Craig's Soil

Now in its eighth edition, this bestselling text continues to blend clarity of explanation with depth of coverage to present students with the fundamental principles of soil mechanics. From the foundations of the subject through to its application in practice, *Craig's Soil Mechanics* provides an indispensable companion to undergraduate courses and beyond.

Revised and fully reworked throughout, several chapters have been significantly extended or had fresh topics added to ensure this new edition reflects more than ever the demands of civil engineering today.

New to this edition:

- Rewritten throughout in line with **Eurocode 7**, with reference to other international standards.
- **Restructuring of the book into two major sections** dealing with both the basic concepts and theories in soil mechanics and the application of these concepts within geotechnical engineering design.
- **Brand new topics** include limit analysis techniques, in-situ testing and foundation systems, plus additional material on seepage, soil stiffness, the critical state concept and foundation design.
- **Enhanced pedagogy** including a comprehensive glossary of terms, start-of-chapter learning objectives, end-of-chapter summaries, and visual examples of real-life engineering equipment to help students approaching the subject for the first time.
- An **extensive companion website** comprising innovative spreadsheet tools for tackling complex problems, digital datasets to accompany worked examples and problems, solutions to end-of-chapter problems, weblinks, extended case studies, and more.

Craig's Soil Mechanics is the definitive text for civil engineering students worldwide.

J. A. Knappett is a lecturer in Civil Engineering at the University of Dundee, UK.

R. F. Craig is a former lecturer in Civil Engineering at the University of Dundee, UK.

Craig's Soil Mechanics

Eighth edition

J. A. Knappett and R. F. Craig

Spon Press
an imprint of Taylor & Francis
LONDON AND NEW YORK

First published 1974 by E & FN Spon, an imprint of Chapman & Hall
Second edition 1978
Third edition 1983
Fourth edition 1987
Fifth edition 1992
Sixth edition 1997
Seventh edition 2004

This edition published 2012
by Spon Press
2 Park Square, Milton Park, Abingdon, Oxon OX14 4RN

Simultaneously published in the USA and Canada
by Spon Press
711 Third Avenue, New York, NY 10017

Spon Press is an imprint of the Taylor & Francis Group, an informa business

© 1974, 1978, 1983, 1987, 1992, 1997, 2004 R. F. Craig
© 2012 J. A. Knappett

The right of J. A. Knappett to be identified as the author of this work has been asserted by him in accordance with sections 77 and 78 of the Copyright, Designs and Patents Act 1988.

Trademark notice: Product or corporate names may be trademarks or registered trademarks, and are used only for identification and explanation without intent to infringe.

All rights reserved. No part of this book may be reprinted or reproduced or utilised in any form or by any electronic, mechanical, or other means, now known or hereafter invented, including photocopying and recording, or in any information storage or retrieval system, without permission in writing from the publishers.

British Library Cataloguing in Publication Data
A catalogue record for this book is available from the British Library

Library of Congress Cataloging in Publication Data
Knappett, Jonathan.
 Craig's soil mechanics / J. A. Knappett and R. F. Craig. – 8th ed.
   p. cm.
   Previous ed. under : Soil mechanics / R.F. Craig.
   Includes bibliographical references and index.
   1. Soil mechanics. I. Craig, R. F. (Robert F.) II. Craig, R. F. (Robert F.) Soil mechanics. III. Title. IV. Title: Soil mechanics.
   TA710.C685 2012
   624.1′5136–dc23   2011033710

ISBN: 978-0-415-56125-9 (hbk)
ISBN: 978-0-415-56126-6 (pbk)
ISBN: 978-0-203-86524-8 (ebk)

Typeset in Times and Frutiger
by Wearset Ltd, Boldon, Tyne and Wear

Printed in Great Britain by Bell & Bain Ltd., Glasgow

To Lis, for her unending support, patience and inspiration.

Contents

List of figures	**xii**
List of tables	**xxii**
Preface	**xxiv**

Part 1 Development of a mechanical model for soil — 1

1 Basic characteristics of soils — 3
Learning outcomes — 3
- 1.1 The origin of soils — 3
- 1.2 The nature of soils — 6
- 1.3 Plasticity of fine-grained soils — 10
- 1.4 Particle size analysis — 13
- 1.5 Soil description and classification — 14
- 1.6 Phase relationships — 22
- 1.7 Soil compaction — 26
- Summary — 35
- Problems — 35
- References — 36
- Further reading — 37

2 Seepage — 39
Learning outcomes — 39
- 2.1 Soil water — 39
- 2.2 Permeability and testing — 41
- 2.3 Seepage theory — 46
- 2.4 Flow nets — 51
- 2.5 Anisotropic soil conditions — 57
- 2.6 Non-homogeneous soil conditions — 59
- 2.7 Numerical solution using the Finite Difference Method — 60
- 2.8 Transfer condition — 63
- 2.9 Seepage through embankment dams — 64
- 2.10 Filter design — 73
- Summary — 74

		Problems	74
		References	77
		Further reading	78

3 Effective stress — 79

Learning outcomes — 79

- 3.1 Introduction — 79
- 3.2 The principle of effective stress — 80
- 3.3 Numerical solution using the Finite Difference Method — 83
- 3.4 Response of effective stress to a change in total stress — 83
- 3.5 Effective stress in partially saturated soils — 87
- 3.6 Influence of seepage on effective stress — 87
- 3.7 Liquefaction — 91
- Summary — 98
- Problems — 98
- References — 100
- Further reading — 100

4 Consolidation — 101

Learning outcomes — 101

- 4.1 Introduction — 101
- 4.2 The oedometer test — 102
- 4.3 Consolidation settlement — 109
- 4.4 Degree of consolidation — 112
- 4.5 Terzaghi's theory of one-dimensional consolidation — 115
- 4.6 Determination of coefficient of consolidation — 121
- 4.7 Secondary compression — 126
- 4.8 Numerical solution using the Finite Difference Method — 127
- 4.9 Correction for construction period — 131
- 4.10 Vertical drains — 136
- 4.11 Pre-loading — 140
- Summary — 142
- Problems — 142
- References — 143
- Further reading — 144

5 Soil behaviour in shear — 145

Learning outcomes — 145

- 5.1 An introduction to continuum mechanics — 145
- 5.2 Simple models of soil elasticity — 149
- 5.3 Simple models of soil plasticity — 152
- 5.4 Laboratory shear tests — 156
- 5.5 Shear strength of coarse-grained soils — 168
- 5.6 Shear strength of saturated fine-grained soils — 174

	5.7	The critical state framework	183
	5.8	Residual strength	188
	5.9	Estimating strength parameters from index tests	189
		Summary	195
		Problems	196
		References	197
		Further reading	199

6 Ground investigation — 201

Learning outcomes — 201

- 6.1 Introduction — 201
- 6.2 Methods of intrusive investigation — 203
- 6.3 Sampling — 210
- 6.4 Selection of laboratory test method(s) — 215
- 6.5 Borehole logs — 216
- 6.6 Cone Penetration Testing (CPT) — 218
- 6.7 Geophysical methods — 222
- 6.8 Contaminated ground — 227
- Summary — 228
- References — 229
- Further reading — 229

7 In-situ testing — 231

Learning outcomes — 231

- 7.1 Introduction — 231
- 7.2 Standard Penetration Test (SPT) — 232
- 7.3 Field Vane Test (FVT) — 236
- 7.4 Pressuremeter Test (PMT) — 240
- 7.5 Cone Penetration Test (CPT) — 252
- 7.6 Selection of in-situ test method(s) — 260
- Summary — 261
- Problems — 262
- References — 265
- Further reading — 266

Part 2 Applications in geotechnical engineering — 267

8 Shallow foundations — 269

Learning outcomes — 269

- 8.1 Introduction — 269
- 8.2 Bearing capacity and limit analysis — 271
- 8.3 Bearing capacity in undrained materials — 273
- 8.4 Bearing capacity in drained materials — 285
- 8.5 Stresses beneath shallow foundations — 295

	8.6	Settlements from elastic theory	300
	8.7	Settlements from consolidation theory	304
	8.8	Settlement from in-situ test data	311
	8.9	Limit state design	316
		Summary	323
		Problems	324
		References	325
		Further reading	326

9 Deep foundations — 327

Learning outcomes — 327

9.1	Introduction	327
9.2	Pile resistance under compressive loads	331
9.3	Pile resistance from in-situ test data	340
9.4	Settlement of piles	341
9.5	Piles under tensile loads	349
9.6	Load testing	350
9.7	Pile groups	353
9.8	Negative skin friction	358
	Summary	359
	Problems	359
	References	361
	Further reading	362

10 Advanced foundation topics — 365

Learning outcomes — 365

10.1	Introduction	365
10.2	Foundation systems	366
10.3	Shallow foundations under combined loading	380
10.4	Deep foundations under combined loading	389
	Summary	398
	Problems	399
	References	400
	Further reading	401

11 Retaining structures — 403

Learning outcomes — 403

11.1	Introduction	403
11.2	Limiting earth pressures from limit analysis	404
11.3	Earth pressure at rest	415
11.4	Gravity retaining structures	418
11.5	Coulomb's theory of earth pressure	429
11.6	Backfilling and compaction-induced earth pressures	434
11.7	Embedded walls	436

	11.8	Ground anchorages	447
	11.9	Braced excavations	452
	11.10	Diaphragm walls	456
	11.11	Reinforced soil	458
		Summary	460
		Problems	461
		References	464
		Further reading	465

12 Stability of self-supporting soil masses — 467
Learning outcomes — 467

	12.1	Introduction	467
	12.2	Vertical cuttings and trenches	468
	12.3	Slopes	472
	12.4	Embankment dams	487
	12.5	An introduction to tunnels	490
		Summary	495
		Problems	496
		References	498
		Further reading	499

13 Illustrative cases — 501
Learning outcomes — 501

	13.1	Introduction	501
	13.2	Selection of characteristic values	502
	13.3	Field instrumentation	506
	13.4	The observational method	514
	13.5	Illustrative cases	515
		Summary	517
		References	517
		Further reading	518

Principal symbols — 519
Glossary — 527
Index — 543

Figures

1.1	The rock cycle	4
1.2	Particle size ranges	4
1.3	Common depositional environments: (a) glacial, (b) fluvial, (c) desert	5
1.4	Particle size distributions of sediments from different depositional environments	5
1.5	Typical ground profile in the West Midlands, UK	6
1.6	Single grain structure	7
1.7	Clay minerals: basic units	8
1.8	Clay minerals: (a) kaolinite, (b) illite, and (c) montmorillonite	9
1.9	Clay structures: (a) dispersed, (b) flocculated, (c) bookhouse, (d) turbostratic, (e) example of a natural clay	9
1.10	Consistency limits for fine soils	11
1.11	Laboratory apparatus for determining liquid limit: (a) fall-cone, (b) Casagrande apparatus	12
1.12	Plasticity chart: British system (BS 1377–2: 1990)	19
1.13	Particle size distribution curves (Example 1.1)	21
1.14	Determination of liquid limit (Example 1.1)	22
1.15	Phase diagrams	23
1.16	Dry density–water content relationship	27
1.17	Dry density–water content curves for different compactive efforts	28
1.18	Dry density–water content curves for a range of soil types	29
1.19	Performance envelopes of various compaction methods for standard soil types: (a) PSD curves of soils, (b) Soil E, (c) Soil F, (d) Soil G, (e) Soil H	31–33
1.20	Moisture condition test	34
2.1	Terminology used to describe groundwater conditions	40
2.2	Laboratory permeability tests: (a) constant head, and (b) falling head	43
2.3	Well pumping tests: (a) unconfined stratum, and (b) confined stratum	45
2.4	Borehole tests: (a) constant-head, (b) variable-head, (c) extension of borehole to prevent clogging, (d) measurement of vertical permeability in anisotropic soil, and (e) measurement of in-situ seepage	47
2.5	Seepage through a soil element	48
2.6	Seepage between two flow lines	50
2.7	Flow lines and equipotentials	50

2.8	Flow net construction: (a) section, (b) boundary conditions, (c) final flow net including a check of the 'square-ness' of the curvilinear squares, and (d) hydraulic gradients inferred from flow net	53
2.9	Example 2.1	54
2.10	Example 2.2	56
2.11	Permeability ellipse	58
2.12	Elemental flow net field	59
2.13	Non-homogeneous soil conditions	59
2.14	Determination of head at an FDM node	61
2.15	Example 2.3	62
2.16	Transfer condition	63
2.17	Homogeneous embankment dam section	65
2.18	Failure of the Teton Dam, 1976	65
2.19	Conformal transformation $r=w^2$: (a) w plane, and (b) r plane	67
2.20	Transformation for embankment dam section: (a) w plane, and (b) r plane	68
2.21	Flow net for embankment dam section	68
2.22	Downstream correction to basic parabola	69
2.23	Example 2.4	69
2.24	(a) Central core and chimney drain, (b) grout curtain, and (c) impermeable upstream blanket	72
2.25	Example 2.5	72
2.26	Problem 2.2	75
2.27	Problem 2.3	75
2.28	Problem 2.4	76
2.29	Problem 2.5	76
2.30	Problem 2.6	76
2.31	Problem 2.7	77
2.32	Problem 2.8	77
3.1	Interpretation of effective stress	81
3.2	Example 3.1	82
3.3	Consolidation analogy	85
3.4	Example 3.2	86
3.5	Relationship of fitting parameter κ to soil plasticity	88
3.6	Partially saturated soil	88
3.7	Forces under seepage conditions	90
3.8	Upward seepage adjacent to sheet piling: (a) determination of parameters from flow net, (b) force diagram, and (c) use of a filter to suppress heave	92
3.9	Examples 3.3 and 3.4	94
3.10	Foundation failure due to liquefaction, 1964 Niigata earthquake, Japan	97
3.11	Problem 3.7	99
4.1	The oedometer: (a) test apparatus, (b) test arrangement	102
4.2	Phase diagram	103
4.3	Void ratio–effective stress relationship	104
4.4	Determination of preconsolidation pressure	106
4.5	In-situ e–log σ' curve	107
4.6	Example 4.1	109

Figures

4.7	Consolidation settlement	111
4.8	Consolidation settlement: graphical procedure	111
4.9	Example 4.2	112
4.10	Assumed linear e–σ' relationship	113
4.11	Consolidation under an increase in total stress $\Delta\sigma$	114
4.12	Element within a consolidating layer of soil	115
4.13	Isochrones	118
4.14	Relationships between average degree of consolidation and time factor	120
4.15	Initial variations of excess pore water pressure	120
4.16	The log time method	121
4.17	The root time method	123
4.18	Hydraulic oedometer	124
4.19	One-dimensional depth-time Finite Difference mesh	128
4.20	Correction for construction period	131
4.21	Example 4.5	132
4.22	Example 4.5 (contd)	133
4.23	Example 4.6	134
4.24	Vertical drains	136
4.25	Cylindrical blocks	137
4.26	Relationships between average degree of consolidation and time factor for radial drainage	138
4.27	Example 4.7	140
4.28	Application of pre-loading: (a) foundation construction on highly compressible soil, (b) foundation constructed following pre-loading	141
5.1	Two-dimensional state of stress in an element of soil: (a) total stresses, (b) effective stresses	146
5.2	Two-dimensional induced state of strain in an element of soil, due to stresses shown in Figure 5.1	147
5.3	(a) Typical stress–strain relationship for soil, (b) elastic–perfectly plastic model, (c) rigid–perfectly plastic model, and (d) strain hardening and strain softening elastic–plastic models	148
5.4	Non-linear soil shear modulus	151
5.5	(a) Frictional strength along a plane of slip, (b) strength of an assembly of particles along a plane of slip	152
5.6	Mohr–Coulomb failure criterion	153
5.7	Mohr circles for total and effective stresses	155
5.8	Direct shear apparatus: (a) schematic, (b) standard direct shear apparatus	157
5.9	The triaxial apparatus: (a) schematic, (b) a standard triaxial cell	158
5.10	Mohr circles for triaxial stress conditions	161
5.11	Interpretation of strength parameters c' and ϕ' using stress invariants	163
5.12	Soil element under isotropic stress increment	165
5.13	Typical relationship between B and degree of saturation	166
5.14	Stress path triaxial cell	167
5.15	Unconfined compression test interpretation	167
5.16	Shear strength characteristics of coarse-grained soils	169
5.17	Mechanics of dilatancy in coarse-grained soils: (a) initially dense soil, exhibiting dilation, (b) initially loose soil, showing contraction	170

5.18	Determination of peak strengths from direct shear test data	171
5.19	Determination of peak strengths from drained triaxial test data	172
5.20	Example 5.1	173
5.21	Example 5.1: Failure envelopes for (a) loose, and (b) dense sand samples	173
5.22	Consolidation characteristics: (a) one-dimensional, (b) isotropic	175
5.23	Typical results from consolidated–undrained and drained triaxial tests	176
5.24	Failure envelopes and stress paths in triaxial tests for: (a) normally consolidated (NC) clays, (b) overconsolidated (OC) clays, (c) corresponding Mohr–Coulomb failure envelope	177
5.25	Example 5.2	178
5.26	Unconsolidated–undrained triaxial test results for saturated clay	179
5.27	Damage observed following the Rissa quick clay flow-slide	181
5.28	Example 5.3	181
5.29	Example 5.4	183
5.30	Volumetric behaviour of soils during (a) undrained tests, (b) drained tests	184
5.31	Position of the Critical State Line (CSL) in p'–q–v space. The effective stress path in an undrained triaxial test is also shown	185
5.32	Example 5.5	186
5.33	Example 5.5 – determination of critical state parameters from line-fitting	187
5.34	(a) Ring shear test, and (b) residual strength	189
5.35	Correlation of ϕ'_{cv} with index properties for (a) coarse-grained, and (b) fine-grained soils	190
5.36	Correlation of remoulded undrained shear strength c_{ur} with index properties	191
5.37	Correlation of sensitivity S_t with index properties	192
5.38	Use of correlations to estimate the undrained strength of cohesive soils: (a) Gault clay, (b) Bothkennar clay	193
5.39	Correlation of ϕ'_r with index properties for fine-grained soils, showing application to UK slope case studies	194
6.1	(a) Percussion boring rig, (b) shell, (c) clay cutter, and (d) chisel	204
6.2	(a) Short-flight auger, (b) continuous-flight auger, (c) bucket auger, and (d) Iwan (hand) auger	205
6.3	Wash boring	207
6.4	Rotary drilling	207
6.5	Open standpipe piezometer	209
6.6	Piezometer tips	210
6.7	Types of sampling tools: (a) open drive sampler, (b) thin-walled sampler, (c) split-barrel sampler, and (d) stationary piston sampler	212
6.8	(a) Continuous sampler, (b) compressed air sampler	215
6.9	Schematic of Cone Penetrometer Test (CPT) showing standard terminology	219
6.10	Soil behaviour type classification chart based on normalised CPT data	219
6.11	Schematic of piezocone (CPTU)	220
6.12	Soil behaviour type classification chart based on normalised CPTU data	221
6.13	Example showing use of CPTU data to provide ground information	221
6.14	Soil behaviour type classification using the I_c method	222
6.15	Seismic refraction method	224
6.16	(a) Electrical resistivity method, (b) identification of soil layers by sounding	226

Figures

7.1	The SPT test: (a) general arrangement, (b) UK standard hammer system, (c) test procedure	233
7.2	Overburden correction factors for coarse-grained soils	234
7.3	Effect of age on SPT data interpretation in coarse-grained soils	235
7.4	Determination of ϕ'_{max} from SPT data in coarse-grained soils	236
7.5	Estimation of c_u from SPT data in fine-grained soils	237
7.6	The FVT test: (a) general arrangement, (b) vane geometry	238
7.7	Correction factor μ for undrained strength as measured by the FVT	239
7.8	Example 7.1 (a) Oedometer test data, (b) I_p calculated from index test data, (c) OCR from FVT and oedometer data	239
7.9	Basic features of (a) Ménard pressuremeter, and (b) self-boring pressuremeter	241
7.10	Idealised soil response during cavity expansion: (a) compatible displacement field, (b) equilibrium stress field	243
7.11	Pressuremeter interpretation during elastic soil behaviour: (a) constitutive model (linear elasticity), (b) derivation of G and σ_{h0} from measured p and dV/V	244
7.12	Pressuremeter interpretation in elasto-plastic soil: (a) constitutive model (linear elasticity, Mohr–Coulomb plasticity), (b) non-linear characteristics of measured p and dV/V	244
7.13	Determination of undrained shear strength from pressuremeter test data	246
7.14	Direct determination of G and σ_{h0} in fine-grained soils from pressuremeter test data	247
7.15	Example 7.2	248
7.16	Direct determination of G, σ_{h0} and u_0 in coarse-grained soils from pressuremeter test data: (a) uncorrected curve, (b) corrected for pore pressure u_0	249
7.17	Determination of parameter s from pressuremeter test data	250
7.18	Determination of ϕ' and ψ from parameter s	250
7.19	Example 7.3	251
7.20	Determination of I_D from CPT/CPTU data	253
7.21	Determination of ϕ'_{max} from CPT/CPTU data	253
7.22	Database of calibration factors for determination of c_u: (a) N_k, (b) N_{kt}	255
7.23	Determination of OCR from CPTU data	256
7.24	Estimation of K_0 from CPTU data	256
7.25	Example 7.4: CPTU data	257
7.26	Example 7.4: Laboratory test data	257
7.27	Example 7.4: Comparison of c_u and OCR from CPTU and laboratory tests	258
7.28	Example 7.5: CPTU data	259
7.29	Example 7.5: Interpretation of ground properties from CPTU and SPT	260
7.30	Problem 7.2	263
7.31	Problem 7.3	263
7.32	Problem 7.4	264
7.33	Problem 7.5	264
8.1	Concepts related to shallow foundation design: (a) soil–structure interaction under vertical actions, (b) foundation performance and limit state design	270

8.2	Modes of failure: (a) general shear, (b) local shear, and (c) punching shear	272
8.3	Idealised stress–strain relationship in a perfectly plastic material	272
8.4	(a) Simple proposed mechanism, UB-1, (b) slip velocities, (c) dimensions	273
8.5	Construction of hodograph for mechanism UB-1	274
8.6	(a) Refined mechanism UB-2, (b) slip velocities on wedge i, (c) geometry of wedge i, (d) hodograph	276
8.7	(a) Simple proposed stress state LB-1, (b) Mohr circles	278
8.8	(a) Refined stress state LB-2, (b) principal stress rotation across a frictional stress discontinuity, (c) Mohr circles	279
8.9	Stress state LB-2 for shallow foundation on undrained soil	280
8.10	Bearing capacity factors N_c for embedded foundations in undrained soil	282
8.11	(a) Bearing capacity factors N_c for strip foundations of width B on layered undrained soils, (b) shape factors s_c	283
8.12	Bearing capacity factors N_c for strip foundations of width B at the crest of a slope of undrained soil	284
8.13	Factor F_z for strip foundations on non-uniform undrained soil	284
8.14	Conditions along a slip plane in drained material	286
8.15	Upper bound mechanism in drained soil: (a) geometry of mechanism, (b) geometry of logarithmic spiral, (c) hodograph	287
8.16	(a) Stress state, (b) principal stress rotation across a frictional stress discontinuity, (c) Mohr circles	289
8.17	Bearing capacity factors for shallow foundations under drained conditions	292
8.18	Shape factors for shallow foundations under drained conditions: (a) s_q, (b) s_γ	293
8.19	(a) Total stresses induced by point load, (b) variation of vertical total stress induced by point load	296
8.20	Total stresses induced by: (a) line load, (b) strip area carrying a uniform pressure	297
8.21	Vertical stress under a corner of a rectangular area carrying a uniform pressure	298
8.22	Contours of equal vertical stress: (a) under a strip area, (b) under a square area	299
8.23	Example 8.3	299
8.24	Distributions of vertical displacement beneath a flexible area: (a) clay, and (b) sand	301
8.25	Contact pressure distribution beneath a rigid area: (a) clay, and (b) sand	301
8.26	Coefficients μ_0 and μ_1 for vertical displacement	302
8.27	Soil element under major principal stress increment	305
8.28	(a) Effective stresses for in-situ conditions and under a general total stress increment $\Delta\sigma_1$, $\Delta\sigma_3$, (b) stress paths	306
8.29	Settlement coefficient μ_c	308
8.30	Example 8.5	309
8.31	Relationship between depth of influence and foundation width	312
8.32	Distribution of strain influence factor	314
8.33	Example 8.6	315
9.1	Deep foundations	328

Figures

9.2	Determination of shaft resistance	329
9.3	Pile installation: non-displacement piling (CFA)	330
9.4	Principal types of pile: (a) precast RC pile, (b) steel H pile, (c) steel tubular pile (plugged), (d) shell pile, (e) CFA pile, (f) under-reamed bored pile (cast-in-situ)	331
9.5	Determination of N_c and s_c for base capacity in undrained soil	332
9.6	Bearing capacity factor N_q for pile base capacity	333
9.7	Determination of adhesion factor α in undrained soil: (a) displacement piles, (b) non-displacement piles	334
9.8	Interface friction angles δ' for various construction materials	335
9.9	Determination of factor β in drained fine-grained soils (all pile types)	336
9.10	Example 9.2	339
9.11	Equilibrium of soil around a settling pile shaft	343
9.12	T–z method	345
9.13	Example 9.3	347
9.14	Approximate values of E_b for preliminary design purposes	348
9.15	Shaft friction in tension: (a) α for non-displacement piles in fine-grained soil, (b) shaft resistance for non-displacement piles in coarse-grained soil	350
9.16	Static load testing of piles: (a) using kentledge, (b) using reaction piles	351
9.17	Interpretation of pile capacity using Chin's method	352
9.18	Failure modes for pile groups at ULS: (a) mode 1, individual pile failure, (b) mode 2, block failure	354
9.19	Diffraction coefficient F_α	356
9.20	Example 9.4	357
9.21	Negative skin friction	358
9.22	Problem 9.5	360
10.1	Foundation systems: (a) pads/strips, (b) raft, (c) piled (plunge column), (d) piled raft	367
10.2	Differential settlement, angular distortion and tilt	367
10.3	The 'Leaning Tower of Pisa': an example of excessive tilt	369
10.4	Damage to load-bearing masonry walls	370
10.5	Damage to masonry infill walls in framed structures	371
10.6	Normalised differential settlement in rafts	372
10.7	Normalised maximum bending moment at the centre of a raft	373
10.8	Example 10.1	374
10.9	Vertical stiffness and load distribution in a square piled raft ($L_p/D_0 = 25$, $S/D_0 = 5$, $\nu = 0.5$)	376
10.10	Minimisation of differential settlements using settlement-reducing piles	377
10.11	(a) Stress state for V–H loading, undrained soil: (b) Mohr circle in zone 1	381
10.12	Yield surface for a strip foundation on undrained soil under V–H loading	382
10.13	Yield surfaces for a strip foundation on undrained soil under (a) V–H loading, (b) V–H–M loading	383
10.14	(a) Stress state for V–H loading, drained soil, (b) Mohr circle in zone 1	384
10.15	N_q for a strip foundation on drained soil under V–H loading	384
10.16	Yield surfaces for a strip foundation on drained soil under (a) V–H loading, (b) V–H–M loading	386
10.17	Non-dimensional factors F_h and F_θ for foundation stiffness determination	387

10.18	Example 10.3	387
10.19	Lateral loading of unrestrained (individual) piles: (a) 'short' pile, (b) 'long' pile	391
10.20	Design charts for determining the lateral capacity of an unrestrained pile under undrained conditions: (a) 'short' pile, (b) 'long' pile	392
10.21	Design charts for determining the lateral capacity of an unrestrained pile under drained conditions: (a) 'short' pile, (b) 'long' pile	392
10.22	Determination of critical failure mode, unrestrained piles: (a) undrained conditions, (b) drained conditions	392
10.23	Lateral loading of restrained (grouped) piles: (a) 'short' pile, (b) 'intermediate' pile, (c) 'long' pile	393
10.24	Design charts for determining the lateral capacity of a restrained pile under undrained conditions: (a) 'short' and 'intermediate' piles, (b) 'long' pile	394
10.25	Design charts for determining the lateral capacity of a restrained pile under drained conditions: (a) 'short' and 'intermediate' piles, (b) 'long' pile	394
10.26	Determination of critical failure mode, restrained piles: (a) undrained conditions, (b) drained conditions	394
10.27	Yield surface for a pile under V–H loading	395
10.28	Critical length of a pile under lateral loading	396
10.29	Example 10.5	397
11.1	Some applications of retained soil: (a) restraint of unstable soil mass, (b) creation of elevated ground, (c) creation of underground space, (d) temporary excavations	404
11.2	Lower bound stress field: (a) stress conditions under active and passive conditions, (b) Mohr circle, undrained case, (c) Mohr circle, drained case	405
11.3	State of plastic equilibrium	407
11.4	Active and passive Rankine states	407
11.5	Example 11.1	408
11.6	Rotation of principal stresses due to wall roughness and batter angle (only total stresses shown)	410
11.7	Mohr circles for zone 2 soil (adjacent to wall) under undrained conditions: (a) active case, (b) passive case	411
11.8	Mohr circles for zone 2 soil (adjacent to wall) under drained conditions: (a) active case, (b) passive case	413
11.9	Equilibrium of sloping retained soil	414
11.10	Mohr circles for zone 1 soil under active conditions: (a) undrained case, (b) drained case	415
11.11	Estimation of K_0 from ϕ' and OCR, and comparison to in-situ test data	417
11.12	Relationship between lateral strain and lateral pressure coefficient	417
11.13	Minimum deformation conditions to mobilise: (a) active state, (b) passive state	418
11.14	Gravity retaining structures	419
11.15	Failure modes for gravity retaining structures at ULS	420
11.16	Pressure distributions and resultant thrusts: undrained soil	422
11.17	Pressure distributions and resultant thrusts: drained soil	423

Figures

11.18	Example 11.2	424
11.19	Example 11.3	425
11.20	Example 11.4	427
11.21	Coulomb theory: active case with $c'=0$: (a) wedge geometry, (b) force polygon	429
11.22	Coulomb theory: active case with $c'>0$	431
11.23	Example 11.5	433
11.24	Stresses due to a line load	435
11.25	Compaction-induced pressure	436
11.26	Cantilever sheet pile wall	437
11.27	Anchored sheet pile wall: free earth support method	438
11.28	Nicoll Highway collapse, Singapore	439
11.29	Anchored sheet pile wall: pressure distribution under working conditions	440
11.30	Arching effects	440
11.31	Various pore water pressure distributions	441
11.32	Example 11.6	443
11.33	Example 11.7	445
11.34	Anchorage types: (a) plate anchor, (b) ground anchor	448
11.35	Ground anchors: (a) grouted mass formed by pressure injection, (b) grout cylinder, and (c) multiple under-reamed anchor	449
11.36	Example 11.8	451
11.37	Earth pressure envelopes for braced excavations	453
11.38	Base failure in a braced excavation	454
11.39	Envelopes of ground settlement behind excavations	455
11.40	(a) Diaphragm wall, (b) contiguous pile wall, (c) secant pile wall	457
11.41	Reinforced soil-retaining structure: (a) tie-back wedge method, (b) coherent gravity method	459
11.42	Problem 11.2	461
11.43	Problem 11.4	462
11.44	Problem 11.5	462
11.45	Problem 11.8	463
12.1	(a) Mechanism UB-1, (b) hodograph	469
12.2	(a) Stress field LB-1, (b) Mohr circle	470
12.3	Stability of a slurry-supported trench in undrained soil	471
12.4	Stability of a slurry-supported trench in drained soil	471
12.5	Slurry-supported excavations: (a) maximum excavation depth in undrained soil, (b) minimum slurry density to avoid collapse in drained soil ($\phi'=35°$, $n=1$)	472
12.6	Types of slope failure	473
12.7	Rotational slope failure at Holbeck, Yorkshire	474
12.8	Limit equilibrium analysis in undrained soil	475
12.9	Stability numbers for slopes in undrained soil	476
12.10	Example 12.1	476
12.11	The method of slices	478
12.12	Example 12.2	481
12.13	Plane translational slip	483

12.14	Pore pressure dissipation and factor of safety: (a) following excavation (i.e. a cutting), (b) following construction (i.e. an embankment)	486
12.15	Failure beneath an embankment	486
12.16	Horizontal drainage layers	488
12.17	Rapid drawdown conditions	490
12.18	Terminology related to tunnels	490
12.19	Stress conditions in the soil above the tunnel crown	491
12.20	Stability numbers for circular tunnels in undrained soil	492
12.21	(a) Support pressure in drained soil for shallow and deep tunnels ($\sigma'_q=0$), (b) maximum support pressure for use in ULS design ($\sigma'_q=0$)	494
12.22	Settlement trough above an advancing tunnel	494
12.23	Problem 12.2	496
12.24	Problem 12.4	497
12.25	Problem 12.6	497
13.1	Examples of characteristic value determination (for undrained shear strength): (a) uniform glacial till, (b) layered overconsolidated clay, (c) overconsolidated fissured clay	505
13.2	Levelling plug	507
13.3	Measurement of vertical movement: (a) plate and rod, (b) deep settlement probe, (c) rod extensometer, and (d) magnetic extensometer	508
13.4	Hydraulic settlement cell	509
13.5	Measurement of horizontal movement: (a) reference studs, (b) tape extensometer, (c) rod extensometer, and (d) tube extensometer	510
13.6	Vibrating wire strain gauge	511
13.7	Inclinometer: (a) probe and guide tube, (b) method of calculation, and (c) force balance accelerometer	512
13.8	(a) Diaphragm pressure cell, and (b) hydraulic pressure cell	514
13.9	Location of illustrative cases	516

Tables

1.1	Activity of some common clay minerals	11
1.2	Composite types of coarse soil	17
1.3	Compactive state and stiffness of soils	17
1.4	Descriptive terms for soil classification (BS 5930)	20
1.5	Example 1.1	20
1.6	Problem 1.1	35
2.1	Coefficient of permeability (m/s)	41
2.2	Example 2.2	56
2.3	Example 2.2 (contd.)	56
2.4	Downstream correction to basic parabola	69
3.1	Example 3.1	82
4.1	Example 4.1	108
4.2	Example 4.2	112
4.3	Secondary compression characteristics of natural soils	127
4.4	Example 4.4	130
4.5	Example 4.5	133
5.1	Example 5.1	174
5.2	Example 5.2	178
5.3	Example 5.2 (contd.)	178
5.4	Example 5.4	182
5.5	Example 5.4 (contd.)	182
5.6	Example 5.5	187
6.1	Guidance on spacing of ground investigation points (Eurocode 7, Part 2: 2007)	202
6.2	Sample quality related to end use (after EC7–2: 2007)	211
6.3	Derivation of key soil properties from undisturbed samples tested in the laboratory	216
6.4	Sample borehole log	217
6.5	Shear wave velocities of common geotechnical materials	223
6.6	Typical resistivities of common geotechnical materials	227
7.1	SPT correction factor ζ	233
7.2	Common Energy ratios in use worldwide	234
7.3	Example 7.5: SPT data	259
7.4	Derivation of key soil properties from in-situ tests	261

7.5	Problem 7.1	262
8.1	Energy dissipated within the soil mass in mechanism UB-1	275
8.2	Work done by the external pressures, mechanism UB-1	275
8.3	Energy dissipated within the soil mass in mechanism UB-2	277
8.4	Work done by the external pressures, mechanism UB-1	288
8.5	Influence factors (I_σ) for vertical stress due to point load	296
8.6	Influence factors (I_s) for vertical displacement under flexible and rigid areas carrying uniform pressure	301
8.7	Example 8.5	310
8.8	Example 8.6	316
8.9	Selection of partial factors for use in ULS design to EC7	318
8.10	Partial factors on actions for use in ULS design to EC7	318
8.11	Partial factors on material properties for use in ULS design to EC7	318
8.12	Example 8.7	320
8.13	Example 8.8	321
8.14	Example 8.9	322
8.15	Example 8.9 (contd.)	323
9.1	Partial resistance factors for use in ULS pile design to EC7 (piles in compression only)	336
9.2	Soil dependent constants for determining base capacity from SPT data	340
9.3	Soil dependent constants for determining base capacity from CPT data	340
9.4	Correlation factors for determination of characteristic resistance from in-situ tests to EC7	341
9.5	Correlation factors for determination of characteristic resistance from static load tests to EC7	353
9.6	Example 9.4 – calculations for pile type A	357
9.7	Example 9.4 – calculations for pile type B	358
10.1	Angular distortion limits for building structures	368
10.2	Tilt limits for building structures	368
10.3	Partial factors on actions for verification of ULS against uplift according to EC7	378
11.1	Example 11.1	409
11.2	Example 11.3	426
11.3	Example 11.4	428
11.4	Example 11.6	444
11.5	Example 11.7 (case $d=6.0$ m)	447
11.6	Example 11.8	452
12.1	Example 12.2	482
13.1	Coefficients of variation of various soil properties	504
13.2	Example calculations for sub-layering approach	505

Preface

When I was approached by Taylor & Francis to write the new edition of Craig's popular textbook, while I was honoured to be asked, I never realised how much time and effort would be required to meet the high standards set by the previous seven editions. Initially published in 1974, I felt that the time was right for a major update as the book approaches its fortieth year, though I have tried to maintain the clarity and depth of explanation which has been a core feature of previous editions.

All chapters have been updated, several extended, and new chapters added to reflect the demands of today's engineering students and courses. It is still intended primarily to serve the needs of the undergraduate civil engineering student and act as a useful reference through the transition into engineering practice. However, inclusion of some more advanced topics extends the scope of the book, making it suitable to also accompany many post-graduate level courses.

The key changes are as follows:

- **Separation of the material into two major sections**: the first deals with basic concepts and theories in soil mechanics, and the determination of the mechanical properties necessary for geotechnical design, which forms the second part of the book.
- **Extensive electronic resources**: including spreadsheet tools for advanced analysis, digital datasets to accompany worked examples and problems, solutions to end-of-chapter problems, weblinks, instructor resources and more, all available through the Companion Website.
- **New chapter on in-situ testing**: focusing on the parameters that can be reliably determined using each test and interpretation of mechanical properties from digital data based on real sites (which is provided on the Companion Website).
- **New chapters on foundation behaviour and design**: coverage of foundations is now split into three separate sections (shallow foundations, deep foundations and advanced topics), for increased flexibility in course design.
- **Limit state design (to Eurocode 7)**: The chapters on geotechnical design are discussed wholly within a modern generic limit state design framework, rather than the out-dated permissible stress approach. More extensive background is provided on Eurocode 7, which is used in the numerical examples and end-of-chapter problems, to aid the transition from university to the design office.
- **Extended case studies (online)**: building on those in previous editions, but now including application of the limit state design techniques in the book to these real-world problems, to start to build engineering judgement.
- **Inclusion of limit analysis techniques**: With the increasing prevalence and popularity of advanced computer software based on these techniques, I believe it is essential for students to leave university with a basic understanding of the underlying theory to aid their future professional development. This also provides a more rigorous background to the origin of bearing capacity factors and limit pressures, missing from previous editions.

Preface

I am immensely grateful to my colleagues at the University of Dundee for allowing me the time to complete this new edition, and for their constructive comments as it took shape. I would also like to express my gratitude to all those at Taylor & Francis who have helped to make such a daunting task achievable, and thank all those who have allowed reproduction of figures, data and images.

I hope that current and future generations of civil engineers will find this new edition as useful, informative and inspiring as previous generations have found theirs.

Jonathan Knappett
University of Dundee
July 2011

Part 1

Development of a mechanical model for soil

Chapter 1

Basic characteristics of soils

> **Learning outcomes**
>
> After working through the material in this chapter, you should be able to:
>
> 1 Understand how soil deposits are formed and the basic composition and structure of soils at the level of the micro-fabric (Sections 1.1 and 1.2);
> 2 Describe (Sections 1.3 and 1.4) and classify (Section 1.5) soils based on their basic physical characteristics;
> 3 Determine the basic physical characteristics of a soil continuum (i.e. at the level of the macro-fabric, Section 1.6);
> 4 Specify compaction required to produce engineered fill materials with desired continuum properties for use in geotechnical constructions (Section 1.7).

1.1 The origin of soils

To the civil engineer, soil is any uncemented or weakly cemented accumulation of mineral particles formed by the weathering of rocks as part of the rock cycle (Figure 1.1), the void space between the particles containing water and/or air. Weak cementation can be due to carbonates or oxides precipitated between the particles, or due to organic matter. Subsequent deposition and compression of soils, combined with cementation between particles, transforms soils into sedimentary rocks (a process known as **lithification**). If the products of weathering remain at their original location they constitute a **residual soil**. If the products are transported and deposited in a different location they constitute a **transported soil**, the agents of transportation being gravity, wind, water and glaciers. During transportation, the size and shape of particles can undergo change and the particles can be sorted into specific size ranges. Particle sizes in soils can vary from over 100 mm to less than 0.001 mm. In the UK, the size ranges are described as shown in Figure 1.2. In Figure 1.2, the terms 'clay', 'silt' etc. are used to describe only the sizes of particles between specified limits. However, the same terms are also used to describe particular types of soil, classified according to their mechanical behaviour (see Section 1.5).

The type of transportation and subsequent deposition of soil particles has a strong influence on the distribution of particle sizes at a particular location. Some common depositional regimes are shown in Figure 1.3. In glacial regimes, soil material is eroded from underlying rock by the frictional and freeze–thaw action of glaciers. The material, which is typically very varied in particle size from clay to

Development of a mechanical model for soil

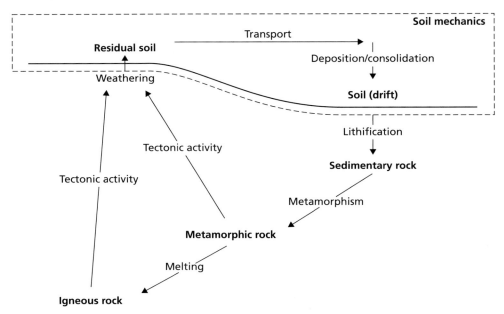

Figure 1.1 The rock cycle.

Figure 1.2 Particle size ranges.

boulder-sized particles, is carried along at the base of the glacier and deposited as the ice melts; the resulting material is known as (**glacial**) **till**. Similar material is also deposited as a terminal moraine at the edge of the glacier. As the glacier melts, moraine is transported in the outwash; it is easier for smaller, lighter particles to be carried in suspension, leading to a gradation in particle size with distance from the glacier as shown in Figure 1.3(a). In warmer temperate climates the chief transporting action is water (i.e. rivers and seas), as shown in Figure 1.3(b). The deposited material is known as **alluvium**, the composition of which depends on the speed of water flow. Faster-flowing rivers can carry larger particles in suspension, resulting in alluvium, which is a mixture of sand and gravel-sized particles, while slower-flowing water will tend to carry only smaller particles. At estuarine locations where rivers meet the sea, material may be deposited as a shelf or **delta**. In arid (desert) environments (Figure 1.3(c)) wind is the key agent of transportation, eroding rock outcrops and forming a **pediment** (the desert floor) of fine wind-blown sediment (**loess**). Towards the coast, a **playa** of temporary evaporating lakes, leaving salt deposits, may also be formed. The large temperature differences between night and day additionally cause thermal weathering of rock outcrops, producing **scree**. These surface processes are geologically very recent, and are referred to as **drift deposits** on geological maps. Soil which has undergone significant compression/consolidation following deposition is typically much older and is referred to as **solid**, alongside rocks, on geological maps.

Basic characteristics of soils

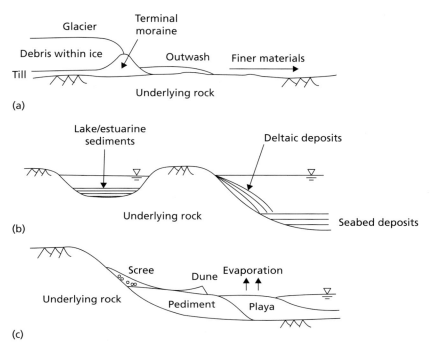

Figure 1.3 Common depositional environments: (a) glacial, (b) fluvial, (c) desert.

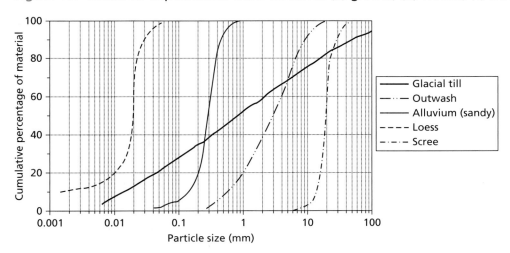

Figure 1.4 Particle size distributions of sediments from different depositional environments.

The relative proportions of different-sized particles within a soil are described as its **particle size distribution** (PSD), and typical curves for materials in different depositional environments are shown in Figure 1.4. The method of determining the PSD of a deposit and its subsequent use in soil classification is described in Sections 1.4 and 1.5.

At a given location, the subsurface materials will be a mixture of rocks and soils, stretching back many hundreds of millions of years in geological time. As a result, it is important to understand the full

Development of a mechanical model for soil

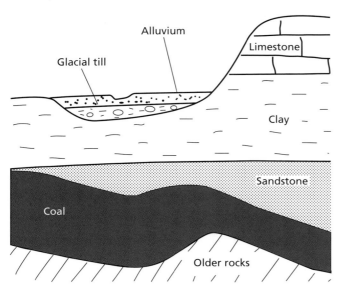

Figure 1.5 Typical ground profile in the West Midlands, UK.

geological history of an area to understand the likely characteristics of the deposits that will be present at the surface, as the depositional regime may have changed significantly over geological time. As an example, the West Midlands in the UK was deltaic in the Carboniferous period (~395–345 million years ago), depositing organic material which subsequently became coal measures. In the subsequent Triassic period (280–225 million years ago), due to a change in sea level sandy materials were deposited which were subsequently lithified to become Bunter sandstone. Mountain building during this period on what is now the European continent caused the existing rock layers to become folded. It was subsequently flooded by the North Sea during the Cretaceous/Jurassic periods (225–136 million years ago), depositing fine particles and carbonate material (Lias clay and Oolitic limestone). The Ice Ages in the Pleistocene period (1.5–2 million years ago) subsequently led to glaciation over all but the southernmost part of the UK, eroding some of the recently deposited softer rocks and depositing glacial till. The subsequent melting of the glaciers created river valleys, which deposited alluvium above the till. The geological history would therefore suggest that the surficial soil conditions are likely to consist of alluvium overlying till/clay overlying stronger rocks, as shown schematically in Figure 1.5. This example demonstrates the importance of engineering geology in understanding ground conditions. A thorough introduction to this topic can be found in Waltham (2002).

1.2 The nature of soils

The destructive process in the formation of soil from rock may be either physical or chemical. The physical process may be erosion by the action of wind, water or glaciers, or disintegration caused by cycles of freezing and thawing in cracks in the rock. The resultant soil particles retain the same mineralogical composition as that of the parent rock (a full description of this is beyond the scope of this text). Particles of this type are described as being of 'bulky' form, and their shape can be indicated by terms such as angular, rounded, flat and elongated. The particles occur in a wide range of sizes, from boulders, through gravels and sands, to the fine rock flour formed by the grinding action of glaciers. The structural arrangement of bulky particles (Figure 1.6) is described as **single grain**, each particle being in direct

Basic characteristics of soils

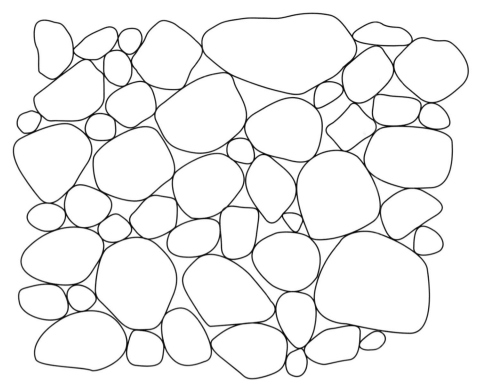

Figure 1.6 Single grain structure.

contact with adjoining particles without there being any bond between them. The state of the particles can be described as dense, medium dense or loose, depending on how they are packed together (see Section 1.5).

Chemical processes result in changes in the mineral form of the parent rock due to the action of water (especially if it contains traces of acid or alkali), oxygen and carbon dioxide. Chemical weathering results in the formation of groups of crystalline particles of **colloidal** size (<0.002 mm) known as clay minerals. The clay mineral kaolinite, for example, is formed by the breakdown of feldspar by the action of water and carbon dioxide. Most clay mineral particles are of 'plate-like' form, having a high specific surface (i.e. a high surface area to mass ratio), with the result that their structure is influenced significantly by surface forces. Long 'needle-shaped' particles can also occur, but are comparatively rare.

The basic structural units of most clay minerals are a silicon–oxygen tetrahedron and an aluminium–hydroxyl octahedron, as illustrated in Figure 1.7(a). There are valency imbalances in both units, resulting in net negative charges. The basic units therefore do not exist in isolation, but combine to form sheet structures. The tetrahedral units combine by the sharing of oxygen ions to form a silica sheet. The octahedral units combine through shared hydroxyl ions to form a gibbsite sheet. The silica sheet retains a net negative charge, but the gibbsite sheet is electrically neutral. Silicon and aluminium may be partially replaced by other elements, this being known as **isomorphous substitution**, resulting in further charge imbalance. The sheet structures are represented symbolically in Figure 1.7(b). Layer structures then form by the bonding of a silica sheet with either one or two gibbsite sheets. Clay mineral particles consist of stacks of these layers, with different forms of bonding between the layers.

The surfaces of clay mineral particles carry residual negative charges, mainly as a result of the isomorphous substitution of silicon or aluminium by ions of lower valency but also due to disassociation of

Development of a mechanical model for soil

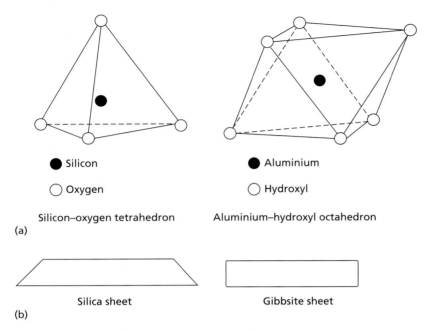

Figure 1.7 Clay minerals: basic units.

hydroxyl ions. Unsatisfied charges due to 'broken bonds' at the edges of particles also occur. The negative charges result in cations present in the water in the void space being attracted to the particles. The cations are not held strongly and, if the nature of the water changes, can be replaced by other cations, a phenomenon referred to as **base exchange**.

Cations are attracted to a clay mineral particle because of the negatively charged surface, but at the same time they tend to move away from each other because of their thermal energy. The net effect is that the cations form a dispersed layer adjacent to the particle, the cation concentration decreasing with increasing distance from the surface until the concentration becomes equal to that in the general mass of water in the void space of the soil as a whole. The term 'double layer' describes the negatively charged particle surface and the dispersed layer of cations. For a given particle, the thickness of the cation layer depends mainly on the valency and concentration of the cations: an increase in valency (due to cation exchange) or an increase in concentration will result in a decrease in layer thickness.

Layers of water molecules are held around a clay mineral particle by hydrogen bonding and (because water molecules are dipolar) by attraction to the negatively charged surfaces. In addition, the exchangeable cations attract water (i.e. they become hydrated). The particle is thus surrounded by a layer of adsorbed water. The water nearest to the particle is strongly held and appears to have a high viscosity, but the viscosity decreases with increasing distance from the particle surface to that of 'free' water at the boundary of the adsorbed layer. Adsorbed water molecules can move relatively freely parallel to the particle surface, but movement perpendicular to the surface is restricted.

The structures of the principal clay minerals are represented in Figure 1.8. Kaolinite consists of a structure based on a single sheet of silica combined with a single sheet of gibbsite. There is very limited isomorphous substitution. The combined silica–gibbsite sheets are held together relatively strongly by hydrogen bonding. A kaolinite particle may consist of over 100 stacks. Illite has a basic structure consisting of a sheet of gibbsite between and combined with two sheets of silica. In the silica sheet, there is partial substitution of silicon by aluminium. The combined sheets are linked together by relatively weak bonding due to non-exchangeable potassium ions held between them. Montmorillonite has the same basic structure as illite. In the gibbsite sheet there is partial substitution of aluminium by magnesium and iron, and in the silica sheet there is again partial substitution of silicon by aluminium. The space between the combined sheets is occupied by water molecules and exchangeable cations other than potassium, resulting in a very weak bond. Considerable swelling of montmorillonite (and therefore of any soil of which it is a part) can occur due to additional water being adsorbed between the combined sheets. This demonstrates that understanding the basic composition of a soil in terms of its mineralogy can provide clues as to the geotechnical problems which may subsequently be encountered.

Forces of repulsion and attraction act between adjacent clay mineral particles. Repulsion occurs between the like charges of the double layers, the force of repulsion depending on the characteristics of the layers. An increase in cation valency or concentration will result in a decrease in repulsive force and vice versa. Attraction between particles is due to short-range van der Waals forces (electrical forces of attraction between neutral molecules), which are independent of the double-layer characteristics, that decrease rapidly with increasing distance between particles. The net inter-particle forces influence the structural form of clay mineral particles on deposition. If there is net repulsion the particles tend to assume a face-to-face orientation, this being referred to as a **dispersed** structure. If, on the other hand, there is net attraction the orientation of the particles tends to be edge-to-face or edge-to-edge, this being referred to as a **flocculated** structure. These structures, involving interaction between single clay mineral particles, are illustrated in Figures 1.9(a) and (b).

In natural clays, which normally contain a significant proportion of larger, bulky particles, the structural arrangement can be extremely complex. Interaction between single clay mineral particles is rare, the tendency being for the formation of elementary **aggregations** of particles with a face-to-face orientation. In turn, these elementary aggregations combine to form larger assemblages, the structure of which

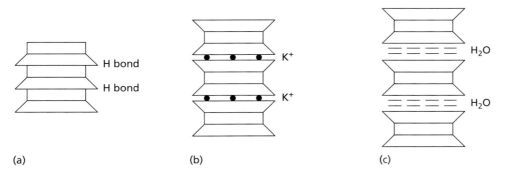

Figure 1.8 Clay minerals: (a) kaolinite, (b) illite, and (c) montmorillonite.

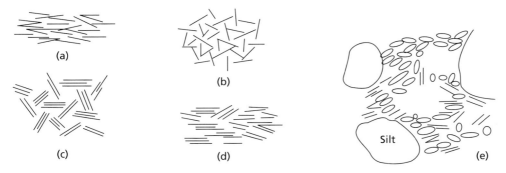

Figure 1.9 Clay structures: (a) dispersed, (b) flocculated, (c) bookhouse, (d) turbostratic, (e) example of a natural clay.

Development of a mechanical model for soil

is influenced by the depositional environment. Two possible forms of particle assemblage, known as the bookhouse and turbostratic structures, are illustrated in Figures 1.9(c) and (d). Assemblages can also occur in the form of connectors or a matrix between larger particles. An example of the structure of a natural clay, in diagrammatical form, is shown in Figure 1.9(e).

If clay mineral particles are present they usually exert a considerable influence on the properties of a soil, an influence out of all proportion to their percentage by weight in the soil. Soils whose properties are influenced mainly by clay and silt size particles are commonly referred to as **fine-grained** (or fine) soils. Those whose properties are influenced mainly by sand and gravel size particles are referred to as **coarse-grained** (or coarse) soils.

1.3 Plasticity of fine-grained soils

Plasticity is an important characteristic in the case of fine-grained soils, the term 'plasticity' describing the ability of a soil to undergo irrecoverable deformation without cracking or crumbling. In general, depending on its **water content** (defined as the ratio of the mass of water in the soil to the mass of solid particles), a soil may exist in one of the liquid, plastic, semi-solid and solid states. If the water content of a soil initially in the liquid state is gradually reduced, the state will change from liquid through plastic and semi-solid, accompanied by gradually reducing volume, until the solid state is reached. The water contents at which the transitions between states occur differ from soil to soil. In the ground, most fine-grained soils exist in the plastic state. Plasticity is due to the presence of a significant content of clay mineral particles (or organic material) in the soil. The void space between such particles is generally very small in size with the result that water is held at negative pressure by capillary tension, allowing the soil to be deformed or moulded. Adsorption of water due to the surface forces on clay mineral particles may contribute to plastic behaviour. Any decrease in water content results in a decrease in cation layer thickness and an increase in the net attractive forces between particles.

The upper and lower limits of the range of water content over which the soil exhibits plastic behaviour are defined as the **liquid limit** (w_L) and the **plastic limit** (w_P), respectively. Above the liquid limit, the soil flows like a liquid (slurry); below the plastic limit, the soil is brittle and crumbly. The water content range itself is defined as the **plasticity index** (I_P), i.e.:

$$I_P = w_L - w_P \tag{1.1}$$

However, the transitions between the different states are gradual, and the liquid and plastic limits must be defined arbitrarily. The natural water content (w) of a soil (adjusted to an equivalent water content of the fraction passing the 425-μm sieve) relative to the liquid and plastic limits can be represented by means of the **liquidity index** (I_L), where

$$I_L = \frac{w - w_P}{I_P} \tag{1.2}$$

The relationship between the different consistency limits is shown in Figure 1.10.

The degree of plasticity of the clay-size fraction of a soil is expressed by the ratio of the plasticity index to the percentage of clay-size particles in the soil (the **clay fraction**): this ratio is called the **activity**. 'Normal' soils have an activity between 0.75 and 1.25, i.e. I_P is approximately equal to the clay fraction. Activity below 0.75 is considered inactive, while soils with activity above 1.25 are considered active. Soils of high activity have a greater change in volume when the water content is changed (i.e. greater swelling when wetted and greater shrinkage when drying). Soils of high activity (e.g. containing a significant amount of montmorillonite) can therefore be particularly damaging to geotechnical works.

Basic characteristics of soils

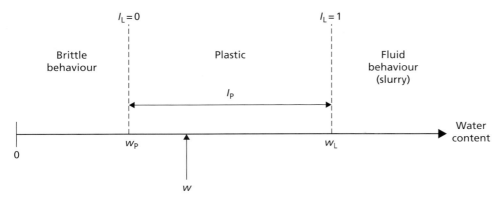

Figure 1.10 Consistency limits for fine soils.

Table 1.1 gives the activity of some common clay minerals, from which it can be seen that activity broadly correlates with the specific surface of the particles (i.e. surface area per unit mass), as this governs the amount of adsorbed water.

The transition between the semi-solid and solid states occurs at the **shrinkage limit**, defined as the water content at which the volume of the soil reaches its lowest value as it dries out.

The liquid and plastic limits are determined by means of arbitrary test procedures. In the UK, these are fully detailed in BS 1377, Part 2 (1990). In Europe CEN ISO/TS 17892–12 (2004) is the current standard, while in the United States ASTM D4318 (2010) is used. These standards all relate to the same basic tests which are described below.

The soil sample is dried sufficiently to enable it to be crumbled and broken up, using a mortar and a rubber pestle, without crushing individual particles; only material passing a 425-μm sieve is typically used in the tests. The apparatus for the liquid limit test consists of a penetrometer (or 'fall-cone') fitted with a 30° cone of stainless steel, 35 mm long: the cone and the sliding shaft to which it is attached have a mass of 80 g. This is shown in Figure 1.11(a). The test soil is mixed with distilled water to form a thick homogeneous paste, and stored for 24 h. Some of the paste is then placed in a cylindrical metal cup, 55 mm internal diameter by 40 mm deep, and levelled off at the rim of the cup to give a smooth surface. The cone is lowered so that it just touches the surface of the soil in the cup, the cone being locked in its support at this stage. The cone is then released for a period of 5 s, and its depth of penetration into the soil is measured. A little more of the soil paste is added to the cup and the test is repeated until a consistent value of penetration has been obtained. (The average of two values within 0.5 mm or of three values within 1.0 mm is taken.) The entire test procedure is repeated at least four times, using the same soil sample but increasing the water content each time by adding distilled water. The penetration values should cover the range of approximately 15–25 mm, the tests proceeding from the drier to the wetter state of the soil. Cone penetration is plotted against water content, and the best straight line fitting the

Table 1.1 Activity of some common clay minerals

Mineral group	Specific surface (m^2/g)[1]	Activity[2]
Kaolinite	10–20	0.3–0.5
Illite	65–100	0.5–1.3
Montmorillonite	Up to 840	4–7

Notes: 1 After Mitchell and Soga (2005). 2 After Day (2001).

plotted points is drawn. This is demonstrated in Example 1.1. The liquid limit is defined as the percentage water content (to the nearest integer) corresponding to a cone penetration of 20 mm. Determination of liquid limit may also be based on a single test (the one-point method), provided the cone penetration is between 15 and 25 mm.

An alternative method for determining the liquid limit uses the Casagrande apparatus (Figure 1.11(b)), which is popular in the United States and other parts of the world (ASTM D4318). A soil paste is placed in a pivoting flat metal cup and divided by cutting a groove. A mechanism enables the cup to be lifted to a height of 10 mm and dropped onto a hard rubber base. The two halves of the soil gradually flow together as the cup is repeatedly dropped The water content of the soil in the cup is then determined; this is plotted against the logarithm of the number of blows, and the best straight line fitting the plotted points is drawn. For this test, the liquid limit is defined as the water content at which 25 blows are required to close the bottom of the groove. It should be noted that the Casagrande method is generally less reliable than the preferred penetrometer method, being more operator dependent and subjective.

For the determination of the plastic limit, the test soil is mixed with distilled water until it becomes sufficiently plastic to be moulded into a ball. Part of the soil sample (approximately 2.5 g) is formed into a thread, approximately 6 mm in diameter, between the first finger and thumb of each hand. The thread is then placed on a glass plate and rolled with the tips of the fingers of one hand until its diameter is reduced to approximately 3 mm: the rolling pressure must be uniform throughout the test. The thread is then remoulded between the fingers (the water content being reduced by the heat of the fingers) and the

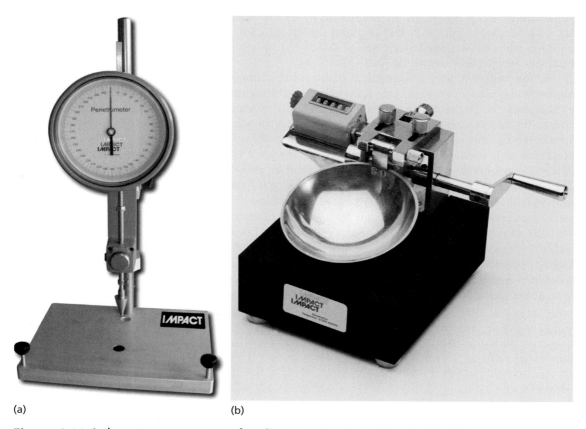

Figure 1.11 Laboratory apparatus for determining liquid limit: (a) fall-cone, (b) Casagrande apparatus (images courtesy of Impact Test Equipment Ltd).

procedure is repeated until the thread of soil shears both longitudinally and transversely when it has been rolled to a diameter of 3 mm. The procedure is repeated using three more parts of the sample, and the percentage water content of all the crumbled soil is determined as a whole. This water content (to the nearest integer) is defined as the plastic limit of the soil. The entire test is repeated using four other sub-samples, and the average taken of the two values of plastic limit: the tests must be repeated if the two values differ by more than 0.5%. Due to the strongly subjective nature of this test, alternative methodologies have recently been proposed for determining w_P, though these are not incorporated within current standards. Further information can be found in Barnes (2009) and Sivakumar *et al.* (2009).

1.4 Particle size analysis

Most soils consist of a graded mixture of particles from two or more size ranges. For example, clay is a type of soil possessing cohesion and plasticity which normally consists of particles in both the clay size and silt size ranges. **Cohesion** is the term used to describe the strength of a clay sample when it is unconfined, being due to negative pressure in the water filling the void space, of very small size, between particles. This strength would be lost if the clay were immersed in a body of water. Cohesion may also be derived from cementation between soil particles. It should be appreciated that all clay-size particles are not necessarily clay mineral particles: the finest rock flour particles may be of clay size.

The particle size analysis of a soil sample involves determining the percentage by mass of particles within the different size ranges. The particle size distribution of a coarse soil can be determined by the method of **sieving**. The soil sample is passed through a series of standard test sieves having successively smaller mesh sizes. The mass of soil retained in each sieve is determined, and the cumulative percentage by mass passing each sieve is calculated. If fine particles are present in the soil, the sample should be treated with a deflocculating agent (e.g. a 4% solution of sodium hexametaphosphate) and washed through the sieves.

The particle size distribution (PSD) of a fine soil or the fine fraction of a coarse soil can be determined by the method of **sedimentation**. This method is based on Stokes' law, which governs the velocity at which spherical particles settle in a suspension: the larger the particles are the greater is the settling velocity, and vice versa. The law does not apply to particles smaller than 0.0002 mm, the settlement of which is influenced by Brownian motion. The size of a particle is given as the diameter of a sphere which would settle at the same velocity as the particle. Initially, the soil sample is pretreated with hydrogen peroxide to remove any organic material. The sample is then made up as a suspension in distilled water to which a deflocculating agent has been added to ensure that all particles settle individually, and placed in a sedimentation tube. From Stokes' law it is possible to calculate the time, t, for particles of a certain 'size', D (the equivalent settling diameter), to settle to a specified depth in the suspension. If, after the calculated time t, a sample of the suspension is drawn off with a pipette at the specified depth below the surface, the sample will contain only particles smaller than the size D at a concentration unchanged from that at the start of sedimentation. If pipette samples are taken at the specified depth at times corresponding to other chosen particle sizes, the particle size distribution can be determined from the masses of the residues. An alternative procedure to pipette sampling is the measurement of the specific gravity of the suspension by means of a special hydrometer, the specific gravity depending on the mass of soil particles in the suspension at the time of measurement. Full details of the determination of particle size distribution by these methods are given in BS 1377–2 (UK), CEN ISO/TS 17892–4 (Europe) and ASTM D6913 (US). Modern optical techniques can also be used to determine the PSD of a coarse soil. Single Particle Optical Sizing (SPOS) works by drawing a stream of dry particles through the beam of a laser diode. As each individual particle passes through the beam it casts a shadow on a light sensor which is proportional to its size (and therefore volume). The optical sizer automatically analyses the sensor output to determine the PSD by volume. Optical methods have been found to

Development of a mechanical model for soil

overestimate particle sizes compared to sieving (White, 2003), though advantages are that the results are repeatable and less operator dependent compared to sieving, and testing requires a much smaller volume of soil.

The particle size distribution of a soil is presented as a curve on a semilogarithmic plot, the ordinates being the percentage by mass of particles smaller than the size given by the abscissa. The flatter the distribution curve, the larger the range of particle sizes in the soil; the steeper the curve, the smaller the size range. A coarse soil is described as **well graded** if there is no excess of particles in any size range and if no intermediate sizes are lacking. In general, a well-graded soil is represented by a smooth, concave distribution curve. A coarse soil is described as **poorly graded** (a) if a high proportion of the particles have sizes within narrow limits (a uniform soil), or (b) if particles of both large and small sizes are present but with a relatively low proportion of particles of intermediate size (a **gap-graded** or step-graded soil). Particle size is represented on a logarithmic scale so that two soils having the same degree of uniformity are represented by curves of the same shape regardless of their positions on the particle size distribution plot. Examples of particle size distribution curves appear in Figure 1.4. The particle size corresponding to any specified percentage value can be read from the particle size distribution curve. The size such that 10% of the particles are smaller than that size is denoted by D_{10}. Other sizes, such as D_{30} and D_{60}, can be defined in a similar way. The size D_{10} is defined as the **effective size**, and can be used to estimate the permeability of the soil (see Chapter 2). The general slope and shape of the distribution curve can be described by means of the **coefficient of uniformity** (C_u) and the **coefficient of curvature** (C_z), defined as follows:

$$C_u = \frac{D_{60}}{D_{10}} \tag{1.3}$$

$$C_z = \frac{D_{30}^2}{D_{60} D_{10}} \tag{1.4}$$

The higher the value of the coefficient of uniformity, the larger the range of particle sizes in the soil. A well-graded soil has a coefficient of curvature between 1 and 3. The sizes D_{15} and D_{85} are commonly used to select appropriate material for granular drains used to drain geotechnical works (see Chapter 2).

1.5 Soil description and classification

It is essential that a standard language should exist for the description of soils. A comprehensive description should include the characteristics of both the soil material and the in-situ soil mass. Material characteristics can be determined from disturbed samples of the soil – i.e. samples having the same particle size distribution as the in-situ soil but in which the in-situ structure has not been preserved. The principal material characteristics are particle size distribution (or grading) and plasticity, from which the soil name can be deduced. Particle size distribution and plasticity properties can be determined either by standard laboratory tests (as described in Sections 1.3 and 1.4) or by simple visual and manual procedures. Secondary material characteristics are the colour of the soil and the shape, texture and composition of the particles. Mass characteristics should ideally be determined in the field, but in many cases they can be detected in undisturbed samples – i.e. samples in which the in-situ soil structure has been essentially preserved. A description of mass characteristics should include an assessment of in-situ compactive state (coarse-grained soils) or stiffness (fine-grained soils), and details of any bedding, discontinuities and weathering. The arrangement of minor geological details, referred to as the soil macro-fabric, should be carefully described, as this can influence the engineering behaviour of the in-situ soil to a considerable extent. Examples of macro-fabric features are thin layers of fine sand and silt in clay, silt-filled fissures in clay, small lenses of clay in sand, organic inclusions, and root holes. The name of the geological formation, if definitely known, should be included in the description; in addition, the type of deposit may be stated (e.g. till, alluvium), as this can indicate, in a general way, the likely behaviour of the soil.

It is important to distinguish between soil description and soil classification. Soil description includes details of both material and mass characteristics, and therefore it is unlikely that any two soils will have identical descriptions. In soil classification, on the other hand, a soil is allocated to one of a limited number of behavioural groups on the basis of material characteristics only. Soil classification is thus independent of the in-situ condition of the soil mass. If the soil is to be employed in its undisturbed condition, for example to support a foundation, a full soil description will be adequate and the addition of the soil classification is discretionary. However, classification is particularly useful if the soil in question is to be used as a construction material when it will be remoulded – for example in an embankment. Engineers can also draw on past experience of the behaviour of soils of similar classification.

Rapid assessment procedures

Both soil description and classification require knowledge of grading and plasticity. This can be determined by the full laboratory procedure using standard tests, as described in Sections 1.3 and 1.4, in which values defining the particle size distribution and the liquid and plastic limits are obtained for the soil in question. Alternatively, grading and plasticity can be assessed using a rapid procedure which involves personal judgements based on the appearance and feel of the soil. The rapid procedure can be used in the field and in other situations where the use of the laboratory procedure is not possible or not justified. In the rapid procedure, the following indicators should be used.

Particles of 0.06 mm, the lower size limit for coarse soils, are just visible to the naked eye, and feel harsh but not gritty when rubbed between the fingers; finer material feels smooth to the touch. The size boundary between sand and gravel is 2 mm, and this represents the largest size of particles which will hold together by capillary attraction when moist. A purely visual judgement must be made as to whether the sample is well graded or poorly graded, this being more difficult for sands than for gravels.

If a predominantly coarse soil contains a significant proportion of fine material, it is important to know whether the fines are essentially plastic or non-plastic (i.e. whether they are predominantly clay or silt, respectively). This can be judged by the extent to which the soil exhibits cohesion and plasticity. A small quantity of the soil, with the largest particles removed, should be moulded together in the hands, adding water if necessary. Cohesion is indicated if the soil, at an appropriate water content, can be moulded into a relatively firm mass. Plasticity is indicated if the soil can be deformed without cracking or crumbling, i.e. without losing cohesion. If cohesion and plasticity are pronounced, then the fines are plastic. If cohesion and plasticity are absent or only weakly indicated, then the fines are essentially non-plastic.

The plasticity of fine soils can be assessed by means of the toughness and dilatancy tests, described below. An assessment of dry strength may also be useful. Any coarse particles, if present, are first removed, and then a small sample of the soil is moulded in the hand to a consistency judged to be just above the plastic limit (i.e. just enough water to mould); water is added or the soil is allowed to dry as necessary. The procedures are then as follows.

Toughness test

A small piece of soil is rolled out into a thread on a flat surface or on the palm of the hand, moulded together, and rolled out again until it has dried sufficiently to break into lumps at a diameter of around 3 mm. In this condition, inorganic clays of high liquid limit are fairly stiff and tough; those of low liquid limit are softer and crumble more easily. Inorganic silts produce a weak and often soft thread, which may be difficult to form and which readily breaks and crumbles.

Dilatancy test

A pat of soil, with sufficient water added to make it soft but not sticky, is placed in the open (horizontal) palm of the hand. The side of the hand is then struck against the other hand several times. Dilatancy is indicated by the appearance of a shiny film of water on the surface of the pat; if the pat is then squeezed or pressed with the fingers, the surface becomes dull as the pat stiffens and eventually crumbles. These reactions are pronounced only for predominantly silt-size material and for very fine sands. Plastic clays give no reaction.

Dry strength test

A pat of soil about 6 mm thick is allowed to dry completely, either naturally or in an oven. The strength of the dry soil is then assessed by breaking and crumbling between the fingers. Inorganic clays have relatively high dry strength; the greater the strength, the higher the liquid limit. Inorganic silts of low liquid limit have little or no dry strength, crumbling easily between the fingers.

Soil description details

A detailed guide to soil description as used in the UK is given in BS 5930 (1999), and the subsequent discussion is based on this standard. In Europe the standard is EN ISO 14688-1 (2002), while in the United States ASTM D2487 (2011) is used. The basic soil types are boulders, cobbles, gravel, sand, silt and clay, defined in terms of the particle size ranges shown in Figure 1.2; added to these are organic clay, silt or sand, and peat. These names are always written in capital letters in a soil description. Mixtures of the basic soil types are referred to as composite types.

A soil is of basic type sand or gravel (these being termed coarse soils) if, after the removal of any cobbles or boulders, over 65% of the material is of sand and gravel sizes. A soil is of basic type silt or clay (termed fine soils) if, after the removal of any cobbles or boulders, over 35% of the material is of silt and clay sizes. However, these percentages should be considered as approximate guidelines, not forming a rigid boundary. Sand and gravel may each be subdivided into coarse, medium and fine fractions as defined in Figure 1.2. The state of sand and gravel can be described as well graded, poorly graded, uniform or gap graded, as defined in Section 1.4. In the case of gravels, particle shape (angular, sub-angular, sub-rounded, rounded, flat, elongated) and surface texture (rough, smooth, polished) can be described if necessary. Particle composition can also be stated. Gravel particles are usually rock fragments (e.g. sandstone, schist). Sand particles usually consist of individual mineral grains (e.g. quartz, feldspar). Fine soils should be described as either silt or clay: terms such as silty clay should not be used.

Organic soils contain a significant proportion of dispersed vegetable matter, which usually produces a distinctive odour and often a dark brown, dark grey or bluish grey colour. Peats consist predominantly of plant remains, usually dark brown or black in colour and with a distinctive odour. If the plant remains are recognisable and retain some strength, the peat is described as fibrous. If the plant remains are recognisable but their strength has been lost, they are pseudo-fibrous. If recognisable plant remains are absent, the peat is described as amorphous. Organic content is measured by burning a sample of soil at a controlled temperature to determine the reduction in mass which corresponds to the organic content. Alternatively, the soil may be treated with hydrogen peroxide (H_2O_2), which also removes the organic content, resulting in a loss of mass.

Composite types of coarse soil are named in Table 1.2, the predominant component being written in capital letters. Fine soils containing 35–65% coarse material are described as sandy and/or gravelly SILT (or CLAY). Deposits containing over 50% of boulders and cobbles are referred to as very coarse, and normally can be described only in excavations and exposures. Mixes of very coarse material with finer soils can be described by combining the descriptions of the two components – e.g. COBBLES with some FINER MATERIAL (sand); gravelly SAND with occasional BOULDERS.

The state of compaction or stiffness of the in-situ soil can be assessed by means of the tests or indications detailed in Table 1.3.

Table 1.2 Composite types of coarse soil

Slightly sandy GRAVEL	Up to 5% sand
Sandy GRAVEL	5–20% sand
Very sandy GRAVEL	Over 20% sand
SAND and GRAVEL	About equal proportions
Very gravelly SAND	Over 20% gravel
Gravelly SAND	5–20% gravel
Slightly gravelly SAND	Up to 5% gravel
Slightly silty SAND (and/or GRAVEL)	Up to 5% silt
Silty SAND (and/or GRAVEL)	5–20% silt
Very silty SAND (and/or GRAVEL)	Over 20% silt
Slightly clayey SAND (and/or GRAVEL)	Up to 5% clay
Clayey SAND (and/or GRAVEL)	5–20% clay
Very clayey SAND (and/or GRAVEL)	Over 20% clay

Note: Terms such as 'Slightly clayey gravelly SAND' (having less than 5% clay and gravel) and 'Silty sandy GRAVEL' (having 5–20% silt and sand) can be used, based on the above proportions of secondary constituents.

Table 1.3 Compactive state and stiffness of soils

Soil group	Term (relative density – Section 1.6)	Field test or indication
Coarse soils	Very loose (0–20%)	Assessed on basis of N-value determined by means of Standard Penetration Test (SPT) – see Chapter 7
	Loose (20–40%)	
	Medium dense (40–60%)	
	Dense (60–80%)	For definition of relative density, see Equation (1.23)
	Very dense (80–100%)	
	Slightly cemented	Visual examination: pick removes soil in lumps which can be abraded
Fine soils	Uncompact	Easily moulded or crushed by the fingers
	Compact	Can be moulded or crushed by strong finger pressure
	Very soft	Finger can easily be pushed in up to 25 mm
	Soft	Finger can be pushed in up to 10 mm
	Firm	Thumb can make impression easily
	Stiff	Thumb can make slight indentation
	Very stiff	Thumb nail can make indentation
	Hard	Thumb nail can make surface scratch
Organic soils	Firm	Fibres already pressed together
	Spongy	Very compressible and open structure
	Plastic	Can be moulded in the hand and smears fingers

Discontinuities such as fissures and shear planes, including their spacings, should be indicated. Bedding features, including their thickness, should be detailed. Alternating layers of varying soil types or with bands or lenses of other materials are described as **interstratified**. Layers of different soil types are described as **interbedded** or **inter-laminated**, their thickness being stated. Bedding surfaces that separate easily are referred to as **partings**. If partings incorporate other material, this should be described.

Some examples of soil description are as follows:

> Dense, reddish-brown, sub-angular, well-graded SAND
> Firm, grey, laminated CLAY with occasional silt partings 0.5–2.0 mm (Alluvium)
> Dense, brown, well graded, very silty SAND and GRAVEL with some COBBLES (Till)
> Stiff, brown, closely fissured CLAY (London Clay)
> Spongy, dark brown, fibrous PEAT (Recent Deposits)

Soil classification systems

General classification systems, in which soils are placed into behavioural groups on the basis of grading and plasticity, have been used for many years. The feature of these systems is that each soil group is denoted by a letter symbol representing main and qualifying terms. The terms and letters used in the UK are detailed in Table 1.4. The boundary between coarse and fine soils is generally taken to be 35% fines (i.e. particles smaller than 0.06 mm). The liquid and plastic limits are used to classify fine soils, employing the plasticity chart shown in Figure 1.12. The axes of the plasticity chart are the plasticity index and liquid limit; therefore, the plasticity characteristics of a particular soil can be represented by a point on the chart. Classification letters are allotted to the soil according to the zone within which the point lies. The chart is divided into five ranges of liquid limit. The four ranges I, H, V and E can be combined as an upper range (U) if closer designation is not required, or if the rapid assessment procedure has been used to assess plasticity. The diagonal line on the chart, known as the **A-line**, should not be regarded as a rigid boundary between clay and silt for purposes of soil description, as opposed to classification. The A-line may be mathematically represented by

$$I_P = 0.73(w_L - 20) \tag{1.5}$$

The letter denoting the dominant size fraction is placed first in the group symbol. If a soil has a significant content of organic matter, the suffix O is added as the last letter of the group symbol. A group symbol may consist of two or more letters, for example:

> SW – well-graded SAND
> SCL – very clayey SAND (clay of low plasticity)
> CIS – sandy CLAY of intermediate plasticity
> MHSO – organic sandy SILT of high plasticity.

The name of the soil group should always be given, as above, in addition to the symbol, the extent of subdivision depending on the particular situation. If the rapid procedure has been used to assess grading and plasticity, the group symbol should be enclosed in brackets to indicate the lower degree of accuracy associated with this procedure.

Basic characteristics of soils

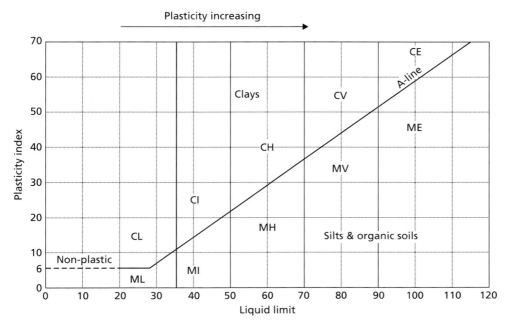

Figure 1.12 Plasticity chart: British system (BS 1377–2: 1990).

The term FINE SOIL or FINES (F) is used when it is not required, or not possible, to differentiate between SILT (M) and CLAY (C). SILT (M) plots below the A-line and CLAY (C) above the A-line on the plasticity chart, i.e. silts exhibit plastic properties over a lower range of water content than clays having the same liquid limit. SILT or CLAY is qualified as gravelly if more than 50% of the coarse fraction is of gravel size, and as sandy if more than 50% of the coarse fraction is of sand size. The alternative term M-SOIL is introduced to describe material which, regardless of its particle size distribution, plots below the A-line on the plasticity chart: the use of this term avoids confusion with soils of predominantly silt size (but with a significant proportion of clay-size particles), which plot above the A-line. Fine soils containing significant amounts of organic matter usually have high to extremely high liquid limits, and plot below the A-line as organic silt. Peats usually have very high or extremely high liquid limits.

Any cobbles or boulders (particles retained on a 63-mm sieve) are removed from the soil before the classification tests are carried out, but their percentages in the total sample should be determined or estimated. Mixtures of soil and cobbles or boulders can be indicated by using the letters Cb (COBBLES) or B (BOULDERS) joined by a + sign to the group symbol for the soil, the dominant component being given first – for example:

> GW + Cb – well-graded GRAVEL with COBBLES
> B + CL – BOULDERS with CLAY of low plasticity.

A similar classification system, known as the Unified Soil Classification System (USCS), was developed in the United States (described in ASTM D2487), but with less detailed subdivisions. As the USCS method is popular in other parts of the world, alternative versions of Figure 1.12 and Table 1.4 are provided on the Companion Website.

Development of a mechanical model for soil

Table 1.4 Descriptive terms for soil classification (BS 5930)

Main terms		Qualifying terms	
GRAVEL	G	Well graded	W
SAND	S	Poorly graded	P
		Uniform	Pu
		Gap graded	Pg
FINE SOIL, FINES	F	Of low plasticity ($w_L < 35$)	L
SILT (M-SOIL)	M	Of intermediate plasticity (w_L 35–50)	I
CLAY	C	Of high plasticity (w_L 50–70)	H
		Of very high plasticity (w_L 70–90)	V
		Of extremely high plasticity ($w_L > 90$)	E
		Of upper plasticity range ($w_L > 35$)	U
PEAT	Pt	Organic (may be a suffix to any group)	O

Example 1.1

The results of particle size analyses of four soils A, B, C and D are shown in Table 1.5. The results of limit tests on soil D are as follows:

TABLE A

Liquid limit:					
Cone penetration (mm)	15.5	18.0	19.4	22.2	24.9
Water content (%)	39.3	40.8	42.1	44.6	45.6
Plastic limit:					
Water content (%)	23.9	24.3			

The fine fraction of soil C has a liquid limit of $I_L = 26$ and a plasticity index of $I_P = 9$.

a Determine the coefficients of uniformity and curvature for soils A, B and C.
b Allot group symbols, with main and qualifying terms to each soil.

Table 1.5 Example 1.1

Sieve	Particle size*	Percentage smaller			
		Soil A	Soil B	Soil C	Soil D
63 mm		100		100	
20 mm		64		76	
6.3 mm		39	100	65	
2 mm		24	98	59	
600 mm		12	90	54	
212 mm		5	9	47	100
63 mm		0	3	34	95
	0.020 mm			23	69
	0.006 mm			14	46
	0.002 mm			7	31

Note: * From sedimentation test.

Solution

The particle size distribution curves are plotted in Figure 1.13. For soils A, B and C, the sizes D_{10}, D_{30} and D_{60} are read from the curves and the values of C_u and C_z are calculated:

TABLE B

Soil	D_{10}	D_{30}	D_{60}	C_U	C_Z
A	0.47	3.5	16	34	1.6
B	0.23	0.30	0.41	1.8	0.95
C	0.003	0.042	2.4	800	0.25

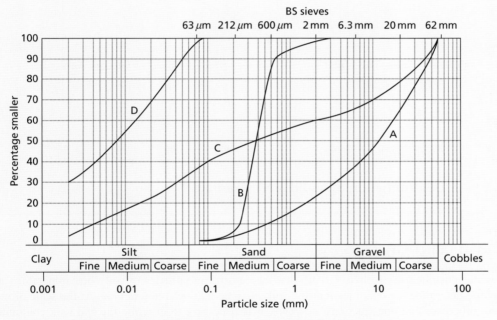

Figure 1.13 Particle size distribution curves (Example 1.1).

For soil D the liquid limit is obtained from Figure 1.14, in which fall-cone penetration is plotted against water content. The percentage water content, to the nearest integer, corresponding to a penetration of 20 mm is the liquid limit, and is 42%. The plastic limit is the average of the two percentage water contents, again to the nearest integer, i.e. 24%. The plasticity index is the difference between the liquid and plastic limits, i.e. 18%.

Soil A consists of 100% coarse material (76% gravel size; 24% sand size) and is classified as GW: well-graded, very sandy GRAVEL.

Soil B consists of 97% coarse material (95% sand size; 2% gravel size) and 3% fines. It is classified as SPu: uniform, slightly silty, medium SAND.

Development of a mechanical model for soil

Soil C comprises 66% coarse material (41% gravel size; 25% sand size) and 34% fines ($w_L = 26$, $I_P = 9$, plotting in the CL zone on the plasticity chart). The classification is GCL: very clayey GRAVEL (clay of low plasticity). This is a till, a glacial deposit having a large range of particle sizes.

Soil D contains 95% fine material: the liquid limit is 42 and the plasticity index is 18, plotting just above the A-line in the CI zone on the plasticity chart. The classification is thus CI: CLAY of intermediate plasticity.

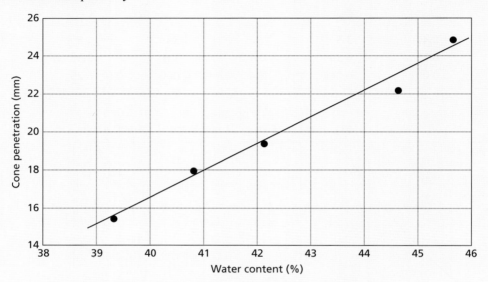

Figure 1.14 Determination of liquid limit (Example 1.1).

1.6 Phase relationships

It has been demonstrated in Sections 1.1–1.5 that the constituent particles of soil, their mineralogy and microstructure determine the classification of a soil into a certain behavioural type. At the scale of most engineering processes and constructions, however, it is necessary to describe the soil as a continuum. Soils can be of either two-phase or three-phase composition. In a completely dry soil there are two phases, namely the solid soil particles and pore air. A fully saturated soil is also two-phase, being composed of solid soil particles and pore water. A partially saturated soil is three-phase, being composed of solid soil particles, pore water and pore air. The components of a soil can be represented by a phase diagram as shown in Figure 1.15(a), from which the following relationships are defined.

The water content (w), or moisture content (m), is the ratio of the mass of water to the mass of solids in the soil, i.e.

$$w = \frac{M_w}{M_s} \tag{1.6}$$

The water content is determined by weighing a sample of the soil and then drying the sample in an oven at a temperature of 105–110°C and re-weighing. Drying should continue until the differences between successive weighings at four-hourly intervals are not greater than 0.1% of the original mass of the sample. A drying period of 24 h is normally adequate for most soils.

Basic characteristics of soils

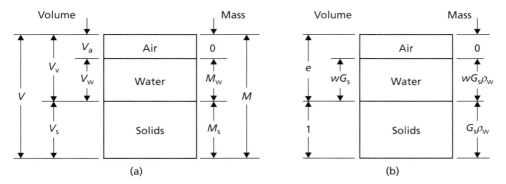

Figure 1.15 Phase diagrams.

The degree of saturation or **saturation ratio** (S_r) is the ratio of the volume of water to the total volume of void space, i.e.

$$S_r = \frac{V_w}{V_v} \tag{1.7}$$

The saturation ratio can range between the limits of zero for a completely dry soil and one (or 100%) for a fully saturated soil.

The **void ratio** (e) is the ratio of the volume of voids to the volume of solids, i.e.

$$e = \frac{V_v}{V_s} \tag{1.8}$$

The **porosity** (n) is the ratio of the volume of voids to the total volume of the soil, i.e.

$$n = \frac{V_v}{V} \tag{1.9}$$

As $V = V_v + V_s$, void ratio and porosity are interrelated as follows:

$$e = \frac{n}{1-n} \tag{1.10}$$

$$n = \frac{e}{1+e} \tag{1.11}$$

The **specific volume** (v) is the total volume of soil which contains a unit volume of solids, i.e.

$$v = \frac{V_v}{V_s} = 1+e \tag{1.12}$$

The **air content** or air voids (A) is the ratio of the volume of air to the total volume of the soil, i.e.

$$A = \frac{V_a}{V} \tag{1.13}$$

The **bulk density** or mass density (ρ) of a soil is the ratio of the total mass to the total volume, i.e.

$$\rho = \frac{M}{V} \tag{1.14}$$

23

Development of a mechanical model for soil

Convenient units for density are kg/m³ or Mg/m³. The density of water (1000 kg/m³ or 1.00 Mg/m³) is denoted by ρ_w.

The **specific gravity** of the soil particles (G_s) is given by

$$G_s = \frac{M_s}{V_s \rho_w} = \frac{\rho_s}{\rho_w} \quad (1.15)$$

where ρ_s is the **particle density**.

From the definition of void ratio, if the volume of solids is 1 unit then the volume of voids is e units. The mass of solids is then $G_s \rho_w$ and, from the definition of water content, the mass of water is $wG_s \rho_w$. The volume of water is thus wG_s. These volumes and masses are represented in Figure 1.15(b). From this figure, the following relationships can then be obtained.

The degree of saturation (definition in Equation 1.7) is

$$S_r = \frac{V_w}{V_v} = \frac{wG_s}{e} \quad (1.16)$$

The air content is the proportion of the total volume occupied by air, i.e.

$$A = \frac{V_a}{V} = \frac{e - wG_s}{1+e} \quad (1.17)$$

or, from Equations 1.11 and 1.16,

$$A = n(1 - S_r) \quad (1.18)$$

From Equation 1.14, the bulk density of a soil is:

$$\rho = \frac{M}{V} = \frac{G_s(1+w)\rho_s}{1+e} \quad (1.19)$$

or, from Equation 1.16,

$$\rho = \frac{G_s + S_r e}{1+e} \rho_w \quad (1.20)$$

Equation 1.20 holds true for any soil. Two special cases that commonly occur, however, are when the soil is fully saturated with either water or air. For a fully saturated soil $S_r = 1$, giving:

$$\rho_{sat} = \frac{G_s + e}{1+e} \rho_w \quad (1.21)$$

For a completely dry soil ($S_r = 0$):

$$\rho_d = \frac{G_s}{1+e} \rho_w \quad (1.22)$$

The **unit weight** or weight density (γ) of a soil is the ratio of the total weight (Mg) to the total volume, i.e.

$$\gamma = \frac{Mg}{V} = \rho g$$

Basic characteristics of soils

Multiplying Equations 1.19 and 1.20 by g then gives

$$\gamma = \frac{G_s(1+w)}{1+e}\gamma_w \tag{1.19a}$$

$$\gamma = \frac{G_s + S_r e}{1+e}\gamma_w \tag{1.20a}$$

where γ_w is the unit weight of water. Convenient units are kN/m³, the unit weight of water being 9.81 kN/m³ (or 10.0 kN/m³ in the case of sea water).

In the case of sands and gravels the **relative density** (I_D) is used to express the relationship between the in-situ void ratio (e), or the void ratio of a sample, and the limiting values e_{max} and e_{min} representing the loosest and densest possible soil packing states respectively. The relative density is defined as

$$I_D = \frac{e_{max} - e}{e_{max} - e_{min}} \tag{1.23}$$

Thus, the relative density of a soil in its densest possible state ($e = e_{min}$) is 1 (or 100%) and in its loosest possible state ($e = e_{max}$) is 0.

The maximum density is determined by compacting a sample underwater in a mould, using a circular steel tamper attached to a vibrating hammer: a 1-l mould is used for sands and a 2.3-l mould for gravels. The soil from the mould is then dried in an oven, enabling the dry density to be determined. The minimum dry density can be determined by one of the following procedures. In the case of sands, a 1-l measuring cylinder is partially filled with a dry sample of mass 1000 g and the top of the cylinder closed with a rubber stopper. The minimum density is achieved by shaking and inverting the cylinder several times, the resulting volume being read from the graduations on the cylinder. In the case of gravels, and sandy gravels, a sample is poured from a height of about 0.5 m into a 2.3-l mould and the resulting dry density determined. Full details of the above tests are given in BS 1377, Part 4 (1990). Void ratio can be calculated from a value of dry density using Equation 1.22. However, the density index can be calculated directly from the maximum, minimum and in-situ values of dry density, avoiding the need to know the value of G_s (see Problem 1.5).

Example 1.2

In its natural condition, a soil sample has a mass of 2290 g and a volume of 1.15×10^{-3} m³. After being completely dried in an oven, the mass of the sample is 2035 g. The value of G_s for the soil is 2.68. Determine the bulk density, unit weight, water content, void ratio, porosity, degree of saturation and air content.

Solution

$$\text{Bulk density, } \rho = \frac{M}{V} = \frac{2.290}{1.15 \times 10^{-3}} = 1990 \text{ kg/m}^3 \, (1.99 \text{ Mg/m}^3)$$

$$\text{Unit weight, } \gamma = \frac{Mg}{V} = 1990 \times 9.8 = 19500 \text{ N/m}^3$$

$$= 19.5 \text{ kN/m}^3$$

$$\text{Water content, } w = \frac{M_w}{M_s} = \frac{2290 - 2035}{2035} = 0.125 \text{ or } 12.5\%$$

From Equation 1.19,

$$\text{Void ratio, } e = G_s(1+w)\frac{\rho_w}{\rho} - 1$$

$$= \left(2.68 \times 1.125 \times \frac{1000}{1990}\right) - 1$$

$$= 1.52 - 1$$

$$= 0.52$$

$$\text{Porosity, } n = \frac{e}{1+e} = \frac{0.52}{1.52} = 0.34 \text{ or } 34\%$$

$$\text{Degree of saturation, } S_r = \frac{wG_s}{e} = \frac{0.125 \times 2.68}{0.52} = 0.645 \text{ or } 64.5\%$$

$$\text{Air content, } A = n(1 - S_r) = 0.34 \times 0.355$$

$$= 0.121 \text{ or } 12.1\%$$

1.7 Soil compaction

Compaction is the process of increasing the density of a soil by packing the particles closer together with a reduction in the volume of air; there is no significant change in the volume of water in the soil. In the construction of fills and embankments, loose soil is typically placed in layers ranging between 75 and 450 mm in thickness, each layer being compacted to a specified standard by means of rollers, vibrators or rammers. In general, the higher the degree of compaction, the higher will be the shear strength and the lower will be the compressibility of the soil (see Chapters 4 and 5). An **engineered fill** is one in which the soil has been selected, placed and compacted to an appropriate specification with the object of achieving a particular engineering performance, generally based on past experience. The aim is to ensure that the resulting fill possesses properties that are adequate for the function of the fill. This is in contrast to non-engineered fills, which have been placed without regard to a subsequent engineering function.

The degree of compaction of a soil is measured in terms of dry density, i.e. the mass of solids only per unit volume of soil. If the bulk density of the soil is ρ and the water content w, then from Equations 1.19 and 1.22 it is apparent that the dry density is given by

$$\rho_d = \frac{\rho}{1+w} \tag{1.24}$$

The dry density of a given soil after compaction depends on the water content and the energy supplied by the compaction equipment (referred to as the **compactive effort**).

Laboratory compaction

The compaction characteristics of a soil can be assessed by means of standard laboratory tests. The soil is compacted in a cylindrical mould using a standard compactive effort. In the **Proctor test**, the volume

Basic characteristics of soils

of the mould is 1-l and the soil (with all particles larger than 20 mm removed) is compacted by a rammer consisting of a 2.5-kg mass falling freely through 300 mm: the soil is compacted in three equal layers, each layer receiving 27 blows with the rammer. In the **modified AASHTO test**, the mould is the same as is used in the above test but the rammer consists of a 4.5-kg mass falling 450 mm; the soil (with all particles larger than 20 mm removed) is compacted in five layers, each layer receiving 27 blows with the rammer. If the sample contains a limited proportion of particles up to 37.5 mm in size, a 2.3-l mould should be used, each layer receiving 62 blows with either the 2.5- or 4.5-kg rammer. In the **vibrating hammer test**, the soil (with all particles larger than 37.5 mm removed) is compacted in three layers in a 2.3-l mould, using a circular tamper fitted in the vibrating hammer, each layer being compacted for a period of 60 s. These tests are detailed in BS1377–4 (UK), EC7–2 (Europe) and, in the US, ASTM D698, D1557 and D7382.

After compaction using one of the three standard methods, the bulk density and water content of the soil are determined and the dry density is calculated. For a given soil the process is repeated at least five times, the water content of the sample being increased each time. Dry density is plotted against water content, and a curve of the form shown in Figure 1.16 is obtained. This curve shows that for a particular method of compaction (i.e. a given compactive effort) there is a particular value of water content, known as the **optimum water content** (w_{opt}), at which a maximum value of dry density is obtained. At low values of water content, most soils tend to be stiff and are difficult to compact. As the water content is increased the soil becomes more workable, facilitating compaction and resulting in higher dry densities. At high water contents, however, the dry density decreases with increasing water content, an increasing proportion of the soil volume being occupied by water.

If all the air in a soil could be expelled by compaction, the soil would be in a state of full saturation and the dry density would be the maximum possible value for the given water content. However, this degree of compaction is unattainable in practice. The maximum possible value of dry density is referred to as the **'zero air voids' dry density** (ρ_{d0}) or the saturation dry density, and can be calculated from the expression:

$$\rho_{d0} = \frac{G_s}{1 + wG_s} \rho_w \tag{1.25}$$

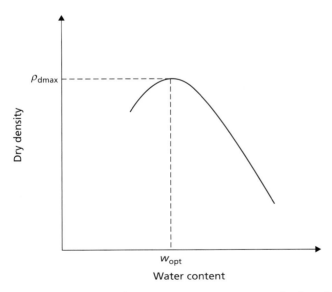

Figure 1.16 Dry density–water content relationship.

Development of a mechanical model for soil

In general, the dry density after compaction at water content w to an air content A can be calculated from the following expression, derived from Equations 1.17 and 1.22:

$$\rho_d = \frac{G_s(1-A)}{1+wG_s}\rho_w \tag{1.26}$$

The calculated relationship between zero air voids dry density (A=0) and water content (for G_s=2.65) is shown in Figure 1.17; the curve is referred to as the zero air voids line or the **saturation line**. The experimental dry density–water content curve for a particular compactive effort must lie completely to the left of the saturation line. The curves relating dry density at air contents of 5% and 10% with water content are also shown in Figure 1.17, the values of dry density being calculated from Equation 1.26. These curves enable the air content at any point on the experimental dry density–water content curve to be determined by inspection.

For a particular soil, different dry density–water content curves are obtained for different compactive efforts. Curves representing the results of tests using the 2.5- and 4.5-kg rammers are shown in Figure 1.17. The curve for the 4.5-kg test is situated above and to the left of the curve for the 2.5-kg test. Thus, a higher compactive effort results in a higher value of maximum dry density and a lower value of optimum water content; however, the values of air content at maximum dry density are approximately equal.

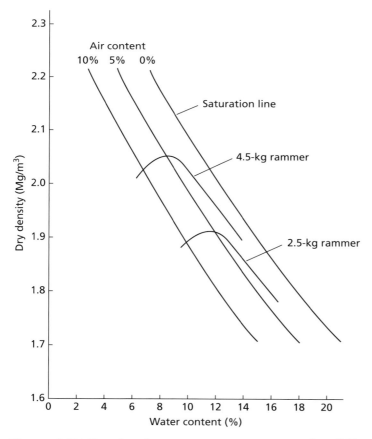

Figure 1.17 Dry density–water content curves for different compactive efforts.

Basic characteristics of soils

The dry density–water content curves for a range of soil types using the same compactive effort (the BS 2.5-kg rammer) are shown in Figure 1.18. In general, coarse soils can be compacted to higher dry densities than fine soils.

Field compaction

The results of laboratory compaction tests are not directly applicable to field compaction because the compactive efforts in the laboratory tests are different, and are applied in a different way, from those produced by field equipment. Further, the laboratory tests are carried out only on material smaller than either 20 or 37.5 mm. However, the maximum dry densities obtained in the laboratory using the 2.5- and 4.5-kg rammers cover the range of dry density normally produced by field compaction equipment.

A minimum number of passes must be made with the chosen compaction equipment to produce the required value of dry density. This number, which depends on the type and mass of the equipment and on the thickness of the soil layer, is usually within the range 3–12. Above a certain number of passes, no significant increase in dry density is obtained. In general, the thicker the soil layer, the heavier the equipment required to produce an adequate degree of compaction.

There are two approaches to the achievement of a satisfactory standard of compaction in the field, known as **method compaction** and **end-product compaction**. In method compaction, the type and mass

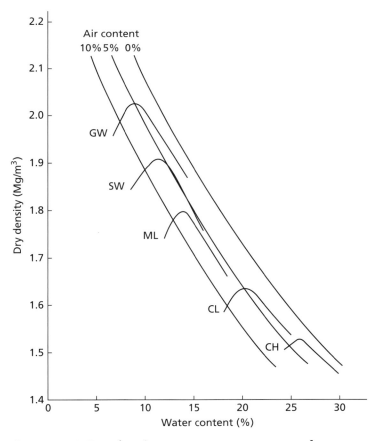

Figure 1.18 Dry density–water content curves for a range of soil types.

of equipment, the layer depth and the number of passes are specified. In the UK, these details are given, for the class of material in question, in the Specification for Highway Works (Highways Agency, 2008). In end-product compaction, the required dry density is specified: the dry density of the compacted fill must be equal to or greater than a stated percentage of the maximum dry density obtained in one of the standard laboratory compaction tests.

Field density tests can be carried out, if considered necessary, to verify the standard of compaction in earthworks, dry density or air content being calculated from measured values of bulk density and water content. A number of methods of measuring bulk density in the field are detailed in BS 1377, Part 4 (1990).

The types of compaction equipment commonly used in the field are described below, and their performance over a range of standard soils is compared in Figure 1.19. Images of all types of plant for identification purposes may be found through the Companion Website.

Smooth-wheeled rollers

These consist of hollow steel drums, the mass of which can be increased by water or sand ballast. They are suitable for most types of soil except uniform sands and silty sands, provided a mixing or kneading action is not required. A smooth surface is produced on the compacted layer, encouraging the run-off of any rainfall but resulting in relatively poor bonding between successive layers; the fill as a whole will therefore tend to be laminated. Smooth-wheeled rollers, and the other types of roller described below, can be either towed or self-propelled.

Pneumatic-tyred rollers

This equipment is suitable for a wide range of coarse and fine soils, but not for uniformly graded material. Wheels are mounted close together on two axles, the rear set overlapping the lines of the front set to ensure complete coverage of the soil surface. The tyres are relatively wide with a flat tread so that the soil is not displaced laterally. This type of roller is also available with a special axle which allows the wheels to wobble, thus preventing the bridging over of low spots. Pneumatic-tyred rollers impart a kneading action to the soil. The finished surface is relatively smooth, resulting in a low degree of bonding between layers. If good bonding is essential, the compacted surface must be scarified between layers. Increased compactive effort can be obtained by increasing the tyre inflation pressure or, less effectively, by adding additional weight (kentledge) to the body of the roller.

Sheepsfoot rollers

A sheepsfoot roller consists of hollow steel drums with numerous tapered or club-shaped feet projecting from their surfaces. The mass of the drums can be increased by ballasting. The arrangement of the feet can vary, but they are usually from 200 to 250 mm in length with an end area of 40–65 cm^2. The feet thus impart a relatively high pressure over a small area. Initially, when the layer of soil is loose, the drums are in contact with the soil surface. Subsequently, as the projecting feet compact below the surface and the soil becomes sufficiently dense to support the high contact pressure, the drums rise above the soil. Sheepsfoot rollers are most suitable for fine soils, both plastic and non-plastic, especially at water contents dry of optimum. They are also suitable for coarse soils with more than 20% of fines. The action of the feet causes significant mixing of the soil, improving its degree of homogeneity, and will break up lumps of stiff material, making the roller particularly suitable for re-compacting excavated clays which tend to be placed in the form of large lumps or peds. Due to the penetration of the feet, excellent bonding is produced between successive soil layers – an important requirement for water-retaining earthworks. Tamping rollers are similar to sheepsfoot rollers but the feet have a larger end area, usually over 100 cm^2, and the total area of the feet exceeds 15% of the surface area of the drums.

Grid rollers

These rollers have a surface consisting of a network of steel bars forming a grid with square holes. Kentledge can be added to the body of the roller. Grid rollers provide high contact pressure but little kneading action, and are suitable for most coarse soils.

Vibratory rollers

These are smooth-wheeled rollers fitted with a power-driven vibration mechanism. They are used for most soil types, and are more efficient if the water content of the soil is slightly wet of optimum. They are particularly effective for coarse soils with little or no fines. The mass of the roller and the frequency of vibration must be matched to the soil type and layer thickness. The lower the speed of the roller, the fewer the number of passes required.

Vibrating plates

This equipment, which is suitable for most soil types, consists of a steel plate with upturned edges, or a curved plate, on which a vibrator is mounted. The unit, under manual guidance, propels itself slowly over the surface of the soil.

Power rammers

Manually controlled power rammers, generally petrol-driven, are used for the compaction of small areas where access is difficult or where the use of larger equipment would not be justified. They are also used extensively for the compaction of backfill in trenches. They do not operate effectively on uniformly graded soils.

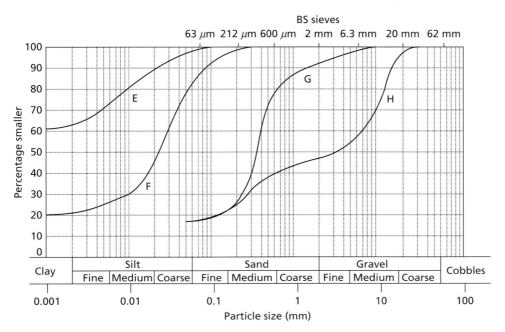

Figure 1.19(a) Performance envelopes of various compaction methods for standard soil types (replotted after Croney and Croney, 1997): PSD curves of soils.

Development of a mechanical model for soil

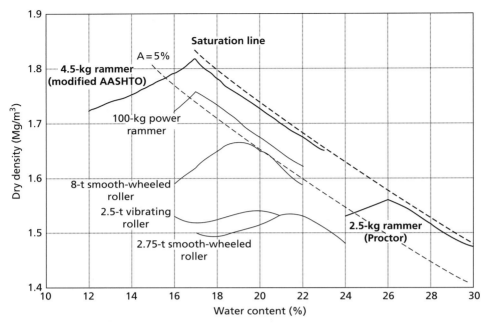

Figure 1.19(b) Performance envelopes of various compaction methods for standard soil types (replotted after Croney and Croney, 1997): Soil E.

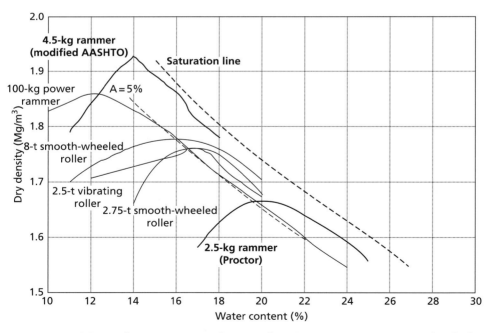

Figure 1.19(c) Performance envelopes of various compaction methods for standard soil types (replotted after Croney and Croney, 1997): Soil F.

Basic characteristics of soils

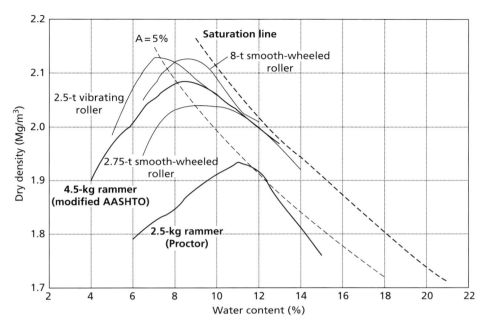

Figure 1.19(d) Performance envelopes of various compaction methods for standard soil types (replotted after Croney and Croney, 1997): Soil G.

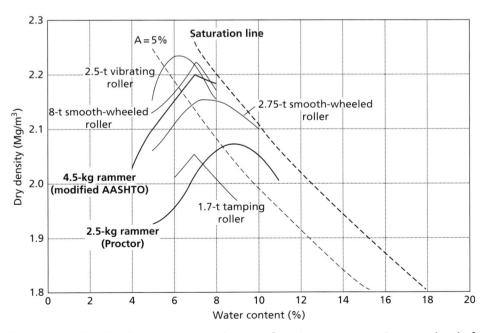

Figure 1.19(e) Performance envelopes of various compaction methods for standard soil types (replotted after Croney and Croney, 1997): Soil H.

Development of a mechanical model for soil

The changed mechanical properties of a compacted soil should be determined using appropriate laboratory tests on samples taken from trial compactions in the field or from soil compacted in the laboratory (e.g. following the standard Proctor test) (see Chapters 4 and 5); alternatively in-situ tests may be conducted on the finished earthworks (Chapter 7).

Moisture condition test

As an alternative to standard compaction tests, the moisture condition test is widely used in the UK. This test, originally developed by the Transport and Road Research Laboratory (now TRL), enables a rapid assessment to be made of the suitability of soils for use as fill materials (Parsons and Boden, 1979). The test does not involve the determination of water content, a cause of delay in obtaining the results of compaction tests due to the necessity of drying the soil. In principle, the test consists of determining the effort required to compact a soil sample (normally 1.5 kg) close to its maximum density. The soil is compacted in a cylindrical mould, having an internal diameter of 10 mm, centred on the base plate of the apparatus. Compaction is imparted by a rammer having a diameter of 97 mm and a mass of 7 kg falling freely from a height of 250 mm. The fall of the rammer is controlled by an adjustable release mechanism and two vertical guide rods. The penetration of the rammer into the mould is measured by means of a scale on the side of the rammer. A fibre disc is placed on top of the soil to prevent extrusion between the rammer and the inside of the mould. Full details are given in BS 1377, Part 4 (1990).

The penetration is measured at various stages of compaction. For a given number of rammer blows (n), the penetration is subtracted from the penetration at four times that number of blows ($4n$). The change in penetration between n and $4n$ blows is plotted against the logarithm (to base 10) of the lesser number of blows (n). A change in penetration of 5 mm is arbitrarily chosen to represent the condition beyond which no significant increase in density occurs. The **moisture condition value** (MCV) is defined as ten times the logarithm of the number of blows corresponding to a change in penetration of 5 mm on the above plot. An example of such a plot is shown in Figure 1.20. For a range of soil types, it has been shown that the relationship between water content and MCV is linear over a substantial range of water content. Details of the soil types for which the test is applicable can be found in Oliphant and Winter (1997).

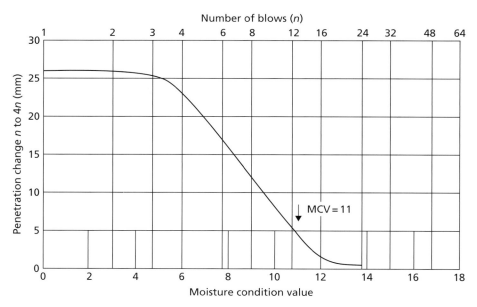

Figure 1.20 Moisture condition test.

Basic characteristics of soils

> ### Summary
> 1 Soil is a particulate material formed of weathered rock. The particles may be single grains in a wide range of sizes (from boulders to silt), or clay minerals (colloidal in size). Soil is typically formed from a mixture of such particles, and the presence of clay minerals may significantly alter the mechanical properties of the soil.
> 2 Soils may be described and classified by their particle size distribution. Fine soils consisting of mainly small particles (e.g. clays and silts) typically exhibit plastic behaviour (e.g. cohesion) which may be defined by the plasticity and liquidity indices. Coarse grained soils generally do not exhibit plastic behaviour.
> 3 At the level of the macro-fabric, all soils may be idealised as a three-phase continuum, the phases being solid particles, water and air. The relative proportions of these phases are controlled by the closeness of particle packing, described by the voids ratio (e), water content (w) and saturation ratio (S_r).
> 4 In addition to being used in their in-situ state, soils may be used as a fill material in geotechnical constructions. Compaction of such soils increases shear strength and reduces compressibility, and is necessary to achieve optimal performance of the fill. Compaction may be quantified using the Proctor or moisture condition tests.

Problems

1.1 The results of particle size analyses and, where appropriate, limit tests on samples of four soils are given in Table 1.6. Allot group symbols and give main and qualifying terms appropriate for each soil.

1.2 A soil has a bulk density of $1.91\,\text{Mg/m}^3$ and a water content of 9.5%. The value of G_s is 2.70. Calculate the void ratio and degree of saturation of the soil. What would be the values of density and water content if the soil were fully saturated at the same void ratio?

Table 1.6 Problem 1.1

BS sieve	Particle size	Percentage smaller			
		Soil I	Soil J	Soil K	Soil L
63 mm					
20 mm		100			
6.3 mm		94	100		
2 mm		69	98		
600 mm		32	88	100	
212 mm		13	67	95	100
63 mm		2	37	73	99
	0.020 mm		22	46	88
	0.006 mm		11	25	71
	0.002 mm		4	13	58
Liquid limit			Non-plastic	32	78
Plastic limit				24	31

1.3 Calculate the dry unit weight and the saturated unit weight of a soil having a void ratio of 0.70 and a value of G_s of 2.72. Calculate also the unit weight and water content at a degree of saturation of 75%.

1.4 A soil specimen is 38 mm in diameter and 76 mm long, and in its natural condition weighs 168.0 g. When dried completely in an oven, the specimen weighs 130.5 g. The value of G_s is 2.73. What is the degree of saturation of the specimen?

1.5 The in-situ dry density of a sand is $1.72\,\text{Mg/m}^3$. The maximum and minimum dry densities, determined by standard laboratory tests, are 1.81 and $1.54\,\text{Mg/m}^3$, respectively. Determine the relative density of the sand.

1.6 Soil has been compacted in an embankment at a bulk density of $2.15\,\text{Mg/m}^3$ and a water content of 12%. The value of G_s is 2.65. Calculate the dry density, void ratio, degree of saturation and air content. Would it be possible to compact the above soil at a water content of 13.5% to a dry density of $2.00\,\text{Mg/m}^3$?

1.7 The following results were obtained from a standard compaction test on a soil:

TABLE C

Mass (g)	2010	2092	2114	2100	2055
Water content (%)	12.8	14.5	15.6	16.8	19.2

The value of G_s is 2.67. Plot the dry density–water content curve, and give the optimum water content and maximum dry density. Plot also the curves of zero, 5% and 10% air content, and give the value of air content at maximum dry density. The volume of the mould is $1000\,\text{cm}^3$.

1.8 Determine the moisture condition value for the soil whose moisture condition test data are given below:

TABLE D

Number of blows	1	2	3	4	6	8	12	16	24	32	64	96	128
Penetration (mm)	15.0	25.2	33.0	38.1	44.7	49.7	57.4	61.0	64.8	66.2	68.2	68.8	69.7

References

ASTM D698 (2007) *Standard Test Methods for Laboratory Compaction Characteristics of Soil Using Standard Effort (12,400 ft lbf/ft³ (600 kN m/m³))*, American Society for Testing and Materials, West Conshohocken, PA.

ASTM D1557 (2009) *Standard Test Methods for Laboratory Compaction Characteristics of Soil Using Modified Effort (56,000 ft lbf/ft³ (2,700 kN m/m³))*, American Society for Testing and Materials, West Conshohocken, PA.

ASTM D2487 (2011) *Standard Practice for Classification of Soils for Engineering Purposes (Unified Soil Classification System)*, American Society for Testing and Materials, West Conshohocken, PA.

ASTM D4318 (2010) *Standard Test Methods for Liquid Limit, Plastic Limit, and Plasticity Index of Soils*, American Society for Testing and Materials, West Conshohocken, PA.

ASTM D6913–04 (2009) *Standard Test Methods for Particle Size Distribution (Gradation) of Soils Using Sieve Analysis*, American Society for Testing and Materials, West Conshohocken, PA.

ASTM D7382 (2008) *Standard Test Methods for Determination of Maximum Dry Unit Weight and Water Content Range for Effective Compaction of Granular Soils Using a Vibrating Hammer*, American Society for Testing and Materials, West Conshohocken, PA.

Barnes, G.E. (2009) An apparatus for the plastic limit and workability of soils, *Proceedings ICE – Geotechnical Engineering*, **162**(3), 175–185.
British Standard 1377 (1990) *Methods of Test for Soils for Civil Engineering Purposes*, British Standards Institution, London.
British Standard 5930 (1999) *Code of Practice for Site Investigations*, British Standards Institution, London.
CEN ISO/TS 17892 (2004) *Geotechnical Investigation and Testing – Laboratory Testing of Soil*, International Organisation for Standardisation, Geneva.
Croney, D. and Croney, P. (1997) *The Performance of Road Pavements* (3rd edn), McGraw Hill, New York, NY.
Day, R.W. (2001) *Soil testing manual*, McGraw Hill, New York, NY.
EC7–2 (2007) *Eurocode 7: Geotechnical Design – Part 2: Ground Investigation and Testing, BS EN 1997–2:2007*, British Standards Institution, London.
Highways Agency (2008) Earthworks, in *Specification for Highway Works*, HMSO, Series 600, London.
Mitchell, J.K. and Soga, K. (2005) *Fundamentals of Soil Behaviour* (3rd edn), John Wiley & Sons, New York, NY.
Oliphant, J. and Winter, M.G. (1997) Limits of use of the moisture condition apparatus, *Proceedings ICE – Transport*, **123**(1), 17–29.
Parsons, A.W. and Boden, J.B. (1979) *The Moisture Condition Test and its Potential Applications in Earthworks*, TRRL Report 522, Crowthorne, Berkshire.
Sivakumar, V., Glynn, D., Cairns, P. and Black, J.A. (2009) A new method of measuring plastic limit of fine materials, *Géotechnique*, **59**(10), 813–823.
Waltham, A.C. (2002) *Foundations of Engineering Geology* (2nd edn), Spon Press, Abingdon, Oxfordshire.
White, D.J. (2003) PSD measurement using the single particle optical sizing (SPOS) method, *Géotechnique*, **53**(3), 317–326.

Further reading

Collins, K. and McGown, A. (1974) The form and function of microfabric features in a variety of natural soils, *Géotechnique*, **24**(2), 223–254.
Provides further information on the structures of soil particles under different depositional regimes.
Grim, R.E. (1962) *Clay Mineralogy*, McGraw-Hill, New York, NY.
Further detail about clay mineralogy in terms of its basic chemistry.
Rowe, P.W. (1972) The relevance of soil fabric to site investigation practice, *Géotechnique*, **22**(2), 195–300.
Presents 35 case studies demonstrating how the depositional/geological history of the soil deposits influences the selection of laboratory tests and their interpretation, and the consequences for geotechnical constructions. Also includes a number of photographs showing a range of different soil types and fabric features to aid interpretation.
Henkel, D.J. (1982) Geology, geomorphology and geotechnics, *Géotechnique*, **32**(3), 175–194.
Presents a series of case studies demonstrating the importance of engineering geology and geomorphological observations in geotechnical engineering.

For further student and instructor resources for this chapter, please visit the Companion Website at www.routledge.com/cw/craig

Chapter 2

Seepage

Learning outcomes

After working through the material in this chapter, you should be able to:

1 Determine the permeability of soils using the results of both laboratory tests and in-situ tests conducted in the field (Sections 2.1 and 2.2);
2 Understand how groundwater flows for a wide range of ground conditions, and determine seepage quantities and pore pressures within the ground (Sections 2.3–2.6);
3 Use computer-based tools for accurately and efficiently solving larger/more complex seepage problems (Section 2.7);
4 Assess seepage through and beneath earthen dams, and understand the design features/remedial methods which may be used to control this (Sections 2.8–2.10).

2.1 Soil water

All soils are permeable materials, water being free to flow through the interconnected pores between the solid particles. It will be shown in Chapters 3–5 that the pressure of the pore water is one of the key parameters governing the strength and stiffness of soils. It is therefore vital that the pressure of the pore water is known both under static conditions and when pore water flow is occurring (this is known as **seepage**).

The pressure of the pore water is measured relative to atmospheric pressure, and the level at which the pressure is atmospheric (i.e. zero) is defined as the **water table** (WT) or the **phreatic surface**. Below the water table the soil is assumed to be fully saturated, although it is likely that, due to the presence of small volumes of entrapped air, the degree of saturation will be marginally below 100%. The level of the water table changes according to climatic conditions, but the level can change also as a consequence of constructional operations. A **perched** water table can occur locally in an **aquitard** (in which water is contained by soil of low permeability, above the normal water table level) or an **aquiclude** (where the surrounding material is impermeable). An example of a perched water table is shown schematically in Figure 2.1. **Artesian** conditions can exist if an inclined soil layer of high permeability is confined locally by an overlying layer of low permeability; the pressure in the artesian

Development of a mechanical model for soil

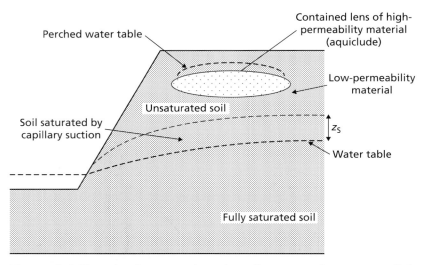

Figure 2.1 Terminology used to describe groundwater conditions.

layer is governed not by the local water table level but by a higher water table level at a distant location where the layer is unconfined.

Below the water table the pore water may be static, the hydrostatic pressure depending on the depth below the water table, or may be seeping through the soil under an **hydraulic gradient**: this chapter is concerned with the second case. Bernoulli's theorem applies to the pore water, but seepage velocities in soils are normally so small that velocity head can be neglected. Thus

$$h = \frac{u}{\gamma_w} + z \tag{2.1}$$

where h is the total head, u the pore water pressure, γ_w the unit weight of water (9.81 kN/m³) and z the elevation head above a chosen datum.

Above the water table, soil can remain saturated, with the pore water being held at negative pressure by capillary tension; the smaller the size of the pores, the higher the water can rise above the water table. The maximum negative pressure which can be sustained by a soil can be estimated using

$$u_c \approx -\frac{4T_s}{eD} \tag{2.2}$$

where T_s is the surface tension of the pore fluid ($=7 \times 10^{-5}$ kN/m for water at 10°C), e is the voids ratio and D is the pore size. As most soils are graded, D is often taken as that at which 10% of material passes on a particle size distribution chart (i.e. D_{10}). The height of the suction zone above the water table may then be estimated by $z_s = u_c/\gamma_w$.

The capillary rise tends to be irregular due to the random pore sizes occurring in a soil. The soil can be almost completely saturated in the lower part of the capillary zone, but in general the degree of saturation decreases with height. When water percolates through the soil from the surface towards the water table, some of this water can be held by surface tension around the points of contact between particles. The negative pressure of water held above the water table results in attractive forces between the particles: this attraction is referred to as soil suction, and is a function of pore size and water content.

2.2 Permeability and testing

In one dimension, water flows through a fully saturated soil in accordance with Darcy's empirical law:

$$v_d = ki \qquad (2.3)$$

or

$$q = v_d A = Aki$$

where q is the volume of water flowing per unit time (also termed **flow rate**), A the cross-sectional area of soil corresponding to the flow q, k the coefficient of permeability (or **hydraulic conductivity**), i the hydraulic gradient and v_d the discharge velocity. The units of the coefficient of permeability are those of velocity (m/s).

The coefficient of permeability depends primarily on the average size of the pores, which in turn is related to the distribution of particle sizes, particle shape and soil structure. In general, the smaller the particles, the smaller is the average size of the pores and the lower is the coefficient of permeability. The presence of a small percentage of fines in a coarse-grained soil results in a value of k significantly lower than the value for the same soil without fines. For a given soil, the coefficient of permeability is a function of void ratio. As soils become denser (i.e. their unit weight goes up), void ratio reduces. Compression of soil will therefore alter its permeability (see Chapter 4). If a soil deposit is **stratified** (layered), the permeability for flow parallel to the direction of stratification is higher than that for flow perpendicular to the direction of stratification. A similar effect may be observed in soils with plate-like particles (e.g. clay) due to alignment of the plates along a single direction. The presence of fissures in a clay results in a much higher value of permeability compared with that of the unfissured material, as the fissures are much larger in size than the pores of the intact material, creating preferential flow paths. This demonstrates the importance of soil fabric in understanding groundwater seepage.

The coefficient of permeability also varies with temperature, upon which the viscosity of the water depends. If the value of k measured at 20°C is taken as 100%, then the values at 10 and 0°C are 77% and 56%, respectively. The coefficient of permeability can also be represented by the equation:

$$k = \frac{\gamma_w}{\eta_w} K$$

where γ_w is the unit weight of water, η_w the viscosity of water and K (units m^2) an absolute coefficient depending only on the characteristics of the soil skeleton.

The values of k for different types of soil are typically within the ranges shown in Table 2.1. For sands, Hazen showed that the approximate value of k is given by

$$k = 10^{-2} D_{10}^2 \quad (\text{m/s}) \qquad (2.4)$$

where D_{10} is in mm.

Table 2.1 Coefficient of permeability (m/s)

1	10⁻¹	10⁻²	10⁻³	10⁻⁴	10⁻⁵	10⁻⁶	10⁻⁷	10⁻⁸	10⁻⁹	10⁻¹⁰
					Desiccated and fissured clays					
Clean gravels		Clean sands and sand–gravel mixtures			Very fine sands, silts and clay-silt laminate			Unfissured clays and clay-silts (>20% clay)		

Development of a mechanical model for soil

On the microscopic scale the water seeping through a soil follows a very tortuous path between the solid particles, but macroscopically the flow path (in one dimension) can be considered as a smooth line. The average velocity at which the water flows through the soil pores is obtained by dividing the volume of water flowing per unit time by the average area of voids (A_v) on a cross-section normal to the macroscopic direction of flow: this velocity is called the seepage velocity (v_s). Thus

$$v_s = \frac{q}{A_v}$$

The porosity of a soil is defined in terms of volume as described by Equation 1.9. However, on average, the porosity can also be expressed as

$$n = \frac{A_v}{A}$$

Hence

$$v_s = \frac{q}{nA} = \frac{v_d}{n}$$

or

$$v_s = \frac{ki}{n} \tag{2.5}$$

Alternatively, Equation 2.5 may be expressed in terms of void ratio, rather than porosity, by substituting for n using Equation 1.11.

Determination of coefficient of permeability

Laboratory methods

The coefficient of permeability for coarse soils can be determined by means of the **constant-head** permeability test (Figure 2.2(a)). The soil specimen, at the appropriate density, is contained in a Perspex cylinder of cross-sectional area A and length l: the specimen rests on a coarse filter or a wire mesh. A steady vertical flow of water, under a constant total head, is maintained through the soil, and the volume of water flowing per unit time (q) is measured. Tappings from the side of the cylinder enable the hydraulic gradient ($i = h/l$) to be measured. Then, from Darcy's law:

$$k = \frac{ql}{Ah}$$

A series of tests should be run, each at a different rate of flow. Prior to running the test, a vacuum is applied to the specimen to ensure that the degree of saturation under flow will be close to 100%. If a high degree of saturation is to be maintained, the water used in the test should be de-aired.

For fine soils, the **falling-head test** (Figure 2.2(b)) should be used. In the case of fine soils undisturbed specimens are normally tested (see Chapter 6), and the containing cylinder in the test may be the sampling tube itself. The length of the specimen is l and the cross-sectional area A. A coarse filter is placed at each end of the specimen, and a standpipe of internal area a is connected to the top of the cylinder. The water drains into a reservoir of constant level. The standpipe is filled with water, and a measurement is made of the time (t_1) for the water level (relative to the water level in the reservoir) to fall from h_0 to h_1. At any intermediate time t, the water level in the standpipe is given by h and its rate of change by $-dh/dt$. At time t, the difference in total head between the top and bottom of the specimen is h. Then, applying Darcy's law:

$$q = Aki$$

$$-a\frac{dh}{dt} = Ak\frac{h}{l}$$

$$\therefore -a\int_{h_0}^{h_1}\frac{dh}{h} = \frac{Ak}{l}\int_0^{t_1} dt \quad (2.6)$$

$$\therefore k = \frac{al}{At_1}\ln\frac{h_0}{h_1}$$

$$= 2.3\frac{al}{At_1}\log\frac{h_0}{h_1}$$

Again, precautions must be taken to ensure that the degree of saturation remains close to 100%. A series of tests should be run using different values of h_0 and h_1 and/or standpipes of different diameters.

The coefficient of permeability of fine soils can also be determined indirectly from the results of consolidation tests (see Chapter 4). Standards governing the implementation of laboratory tests for permeability include BS1377–5 (UK), CEN ISO 17892–11 (Europe) and ASTM D5084 (US).

The reliability of laboratory methods depends on the extent to which the test specimens are representative of the soil mass as a whole. More reliable results can generally be obtained by the in-situ methods described below.

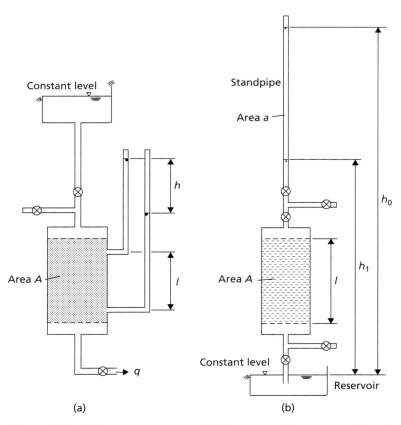

Figure 2.2 Laboratory permeability tests: (a) constant head, and (b) falling head.

Development of a mechanical model for soil

Well pumping test

This method is most suitable for use in homogeneous coarse soil strata. The procedure involves continuous pumping at a constant rate from a well, normally at least 300 mm in diameter, which penetrates to the bottom of the stratum under test. A screen or filter is placed in the bottom of the well to prevent ingress of soil particles. Perforated casing is normally required to support the sides of the well. Steady seepage is established, radially towards the well, resulting in the water table being drawn down to form a cone of depression. Water levels are observed in a number of boreholes spaced on radial lines at various distances from the well. An unconfined stratum of uniform thickness with a (relatively) impermeable lower boundary is shown in Figure 2.3(a), the water table being below the upper surface of the stratum. A confined layer between two impermeable strata is shown in Figure 2.3(b), the original water table being within the overlying stratum. Frequent recordings are made of the water levels in the boreholes, usually by means of an electrical dipper. The test enables the average coefficient of permeability of the soil mass below the cone of depression to be determined.

Analysis is based on the assumption that the hydraulic gradient at any distance r from the centre of the well is constant with depth and is equal to the slope of the water table, i.e.

$$i_r = \frac{dh}{dr}$$

where h is the height of the water table at radius r. This is known as the Dupuit assumption, and is reasonably accurate except at points close to the well.

In the case of an unconfined stratum (Figure 2.3(a)), consider two boreholes located on a radial line at distances r_1 and r_2 from the centre of the well, the respective water levels relative to the bottom of the stratum being h_1 and h_2. At distance r from the well, the area through which seepage takes place is $2\pi rh$, where r and h are variables. Then, applying Darcy's law:

$$q = Aki$$

$$q = 2\pi rhk \frac{dh}{dr}$$

$$\therefore q \int_{r_1}^{r_2} \frac{dr}{r} = 2\pi k \int_{h_1}^{h_2} h\,dh$$

$$\therefore q \ln\left(\frac{r_2}{r_1}\right) = \pi k(h_2^2 - h_1^2) \tag{2.7}$$

$$\therefore k = \frac{2.3q \log(r_2/r_1)}{\pi(h_2^2 - h_1^2)}$$

For a confined stratum of thickness H (Figure 2.3(b)) the area through which seepage takes place is $2\pi rH$, where r is variable and H is constant. Then

$$q = 2\pi rHk \frac{dh}{dr}$$

$$\therefore q \int_{r_1}^{r_2} \frac{dr}{r} = 2\pi Hk \int_{h_1}^{h_2} dh$$

$$\therefore q \ln\left(\frac{r_2}{r_1}\right) = 2\pi Hk(h_2 - h_1) \tag{2.8}$$

$$\therefore k = \frac{2.3q \log(r_2/r_1)}{2\pi H(h_2 - h_1)}$$

Seepage

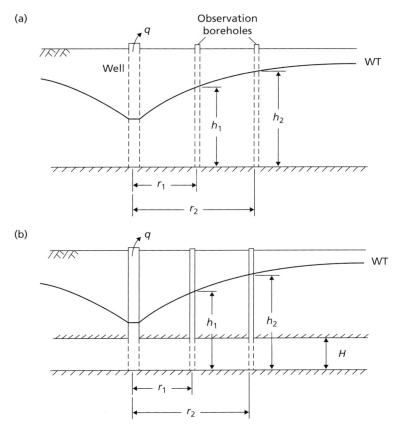

Figure 2.3 Well pumping tests: (a) unconfined stratum, and (b) confined stratum.

Borehole tests

The general principle is that water is either introduced into or pumped out of a borehole which terminates within the stratum in question, the procedures being referred to as inflow and outflow tests, respectively. A hydraulic gradient is thus established, causing seepage either into or out of the soil mass surrounding the borehole, and the rate of flow is measured. In a constant-head test, the water level above the water table is maintained throughout at a given level (Figure 2.4(a)). In a falling-head test, the water level is allowed to fall or rise from its initial position and the time taken for the level to change between two values is recorded (Figure 2.4(b)). The tests indicate the permeability of the soil within a radius of only 1–2 m from the centre of the borehole. Careful boring is essential to avoid disturbance in the soil structure.

A problem in such tests is that clogging of the soil face at the bottom of the borehole tends to occur due to the deposition of sediment from the water. To alleviate the problem the borehole may be extended below the bottom of the casing, as shown in Figure 2.4(c), increasing the area through which seepage takes place. The extension may be uncased in stiff fine soils, or supported by perforated casing in coarse soils.

Expressions for the coefficient of permeability depend on whether the stratum is unconfined or confined, the position of the bottom of the casing within the stratum, and details of the drainage face in the soil. If the soil is anisotropic with respect to permeability and if the borehole extends below the bottom of the casing (Figure 2.4(c)), then the horizontal permeability tends to be measured. If, on the other hand, the casing penetrates below soil level in the bottom of the borehole (Figure 2.4(d)), then vertical

Development of a mechanical model for soil

permeability tends to be measured. General formulae can be written, with the above details being represented by an intake factor (F_i).

For a constant-head test:

$$k = \frac{q}{F_i h_c}$$

For a falling-head test:

$$k = \frac{2.3A}{F_i (t_2 - t_1)} \log \frac{h_1}{h_2} \quad (2.9)$$

where k is the coefficient of permeability, q the rate of flow, h_c the constant head, h_1 the variable head at time t_1, h_2 the variable head at time t_2, and A the cross-sectional area of casing or standpipe. Values of intake factor were originally published by Hvorslev (1951), and are also given in Cedergren (1989).

For the case shown in Figure 2.4(b)

$$F_i = \frac{11R}{2}$$

where R is the radius of the inner radius of the casing, while for Figure 2.4(c)

$$F_i = \frac{2\pi L}{\ln(L/R)}$$

and for Figure 2.4(d)

$$F_i = \frac{11\pi R^2}{2\pi R + 11L}$$

The coefficient of permeability for a coarse soil can also be obtained from in-situ measurements of seepage velocity, using Equation 2.5. The method involves excavating uncased boreholes or trial pits at two points A and B (Figure 2.4(e)), seepage taking place from A towards B. The hydraulic gradient is given by the difference in the steady-state water levels in the boreholes divided by the distance AB. Dye or any other suitable tracer is inserted into borehole A, and the time taken for the dye to appear in borehole B is measured. The seepage velocity is then the distance AB divided by this time. The porosity of the soil can be determined from density tests. Then

$$k = \frac{v_s n}{i}$$

Further information on the implementation of in-situ permeability tests may be found in Clayton *et al.* (1995).

2.3 Seepage theory

The general case of seepage in two dimensions will now be considered. Initially it will be assumed that the soil is homogeneous and isotropic with respect to permeability, the coefficient of permeability being k. In the x–z plane, Darcy's law can be written in the generalised form:

$$v_x = k i_x = -k \frac{\partial h}{\partial x} \quad (2.10a)$$

Seepage

$$v_z = ki_z = -k\frac{\partial h}{\partial z} \tag{2.10b}$$

with the total head h decreasing in the directions of v_x and v_z.

An element of fully saturated soil having dimensions dx, dy and dz in the x, y and z directions, respectively, with flow taking place in the x–z plane only, is shown in Figure 2.5. The components of discharge velocity of water entering the element are v_x and v_z, and the rates of change of discharge velocity in the x and z directions are $\partial v_x/\partial x$ and $\partial v_z/\partial z$, respectively. The volume of water entering the element per unit time is

$$v_x\,\mathrm{d}y\mathrm{d}z + v_z\,\mathrm{d}x\mathrm{d}y$$

and the volume of water leaving per unit time is

$$\left(v_x + \frac{\partial v_x}{\partial x}\mathrm{d}x\right)\mathrm{d}y\mathrm{d}z + \left(v_z + \frac{\partial v_z}{\partial z}\mathrm{d}z\right)\mathrm{d}x\mathrm{d}y$$

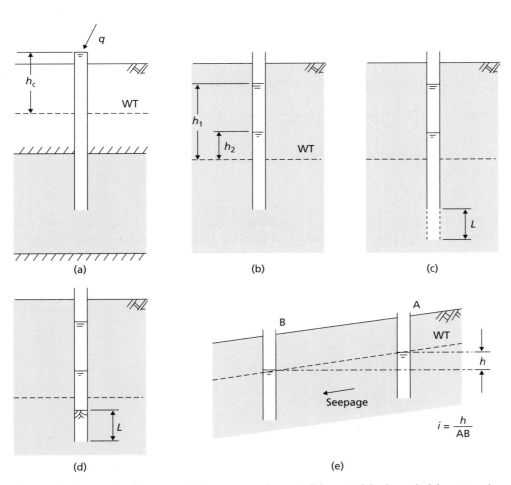

Figure 2.4 Borehole tests: (a) constant-head, (b) variable-head, (c) extension of borehole to prevent clogging, (d) measurement of vertical permeability in anisotropic soil, and (e) measurement of in-situ seepage.

Development of a mechanical model for soil

If the element is undergoing no volume change and if water is assumed to be incompressible, the difference between the volume of water entering the element per unit time and the volume leaving must be zero. Therefore

$$\frac{\partial v_x}{\partial x} + \frac{\partial v_z}{\partial z} = 0 \tag{2.11}$$

Equation 2.11 is the equation of continuity in two dimensions. If, however, the volume of the element is undergoing change, the equation of continuity becomes

$$\left(\frac{\partial v_x}{\partial x} + \frac{\partial v_z}{\partial z}\right) dx\, dy\, dz = \frac{dV}{dt} \tag{2.12}$$

where dV/dt is the volume change per unit time.

Consider, now, the function $\phi(x, z)$, called the potential function, such that

$$\frac{\partial \phi}{\partial x} = v_x = -k \frac{\partial h}{\partial x} \tag{2.13a}$$

$$\frac{\partial \phi}{\partial z} = v_z = -k \frac{\partial h}{\partial z} \tag{2.13b}$$

From Equations 2.11 and 2.13 it is apparent that

$$\frac{\partial^2 \phi}{\partial x^2} + \frac{\partial^2 \phi}{\partial z^2} = 0 \tag{2.14}$$

i.e. the function $\phi(x, z)$ satisfies the Laplace equation.

Integrating Equation 2.13:

$$\phi(x, z) = -kh(x, z) + C$$

where C is a constant. Thus, if the function $\phi(x, z)$ is given a constant value, equal to ϕ_1 (say), it will represent a curve along which the value of total head (h_1) is constant. If the function $\phi(x, z)$ is given a series of constant values, ϕ_1, ϕ_2, ϕ_3 etc., a family of curves is specified along each of which the total head is a constant value (but a different value for each curve). Such curves are called **equipotentials**.

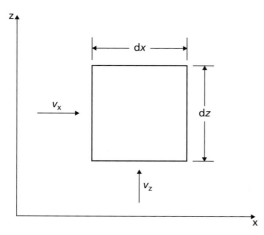

Figure 2.5 Seepage through a soil element.

A second function $\psi(x, z)$, called the flow function, is now introduced, such that

$$-\frac{\partial \psi}{\partial x} = v_z = -k\frac{\partial h}{\partial z} \tag{2.15a}$$

$$\frac{\partial \psi}{\partial x} = v_x = -k\frac{\partial h}{\partial x} \tag{2.15b}$$

It can be shown that this function also satisfies the Laplace equation.

The total differential of the function $\psi(x, z)$ is

$$d\psi = \frac{\partial \psi}{\partial x}dx + \frac{\partial \psi}{\partial z}dz$$
$$= -v_z dx + v_x dz$$

If the function $\psi(x, z)$ is given a constant value ψ_1 then $d\psi = 0$ and

$$\frac{dz}{dx} = \frac{v_z}{v_x} \tag{2.16}$$

Thus, the tangent at any point on the curve represented by

$$\psi(x, z) = \psi_1$$

specifies the direction of the resultant discharge velocity at that point: the curve therefore represents the flow path. If the function $\psi(x, z)$ is given a series of constant values, ψ_1, ψ_2, ψ_3 etc., a second family of curves is specified, each representing a flow path. These curves are called **flow lines**.

Referring to Figure 2.6, the flow per unit time between two flow lines for which the values of the flow function are ψ_1 and ψ_2 is given by

$$\Delta q = \int_{\psi_1}^{\psi_2} (-v_z dx + v_x dz)$$
$$= \int_{\psi_1}^{\psi_2} \left(\frac{\partial \psi}{\partial x}dx + \frac{\partial \psi}{\partial z}dz\right) = \psi_2 - \psi_1$$

Thus, the flow through the 'channel' between the two flow lines is constant.

The total differential of the function $\phi(x, z)$ is

$$d\phi = \frac{\partial \phi}{\partial x}dx + \frac{\partial \phi}{\partial z}dz$$
$$= v_x dx + v_z dz$$

If $\phi(x, z)$ is constant then $d\phi = 0$ and

$$\frac{dz}{dx} = -\frac{v_x}{v_z} \tag{2.17}$$

Comparing Equations 2.16 and 2.17, it is apparent that the flow lines and the equipotentials intersect each other at right angles.

Consider, now, two flow lines ψ_1 and $(\psi_1 + \Delta\psi)$ separated by the distance Δn. The flow lines are intersected orthogonally by two equipotentials ϕ_1 and $(\phi_1 + \Delta\phi)$ separated by the distance Δs, as shown in Figure 2.7. The directions s and n are inclined at angle α to the x and z axes, respectively. At point A the discharge velocity (in direction s) is v_s; the components of v_s in the x and z directions, respectively, are

$$v_x = v_s \cos \alpha$$
$$v_z = v_s \sin \alpha$$

Development of a mechanical model for soil

Now

$$\frac{\partial \phi}{\partial s} = \frac{\partial \phi}{\partial x}\frac{\partial x}{\partial s} + \frac{\partial \phi}{\partial z}\frac{\partial z}{\partial s}$$
$$= v_s \cos^2 \alpha + v_s \sin^2 \alpha = v_s$$

and

$$\frac{\partial \psi}{\partial n} = \frac{\partial \psi}{\partial x}\frac{\partial x}{\partial n} + \frac{\partial \psi}{\partial z}\frac{\partial z}{\partial n}$$
$$= -v_s \sin \alpha (-\sin \alpha) + v_s \cos^2 \alpha = v_s$$

Thus

$$\frac{\partial \psi}{\partial n} = \frac{\partial \phi}{\partial s}$$

or approximately

$$\frac{\Delta \psi}{\Delta n} = \frac{\Delta \phi}{\Delta s} \tag{2.18}$$

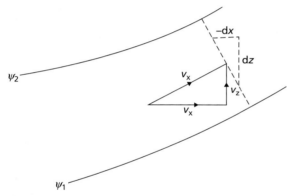

Figure 2.6 Seepage between two flow lines.

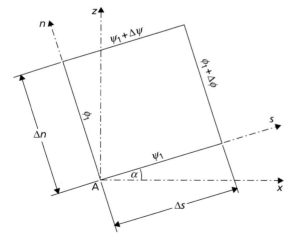

Figure 2.7 Flow lines and equipotentials.

2.4 Flow nets

In principle, for the solution of a practical seepage problem the functions $\phi(x, z)$ and $\psi(x, z)$ must be found for the relevant boundary conditions. The solution is represented by a family of flow lines and a family of equipotentials, constituting what is referred to as a **flow net**. Computer software based on either the finite difference or finite element methods is widely available for the solution of seepage problems. Williams *et al.* (1993) described how solutions can be obtained using the finite difference form of the Laplace equation by means of a spreadsheet. This method will be described in Section 2.7, with electronic resources available on the Companion Website to accompany the material discussed in the remainder of this chapter. Relatively simple problems can be solved by the trial-and-error sketching of the flow net, the general form of which can be deduced from consideration of the boundary conditions. Flow net sketching leads to a greater understanding of seepage principles. However, for problems in which the geometry becomes complex and there are zones of different permeability throughout the flow region, use of the finite difference method is usually necessary.

The fundamental condition to be satisfied in a flow net is that every intersection between a flow line and an equipotential must be at right angles. In addition, it is convenient to construct the flow net such that $\Delta\psi$ has the same value between any two adjacent flow lines and $\Delta\phi$ has the same value between any two adjacent equipotentials. It is also convenient to make $\Delta s = \Delta n$ in Equation 2.18, i.e. the flow lines and equipotentials form curvilinear squares throughout the flow net. Then, for any curvilinear square

$$\Delta\psi = \Delta\phi$$

Now, $\Delta\psi = \Delta q$ and $\Delta\phi = k\Delta h$, therefore:

$$\Delta q = k\Delta h \tag{2.19}$$

For the entire flow net, h is the difference in total head between the first and last equipotentials, N_d the number of equipotential drops, each representing the same total head loss Δh, and N_f the number of **flow channels**, each carrying the same flow Δq. Then,

$$\Delta h = \frac{h}{N_d} \tag{2.20}$$

and

$$q = N_f \Delta q$$

Hence, from Equation 2.19

$$q = kh\frac{N_f}{N_d} \tag{2.21}$$

Equation 2.21 gives the total volume of water flowing per unit time (per unit dimension in the y-direction) and is a function of the ratio N_f/N_d.

Between two adjacent equipotentials the hydraulic gradient is given by

$$i = \frac{\Delta h}{\Delta s} \tag{2.22}$$

Development of a mechanical model for soil

Example of a flow net

As an illustration, the flow net for the problem detailed in Figure 2.8(a) will be considered. The figure shows a line of sheet piling driven 6.00 m into a stratum of soil 8.60 m thick, underlain by an impermeable stratum. On one side of the piling the depth of water is 4.50 m; on the other side the depth of water (reduced by pumping) is 0.50 m. The soil has a permeability of 1.5×10^{-5} m/s.

The first step is to consider the boundary conditions of the flow region (Figure 2.8(b)). At every point on the boundary AB the total head is constant, so AB is an equipotential; similarly, CD is an equipotential. The datum to which total head is referred may be chosen to be at any level, but in seepage problems it is convenient to select the downstream water level as datum. Then, the total head on equipotential CD is zero after Equation 2.1 (pressure head 0.50 m; elevation head –0.50 m) and the total head on equipotential AB is 4.00 m (pressure head 4.50 m; elevation head –0.50 m). From point B, water must flow down the upstream face BE of the piling, round the tip E and up the downstream face EC. Water from point F must flow along the impermeable surface FG. Thus, BEC and FG are flow lines. The other flow lines must lie between the extremes of BEC and FG, and the other equipotentials must lie between AB and CD. As the flow region is symmetric on either side of the sheet piling, when flowline BEC reaches point E, halfway between AB and CD (i.e. at the toe of the sheet piling), the total head must be half way between the values along AB and CD. This principle also applies to flowline FG, such that a third vertical equipotential can be drawn from point E, as shown in Figure 2.8(b).

The number of equipotential drops should then be selected. Any number may be selected, however it is convenient to use a value of N_d which divides precisely into the total change in head through the flow region. In this example $N_d = 8$ is chosen so that each equipotential will represent a drop in head of 0.5 m. The choice of N_d has a direct influence on the value of N_f. As N_d is increased, the equipotentials get closer together such that, to get a 'square' flow net, the flow channels will also have to be closer together (i.e. more flow lines will need to be plotted). This will lead to a finer net with greater detail in the distribution of seepage pressures; however, the total flow quantity will be unchanged. Figure 2.8(c) shows the flow net for $N_d = 8$ and $N_f = 3$. These parameters for this particular example give a 'square' flow net and a whole number of flow channels. This should be formed by trial and error: a first attempt should be made and the positions of the flow lines and equipotentials (and even N_f and N_d) should then be adjusted as necessary until a satisfactory flow net is achieved. A satisfactory flow net should satisfy the following conditions:

- All intersections between flow lines and equipotentials should be at 90°.
- The curvilinear squares must be square – in Figure 2.8(c), the 'square-ness' of the flow net has been checked by inscribing a circle within each square. The flow net is acceptable if the circle just touches the edges of the curvilinear square (i.e. there are no rectangular elements).

Due to the symmetry within the flow region, the equipotentials and flow lines may be drawn in half of the problem and then reflected about the line of symmetry (i.e. the sheet piling). In constructing a flownet it is a mistake to draw too many flow lines; typically, three to five flow channels are sufficient, depending on the geometry of the problem and the value of N_d which is most convenient.

In the flow net in Figure 2.8(c) the number of flow channels is three and the number of equipotential drops is eight; thus the ratio N_f/N_d is 0.375. The loss in total head between any two adjacent equipotentials is

$$\Delta h = \frac{h}{N_d} = \frac{4.00}{8} = 0.5 \text{ m}$$

The total volume of water flowing under the piling per unit time per unit length of piling is given by

$$q = kh\frac{N_f}{N_d} = 1.5 \times 10^{-5} \times 4.00 \times 0.375$$

$$= 2.25 \times 10^{-5} \text{ m}^3/\text{s}$$

Seepage

A piezometer tube is shown at a point P on the equipotential with total head $h = 1.00$ m, i.e. the water level in the tube is 1.00 m above the datum. The point P is at a distance $z_P = 6$ m below the datum, i.e. the elevation head is $-z_P$. The pore water pressure at P can then be calculated from Bernoulli's theorem:

$$u_P = \gamma_w \left[h_P - (-z_P) \right]$$
$$= \gamma_w (h_P + z_P)$$
$$= 9.81 \times (1 + 6)$$
$$= 68.7 \text{ kPa}$$

The hydraulic gradient across any square in the flow net involves measuring the average dimension of the square (Equation 2.22). The highest hydraulic gradient (and hence the highest seepage velocity) occurs across the smallest square, and vice versa. The dimension Δs has been estimated by measuring the diameter of the circles in Figure 2.8(c). The hydraulic gradients across each square are shown using a quiver plot in Figure 2.8(d) in which the length of the arrows is proportional to the magnitude of the hydraulic gradient.

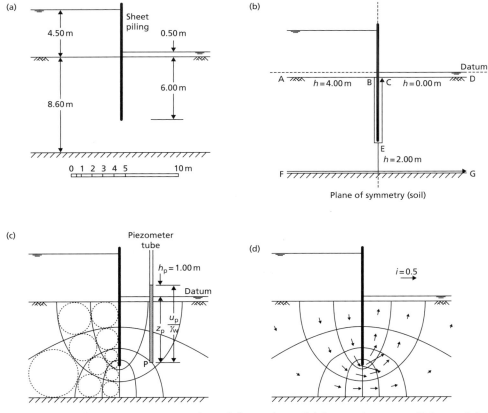

Figure 2.8 Flow net construction: (a) section, (b) boundary conditions, (c) final flow net including a check of the 'square-ness' of the curvilinear squares, and (d) hydraulic gradients inferred from flow net.

Development of a mechanical model for soil

Example 2.1

A river bed consists of a layer of sand 8.25 m thick overlying impermeable rock; the depth of water is 2.50 m. A long cofferdam 5.50 m wide is formed by driving two lines of sheet piling to a depth of 6.00 m below the level of the river bed, and excavation to a depth of 2.00 m below bed level is carried out within the cofferdam. The water level within the cofferdam is kept at excavation level by pumping. If the flow of water into the cofferdam is 0.25 m³/h per unit length, what is the coefficient of permeability of the sand? What is the hydraulic gradient immediately below the excavated surface?

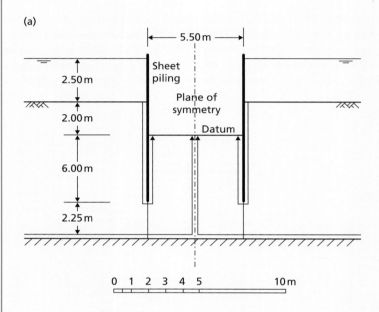

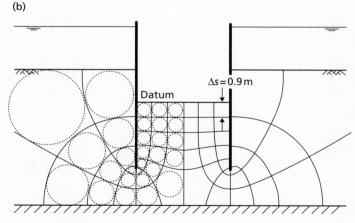

Figure 2.9 Example 2.1.

Solution

The section and boundary conditions appear in Figure 2.9(a) and the flow net is shown in Figure 2.9(b). In the flow net there are six flow channels (three per side) and ten equipotential drops. The total head loss is 4.50 m. The coefficient of permeability is given by

$$k = \frac{q}{h(N_f / N_d)}$$

$$= \frac{0.25}{4.50 \times 6/10 \times 60^2} = 2.6 \times 10^{-5} \text{ m/s}$$

The distance (Δs) between the last two equipotentials is measured as 0.9 m. The required hydraulic gradient is given by

$$i = \frac{\Delta h}{\Delta s}$$

$$= \frac{4.50}{10 \times 0.9} = 0.50$$

Example 2.2

The section through a dam spillway is shown in Figure 2.10. Determine the quantity of seepage under the dam and plot the distributions of uplift pressure on the base of the dam, and the net distribution of water pressure on the cut-off wall at the upstream end of the spillway. The coefficient of permeability of the foundation soil is 2.5×10^{-5} m/s.

Solution

The flow net is shown in Figure 2.10. The downstream water level (ground surface) is selected as datum. Between the upstream and downstream equipotentials the total head loss is 5.00 m. In the flow net there are three flow channels and ten equipotential drops. The seepage is given by

$$q = kh\frac{N_f}{N_d} = 2.5 \times 10^{-5} \times 5.00 \times \frac{3}{10}$$

$$= 3.75 \times 10^{-5} \text{ m}^3/\text{s}$$

This inflow rate is per metre length of the cofferdam. The pore water pressures acting on the base of the spillway are calculated at the points of intersection of the equipotentials with the base of the spillway. The total head at each point is obtained from the flow net, and the elevation head from the section. The calculations are shown in Table 2.2, and the pressure diagram is plotted in Figure 2.10.

The water pressures acting on the cut-off wall are calculated on both the back (h_b) and front (h_f) of the wall at the points of intersection of the equipotentials with the wall. The net pressure acting on the back face of the wall is therefore

$$u_{net} = u_b - u_f = \left(\frac{h_b - z}{\gamma_w}\right) - \left(\frac{h_f - z}{\gamma_w}\right)$$

Development of a mechanical model for soil

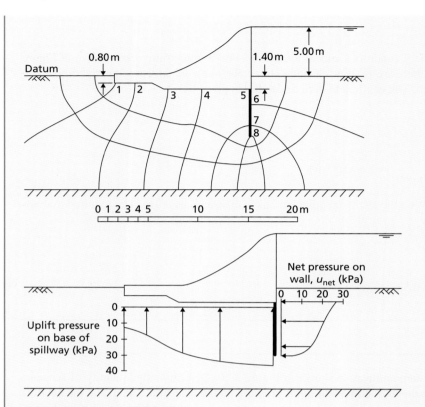

Figure 2.10 Example 2.2.

The calculations are shown in Table 2.3, and the pressure diagram is plotted in Figure 2.10. The levels (z) of points 5–8 in Table 2.3 were found by scaling from the diagram.

Table 2.2 Example 2.2

Point	h(m)	z(m)	$h-z$(m)	$u = \gamma_w(h-z)$ (kPa)
1	0.50	−0.80	1.30	12.8
2	1.00	−0.80	1.80	17.7
3	1.50	−1.40	2.90	28.4
4	2.00	−1.40	3.40	33.4
5	2.30	−1.40	3.70	36.3

Table 2.3 Example 2.2 (contd.)

Level	z(m)	h_b(m)	u_b/γ_w(m)	h_f(m)	u_f/γ_w(m)	$u_b - u_f$ (kPa)
5	−1.40	5.00	6.40	2.28	3.68	26.7
6	−3.07	4.50	7.57	2.37	5.44	20.9
7	−5.20	4.00	9.20	2.50	7.70	14.7
8	−6.00	3.50	9.50	3.00	9.00	4.9

2.5 Anisotropic soil conditions

It will now be assumed that the soil, although homogeneous, is anisotropic with respect to permeability. Most natural soil deposits are anisotropic, with the coefficient of permeability having a maximum value in the direction of stratification and a minimum value in the direction normal to that of stratification; these directions are denoted by x and z, respectively, i.e.

$$k_x = k_{max} \text{ and } k_z = k_{min}$$

In this case, the generalised form of Darcy's law is

$$v_x = k_x i_x = -k_x \frac{\partial h}{\partial x} \quad (2.23a)$$

$$v_z = k_z i_z = -k_z \frac{\partial h}{\partial z} \quad (2.23b)$$

Also, in any direction s, inclined at angle α to the x direction, the coefficient of permeability is defined by the equation

$$v_s = -k_s \frac{\partial h}{\partial s}$$

Now

$$\frac{\partial h}{\partial s} = \frac{\partial h}{\partial x} \frac{\partial x}{\partial s} + \frac{\partial h}{\partial z} \frac{\partial z}{\partial s}$$

i.e.

$$\frac{v_s}{k_s} = \frac{v_x}{k_x} \cos\alpha + \frac{v_z}{k_z} \sin\alpha$$

The components of discharge velocity are also related as follows:

$$v_x = v_s \cos\alpha$$
$$v_z = v_s \sin\alpha$$

Hence

$$\frac{1}{k_s} = \frac{\cos^2\alpha}{k_x} + \frac{\sin^2\alpha}{k_z}$$

or

$$\frac{s^2}{k_s} = \frac{x^2}{k_x} + \frac{z^2}{k_z} \quad (2.24)$$

The directional variation of permeability is thus described by Equation 2.24, which represents the ellipse shown in Figure 2.11.

Given the generalised form of Darcy's law (Equation 2.23), the equation of continuity (2.11) can be written:

$$k_x \frac{\partial^2 h}{\partial x^2} + k_z \frac{\partial^2 h}{\partial z^2} = 0 \quad (2.25)$$

Development of a mechanical model for soil

or

$$\frac{\partial^2 h}{(k_z/k_x)\partial x^2} + \frac{\partial^2 h}{\partial z^2} = 0$$

Substituting

$$x_t = x\sqrt{\frac{k_z}{k_x}} \tag{2.26}$$

the equation of continuity becomes

$$\frac{\partial^2 h}{\partial x_t^2} + \frac{\partial^2 h}{\partial z^2} = 0 \tag{2.27}$$

which is the equation of continuity for an isotropic soil in an x_t–z plane.

Thus, Equation 2.26 defines a scale factor which can be applied in the x direction to transform a given anisotropic flow region into a fictitious isotropic flow region in which the Laplace equation is valid. Once the flow net (representing the solution of the Laplace equation) has been drawn for the transformed section, the flow net for the natural section can be obtained by applying the inverse of the scaling factor. Essential data, however, can normally be obtained from the transformed section. The necessary transformation could also be made in the z direction.

The value of coefficient of permeability applying to the transformed section, referred to as the equivalent isotropic coefficient, is

$$k' = \sqrt{(k_x k_z)} \tag{2.28}$$

A formal proof of Equation 2.28 has been given by Vreedenburgh (1936). The validity of Equation 2.28 can be demonstrated by considering an elemental flow net field through which flow is in the x direction. The flow net field is drawn to the transformed and natural scales in Figure 2.12, the transformation being in the x direction. The discharge velocity v_x can be expressed in terms of either k' (transformed section) or k_x (natural section), i.e.

$$v_x = -k'\frac{\partial h}{\partial x_t} = -k_x \frac{\partial h}{\partial x}$$

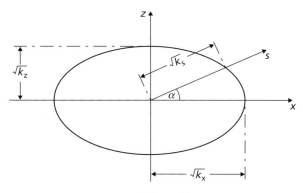

Figure 2.11 Permeability ellipse.

where

$$\frac{\partial h}{\partial x_t} = \frac{\partial h}{\sqrt{(k_z/k_x)}\partial x}$$

Thus

$$k' = k_x\sqrt{\frac{k_z}{k_x}} = \sqrt{(k_x k_z)}$$

2.6 Non-homogeneous soil conditions

Two isotropic soil layers of thicknesses H_1 and H_2 are shown in Figure 2.13, the respective coefficients of permeability being k_1 and k_2; the boundary between the layers is horizontal. (If the layers are anisotropic, k_1 and k_2 represent the equivalent isotropic coefficients for the layers.) The two layers can be approximated as a single homogeneous anisotropic layer of thickness (H_1+H_2) in which the coefficients in the directions parallel and normal to that of stratification are $\bar{k}_x$ and $\bar{k}_z$, respectively.

For one-dimensional seepage in the horizontal direction, the equipotentials in each layer are vertical. If h_1 and h_2 represent total head at any point in the respective layers, then for a common point on the boundary $h_1=h_2$. Therefore, any vertical line through the two layers represents a common equipotential. Thus, the hydraulic gradients in the two layers, and in the equivalent single layer, are equal; the equal hydraulic gradients are denoted by i_x. The total horizontal flow per unit time is then given by

$$\bar{q}_x = (H_1+H_2)\bar{k}_x i_x = (H_1 k_1 + H_2 k_2)i_x$$
$$\therefore \bar{k}_x = \frac{H_1 k_1 + H_2 k_2}{H_1 + H_2} \qquad (2.29)$$

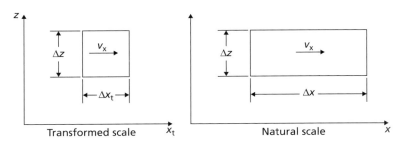

Figure 2.12 Elemental flow net field.

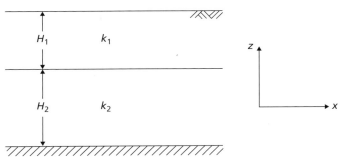

Figure 2.13 Non-homogeneous soil conditions.

Development of a mechanical model for soil

For one-dimensional seepage in the vertical direction, the discharge velocities in each layer, and in the equivalent single layer, must be equal if the requirement of continuity is to be satisfied. Thus

$$v_z = \bar{k}_z \bar{i}_z = k_1 i_1 = k_2 i_2$$

where $\bar{i}_z$ is the average hydraulic gradient over the depth $(H_1 + H_2)$. Therefore

$$i_1 = \frac{\bar{k}_z}{k_1} \bar{i}_z \quad \text{and} \quad i_2 = \frac{\bar{k}_z}{k_2} \bar{i}_z$$

Now the loss in total head over the depth $(H_1 + H_2)$ is equal to the sum of the losses in total head in the individual layers, i.e.

$$\begin{aligned} \bar{i}_z (H_1 + H_2) &= i_1 H_1 + i_2 H_2 \\ &= \bar{k}_z \bar{i}_z \left(\frac{H_1}{k_1} + \frac{H_2}{k_2} \right) \\ \therefore \bar{k}_z &= \frac{H_1 + H_2}{\left(\dfrac{H_1}{k_1}\right) + \left(\dfrac{H_2}{k_2}\right)} \end{aligned} \qquad (2.30)$$

Similar expressions for $\bar{k}_x$ and $\bar{k}_z$ apply in the case of any number of soil layers. It can be shown that $\bar{k}_x$ must always be greater than $\bar{k}_z$, i.e. seepage can occur more readily in the direction parallel to stratification than in the direction perpendicular to stratification.

2.7 Numerical solution using the Finite Difference Method

Although flow net sketches are useful for estimating seepage-induced pore pressures and volumetric flow rates for simple problems, a great deal of practice is required to produce reliable results. As an alternative, spreadsheets (which are almost universally available to practising engineers) may be used to determine the seepage pressures more rapidly and reliably. A spreadsheet-based tool for solving a wide range of problems may be found on the Companion Website which accompanies the material in this chapter. Spreadsheets analyse seepage problems by solving Laplace's equation using the **Finite Difference Method** (FDM). The problem is first discretised into a mesh of regularly-spaced nodes representing the soil in the problem. If steady-state seepage is occurring through a given region of soil with isotopic permeability k, the total head at a general node within the soil is the average of the values of head at the four connecting nodes, as shown in Figure 2.14:

$$h_0 = \frac{1}{4}(h_1 + h_2 + h_3 + h_4) \qquad (2.31)$$

Modified forms of Equation 2.31 can be derived for impervious boundaries and for determining the values of head either side of thin impermeable inclusions such as sheet piling. The analysis technique may additionally be used to study anisotropic soils at the transformed scale for soil of permeability k' as defined at the end of Section 2.5. It is also possible to modify Equation 2.31 to model nodes at the boundary between two soil layers of different isotropic permeabilities (k_1, k_2) as discussed in Section 2.6.

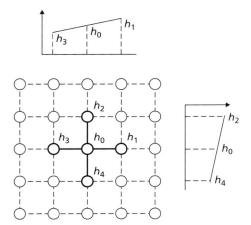

Figure 2.14 Determination of head at an FDM node.

The Companion Website contains a spreadsheet analysis tool (Seepage_CSM8.xls) that can be used for the solution of a wide range of seepage problems. Each cell is used to represent a node, with the value of the cell equal to the total head, h_0 for that node. A library of equations for h_0 for a range of different boundary conditions is included within this spreadsheet, which may be copied as necessary to build up a complete and detailed model of a wide range of problems. A more detailed description of the boundary conditions which may be used and their formulation are given in the User's Manual, which may also be found on the Companion Website. After the formulae have been copied in for the relevant nodes, the values of head are entered on the recharge and discharge boundaries and the calculations are then conducted iteratively until further iterations give a negligible change in the head distribution. This iterative process is entirely automated within the spreadsheet. On a modern computer, this should take a matter of seconds for the problems addressed in this chapter. The use of this spreadsheet will be demonstrated in the following example, which utilises all of the boundary conditions included in the spreadsheet.

Example 2.3

A deep excavation is to be made next to an existing masonry tunnel carrying underground railway lines, as shown in Figure 2.15. The surrounding soil is layered, with isotropic permeabilities as shown in the figure. Calculate the pore water pressure distribution around the tunnel, and find also the flow-rate of water into the excavation.

Solution

Given the geometry shown in Figure 2.15, a grid spacing of 1 m is chosen in the horizontal and vertical directions, giving the nodal layout shown in the figure. The appropriate formulae are then entered in the cells representing each node, as shown in the User's Manual on the Companion Website. The datum is set at the level of the excavation. The resulting total head distribution is shown in Figure 2.15, and by applying Equation 2.1 the pore water pressure distribution around the tunnel can be plotted.

Development of a mechanical model for soil

The flow rate of water into the excavation can be found by considering the flow between the eight adjacent nodes on the discharge boundary. Considering the nodes next to the sheet pile wall, the change in head between the last two nodes $\Delta h = 0.47$. This is repeated along the discharge boundary, and the average Δh between each set of nodes is computed. Adapting Equation 2.19, and noting that the soil on the discharge boundary has permeability k_1, the flow rate is then given by:

$$q = k_1 \sum \Delta h$$

$$= 3.3 \times 10^{-9}\, \text{m}^3/\text{s}$$

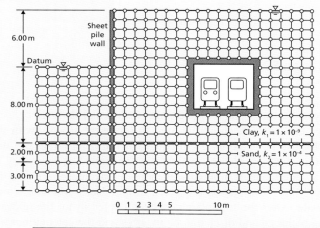

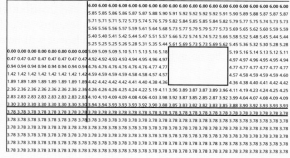

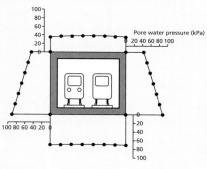

Figure 2.15 Example 2.3.

2.8 Transfer condition

Consideration is now given to the condition which must be satisfied when seepage takes place diagonally across the boundary between two isotropic soils 1 and 2 having coefficients of permeability k_1 and k_2, respectively. The direction of seepage approaching a point B on the boundary ABC is at angle α_1 to the normal at B, as shown in Figure 2.16; the discharge velocity approaching B is v_1. The components of v_1 along the boundary and normal to the boundary are v_{1s} and v_{1n} respectively. The direction of seepage leaving point B is at angle α_2 to the normal, as shown; the discharge velocity leaving B is v_2. The components of v_2 are v_{2s} and v_{2n}.

For soils 1 and 2 respectively

$$\phi_1 = -k_1 h_1 \text{ and } \phi_2 = -k_2 h_2$$

At the common point B, $h_1 = h_2$; therefore

$$\frac{\phi_1}{k_1} = \frac{\phi_2}{k_2}$$

Differentiating with respect to s, the direction along the boundary:

$$\frac{1}{k_1}\frac{\partial \phi_1}{\partial s} = \frac{1}{k_2}\frac{\partial \phi_2}{\partial s}$$

i.e.

$$\frac{v_{1s}}{k_1} = \frac{v_{2s}}{k_2}$$

For continuity of flow across the boundary the normal components of discharge velocity must be equal, i.e.

$$v_{1n} = v_{2n}$$

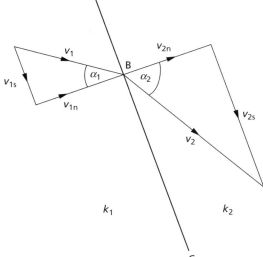

Figure 2.16 Transfer condition.

Development of a mechanical model for soil

Therefore

$$\frac{1}{k_1}\frac{v_{1s}}{v_{1n}} = \frac{1}{k_2}\frac{v_{2s}}{v_{2n}}$$

Hence it follows that

$$\frac{\tan \alpha_1}{\tan \alpha_2} = \frac{k_1}{k_2} \tag{2.32}$$

Equation 2.32 specifies the change in direction of the flow line passing through point B. This equation must be satisfied on the boundary by every flow line crossing the boundary.

Equation 2.18 can be written as

$$\Delta \psi = \frac{\Delta n}{\Delta s}\Delta \phi$$

i.e.

$$\Delta q = \frac{\Delta n}{\Delta s}k\Delta \phi$$

If Δq and Δh are each to have the same values on both sides of the boundary, then

$$\left(\frac{\Delta n}{\Delta s}\right)_1 k_1 = \left(\frac{\Delta n}{\Delta s}\right)_2 k_2$$

and it is clear that curvilinear squares are possible only in one soil. If

$$\left(\frac{\Delta n}{\Delta s}\right)_1 = 1$$

then

$$\left(\frac{\Delta n}{\Delta s}\right)_2 = \frac{k_1}{k_2} \tag{2.33}$$

If the permeability ratio is less than $1/10$, it is unlikely that the part of the flow net in the soil of higher permeability needs to be considered.

2.9 Seepage through embankment dams

This problem is an example of unconfined seepage, one boundary of the flow region being a phreatic surface on which the pressure is atmospheric. In section, the phreatic surface constitutes the top flow line and its position must be estimated before the flow net can be drawn.

Consider the case of a homogeneous isotropic embankment dam on an impermeable foundation, as shown in Figure 2.17. The impermeable boundary BA is a flow line, and CD is the required top flow line. At every point on the upstream slope BC the total head is constant (u/γ_w and z varying from point to point but their sum remaining constant); therefore, BC is an equipotential. If the downstream water level is taken as datum, then the total head on equipotential BC is equal to h, the difference between the upstream and downstream water levels. The discharge surface AD, for the case shown in Figure 2.17 only, is the equipotential for zero total head. At every point on the top flow line the pressure is zero (atmospheric), so total head is equal to elevation head and there must be equal vertical intervals Δz between the points of intersection between successive equipotentials and the top flow line.

Seepage

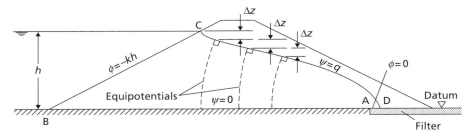

Figure 2.17 Homogeneous embankment dam section.

A suitable filter must always be constructed at the discharge surface in an embankment dam. The function of the filter is to keep the seepage entirely within the dam; water seeping out onto the downstream slope would result in the gradual erosion of the slope. The consequences of such erosion can be severe. In 1976, a leak was noticed close to one of the abutments of the Teton Dam in Idaho, USA. Seepage was subsequently observed through the downstream face. Within two hours of this occurring the dam failed catastrophically, causing extensive flooding, as shown in Figure 2.18. The direct and indirect costs of this failure were estimated at close to $1 billion. A horizontal under-filter is shown in Figure 2.17. Other possible forms of filter are illustrated in Figures 2.22(a) and (b); in these two cases the discharge surface AD is neither a flow line nor an equipotential since there are components of discharge velocity both normal and tangential to AD.

The boundary conditions of the flow region ABCD in Figure 2.17 can be written as follows:

Equipotential BC: $\phi = -kh$

Equipotential AD: $\phi = 0$

Flow line CD: $\psi = q$ (also, $\phi = -kz$)

Flow line BA: $\psi = 0$

Figure 2.18 Failure of the Teton Dam, 1976 (photo courtesy of the Bureau of Reclamation).

Development of a mechanical model for soil

The conformal transformation $r = w^2$

Complex variable theory can be used to obtain a solution to the embankment dam problem. Let the complex number $w = \phi + i\psi$ be an analytic function of $r = x + iz$. Consider the function

$$r = w^2$$

Thus

$$(x + iz) = (\phi + i\psi)^2$$
$$= (\phi^2 + 2i\phi\psi - \psi^2)$$

Equating real and imaginary parts:

$$x = \phi^2 - \psi^2 \tag{2.34}$$

$$z = 2\phi\psi \tag{2.35}$$

Equations 2.34 and 2.35 govern the transformation of points between the r and w planes.

Consider the transformation of the straight lines $\psi = n$, where $n = 0, 1, 2, 3$ (Figure 2.19(a)). From Equation 2.35

$$\phi = \frac{z}{2n}$$

and Equation 2.34 becomes

$$x = \frac{z^2}{4n^2} - n^2 \tag{2.36}$$

Equation 2.36 represents a family of confocal parabolas. For positive values of z the parabolas for the specified values of n are plotted in Figure 2.19(b).

Consider also the transformation of the straight lines $\phi = m$, where $m = 0, 1, 2, \ldots, 6$ (Figure 2.19(a)). From Equation 2.35

$$\psi = \frac{z}{2m}$$

and Equation 2.34 becomes

$$x = m^2 - \frac{z^2}{4m^2} \tag{2.37}$$

Equation 2.37 represents a family of confocal parabolas conjugate with the parabolas represented by Equation 2.36. For positive values of z the parabolas for the specified values of m are plotted in Figure 2.19(b). The two families of parabolas satisfy the requirements of a flow net.

Application to embankment dam sections

The flow region in the w plane satisfying the boundary conditions for the section (Figure 2.17) is shown in Figure 2.20(a). In this case, the transformation function

$$r = Cw^2$$

Seepage

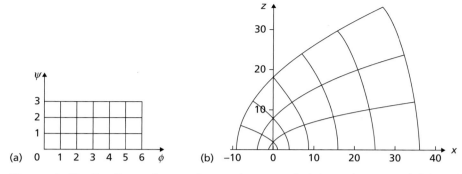

Figure 2.19 Conformal transformation $r=w^2$: (a) w plane, and (b) r plane.

will be used, where C is a constant. Equations 2.34 and 2.35 then become

$$x = C(\phi^2 - \psi^2)$$
$$z = 2C\phi\psi$$

The equation of the top flow line can be derived by substituting the conditions

$$\psi = q$$
$$\phi = -kz$$

Thus

$$z = -2Ckzq$$
$$\therefore C = -\frac{1}{2kq}$$

Hence

$$x = -\frac{1}{2kq}(k^2z^2 - q^2)$$
$$x = \frac{1}{2}\left(\frac{q}{k} - \frac{k}{q}z^2\right)$$

(2.38)

The curve represented by Equation 2.38 is referred to as Kozeny's basic parabola and is shown in Figure 2.20(b), the origin and focus both being at A.

When $z=0$, the value of x is given by

$$x_0 = \frac{q}{2k}$$
$$\therefore q = 2kx_0$$

(2.39)

where $2x_0$ is the directrix distance of the basic parabola. When $x=0$, the value of z is given by

$$z_0 = \frac{q}{k} = 2x_0$$

Substituting Equation 2.39 in Equation 2.38 yields

$$x = x_0 - \frac{z^2}{4x_0}$$

(2.40)

Development of a mechanical model for soil

The basic parabola can be drawn using Equation 2.40, provided the coordinates of one point on the parabola are known initially.

An inconsistency arises due to the fact that the conformal transformation of the straight line $\phi=-kh$ (representing the upstream equipotential) is a parabola, whereas the upstream equipotential in the embankment dam section is the upstream slope. Based on an extensive study of the problem, Casagrande (1940) recommended that the initial point of the basic parabola should be taken at G (Figure 2.21) where GC=0.3HC. The coordinates of point G, substituted in Equation 2.40, enable the value of x_0 to be determined; the basic parabola can then be plotted. The top flow line must intersect the upstream slope at right angles; a correction CJ must therefore be made (using personal judgement) to the basic parabola. The flow net can then be completed as shown in Figure 2.21.

If the discharge surface AD is not horizontal, as in the cases shown in Figure 2.22, a further correction KD to the basic parabola is required. The angle β is used to describe the direction of the discharge surface relative to AB. The correction can be made with the aid of values of the ratio MD/MA=$\Delta a/a$, given by Casagrande for the range of values of β (Table 2.4).

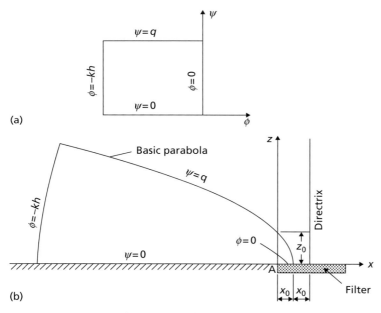

Figure 2.20 Transformation for embankment dam section: (a) w plane, and (b) r plane.

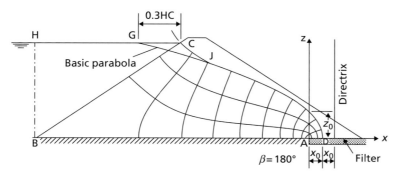

Figure 2.21 Flow net for embankment dam section.

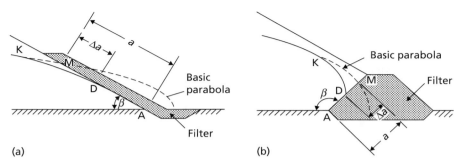

(a) (b)

Figure 2.22 Downstream correction to basic parabola.

Table 2.4 Downstream correction to basic parabola. Reproduced from A. Casagrande (1940) 'Seepage through dams', in *Contributions to Soil Mechanics 1925–1940*, by permission of the Boston Society of Civil Engineers

β	30°	60°	90°	120°	150°	180°
$\Delta a/a$	(0.36)	0.32	0.26	0.18	0.10	0

Example 2.4

A homogeneous anisotropic embankment dam section is detailed in Figure 2.23(a), the coefficients of permeability in the x and z directions being 4.5×10^{-8} and 1.6×10^{-8} m/s, respectively. Construct the flow net and determine the quantity of seepage through the dam. What is the pore water pressure at point P?

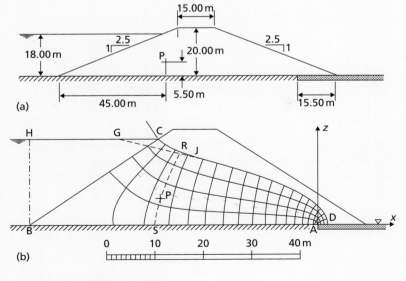

Figure 2.23 Example 2.4.

Solution

The scale factor for transformation in the x direction is

$$\sqrt{\frac{k_z}{k_x}} = \sqrt{\frac{1.6}{4.5}} = 0.60$$

The equivalent isotropic permeability is

$$k' = \sqrt{(k_x k_z)}$$
$$= \sqrt{(4.5 \times 1.6)} \times 10^{-8} = 2.7 \times 10^{-8} \, \text{m/s}$$

The section is drawn to the transformed scale in Figure 2.23(b). The focus of the basic parabola is at point A. The basic parabola passes through point G such that

$$GC = 0.3 HC = 0.3 \times 27.00 = 8.10 \, \text{m}$$

i.e. the coordinates of G are

$$x = -40.80, \; z = +18.00$$

Substituting these coordinates in Equation 2.40:

$$-40.80 = x_0 - \frac{18.00^2}{4x_0}$$

Hence

$$x_0 = 1.90 \, \text{m}$$

Using Equation 2.40, the coordinates of a number of points on the basic parabola are now calculated:

TABLE E

x	1.90	0	−5.00	−10.00	−20.00	−30.00
z	0	3.80	7.24	9.51	12.90	15.57

The basic parabola is plotted in Figure 2.23(b). The upstream correction is made (JC) and the flow net completed, ensuring that there are equal vertical intervals between the points of intersection of successive equipotentials with the top flow line. In the flow net there are four flow channels and 18 equipotential drops. Hence, the quantity of seepage (per unit length) is

$$q = k'h \frac{N_f}{N_d}$$

$$= 2.7 \times 10^{-8} \times 18 \times \frac{4}{18} = 1.1 \times 10^{-7} \, \text{m}^3/\text{s}$$

The quantity of seepage can also be determined from Equation 2.39 (without the necessity of drawing the flow net):

$$q = 2k'x_0$$
$$= 2 \times 2.7 \times 10^{-8} \times 1.90 = 1.0 \times 10^{-7} \, \text{m}^3/\text{s}$$

To determine the pore water pressure at P, Level AD is first selected as datum. An equipotential RS is drawn through point P (transformed position). By inspection, the total head at P is 15.60 m. At P the elevation head is 5.50 m, so the pressure head is 10.10 m and the pore water pressure is

$$u_P = 9.81 \times 10.10 = 99 \, \text{kPa}$$

Alternatively, the pressure head at P is given directly by the vertical distance of P below the point of intersection (R) of equipotential RS with the top flow line.

Seepage control in embankment dams

The design of an embankment dam section and, where possible, the choice of soils are aimed at reducing or eliminating the detrimental effects of seeping water. Where high hydraulic gradients exist there is a possibility that the seeping water may cause internal erosion within the dam, especially if the soil is poorly compacted. Erosion can work its way back into the embankment, creating voids in the form of channels or 'pipes', and thus impairing the stability of the dam. This form of erosion is referred to as **piping**.

A section with a central core of low permeability, aimed at reducing the volume of seepage, is shown in Figure 2.24(a). Practically all the total head is lost in the core, and if the core is narrow, high hydraulic gradients will result. There is a particular danger of erosion at the boundary between the core and the adjacent soil (of higher permeability) under a high exit gradient from the core. Protection against this danger can be given by means of a 'chimney' drain (Figure 2.24(a)) at the downstream boundary of the core. The drain, designed as a filter (see Section 2.10) to provide a barrier to soil particles from the core, also serves as an interceptor, keeping the downstream slope in an unsaturated state.

Most embankment dam sections are non-homogeneous owing to zones of different soil types, making the construction of the flow net more difficult. The basic parabola construction for the top flow line applies only to homogeneous sections, but the condition that there must be equal vertical distances between the points of intersection of equipotentials with the top flow line applies equally to a non-homogeneous section. The transfer condition (Equation 2.32) must be satisfied at all zone boundaries. In the case of a section with a central core of low permeability, the application of Equation 2.32 means that the lower the permeability ratio, the lower the position of the top flow line in the downstream zone (in the absence of a chimney drain).

If the foundation soil is more permeable than the dam, the control of **underseepage** is essential. Underseepage can be virtually eliminated by means of an 'impermeable' cut-off such as a grout curtain (Figure 2.24(b)). Another form of cut-off is the concrete diaphragm wall (see Section 11.10). Any measure designed to lengthen the seepage path, such as an impermeable upstream blanket (Figure 2.24(c)), will result in a partial reduction in underseepage.

An excellent treatment of seepage control is given by Cedergren (1989).

Development of a mechanical model for soil

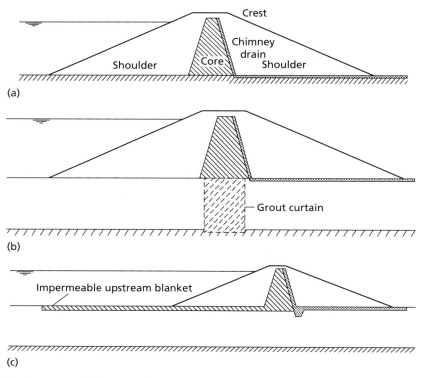

Figure 2.24 (a) Central core and chimney drain, (b) grout curtain, and (c) impermeable upstream blanket.

Example 2.5

Consider the concrete dam spillway from Example 2.2 (Figure 2.10). Determine the effect of the length of the cut-off wall on the reduction in seepage flow beneath the spillway. Determine also the reduction in seepage flow due to an impermeable upstream blanket of length L_b, and compare the efficacy of the two methods of seepage control.

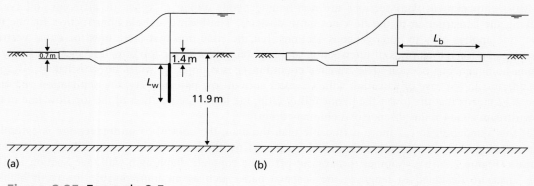

Figure 2.25 Example 2.5.

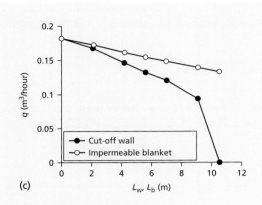
(c)

Figure 2.25 cont.

Solution

The seepage flow can be determined using the spreadsheet tool on the Companion Website. Repeating the calculations for different lengths, L_w, of cut-off wall (Figure 2.25(a)) and separately for different lengths of impermeable blanket, L_b, as shown in Figure 2.25(b), the results shown in the figure can be determined. By comparing the two methods, it can be seen that for thin layers such as in this problem, cut-off walls are generally more effective at reducing seepage and typically require a lower volume of material (and therefore lower cost) to achieve the same reduction in seepage flow.

2.10 Filter design

Filters used to control seepage must satisfy certain fundamental requirements. The pores must be small enough to prevent particles from being carried in from the adjacent soil. The permeability must simultaneously be high enough to ensure the free drainage of water entering the filter. The capacity of a filter should be such that it does not become fully saturated. In the case of an embankment dam, a filter placed downstream from the core should be capable of controlling and sealing any leak which develops through the core as a result of internal erosion. The filter must also remain stable under the abnormally high hydraulic gradient which is liable to develop adjacent to such a leak.

Based on extensive laboratory tests by Sherard *et al.* (1984a, 1984b) and on design experience, it has been shown that filter performance can be related to the size D_{15} obtained from the particle size distribution curve of the filter material. Average pore size, which is largely governed by the smaller particles in the filter, is well represented by D_{15}. A filter of uniform grading will trap all particles larger than around $0.11D_{15}$; particles smaller than this size will be carried through the filter in suspension in the seeping water. The characteristics of the adjacent soil, in respect of its retention by the filter, can be represented by the size D_{85} for that soil. The following criterion has been recommended for satisfactory filter performance:

$$\frac{(D_{15})_f}{(D_{85})_s} < 5 \tag{2.41}$$

where $(D_{15})_f$ and $(D_{85})_s$ refer to the filter and the adjacent (upstream) soil, respectively. However, in the case of filters for fine soils the following limit is recommended for the filter material:

$$D_{15} \leq 0.5\,\text{mm}$$

Development of a mechanical model for soil

Care must be taken to avoid segregation of the component particles of the filter during construction.

To ensure that the permeability of the filter is high enough to allow free drainage, it is recommended that

$$\frac{(D_{15})_f}{(D_{15})_s} > 5 \tag{2.42}$$

Graded filters comprising two (or more) layers with different gradings can also be used, the finer layer being on the upstream side. The above criterion (Equation 2.41) would also be applied to the component layers of the filter.

> **Summary**
>
> 1. The permeability of soil is strongly affected by the size of the voids. As a result, the permeability of fine soils may be many orders of magnitude smaller than those for coarse soils. Falling head tests are commonly used to measure permeability in fine soils, and constant head tests are used for coarse soils. These may be conducted in the laboratory on undisturbed samples removed from the ground (see Chapter 6) or in-situ.
> 2. Groundwater will flow wherever a hydraulic gradient exists. For flow in two-dimensions, a flow net may be used to determine the distribution of total head, pore pressure and seepage quantity. This technique can also account for soils which are layered or anisotropic in permeability, both of which parameters significantly affect seepage.
> 3. Complex or large seepage problems, for which flow net sketching would prove impractical, can be accurately and efficiently solved using the finite difference method. A spreadsheet implementation of this method is available on the Companion Website.
> 4. Seepage through earth dams is more complex as it is unconfined. The conformal transformation provides a straightforward and efficient method for determining flow quantities and developing the flow net for such a case. Seepage through and beneath earth dams, which may affect their stability, may be controlled using a range of techniques, including a low permeability core, cut-off walls, or impermeable blankets. The efficacy of these methods may be determined using the techniques outlined in the chapter.

Problems

2.1 In a falling-head permeability test the initial head of 1.00 m dropped to 0.35 m in 3 h, the diameter of the standpipe being 5 mm. The soil specimen was 200 mm long by 100 mm in diameter. Calculate the coefficient of permeability of the soil.

2.2 The section through part of a cofferdam is shown in Figure 2.26, the coefficient of permeability of the soil being 2.0×10^{-6} m/s. Draw the flow net and determine the quantity of seepage.

2.3 The section through a long cofferdam is shown in Figure 2.27, the coefficient of permeability of the soil being 4.0×10^{-7} m/s. Draw the flow net and determine the quantity of seepage entering the cofferdam.

2.4 The section through a sheet pile wall along a tidal estuary is given in Figure 2.28. At low tide the depth of water in front of the wall is 4.00 m; the water table behind the wall lags 2.50 m behind tidal level. Plot the net distribution of water pressure on the piling.

2.5 Details of an excavation adjacent to a canal are shown in Figure 2.29. Determine the quantity of seepage into the excavation if the coefficient of permeability is 4.5×10^{-5} m/s.

2.6 The dam shown in section in Figure 2.30 is located on anisotropic soil. The coefficients of permeability in the x and z directions are 5.0×10^{-7} and 1.8×10^{-7} m/s, respectively. Determine the quantity of seepage under the dam.

2.7 Determine the quantity of seepage under the dam shown in section in Figure 2.31. Both layers of soil are isotropic, the coefficients of permeability of the upper and lower layers being 2.0×10^{-6} and 1.6×10^{-5} m/s, respectively.

2.8 An embankment dam is shown in section in Figure 2.32, the coefficients of permeability in the horizontal and vertical directions being 7.5×10^{-6} and 2.7×10^{-6} m/s, respectively. Construct the top flow line and determine the quantity of seepage through the dam.

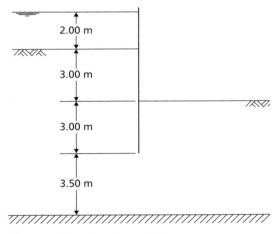

Figure 2.26 Problem 2.2.

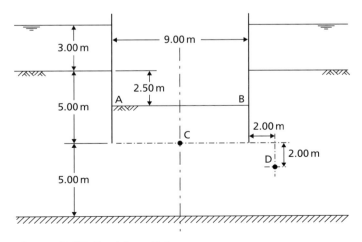

Figure 2.27 Problem 2.3.

Development of a mechanical model for soil

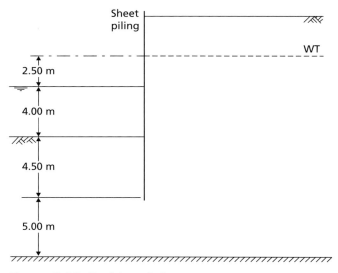

Figure 2.28 Problem 2.4.

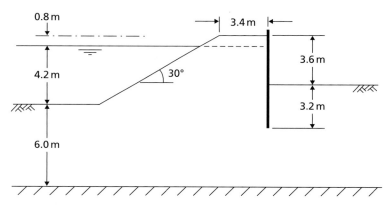

Figure 2.29 Problem 2.5.

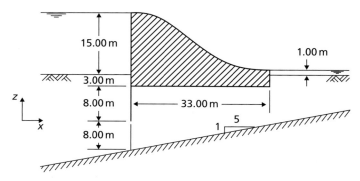

Figure 2.30 Problem 2.6.

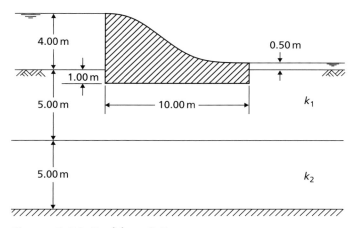

Figure 2.31 Problem 2.7.

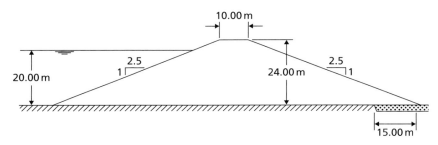

Figure 2.32 Problem 2.8.

References

ASTM D5084 (2010) *Standard Test Methods for Measurement of Hydraulic Conductivity of Saturated Porous Materials Using a Flexible Wall Permeameter*, American Society for Testing and Materials, West Conshohocken, PA.

British Standard 1377 (1990) *Methods of Test for Soils for Civil Engineering Purposes*, British Standards Institution, London.

CEN ISO/TS 17892 (2004) *Geotechnical Investigation and Testing – Laboratory Testing of Soil*, International Organisation for Standardisation, Geneva.

Casagrande, A. (1940) Seepage through dams, in *Contributions to Soil Mechanics 1925–1940*, Boston Society of Civil Engineers, Boston, MA, pp. 295–336.

Cedergren, H.R. (1989) *Seepage, Drainage and Flow Nets* (3rd edn), John Wiley & Sons, New York, NY.

Clayton, C.R.I., Matthews, M.C. and Simons, N.E. (1995) *Site Investigation* (2nd edn), Blackwell, London.

Hvorslev, M.J. (1951) *Time Lag and Soil Permeability in Ground-Water Observations*, Bulletin No. 36, Waterways Experimental Station, US Corps of Engineers, Vicksburg, MS.

Sherard, J.L., Dunnigan, L.P. and Talbot, J.R. (1984a) Basic properties of sand and gravel filters, *Journal of the ASCE*, **110**(GT6), 684–700.

Sherard, J.L., Dunnigan, L.P. and Talbot, J.R. (1984b) Filters for silts and clays, *Journal of the ASCE*, **110**(GT6), 701–718.

Vreedenburgh, C.G.F. (1936) On the steady flow of water percolating through soils with homogeneous-anisotropic permeability, in *Proceedings of the 1st International Conference on SMFE, Cambridge, MA*, Vol. 1.

Williams, B.P., Smyrell, A.G. and Lewis, P.J. (1993) Flownet diagrams – the use of finite differences and a spreadsheet to determine potential heads, *Ground Engineering*, **25**(5), 32–38.

Further reading

Cedergren, H.R. (1989) *Seepage, Drainage and Flow Nets* (3rd edn), John Wiley & Sons, New York, NY.

This is still the definitive text on seepage, particularly regarding flow net construction. The book also includes case histories showing the application of flow net techniques to real problems.

Preene, M., Roberts, T.O.L., Powrie, W. and Dyer, M.R. (2000) *Groundwater Control – Design and Practice*, CIRIA Publication C515, CIRIA, London.

This text covers groundwater control in more detail (it has only been touched on here). Also a valuable source of practical guidance.

For further student and instructor resources for this chapter, please visit the Companion Website at www.routledge.com/cw/craig

Chapter 3

Effective stress

> **Learning outcomes**
>
> After working through the material in this chapter, you should be able to:
>
> 1 Understand how total stress, pore water pressure and effective stress are related and the importance of effective stress in soil mechanics (Sections 3.1, 3.2 and 3.4);
> 2 Determine the effective stress state within the ground, both under hydrostatic conditions and when seepage is occurring (Sections 3.3 and 3.6);
> 3 Describe the phenomenon of liquefaction, and determine the hydraulic conditions within the groundwater under which liquefaction will occur (Section 3.7).

3.1 Introduction

A soil can be visualised as a skeleton of solid particles enclosing continuous voids which contain water and/or air. For the range of stresses usually encountered in practice, the individual solid particles and water can be considered incompressible; air, on the other hand, is highly compressible. The volume of the soil skeleton as a whole can change due to rearrangement of the soil particles into new positions, mainly by rolling and sliding, with a corresponding change in the forces acting between particles. The actual compressibility of the soil skeleton will depend on the structural arrangement of the solid particles, i.e. the void ratio, e. In a fully saturated soil, since water is considered to be incompressible, a reduction in volume is possible only if some of the water can escape from the voids. In a dry or a partially saturated soil a reduction in volume is always possible, due to compression of the air in the voids, provided there is scope for particle rearrangement (i.e. the soil is not already in its densest possible state, $e > e_{min}$).

Shear stress can be resisted only by the skeleton of solid particles, by means of reaction forces developed at the interparticle contacts. Normal stress may similarly be resisted by the soil skeleton through an increase in the interparticle forces. If the soil is fully saturated, the water filling the voids can also withstand normal stress by an increase in pore water pressure.

3.2 The principle of effective stress

The importance of the forces transmitted through the soil skeleton from particle to particle was recognised by Terzaghi (1943), who presented his **Principle of Effective Stress**, an intuitive relationship based on experimental data. The principle applies only to fully saturated soils, and relates the following three stresses:

1. the **total normal stress** (σ) on a plane within the soil mass, being the force per unit area transmitted in a normal direction across the plane, imagining the soil to be a solid (single-phase) material;
2. the **pore water pressure** (u), being the pressure of the water filling the void space between the solid particles;
3. the **effective normal stress** (σ') on the plane, representing the stress transmitted through the soil skeleton only (i.e. due to interparticle forces).

The relationship is:

$$\sigma = \sigma' + u \tag{3.1}$$

The principle can be represented by the following physical model. Consider a 'plane' XX in a fully saturated soil, passing through points of interparticle contact only, as shown in Figure 3.1. The wavy plane XX is really indistinguishable from a true plane on the mass scale due to the relatively small size of individual soil particles. A normal force P applied over an area A may be resisted partly by interparticle forces and partly by the pressure in the pore water. The interparticle forces are very random in both magnitude and direction throughout the soil mass, but at every point of contact on the wavy plane may be split into components normal and tangential to the direction of the true plane to which XX approximates; the normal and tangential components are N' and T, respectively. Then, the effective normal stress is approximated as the sum of all the components N' within the area A, divided by the area A, i.e.

$$\sigma' = \frac{\Sigma N'}{A} \tag{3.2}$$

The total normal stress is given by

$$\sigma = \frac{P}{A} \tag{3.3}$$

If point contact is assumed between the particles, the pore water pressure will act on the plane over the entire area A. Then, for equilibrium in the direction normal to XX

$$P = \Sigma N' + uA$$

or

$$\frac{P}{A} = \frac{\Sigma N'}{A} + u$$

i.e.

$$\sigma = \sigma' + u$$

The pore water pressure which acts equally in every direction will act on the entire surface of any particle, but is assumed not to change the volume of the particle (i.e. the soil particles themselves are incompressible); also, the pore water pressure does not cause particles to be pressed together. The error involved in

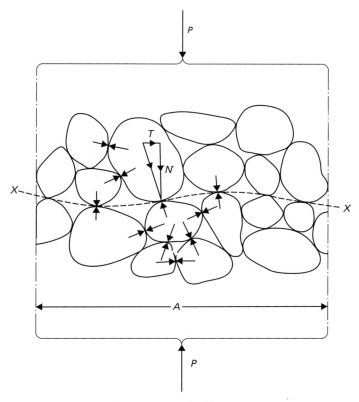

Figure 3.1 Interpretation of effective stress.

assuming point contact between particles is negligible in soils, the actual contact area a normally being between 1% and 3% of the cross-sectional area A. It should be understood that σ' does not represent the true contact stress between two particles, which would be the random but very much higher stress N'/a.

Effective vertical stress due to self-weight of soil

Consider a soil mass having a horizontal surface and with the water table at surface level. The total vertical stress (i.e. the total normal stress on a horizontal plane) σ_v at depth z is equal to the weight of all material (solids + water) per unit area above that depth, i.e.

$$\sigma_v = \gamma_{sat} z$$

The pore water pressure at any depth will be hydrostatic since the void space between the solid particles is continuous, so at depth z

$$u = \gamma_w z$$

Hence, from Equation 3.1 the effective vertical stress at depth z in this case will be

$$\sigma'_v = \sigma_v - u$$
$$= (\gamma_{sat} - \gamma_w) z$$

Example 3.1

A layer of saturated clay 4 m thick is overlain by sand 5 m deep, the water table being 3 m below the surface, as shown in Figure 3.2. The saturated unit weights of the clay and sand are 19 and 20 kN/m³, respectively; above the water table the (dry) unit weight of the sand is 17 kN/m³. Plot the values of total vertical stress and effective vertical stress against depth. If sand to a height of 1 m above the water table is saturated with capillary water, how are the above stresses affected?

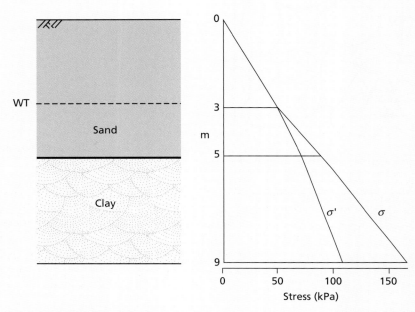

Figure 3.2 Example 3.1.

Solution

The total vertical stress is the weight of all material (solids + water) per unit area above the depth in question. Pore water pressure is the hydrostatic pressure corresponding to the depth below the water table. The effective vertical stress is the difference between the total vertical stress and the pore water pressure at the same depth. The stresses need only be calculated at depths where there is a change in unit weight (Table 3.1).

Table 3.1 Example 3.1

Depth (m)	σ_v (kPa)		u (kPa)		$\sigma'_v = \sigma_v - u$ (kPa)
3	3×17	=51.0	0		51.0
5	(3×17)+(2×20)	=91.0	2×9.81	=19.6	71.4
9	(3×17)+(2×20)+(4×19)	=167.0	6×9.81	=58.9	108.1

Effective stress

In all cases the stresses would normally be rounded off to the nearest whole number. The stresses are plotted against depth in Figure 3.2.

From Section 2.1, the water table is the level at which pore water pressure is atmospheric (i.e. $u=0$). Above the water table, water is held under negative pressure and, even if the soil is saturated above the water table, does not contribute to hydrostatic pressure below the water table. The only effect of the 1-m capillary rise, therefore, is to increase the total unit weight of the sand between 2 and 3 m depth from 17 to 20 kN/m³, an increase of 3 kN/m³. Both total and effective vertical stresses below 3 m depth are therefore increased by the constant amount $3 \times 1 = 3.0$ kPa, pore water pressures being unchanged.

3.3 Numerical solution using the Finite Difference Method

The calculation of the effective stress profile within the ground (i.e. variation with depth) under hydrostatic groundwater conditions is amenable to analysis using spreadsheets. In this way, the repeated calculations required to define the total stress pore pressure and effective stress over a fine depth scale can be automated for rapid analysis.

The soil body in question is divided into thin slices, i.e. a one-dimensional finite difference grid. The total stress due to each slice is then calculated as the product of its thickness and saturated or dry unit weight as appropriate. The total stress at any depth is then found as the sum of the total stress increments due to each of the slices above that depth. By subdividing the problem in this way, it is straightforward to tackle even problems with many layers of different soils. Because the pore fluid is continuous through the interconnected pore structure, the pore water pressure at any depth can be found from the depth below the water table (hydrostatic distribution), suitably modified to account for artesian or seepage-induced pressure (see Chapter 2) as necessary. The effective stress at any depth is then found by a simple application of Equation 3.1.

A spreadsheet tool for undertaking such analysis, Stress_CSM8.xls, may be found on the Companion Website. This is able to address a wide range of problems including up to ten layers of soil with different unit weights, a variable water table (either below or above the ground surface), surcharge loading at the ground surface, and confined layers under artesian fluid pressure (see Section 2.1).

3.4 Response of effective stress to a change in total stress

As an illustration of how effective stress responds to a change in total stress, consider the case of a fully saturated soil subject to an increase in total vertical stress $\Delta\sigma$ and in which the lateral strain is zero, volume change being entirely due to deformation of the soil in the vertical direction. This condition may be assumed in practice when there is a change in total vertical stress over an area which is large compared with the thickness of the soil layer in question.

It is assumed initially that the pore water pressure is constant at a value governed by a constant position of the water table. This initial value is called the **static pore water pressure** (u_s). When the total vertical stress is increased, the solid particles immediately try to take up new positions closer together. However, if water is incompressible and the soil is laterally confined, no such particle rearrangement, and therefore no increase in the interparticle forces, is possible unless some of the pore water can escape. Since it takes time for the pore water to escape by seepage, the pore water pressure is increased above the static value immediately after the increase in total stress takes place. The component of pore water pressure above the static value is known as the **excess pore water pressure** (u_e). This increase in pore water pressure will be equal to the increase in total vertical stress, i.e. the increase in total vertical stress is carried initially entirely by the pore water ($u_e = \Delta\sigma$). Note that if the lateral strain were not zero, some degree of particle rearrangement would be possible, resulting in an immediate increase in effective vertical stress, and the increase in pore water pressure would be less than the increase in total vertical stress by Terzaghi's Principle.

The increase in pore water pressure causes a hydraulic pressure gradient, resulting in transient flow of pore water (i.e. seepage, see Chapter 2) towards a free-draining boundary of the soil layer. This flow or drainage will continue until the pore water pressure again becomes equal to the value governed by the position of the water table, i.e. until it returns to its static value. It is possible, however, that the position of the water table will have changed during the time necessary for drainage to take place, so that the datum against which excess pore water pressure is measured will have changed. In such cases, the excess pore water pressure should be expressed with reference to the static value governed by the new water table position. At any time during drainage, the overall pore water pressure (u) is equal to the sum of the static and excess components, i.e.

$$u = u_s + u_e \tag{3.4}$$

The reduction of excess pore water pressure as drainage takes place is described as **dissipation**, and when this has been completed (i.e. when $u_e = 0$ and $u = u_s$) the soil is said to be in the **drained** condition. Prior to dissipation, with the excess pore water pressure at its initial value, the soil is said to be in the **undrained** condition. It should be noted that the term 'drained' does not mean that all of the water has flowed out of the soil pores; it means that there is no stress-induced (excess) pressure in the pore water. The soil remains fully saturated throughout the process of dissipation.

As drainage of pore water takes place the solid particles become free to take up new positions, with a resulting increase in the interparticle forces. In other words, as the excess pore water pressure dissipates, the effective vertical stress increases, accompanied by a corresponding reduction in volume. When dissipation of excess pore water pressure is complete, the increment of total vertical stress will be carried entirely by the soil skeleton. The time taken for drainage to be completed depends on the permeability of the soil. In soils of low permeability, drainage will be slow; in soils of high permeability, drainage will be rapid. The whole process is referred to as **consolidation**. With deformation taking place in one direction only (vertical as described here), consolidation is described as one-dimensional. This process will be described in greater detail in Chapter 4.

When a soil is subject to a reduction in total normal stress the scope for volume increase is limited, because particle rearrangement due to total stress increase is largely irreversible. As a result of increase in the interparticle forces there will be small elastic strains (normally ignored) in the solid particles, especially around the contact areas, and if clay mineral particles are present in the soil they may experience bending. In addition, the adsorbed water surrounding clay mineral particles will experience recoverable compression due to increases in interparticle forces, especially if there is face-to-face orientation of the particles. When a decrease in total normal stress takes place in a soil there will thus be a tendency for the soil skeleton to expand to a limited extent, especially so in soils containing an appreciable proportion of clay mineral particles. As a result, the pore water pressure will initially be reduced and the excess pore water pressure will be negative. The pore water pressure will gradually increase to the static value, flow taking place into the soil, accompanied by a corresponding reduction in effective normal stress and increase in volume. This process is known as **swelling**.

Under seepage (as opposed to static) conditions, the excess pore water pressure due to a change in total stress is the value above or below the **steady-state seepage pore water pressure** (u_{ss}), which is determined, at the point in question, from the appropriate flow net (see Chapter 2).

Effective stress

Consolidation analogy

The mechanics of the one-dimensional consolidation process can be represented by means of a simple analogy. Figure 3.3(a) shows a spring inside a cylinder filled with water, and a piston, fitted with a valve, on top of the spring. It is assumed that there can be no leakage between the piston and the cylinder, and no friction. The spring represents the compressible soil skeleton, the water in the cylinder the pore water, and the bore diameter of the valve the permeability of the soil. The cylinder itself simulates the condition of no lateral strain in the soil.

Suppose a load is now placed on the piston with the valve closed, as in Figure 3.3(b). Assuming water to be incompressible, the piston will not move as long as the valve is closed, with the result that no load can be transmitted to the spring; the load will be carried by the water, the increase in pressure in the water being equal to the load divided by the piston area. This situation with the valve closed corresponds to the undrained condition in the soil.

If the valve is now opened, water will be forced out through the valve at a rate governed by the bore diameter. This will allow the piston to move and the spring to be compressed as load is gradually transferred to it. This situation is shown in Figure 3.3(c). At any time, the increase in load on the spring will be directly proportional to the reduction in pressure in the water. Eventually, as shown in Figure 3.3(d), all the load will be carried by the spring and the piston will come to rest, this corresponding to the drained condition in the soil. At any time, the load carried by the spring represents the effective normal stress in the soil, the pressure of the water in the cylinder represents the pore water pressure, and the load on the piston represents the total normal stress. The movement of the piston represents the change in volume of the soil, and is governed by the compressibility of the spring (the equivalent of the compressibility of the soil skeleton, see Chapter 4). The piston and spring analogy represents only an element of soil, since the stress conditions vary from point to point throughout a soil mass.

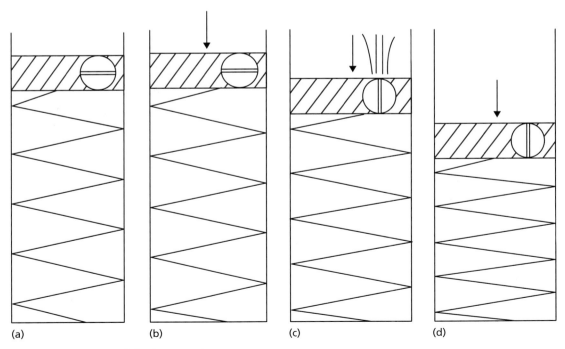

Figure 3.3 Consolidation analogy.

Example 3.2

A 5-m depth of sand overlies a 6-m thick layer of clay, the water table being at the surface; the permeability of the clay is very low. The saturated unit weight of the sand is $19\,\text{kN/m}^3$ and that of the clay is $20\,\text{kN/m}^3$. A 4-m depth of fill material of unit weight $20\,\text{kN/m}^3$ is placed on the surface over an extensive area. Determine the effective vertical stress at the centre of the clay layer (a) immediately after the fill has been placed, assuming this to take place rapidly, and (b) many years after the fill has been placed.

Solution

The soil profile is shown in Figure 3.4. Since the fill covers an extensive area, it can be assumed that the condition of zero lateral strain applies. As the permeability of the clay is very low, dissipation of excess pore water pressure will be very slow; immediately after the rapid placement of the fill, no appreciable dissipation will have taken place. The initial stresses and pore water pressure at the centre of the clay layer are

$$\sigma_v = (5\times 19)+(3\times 20) = 155\,\text{kPa}$$
$$u_s = 8\times 9.81 = 78\,\text{kPa}$$
$$\sigma'_v = \sigma_v - u_s = 77\,\text{kPa}$$

Many years after placement of the fill, dissipation of excess pore water pressure should be essentially complete such that the increment of total stress from the fill is entirely carried by the soil skeleton (effective stress). The effective vertical stress at the centre of the clay layer is then

$$\sigma'_v = 77+(4\times 20) = 157\,\text{kPa}$$

Immediately after the fill has been placed, the total vertical stress at the centre of the clay increases by $80\,\text{kPa}$ due to the weight of the fill. Since the clay is saturated and there is no lateral strain, there will be a corresponding increase in pore water pressure of $u_e = 80\,\text{kPa}$ (the initial excess pore water pressure). The static pore water pressure was calculated previously as $u_s = 78\,\text{kPa}$. Immediately after placement, the pore water pressure therefore increases from 78 to $158\,\text{kPa}$, and then during subsequent consolidation gradually decreases again to $78\,\text{kPa}$, accompanied by the gradual increase of effective vertical stress from 77 to $157\,\text{kPa}$.

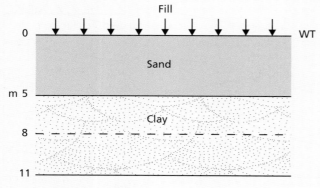

Figure 3.4 Example 3.2.

3.5 Effective stress in partially saturated soils

In the case of partially saturated soils, part of the void space is occupied by water and part by air. The pore water pressure (u_w) must always be less than the pore air pressure (u_a) due to surface tension. Unless the degree of saturation is close to unity, the pore air will form continuous channels through the soil and the pore water will be concentrated in the regions around the interparticle contacts. The boundaries between pore water and pore air will be in the form of menisci whose radii will depend on the size of the pore spaces within the soil. Part of any wavy plane through the soil will therefore pass part through water, and part through air.

Bishop (1959) proposed the following effective stress equation for partially saturated soils:

$$\sigma = \sigma' + u_a - \chi(u_a - u_w) \tag{3.5}$$

where χ is a parameter, to be determined experimentally, related primarily to the degree of saturation of the soil. The term ($\sigma' - u_a$) is also described as the **net stress**, while ($u_a - u_w$) is a measure of the suction in the soil.

Vanapalli and Fredlund (2000) conducted a series of laboratory triaxial tests (this test is described in Chapter 5) on five different soils, and found that

$$\chi = (S_r)^\kappa \tag{3.6}$$

where κ is a fitting parameter that is predominantly a function of plasticity index (I_p) as shown in Figure 3.5. Note that for non-plastic (coarse) soils, $\kappa = 1$. For a fully saturated soil ($S_r = 1$), $\chi = 1$; and for a completely dry soil ($S_r = 0$), $\chi = 0$. Equation 3.5 thus reduces to Equation 3.1 when $S_r = 1$. The value of χ is also influenced, to a lesser extent, by the soil structure and the way the particular degree of saturation was brought about.

A physical model may be considered in which the parameter χ is interpreted as the average proportion of any cross-section which passes through water. Then, across a given section of gross area A (Figure 3.6), total force is given by the equation

$$\sigma A = \sigma' A + u_w \chi A + u_a(1 - \chi)A \tag{3.7}$$

which leads to Equation 3.5.

If the degree of saturation of the soil is close to unity, it is likely that the pore air will exist in the form of bubbles within the pore water and it is possible to draw a wavy plane through pore water only. The soil can then be considered as a fully saturated soil, but with the pore water having some degree of compressibility due to the presence of the air bubbles. This is reasonable for most common applications in temperate climates (such as the UK). Equation 3.1 may then represent effective stress with sufficient accuracy for most practical purposes. A notable application where it is of greater importance to understand the behaviour of partially saturated soil is the stability of slopes under seasonal groundwater changes.

3.6 Influence of seepage on effective stress

When water is seeping through the pores of a soil, total head is dissipated as viscous friction producing a frictional drag, acting in the direction of flow, on the solid particles. A transfer of energy thus takes place from the water to the solid particles, and the force corresponding to this energy transfer is called **seepage**

Development of a mechanical model for soil

force. Seepage force acts on the particles of a soil in addition to gravitational force, and the combination of the forces on a soil mass due to gravity and seeping water is called the **resultant body force**. It is the resultant body force that governs the effective normal stress on a plane within a soil mass through which seepage is taking place.

Consider a point in a soil mass where the direction of seepage is at angle θ below the horizontal. A square element ABCD of dimension L (unit dimension normal to the page) is centred at the above point with sides parallel and normal to the direction of seepage, as shown in Figure 3.7(a) – i.e. the square element can be considered as a flow net element (curvilinear square). Let the drop in total head between the sides AD and BC be Δh. Consider the pore water pressures on the boundaries of the element, taking the value of pore water pressure at point A as u_A. The difference in pore water pressure between A and D

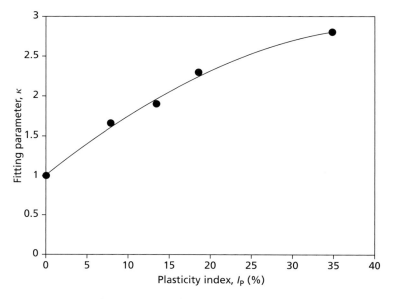

Figure 3.5 Relationship of fitting parameter κ to soil plasticity (re-plotted after Vanapalli and Fredlund, 2000).

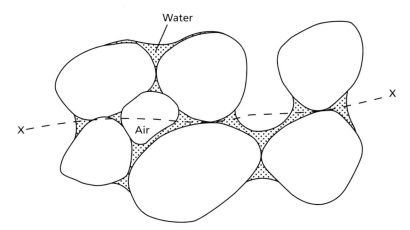

Figure 3.6 Partially saturated soil.

Effective stress

is due only to the difference in elevation head between A and D, the total head being the same at A and D. However, the difference in pore water pressure between A and either B or C is due to the difference in elevation head and the difference in total head between A and either B or C. If the total head at A is h_A and the elevation head at the same point is z_A, applying Equation 2.1 gives:

$$u_A = [h_A - z_A]\gamma_w$$

$$u_B = [(h_A - \Delta h) - (z_A - L\sin\theta)]\gamma_w$$

$$u_C = [(h_A - \Delta h) - (z_A - L\sin\theta - L\cos\theta)]\gamma_w$$

$$u_D = [(h_A) - (z_A - L\cos\theta)]\gamma_w$$

The following pressure differences can now be established:

$$u_B - u_A = u_C - u_D = [-\Delta h + L\sin\theta]\gamma_w$$

$$u_D - u_A = u_C - u_B = [L\cos\theta]\gamma_w$$

These values are plotted in Figure 3.7(b), giving the distribution diagrams of net pressure across the element in directions parallel and normal to the direction of flow as shown.

Therefore, the force on BC due to pore water pressure acting on the boundaries of the element, called the boundary water force, is given by

$$\gamma_w(-\Delta h + L\sin\theta)L$$

or

$$\gamma_w L^2 \sin\theta - \Delta h \gamma_w L$$

and the boundary water force on CD by

$$\gamma_w L^2 \cos\theta$$

If there were no seepage, i.e. if the pore water were static, the value of Δh would be zero, the forces on BC and CD would be $\gamma_w L^2 \sin\theta$ and $\gamma_w L^2 \cos\theta$, respectively, and their resultant would be $\gamma_w L^2$ acting in the vertical direction. The force $\Delta h\, \gamma_w L$ represents the only difference between the static and seepage cases and is therefore the seepage force (J), acting in the direction of flow (in this case normal to BC).

Now, the average hydraulic gradient across the element is given by

$$i = \frac{\Delta h}{L}$$

hence,

$$J = \Delta h \gamma_w L = \frac{\Delta h}{L} \gamma_w L^2 = i\gamma_w L^2$$

or

$$J = i\gamma_w V \qquad (3.8)$$

where V is the volume of the soil element.

Development of a mechanical model for soil

The seepage pressure (j) is defined as the seepage force per unit volume, i.e.

$$j = i\gamma_w \quad (3.9)$$

It should be noted that j (and hence J) depends only on the value of the hydraulic gradient.

All the forces, both gravitational and forces due to seeping water, acting on the element ABCD, may be represented in the vector diagram shown in Figure 3.7(c). The magnitude of the forces shown are summarised below.

> Total weight of the element = $\gamma_{sat} L^2$ = vector ab
> Boundary water force on CD = $\gamma_w L^2 \cos\theta$ = vector bd
> Boundary water force on BC = $\gamma_w L^2 \sin\theta - \Delta h\gamma_w L$ = vector de
> Resultant body force = vector ea

The resultant body force can be obtained as the vector summation of $ab + bd + de$, as shown in Figure 3.7(c). The seepage force $J = \Delta h\gamma_w L$ is represented by the dashed line in Figure 3.7(c), i.e. vector ce. Considering the triangle cbd:

$$bc = \gamma_w L^2$$

$$ac = ab - bc = (\gamma_{sat} - \gamma_w)L^2 = \gamma' L^2$$

Applying the cosine rule to triangle ace, and recognising that angle $ace = 90 + \theta$ degrees, it can be shown that the magnitude of the resultant body force (length of $|ea|$) is given by:

$$|ea| = \sqrt{(\gamma' L^2)^2 + (\Delta h\gamma_w L)^2 + (2\gamma' L^2 \Delta h\gamma_w \sin\theta)} \quad (3.10)$$

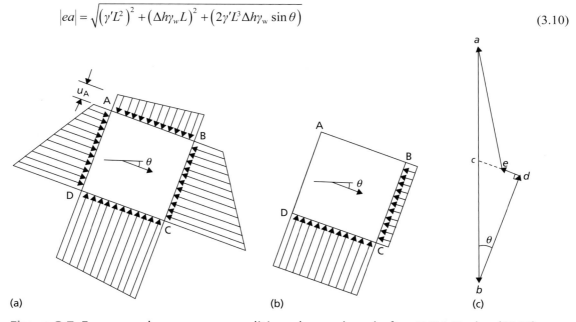

Figure 3.7 Forces under seepage conditions (reproduced after D.W. Taylor (1948) *Fundamentals of Soil Mechanics*, © John Wiley & Sons Inc., by permission).

Effective stress

This resultant body force acts at an angle to the vertical described by angle *bae*. Applying the sine rule to triangle *ace* gives:

$$\angle bae = \sin^{-1}\left(\frac{\Delta h \gamma_w L}{ea}\cos\theta\right) \quad (3.11)$$

Only the resultant body force contributes to effective stress. A component of seepage force acting vertically upwards will therefore reduce a vertical effective stress component from the static value. A component of seepage force acting vertically downwards will increase a vertical effective stress component from the static value.

3.7 Liquefaction

Seepage-induced liquefaction

Consider the special case of seepage vertically upwards ($\theta = -90°$). The vector *ce* in Figure 3.7(c) would then be vertically upwards, and if the hydraulic gradient were high enough the resultant body force would be zero. The value of hydraulic gradient corresponding to zero resultant body force is called the **critical hydraulic gradient** (i_{cr}). Substituting $|ea|=0$ and $\theta = -90°$ into Equation 3.10 gives

$$0 = (\gamma' L^2)^2 + (\Delta h \gamma_w L)^2 - 2(\gamma' L^3 \Delta h \gamma_w)$$
$$= (\gamma' L^2 - \Delta h \gamma_w L)^2$$
$$\therefore i_{cr} = \frac{\Delta h}{L} = \frac{\gamma'}{\gamma_w}$$

Therefore

$$i_{cr} = \frac{\gamma'}{\gamma_w} = \frac{G_s - 1}{1 + e} \quad (3.12)$$

The ratio γ'/γ_w, and hence the critical hydraulic gradient, is approximately 1.0 for most soils.

When the hydraulic gradient is i_{cr}, the effective normal stress on any plane will be zero, gravitational forces having been cancelled out by upward seepage forces. In the case of sands, the contact forces between particles will be zero and the soil will have no strength. The soil is then said to be **liquefied**, and if the critical gradient is exceeded the surface will appear to be 'boiling' as the particles are moved around in the upward flow of water. It should be realised that 'quicksand' is not a special type of soil, but simply sand through which there is an upward flow of water under a hydraulic gradient equal to or exceeding i_{cr}. In the case of clays, liquefaction may not necessarily result when the hydraulic gradient reaches the critical value given by Equation 3.12.

Conditions adjacent to sheet piling

High upward hydraulic gradients may be experienced in the soil adjacent to the downstream face of a sheet pile wall. Figure 3.8 shows part of the flow net for seepage under a sheet pile wall, the embedded length on the downstream side being *d*. A mass of soil adjacent to the piling may become unstable and be unable to support the wall. Model tests have shown that failure is likely to occur within a soil mass of approximate dimensions $d \times d/2$ in section (ABCD in Figure 3.8). Failure first shows in the form of a rise or **heave** at the surface, associated with an expansion of the soil which results in an increase in permeability. This in turn leads to increased flow, surface 'boiling' in the case of sands, and complete failure of the wall.

Development of a mechanical model for soil

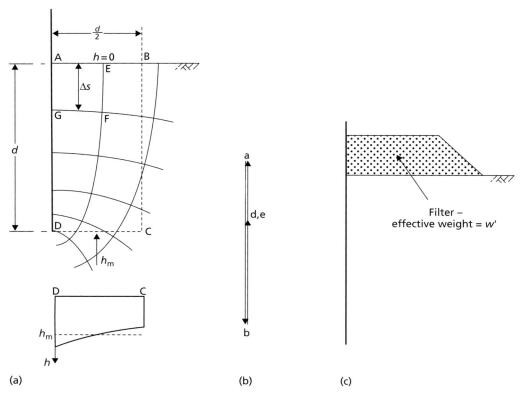

Figure 3.8 Upward seepage adjacent to sheet piling: (a) determination of parameters from flow net, (b) force diagram, and (c) use of a filter to suppress heave.

The variation of total head on the lower boundary CD of the soil mass can be obtained from the flow net equipotentials, or directly from the results of analysis using the FDM, such as the spreadsheet analysis tool described in Chapter 2. For the purpose of analysis, however, it is usually sufficient to determine the average total head h_m by inspection. The total head on the upper boundary AB is zero. The average hydraulic gradient is given by

$$i_m = \frac{h_m}{d}$$

Since failure due to heaving may be expected when the hydraulic gradient becomes i_{cr}, the factor of safety (F) against heaving may be expressed as

$$F = \frac{i_{cr}}{i_m} \qquad (3.13)$$

In the case of sands, a factor of safety can also be obtained with respect to 'boiling' at the surface. The exit hydraulic gradient (i_e) can be determined by measuring the dimension Δs of the flow net field AEFG adjacent to the piling:

$$i_e = \frac{\Delta h}{\Delta s}$$

where Δh is the drop in total head between equipotentials GF and AE. Then, the factor of safety is

$$F = \frac{i_{cr}}{i_e} \qquad (3.14)$$

There is unlikely to be any appreciable difference between the values of F given by Equations 3.13 and 3.14.

The sheet pile wall problem shown in Figure 3.8 can also be used to illustrate the graphical method for determining seepage forces outlined in Section 3.6:

> Total weight of mass ABCD = vector $ab = \frac{1}{2}\gamma_{sat} d^2$
> Average total head on CD = h_m
> Elevation head on CD = $-d$
> Average pore water pressure on CD = $(h_m + d)\gamma_w$
> Boundary water force on CD = vector $bd = \frac{d}{2}(h_m + d)\gamma_w$
> Boundary water force on BC = vector $de = 0$ (as seepage is vertically upwards).

Resultant body force of ABCD = $ab - bd - de$

$$= \frac{1}{2}\gamma_{sat} d^2 - \frac{d}{2}(h_m + d)\gamma_w - 0$$

$$= \frac{1}{2}(\gamma' + \gamma_w)d^2 - \frac{1}{2}(h_m d + d^2)\gamma_w$$

$$= \frac{1}{2}\gamma' d^2 - \frac{1}{2}h_m \gamma_w d$$

The resultant body force will be zero, leading to heave, when

$$\frac{1}{2}h_m \gamma_w d = \frac{1}{2}\gamma' d^2$$

The factor of safety can then be expressed as

$$F = \frac{\frac{1}{2}\gamma' d^2}{\frac{1}{2}h_m \gamma_w d} = \frac{\gamma' d}{h_m \gamma_w} = \frac{i_{cr}}{i_m}$$

If the factor of safety against heave is considered inadequate, the embedded length d may be increased or a surcharge load in the form of a filter may be placed on the surface AB, the filter being designed to prevent entry of soil particles, following the recommendations of Section 2.10. Such a filter is shown in Figure 3.8(c). If the effective weight of the filter per unit area is w', then the factor of safety becomes

$$F = \frac{\gamma' d + w'}{h_m \gamma_w}$$

Example 3.3

The flow net for seepage under a sheet pile wall is shown in Figure 3.9(a), the saturated unit weight of the soil being 20 kN/m³. Determine the values of effective vertical stress at points A and B.

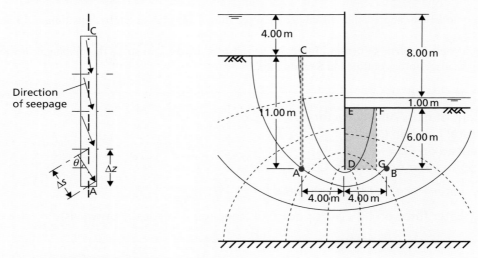

Figure 3.9 Examples 3.3 and 3.4.

Solution

First, consider the column of saturated soil of unit area between A and the soil surface at C. The total weight of the column is $11\gamma_{sat}$ (220 kN). Due to the change in level of the equipotentials across the column, the boundary water forces on the sides of the column will not be equal, although in this case the difference will be small. There is thus a net horizontal boundary water force on the column. However, as the effective vertical stress is to be calculated, only the vertical component of the resultant body force is required and the net horizontal boundary water force need not be considered. The vertical component of the boundary water force on the top surface of the column is due only to the depth of water above C, and is $4\gamma_w$ (39 kN). The boundary water force on the bottom surface of the column must be determined; in this example the FDM spreadsheet described in Section 2.7 is used as the required values of total head can be directly read from the spreadsheet, though this may equally be determined from a hand-drawn flow net. The calculated total head distribution may be found in Seepage_CSM8.xls on the Companion Website.

Total head at A, $h_A = 5.2$ m
Elevation head at A, $zh_A = -7.0$ m
Pore water pressure at A, $u_A = \gamma_w(h_A - z_A) = 9.81(5.2 + 7.0) = 120$ kPa
i.e. boundary water force on bottom surface = 120 kN
Net vertical boundary water force = 120 − 39 = 81 kN
Total weight of the column = 220 kN
Vertical component of resultant body force = 220 − 81 = 139 kN
i.e. effective vertical stress at A = 139 kPa.

Effective stress

It should be realised that the same result would be obtained by the direct application of the effective stress equation, the total vertical stress at A being the weight of saturated soil and water, per unit area, above A. Thus

$$\sigma_A = 11\gamma_{sat} + 4\gamma_w = 220 + 39 = 259 \, \text{kPa}$$
$$u_A = 120 \, \text{kPa}$$
$$\sigma'_A = \sigma_A - u_A = 259 - 120 = 139 \, \text{kPa}$$

The only difference in concept is that the boundary water force per unit area on top of the column of saturated soil AC contributes to the total vertical stress at A. Similarly at B

$$\sigma_B = 6\gamma_{sat} + 1\gamma_w = 120 + 9.81 = 130 \, \text{kPa}$$
$$h_B = 1.7 \, \text{m}$$
$$z_B = -7.0 \, \text{m}$$
$$u_B = \gamma_w (h_B - z_B) = 9.81(1.7 + 7.0) = 85 \, \text{kPa}$$
$$\sigma'_B = \sigma_B - u_B = 130 - 85 = 45 \, \text{kPa}$$

Example 3.4

Example 3.4

Using the flow net in Figure 3.9(a), determine the factor of safety against failure by heaving adjacent to the downstream face of the piling. The saturated unit weight of the soil is $20 \, \text{kN/m}^3$.

Solution

The stability of the soil mass DEFG in Figure 3.9(a), 6 m by 3 m in section, will be analysed. By inspection of the flow net or from Seepage_CSM8.xls on the Companion Website, the average value of total head on the base DG is given by

$$h_m = 2.6 \, \text{m}$$

The average hydraulic gradient between DG and the soil surface EF is

$$i_m = \frac{2.6}{6} = 0.43$$

Critical hydraulic gradient, $i_{cr} = \dfrac{\gamma'}{\gamma_w} = \dfrac{10.2}{9.8} = 1.04$

Factor of safety, $F = \dfrac{i_{cr}}{i_m} = \dfrac{1.04}{0.43} = 2.4$

Development of a mechanical model for soil

Dynamic/seismic liquefaction

In the previous examples, seepage-induced or **static liquefaction** of soil has been discussed – that is to say, situations where the effective stress within the soil is reduced to zero as a result of high pore water pressures due to seepage. Pore water pressure may also be increased due to dynamic loading of soil. As soil is sheared cyclically it has a tendency to contract, reducing the void ratio e. If this shearing and resulting contraction happens rapidly, then there may not be sufficient time for the pore water to escape from the voids, such that the reduction in volume will lead to an increase in pore water pressure due to the incompressibility of water.

Consider a uniform layer of fully saturated soil with the water table at the surface. The total stress at any depth z within the soil is

$$\sigma_v = \gamma_{sat} z$$

The soil will liquefy at this depth when the effective stress becomes zero. By Terzaghi's Principle (Equation 3.1) this will occur when $u = \sigma_v$. From Equation 3.4, the pore water pressure u is made up of two components: hydrostatic pressure u_s (present initially before the soil is loaded) and an excess component u_e (which is induced by the dynamic load). Therefore, the critical excess pore water pressure at the onset of liquefaction (u_{eL}) is given by

$$u = \sigma_v$$
$$u_s + u_{eL} = \gamma_{sat} z$$
$$\gamma_w z + u_{eL} = \gamma_{sat} z$$

$$u_{eL} = \gamma' z \qquad (3.15)$$

i.e. for soil to liquefy, the excess pore water pressure must be equal to the initial effective stress in the ground (prior to application of the dynamic load). Furthermore, considering a datum at the surface of the soil, from Equation 2.1

$$u = \gamma_w (h + z)$$
$$\therefore h = \frac{(\gamma_{sat} - \gamma_w)}{\gamma_w} z$$
$$\frac{h}{z} = \frac{\gamma'}{\gamma_w}$$

This demonstrates that there will be a positive hydraulic gradient h/z between the soil at depth z and the surface (i.e. vertically upwards), when liquefaction has been achieved. This is the same as the critical hydraulic gradient defined by Equation 3.12 for seepage-induced liquefaction.

Excess pore water pressure rise due to volumetric contraction may be induced by vibrating loads from direct sources on or in the soil, or may be induced due to cyclic ground motion during an earthquake. An example of the former case is a shallow foundation for a piece of machinery such as a power station turbine. In this case, the cyclic straining (and therefore volumetric contraction and induced excess pore pressure) in the soil will generally decrease with distance from the source. From Equation 3.15, it can be seen that the amount of excess pore water pressure required to initiate liquefaction increases with depth. As a result, any liquefaction will be concentrated towards the surface of the ground, close to the source.

Effective stress

In the case of an earthquake, ground motion is induced as a result of powerful stress waves which are transmitted from within the Earth's crust (i.e. far beneath the soil). As a result, liquefaction may extend to much greater depths. Combining Equations 3.12 and 3.15

$$u_{eL} = i_{cr}\gamma_w z = \frac{\gamma_w (G_s - 1)}{1+e} z \tag{3.16}$$

The soil towards the surface is often at a lower density (higher e) than the soil beneath. Combined with the shallow depth z, it is clear that, under earthquake shaking, liquefaction will start at the ground surface and move downwards as shaking continues, requiring larger excess pore water pressures at depth to liquefy the deeper layers. It is also clear from Equation 3.16 that looser soils at high e will require lower excess pore water pressures to cause liquefaction. Soils with high voids ratio also have a higher potential for densification when shaken (towards e_{min}, the densest possible state), so loose soils are particularly vulnerable to liquefaction. Indeed, strong earthquakes may fully liquefy layers of loose soil many metres thick.

It will be demonstrated in Chapter 5 that the shear strength of a soil (which resists applied loads due to foundations and other geotechnical constructions) is proportional to the effective stresses within the ground. It is therefore clear that the occurrence of liquefaction ($\sigma'_v = 0$) can lead to significant damage to structures – an example, observed during the 1964 Niigata earthquake in Japan, is shown in Figure 3.10.

Figure 3.10 Foundation failure due to liquefaction, 1964 Niigata earthquake, Japan.

Development of a mechanical model for soil

> **Summary**
>
> 1. Total stress is used to define the applied stresses on an element of soil (both due to external applied loads and due to self weight). Soils support total stresses through a combination of effective stress due to interparticle contact and pore water pressure in the voids. This is known as Terzaghi's Principle (Equation 3.1).
> 2. Under hydrostatic conditions, the effective stress state at any depth within the ground can be found from knowledge of the unit weight of the soil layers and the location of the water table. If seepage is occurring, a flow net or finite difference mesh can be used to determine the pore water pressures at any point within the ground, with effective stress subsequently being found using Terzaghi's Principle.
> 3. A consequence of Terzaghi's Principle is that if there is significant excess pore water pressure developed in the ground, the soil skeleton may become unloaded (zero effective stress). This condition is known as liquefaction and may occur due to seepage, or due to dynamic external loads which cause the soil to contract rapidly. Seepage-induced liquefaction can lead to uplift or boiling of soil along the downstream face of a sheet piled excavation, and subsequent failure of the excavation.

Problems

3.1 A river is 2 m deep. The river bed consists of a depth of sand of saturated unit weight 20 kN/m³. What is the effective vertical stress 5 m below the top of the sand?

3.2 The North Sea is 200 m deep. The sea bed consists of a depth of sand of saturated unit weight 20 kN/m³. What is the effective vertical stress 5 m below the top of the sand? Compare your answer to the value found in 3.1 – how does the water level above the ground surface affect the stresses within the ground?

3.3 A layer of clay 4 m thick lies between two layers of sand each 4 m thick, the top of the upper layer of sand being ground level. The water table is 2 m below ground level but the lower layer of sand is under artesian pressure, the piezometric surface being 4 m above ground level. The saturated unit weight of the clay is 20 kN/m³ and that of the sand 19 kN/m³; above the water table the unit weight of the sand is 16.5 kN/m³. Calculate the effective vertical stresses at the top and bottom of the clay layer.

3.4 In a deposit of fine sand the water table is 3.5 m below the surface, but sand to a height of 1.0 m above the water table is saturated by capillary water; above this height the sand may be assumed to be dry. The saturated and dry unit weights, respectively, are 20 and 16 kN/m³. Calculate the effective vertical stress in the sand 8 m below the surface.

3.5 A layer of sand extends from ground level to a depth of 9 m and overlies a layer of clay, of very low permeability, 6 m thick. The water table is 6 m below the surface of the sand. The saturated unit weight of the sand is 19 kN/m³ and that of the clay 20 kN/m³; the unit weight of the sand above the water table is 16 kN/m³. Over a short period of time the water table rises by 3 m, and is expected to remain permanently at this new level. Determine the effective vertical stress at depths of 8 and 12 m below ground level (a) immediately after the rise of the water table, and (b) several years after the rise of the water table.

Effective stress

3.6 An element of soil with sides horizontal and vertical measures 1 m in each direction. Water is seeping through the element in a direction inclined upwards at 30° above the horizontal under a hydraulic gradient of 0.35. The saturated unit weight of the soil is 21 kN/m³. Draw a force diagram to scale showing the following: total and effective weights, resultant boundary water force, seepage force. What is the magnitude and direction of the resultant body force?

3.7 For the seepage situations shown in Figure 3.11, determine the effective normal stress on plane XX in each case, (a) by considering pore water pressure and (b) by considering seepage pressure. The saturated unit weight of the soil is 20 kN/m³.

3.8 The section through a long cofferdam is shown in Figure 2.27, the saturated unit weight of the soil being 20 kN/m³. Determine the factor of safety against 'boiling' at the surface AB, and the values of effective vertical stress at C and D.

3.9 The section through part of a cofferdam is shown in Figure 2.26, the saturated unit weight of the soil being 19.5 kN/m³. Determine the factor of safety against heave failure in the excavation adjacent to the sheet piling. What depth of filter (unit weight 21 kN/m³) would be required to ensure a factor of safety of 3.0?

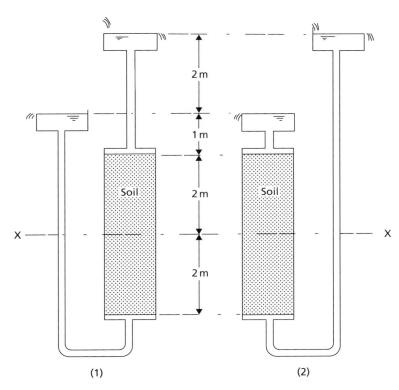

Figure 3.11 Problem 3.7.

References

Bishop, A.W. (1959) The principle of effective stress, *Tekniche Ukeblad*, **39**, 4–16.
Taylor, D.W. (1948) *Fundamentals of Soil Mechanics*, John Wiley & Sons, New York, NY.
Terzaghi, K. (1943) *Theoretical Soil Mechanics*, John Wiley & Sons, New York, NY.
Vanapalli, S.K. and Fredlund, D.G. (2000) Comparison of different procedures to predict unsaturated soil shear strength, *Proceedings of Geo-Denver 2000, ASCE Geotechnical Special Publication*, **99**, 195–209.

Further reading

Rojas, E. (2008a) Equivalent Stress Equation for Unsaturated Soils. I: Equivalent Stress, *International Journal of Geomechanics*, **8**(5), 285–290.
Rojas, E. (2008b) Equivalent Stress Equation for Unsaturated Soils. II: Solid-Porous Model, *International Journal of Geomechanics*, **8**(5), 291–299.
These companion papers describe in detail the development of a strength model for unsaturated soil, aiming to address one of the key questions in geotechnical engineering for which a satisfactory and widely accepted answer is still unclear.

For further student and instructor resources for this chapter, please visit the Companion Website at www.routledge.com/cw/craig

Chapter 4

Consolidation

> **Learning outcomes**
>
> After working through the material in this chapter, you should be able to:
>
> 1 Understand the behaviour of soil during consolidation (drainage of pore water pressure), and determine the mechanical properties which characterise this behaviour from laboratory testing (Sections 4.1–4.2 and 4.6–4.7);
> 2 Calculate ground settlements as a function of time due to consolidation both analytically and using computer-based tools for more complex problems (Sections 4.3–4.5 and 4.8–4.9);
> 3 Design a remedial scheme of vertical drains to speed-up consolidation and meet specified performance criteria (Section 4.10).

4.1 Introduction

As explained in Chapter 3, **consolidation** is the gradual reduction in volume of a fully saturated soil of low permeability due to change of effective stress. This may be as a result of drainage of some of the pore water, the process continuing until the excess pore water pressure set up by an increase in total stress has completely dissipated; consolidation may also occur due to a reduction in pore water pressure, e.g. from groundwater pumping or well abstraction (see Example 4.5). The simplest case is that of one-dimensional consolidation, in which the stress increment is applied in one direction only (usually vertical) with a condition of zero lateral strain being implicit. The process of **swelling**, the reverse of consolidation, is the gradual increase in volume of a soil under negative excess pore water pressure.

Consolidation settlement is the vertical displacement of the soil surface corresponding to the volume change at any stage of the consolidation process. Consolidation settlement will result, for example, if a structure (imposing additional total stress) is built over a layer of saturated clay, or if the water table is lowered permanently in a stratum overlying a clay layer. On the other hand, if an excavation (reduction in total stress) is made in a saturated clay, **heave** (upward displacement) will result in the bottom of the excavation due to swelling of the clay. In cases in which significant lateral strain takes place there will be an immediate settlement due to deformation of the soil under undrained conditions, in addition to consolidation settlement. The determination of immediate settlement will be discussed further in Chapter 8. This chapter is concerned with the prediction of both the magnitude and the rate of consolidation settlement

Development of a mechanical model for soil

under one-dimensional conditions (i.e. where the soil deforms only in the vertical direction). This is extended to the case when the soil can strain laterally (such as beneath a foundation) in Section 8.7.

The progress of consolidation in-situ can be monitored by installing piezometers to record the change in pore water pressure with time (these are described in Chapter 6). The magnitude of settlement can be measured by recording the levels of suitable reference points on a structure or in the ground: precise levelling is essential, working from a benchmark which is not subject to even the slightest settlement. Every opportunity should be taken of obtaining settlement data in the field, as it is only through such measurements that the adequacy of theoretical methods can be assessed.

4.2 The oedometer test

The characteristics of a soil during one-dimensional consolidation or swelling can be determined by means of the oedometer test. Figure 4.1 shows diagrammatically a cross-section through an oedometer. The test specimen is in the form of a disc of soil, held inside a metal ring and lying between two porous stones. The upper porous stone, which can move inside the ring with a small clearance, is fixed below a metal loading cap through which pressure can be applied to the specimen. The whole assembly sits in an open cell of water to which the pore water in the specimen has free access. The ring confining the specimen may be either fixed (clamped to the body of the cell) or floating (being free to move vertically); the inside of the ring should have a smooth polished surface to reduce side friction. The confining ring imposes a condition of zero lateral strain on the specimen. The compression of the specimen under pressure is measured by means of a dial gauge or electronic displacement transducer operating on the loading cap.

The test procedure has been standardised in CEN ISO/TS17892–5 (Europe) and ASTM D2435 (US), though BS 1377, Part 5 remains current in the UK, which specifies that the oedometer shall be of the fixed ring type. The initial pressure (total stress) applied will depend on the type of soil; following this, a sequence of pressures is applied to the specimen, each being double the previous value. Each pressure is normally maintained for a period of 24 h (in exceptional cases a period of 48 h may be required), compression readings being observed at suitable intervals during this period. At the end of the increment period, when the excess pore water pressure has completely dissipated, the applied total stress equals the effective vertical

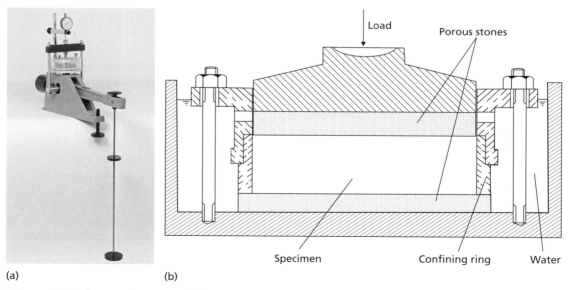

(a)   (b)

Figure 4.1 The oedometer: (a) test apparatus, (b) test arrangement (image courtesy of Impact Test Equipment Ltd.).

stress in the specimen. The results are presented by plotting the thickness (or percentage change in thickness) of the specimen or the void ratio at the end of each increment period against the corresponding effective stress. The effective stress may be plotted to either a natural or a logarithmic scale, though the latter is normally adopted due to the reduction in volume change in a given increment as total stress increases. If desired, the expansion of the specimen can additionally be measured under successive decreases in applied pressure to observe the swelling behaviour. However, even if the swelling characteristics of the soil are not required, the expansion of the specimen due to the removal of the final pressure should be measured.

Eurocode 7, Part 2 recommends that a minimum of two tests are conducted in a given soil stratum; this value should be doubled if there is considerable discrepancy in the measured compressibility, especially if there is little or no previous experience relating to the soil in question.

The void ratio at the end of each increment period can be calculated from the displacement readings and either the water content or the dry weight of the specimen at the end of the test. Referring to the phase diagram in Figure 4.2, the two methods of calculation are as follows:

1. Water content measured at end of test $= w_1$
   Void ratio at end of test $= e_1 = w_1 G_s$ (assuming $S_r = 100\%$)
   Thickness of specimen at start of test $= H_0$
   Change in thickness during test $= \Delta H$
   Void ratio at start of test $= e_0 = e_1 + \Delta e$ where

$$\frac{\Delta e}{\Delta H} = \frac{1 + e_0}{H_0} \tag{4.1}$$

   In the same way Δe can be calculated up to the end of any increment period.

2. Dry weight measured at end of test $= M_s$ (i.e. mass of solids)
   Thickness at end of any increment period $= H_1$
   Area of specimen $= A$
   Equivalent thickness of solids $= H_s = M_s/AG_s\rho w$
   Void ratio,

$$e_1 = \frac{H_1 - H_s}{H_s} = \frac{H_1}{H_s} - 1 \tag{4.2}$$

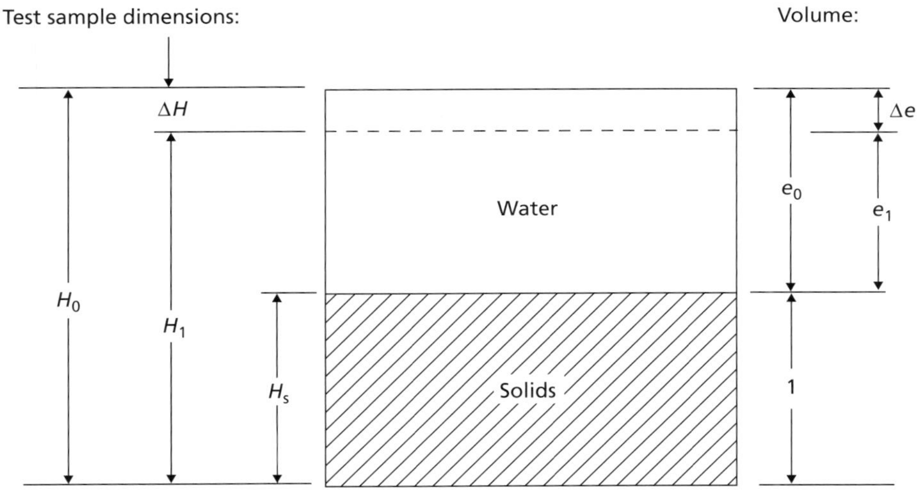

Figure 4.2 Phase diagram.

Development of a mechanical model for soil

Stress history

The relationship between void ratio and effective stress depends on the **stress history** of the soil. If the present effective stress is the maximum to which the soil has ever been subjected, the clay is said to be **normally consolidated**. If, on the other hand, the effective stress at some time in the past has been greater than the present value, the soil is said to be **overconsolidated**. The maximum value of effective stress in the past divided by the present value is defined as the **overconsolidation ratio** (OCR). A normally consolidated soil thus has an overconsolidation ratio of unity; an overconsolidated soil has an overconsolidation ratio greater than unity. The overconsolidation ratio can never be less than one.

Most soils will initially be formed by sedimentation of particles, which leads to gradual consolidation under increasing self-weight. Under these conditions, the effective stresses within the soil will be constantly increasing as deposition continues, and the soil will therefore be normally consolidated. Seabed or riverbed soils are common examples of soils which are naturally in a normally consolidated state (or close to it). Overconsolidation is usually the result of geological factors – for example, the erosion of overburden (due to glacier motion, wind, wave or ocean currents), the melting of ice sheets (and therefore reduction in stress) after glaciation, or permanent rise of the water table. Overconsolidation may also occur due to man-made processes: for example, the demolition of an old structure to redevelop the land will remove the total stresses that were applied by the building's foundations causing heave, such that, for the redevelopment, the soil will initially be overconsolidated.

Compressibility characteristics

Typical plots of void ratio (e) after consolidation against effective stress (σ') for a saturated soil are shown in Figure 4.3, the plots showing an initial compression followed by unloading and recompression. The shapes of the curves are related to the stress history of the soil. The e–$\log \sigma'$ relationship for a normally consolidated soil is linear (or nearly so), and is called the **virgin (one-dimensional) compression line** (1DCL). During compression along this line, permanent (irreversible) changes in soil structure continuously take place and the soil does not revert to the original structure during expansion. If a soil is overconsolidated, its state will be represented by a point on the expansion or recompression parts of the e–$\log \sigma'$ plot. The changes in soil structure along this line are almost wholly recoverable as shown in Figure 4.3.

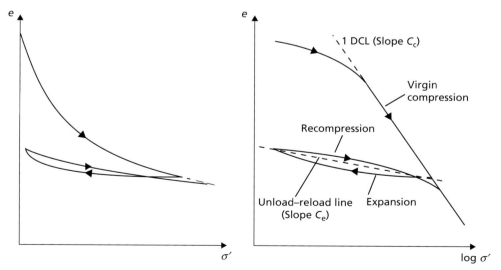

Figure 4.3 Void ratio–effective stress relationship.

Consolidation

The recompression curve ultimately rejoins the virgin compression line: further compression then occurs along the virgin line. The plots show that a soil in the overconsolidated state will be much less compressible than that in a normally consolidated state.

The compressibility of the soil can be quantified by one of the following coefficients.

1. The **coefficient of volume compressibility** (m_v), defined as the volume change per unit volume per unit increase in effective stress (i.e. ratio of volumetric strain to applied stress). The units of m_v are the inverse of stiffness (m²/MN). The volume change may be expressed in terms of either void ratio or specimen thickness. If, for an increase in effective stress from σ'_0 to σ'_1, the void ratio decreases from e_0 to e_1, then

$$m_v = \frac{1}{1+e_0}\left(\frac{e_0 - e_1}{\sigma'_1 - \sigma'_0}\right) \tag{4.3}$$

$$m_v = \frac{1}{H_0}\left(\frac{H_0 - H_1}{\sigma'_1 - \sigma'_0}\right) \tag{4.4}$$

The value of m_v for a particular soil is not constant but depends on the stress range over which it is calculated, as this parameter appears in the denominator of Equations 4.3 and 4.4. Most test standards specify a single value of the coefficient m_v calculated for a stress increment of 100 kN/m² in excess of the in-situ vertical effective stress of the soil sample at the depth it was sampled from (also termed the effective **overburden pressure**), although the coefficient may also be calculated, if required, for any other stress range, selected to represent the expected stress changes due to a particular geotechnical construction.

2. The **constrained modulus** (also called one-dimensional elastic modulus) E'_{oed} is the reciprocal of m_v (i.e. having units of stiffness, MN/m² = MPa) where:

$$E'_{oed} = \frac{1}{m_v} \tag{4.5}$$

3. The **compression index** (C_c) is the slope of the 1DCL, which is a straight line on the e–log σ' plot, and is dimensionless. For any two points on the linear portion of the plot

$$C_C = \frac{e_0 - e_1}{\log(\sigma'_1/\sigma'_0)} \tag{4.6}$$

The expansion part of the e–log σ' plot can be approximated to a straight line, the slope of which is referred to as the **expansion index** C_e (also called swell-back index). The expansion index is usually many times less than the compression index (as indicated in Figure 4.3).

It should be noted that although C_c and C_e represent negative gradients on the e–log σ' plot, their value is always given as positive (i.e. they represent the magnitude of the gradients).

Preconsolidation pressure

Casagrande (1936) proposed an empirical construction to obtain, from the e–log σ' curve for an overconsolidated soil, the maximum effective vertical stress that has acted on the soil in the past, referred to as the **preconsolidation pressure** (σ'_{max}). This parameter may be used to determine the in-situ OCR for the soil tested:

$$OCR = \frac{\sigma'_{max}}{\sigma'_{v0}} \tag{4.7}$$

105

Development of a mechanical model for soil

where σ'_{v0} is the in-situ vertical effective stress of the soil sample at the depth it was sampled from (effective overburden pressure), which may be calculated using the methods outlined in Chapter 3.

Figure 4.4 shows a typical e–$\log \sigma'$ curve for a specimen of soil which is initially overconsolidated. The initial curve (AB) and subsequent transition to a linear compression (BC) indicates that the soil is undergoing recompression in the oedometer, having at some stage in its history undergone swelling. Swelling of the soil in-situ may, for example, have been due to melting of ice sheets, erosion of overburden, or a rise in water table level. The construction for estimating the preconsolidation pressure consists of the following steps:

1. Produce back the straight-line part (BC) of the curve;
2. Determine the point (D) of maximum curvature on the recompression part (AB) of the curve;
3. Draw a horizontal line through D;
4. Draw the tangent to the curve at D and bisect the angle between the tangent and the horizontal through D;
5. The vertical through the point of intersection of the bisector and CB produced gives the approximate value of the preconsolidation pressure.

Whenever possible, the preconsolidation pressure for an overconsolidated clay should not be exceeded in construction. Compression will not usually be great if the effective vertical stress remains below σ'_{max}, as the soil will be always on the unload–reload part of the compression curve. Only if σ'_{max} is exceeded will compression be large. This is the key principle behind **preloading**, which is a technique used to reduce the compressibility of soils to make them more suitable for use in foundations; this is discussed in Section 4.11.

In-situ e–log σ´ curve

Due to the effects of sampling (see Chapter 6) and test preparation, the specimen in an oedometer test will be slightly disturbed. It has been shown that an increase in the degree of specimen disturbance

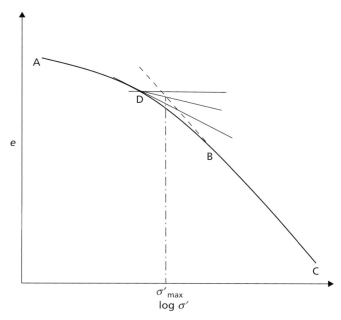

Figure 4.4 Determination of preconsolidation pressure.

Consolidation

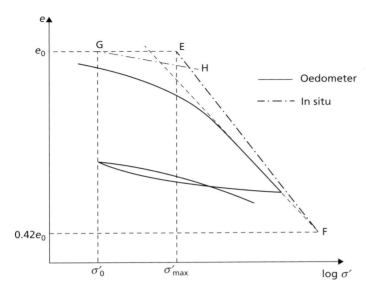

Figure 4.5 In-situ e–log σ' curve.

results in a slight decrease in the slope of the virgin compression line. It can therefore be expected that the slope of the line representing virgin compression of the in-situ soil will be slightly greater than the slope of the virgin line obtained in a laboratory test.

No appreciable error will be involved in taking the in-situ void ratio as being equal to the void ratio (e_0) at the start of the laboratory test. Schmertmann (1953) pointed out that the laboratory virgin line may be expected to intersect the in-situ virgin line at a void ratio of approximately 0.42 times the initial void ratio. Thus the in-situ virgin line can be taken as the line EF in Figure 4.5 where the coordinates of E are log σ'_{max} and e_0, and F is the point on the laboratory virgin line at a void ratio of $0.42e_0$.

In the case of overconsolidated clays, the in-situ condition is represented by the point (G) having coordinates σ'_0 and e_0, where σ'_{max} is the present effective overburden pressure. The in-situ recompression curve can be approximated to the straight line GH parallel to the mean slope of the laboratory recompression curve.

Example 4.1

The following compression readings were obtained in an oedometer test on a specimen of saturated clay ($G_s = 2.73$):

TABLE F

Pressure (kPa)	0	54	107	214	429	858	1716	3432	0
Dial gauge after 24 h (mm)	5.000	4.747	4.493	4.108	3.449	2.608	1.676	0.737	1.480

The initial thickness of the specimen was 19.0 mm, and at the end of the test the water content was 19.8%. Plot the e–log σ' curve and determine the preconsolidation pressure. Determine the values of m_v for the stress increments 100–200 and 1000–1500 kPa. What is the value of C_c for the latter increment?

Solution

Void ratio at end of test $= e_1 = w_1 G_s = 0.198 \times 2.73 = 0.541$

Void ratio at start of test $= e_0 = e_1 + \Delta e$

Now

$$\frac{\Delta e}{\Delta H} = \frac{1+e_0}{H_0} = \frac{1+e_1+\Delta e}{H_0}$$

i.e.

$$\frac{\Delta e}{3.520} = \frac{1.541 + \Delta e}{19.0}$$

$$\Delta e = 0.350$$

$$e_0 = 0.541 + 0.350 = 0.891$$

In general, the relationship between Δe and ΔH is given by

$$\frac{\Delta e}{\Delta H} = \frac{1.891}{19.0}$$

i.e. $\Delta e = 0.0996 \, \Delta H$, and can be used to obtain the void ratio at the end of each increment period (see Table 4.1). The e–log σ' curve using these values is shown in Figure 4.6. Using Casagrande's construction, the value of the preconsolidation pressure is 325 kPa.

Table 4.1 Example 4.1

Pressure (kPa)	ΔH (mm)	Δe	e
0	0	0	0.891
54	0.253	0.025	0.866
107	0.507	0.050	0.841
214	0.892	0.089	0.802
429	1.551	0.154	0.737
858	2.392	0.238	0.653
1716	3.324	0.331	0.560
3432	4.263	0.424	0.467
0	3.520	0.350	0.541

$$m_v = \frac{1}{1+e_0} \cdot \frac{e_0 - e_1}{\sigma'_1 - \sigma'_0}$$

For $\sigma'_0 = 100$ kPa and $\sigma'_1 = 200$ kPa

$$e_0 = 0.845 \text{ and } e_1 = 0.808$$

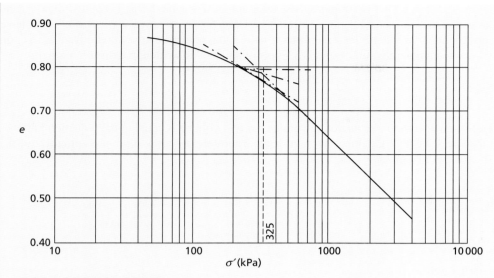

Figure 4.6 Example 4.1.

and therefore

$$m_v = \frac{1}{1.845} \times \frac{0.037}{100} = 2.0 \times 10^{-4} \text{ m}^2/\text{kN} = 0.20 \text{ m}^2/\text{MN}$$

For $\sigma'_0 = 1000$ kPa and $\sigma'_1 = 1500$ kPa

$e_0 = 0.632$ and $e_1 = 0.577$

and therefore

$$m_v = \frac{1}{1.845} \times \frac{0.055}{500} = 6.7 \times 10^{-5} \text{ m}^2/\text{kN} = 0.07 \text{ m}^2/\text{MN}$$

and

$$C_c = \frac{0.632 - 0.557}{\log(1500/1000)} = \frac{0.055}{0.176} = 0.31$$

Note that C_c will be the same for any stress range on the linear part of the e–$\log \sigma'$ curve; m_v will vary according to the stress range, even for ranges on the linear part of the curve.

4.3 Consolidation settlement

In order to estimate one-dimensional consolidation settlement, the value of either the coefficient of volume compressibility or the compression index is required. Consider a layer of saturated soil of thickness H. Due to construction, the total vertical stress in an elemental layer of thickness dz at depth z is increased by a **surcharge pressure** $\Delta\sigma$ (Figure 4.7). It is assumed that the condition of zero lateral strain

Development of a mechanical model for soil

applies within the soil layer. This is an appropriate assumption if the surcharge is applied over a wide area. After the completion of consolidation, an equal increase $\Delta\sigma'$ in effective vertical stress will have taken place corresponding to a stress increase from σ'_0 to σ'_1 and a reduction in void ratio from e_0 to e_1 on the e–σ' curve. The reduction in volume per unit volume of soil can be written in terms of void ratio:

$$\frac{\Delta V}{V_0} = \frac{e_0 - e_1}{1 + e_0}$$

Since the lateral strain is zero, the reduction in volume per unit volume is equal to the reduction in thickness per unit thickness, i.e. the settlement per unit depth. Therefore, by proportion, the settlement of the layer of thickness dz will be given by

$$ds_{oed} = \frac{e_0 - e_1}{1 + e_0} dz$$

$$= \left(\frac{e_0 - e_1}{\sigma'_1 - \sigma'_0}\right)\left(\frac{\sigma'_1 - \sigma'_0}{1 + e_0}\right) dz$$

$$= m_v \Delta\sigma' dz$$

where s_{oed} = consolidation settlement (one dimensional).

The settlement of the layer of thickness H is given by integrating the incremental change ds_{oed} over the height of the layer

$$s_{oed} = \int_0^H m_v \Delta\sigma' dz \tag{4.8}$$

If m_v and $\Delta\sigma'$ are constant with depth, then

$$s_{oed} = m_v \Delta\sigma' H \tag{4.9}$$

Substituting Equation 4.3 for m_v, Equation 4.9 may also be written as

$$s_{oed} = \frac{e_0 - e_1}{1 + e_0} H \tag{4.10}$$

or, in the case of a normally consolidated soil, by substituting Equation 4.6, Equation 4.9 is rewritten as

$$s_{oed} = \frac{C_c \log(\sigma'_1/\sigma'_0)}{1 + e_0} H \tag{4.11}$$

In order to take into account the variation of m_v and/or $\Delta\sigma'$ with depth (i.e. non-uniform soil), the graphical procedure shown in Figure 4.8 can be used to determine s_{oed}. The variations of initial effective vertical stress (σ'_0) and effective vertical stress increment ($\Delta\sigma'$) over the depth of the layer are represented in Figure 4.8(a); the variation of m_v is represented in Figure 4.8(b). The curve in Figure 4.8(c) represents the variation with depth of the dimensionless product $m_v\Delta\sigma'$, and the area under this curve is the settlement of the layer. Alternatively, the layer can be divided into a suitable number of sub-layers and the product $m_v\Delta\sigma'$ evaluated at the centre of each sub-layer: each product $m_v\Delta\sigma'$ is then multiplied by the appropriate sub-layer thickness to give the sub-layer settlement. The settlement of the whole layer is equal to the sum of the sub-layer settlements. The sub-layer technique may also be used to analyse layered soil profiles where the soil layers have very different compressional characteristics.

Consolidation

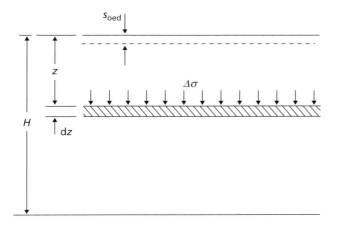
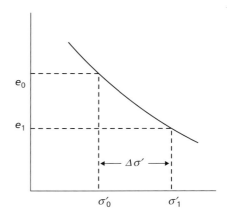

Figure 4.7 Consolidation settlement.

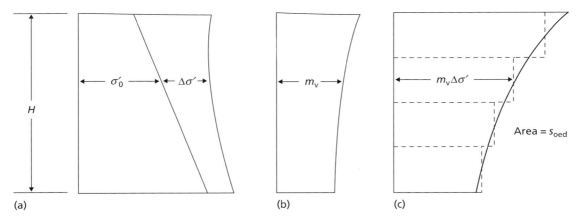

Figure 4.8 Consolidation settlement: graphical procedure.

Example 4.2

A long embankment 30 m wide is to be built on layered ground as shown in Figure 4.9. The net vertical pressure applied by the embankment (assumed to be uniformly distributed) is 90 kPa. The soil profile and stress distribution beneath the centre of the embankment are as shown in Figure 4.9 (the methods used to determine such a distribution are outlined in Section 8.5). The value of m_v for the upper clay is 0.35 m²/MN, and for the lower clay $m_v = 0.13$ m²/MN. The permeabilities of the clays are 10^{-10} m/s and 10^{-11} m/s for the upper and lower soils respectively. Determine the final settlement under the centre of the embankment due to consolidation.

Development of a mechanical model for soil

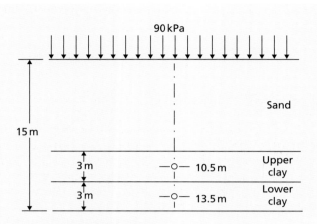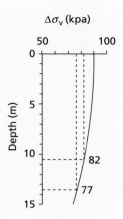

Figure 4.9 Example 4.2.

Solution

The clay layers are thin relative to the width of the applied surcharge from the embankment, therefore it can be assumed that consolidation is approximately one-dimensional. Considering the stresses shown in Figure 4.9, it will be sufficiently accurate to treat each clay layer as a single sub-layer. For one-dimensional consolidation $\Delta\sigma' = \Delta\sigma$, with the stress increments at the middle of each layer indicated in Figure 4.9. Equation 4.9 is then applied to determine the settlement of each of the layers, with H being the layer thickness ($=3\,\text{m}$), and these are combined to give the total consolidation settlement. The calculations are shown in Table 4.2.

Table 4.2 Example 4.2

Sublayer	z (m)	$\Delta\sigma'$ (kPa)	m_v (m²/MN)	H (m)	S_{oed} (mm)
1 (Upper clay)	10.5	82	0.35	3	86.1
2 (Lower clay)	13.5	77	0.13	3	30.0
					116.1

4.4 Degree of consolidation

For an element of soil at a particular depth z in a layer of soil, the progress of the consolidation process under a particular total stress increment can be expressed in terms of void ratio as follows:

$$U_v = \frac{e_0 - e}{e_0 - e_1}$$

where U_v is defined as the **degree of consolidation**, at a particular instant of time, at depth z ($0 \leq U_v \leq 1$); e_0 = void ratio before the start of consolidation; e_1 = void ratio at the end of consolidation and e = void ratio, at the time in question, during consolidation. A value of $U_v = 0$ means that consolidation has not yet begun (i.e. $e = e_0$); $U_v = 1$ implies that consolidation is complete (i.e. $e = e_1$)

Consolidation

If the e–σ' curve is assumed to be linear over the stress range in question, as shown in Figure 4.10, the degree of consolidation can be expressed in terms of σ':

$$U_v = \frac{\sigma' - \sigma'_0}{\sigma'_1 - \sigma'_0}$$

Suppose that the total vertical stress in the soil at the depth z is increased from σ_0 to σ_1 and there is no lateral strain. With reference to Figure 4.11, immediately after the increase takes place, although the total stress has increased to σ_1, the effective vertical stress will still be σ'_0 (undrained conditions); only after the completion of consolidation will the effective stress become σ'_1 (drained conditions). During consolidation, the increase in effective vertical stress is numerically equal to the decrease in excess pore water pressure. If σ' and u_e are, respectively, the values of effective stress and excess pore water pressure at any time during the consolidation process and if u_i is the initial excess pore water pressure (i.e. the value immediately after the increase in total stress), then, referring to Figure 4.10:

$$\sigma'_1 = \sigma'_0 + u_i = \sigma' + u_e$$

The degree of consolidation can then be expressed as

$$U_v = \frac{u_i - u_e}{u_i} = 1 - \frac{u_e}{u_i} \tag{4.12}$$

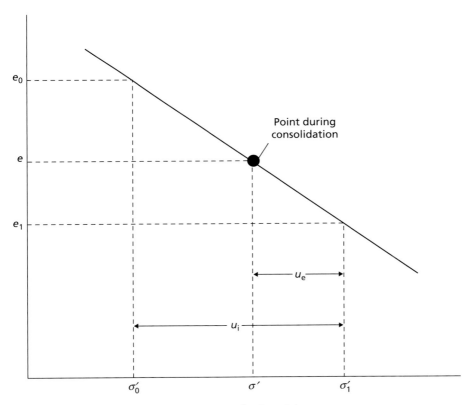

Figure 4.10 Assumed linear e–σ' relationship.

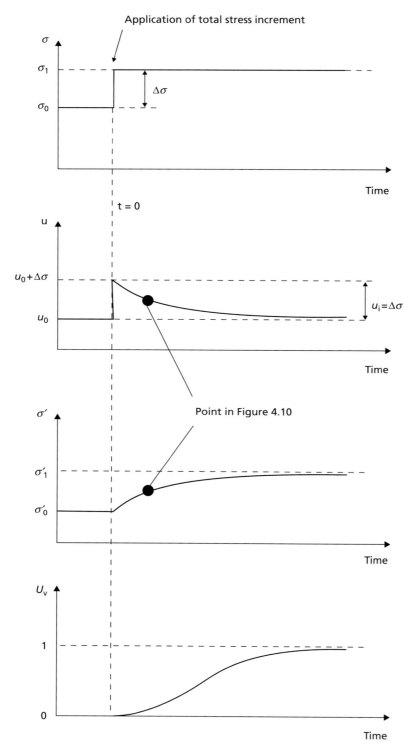

Figure 4.11 Consolidation under an increase in total stress $\Delta\sigma$.

4.5 Terzaghi's theory of one-dimensional consolidation

Terzaghi (1943) developed an analytical model for determining the degree of consolidation within soil at any time, t. The assumptions made in the theory are:

1. The soil is homogeneous.
2. The soil is fully saturated.
3. The solid particles and water are incompressible.
4. Compression and flow are one-dimensional (vertical).
5. Strains are small.
6. Darcy's law is valid at all hydraulic gradients.
7. The coefficient of permeability and the coefficient of volume compressibility remain constant throughout the process.
8. There is a unique relationship, independent of time, between void ratio and effective stress.

Regarding assumption 6, there is evidence of deviation from Darcy's law at low hydraulic gradients. Regarding assumption 7, the coefficient of permeability decreases as the void ratio decreases during consolidation (Al-Tabbaa and Wood, 1987). The coefficient of volume compressibility also decreases during consolidation since the e–σ' relationship is non-linear. However, for small stress increments assumption 7 is reasonable. The main limitations of Terzaghi's theory (apart from its one-dimensional nature) arise from assumption 8. Experimental results show that the relationship between void ratio and effective stress is not independent of time.

The theory relates the following three quantities.

1. the excess pore water pressure (u_e);
2. the depth (z) below the top of the soil layer;
3. the time (t) from the instantaneous application of a total stress increment.

Consider an element having dimensions dx, dy and dz within a soil layer of thickness $2d$, as shown in Figure 4.12. An increment of total vertical stress $\Delta\sigma$ is applied to the element.

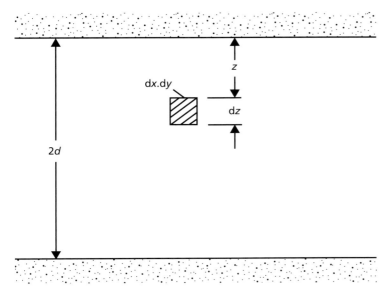

Figure 4.12 Element within a consolidating layer of soil.

Development of a mechanical model for soil

The flow velocity through the element is given by Darcy's law as

$$v_z = ki_z = -k\frac{\partial h}{\partial z}$$

Since at a fixed position z any change in total head (h) is due only to a change in pore water pressure:

$$v_z = -\frac{k}{\gamma_w}\frac{\partial u_e}{\partial z}$$

The condition of continuity (Equation 2.12) can therefore be expressed as

$$-\frac{k}{\gamma_w}\frac{\partial^2 u_e}{\partial z^2}dx\,dy\,dz = \frac{dV}{dt} \tag{4.13}$$

The rate of volume change can be expressed in terms of m_v:

$$\frac{dV}{dt} = m_v\frac{\partial \sigma'}{\partial t}dx\,dy\,dz$$

The total stress increment is gradually transferred to the soil skeleton, increasing effective stress, as the excess pore water pressure decreases. Hence the rate of volume change can be expressed as

$$\frac{dV}{dt} = -m_v\frac{\partial u_e}{\partial t}dx\,dy\,dz \tag{4.14}$$

Combining Equations 4.13 and 4.14,

$$m_v\frac{\partial u_e}{\partial t} = \frac{k}{\gamma_w}\frac{\partial^2 u_e}{\partial z^2}$$

or

$$\frac{\partial u_e}{\partial t} = c_v\frac{\partial^2 u_e}{\partial z^2} \tag{4.15}$$

This is the differential equation of consolidation, in which

$$c_v = \frac{k}{m_v\gamma_w} \tag{4.16}$$

c_v being defined as the **coefficient of consolidation**, with a suitable unit being m²/year. Since k and m_v are assumed as constants (assumption 7), c_v is constant during consolidation.

Solution of the consolidation equation

The total stress increment is assumed to be applied instantaneously, as in Figure 4.11. At zero time, therefore, the increment will be carried entirely by the pore water, i.e. the initial value of excess pore water pressure (u_i) is equal to $\Delta\sigma$ and the initial condition is

$$u_e = u_i \text{ for } 0 \leq z \leq 2d \text{ when } t = 0$$

The upper and lower boundaries of the soil layer are assumed to be free-draining, the permeability of the soil adjacent to each boundary being very high compared to that of the soil. Water therefore drains from the centre of the soil element to the upper and lower boundaries simultaneously so that the drainage path length is d if the soil is of thickness $2d$. Thus the boundary conditions at any time after the application of $\Delta\sigma$ are

$$u_e = 0 \text{ for } z = 0 \text{ and } z = 2d \text{ when } t > 0$$

The solution for the excess pore water pressure at depth z after time t from Equation 4.15 is

$$u_e = \sum_{n=1}^{n=\infty} \left(\frac{1}{d} \int_0^{2d} u_i \sin \frac{n\pi z}{2d} dz \right) \left(\sin \frac{n\pi z}{2d} \right) \exp\left(-\frac{n^2 \pi^2 c_v t}{4d^2} \right) \quad (4.17)$$

where u_i = initial excess pore water pressure, in general a function of z. For the particular case in which u_i is constant throughout the clay layer:

$$u_e = \sum_{n=1}^{n=\infty} \frac{2u_i}{n\pi}(1 - \cos n\pi) \left(\sin \frac{n\pi z}{2d} \right) \exp\left(-\frac{n^2 \pi^2 c_v t}{4d^2} \right) \quad (4.18)$$

When n is even, $(1 - \cos n\pi) = 0$, and when n is odd, $(1 - \cos n\pi) = 2$. Only odd values of n are therefore relevant, and it is convenient to make the substitutions

$$n = 2m + 1$$

and

$$M = \frac{\pi}{2}(2m + 1)$$

It is also convenient to substitute

$$T_v = \frac{c_v t}{d^2} \quad (4.19)$$

a dimensionless number called the **time factor**. Equation 4.18 then becomes

$$u_e = \sum_{m=0}^{m=\infty} \frac{2u_i}{M} \left(\sin \frac{Mz}{d} \right) \exp(-M^2 T_v) \quad (4.20)$$

The progress of consolidation can be shown by plotting a series of curves of u_e against z for different values of t. Such curves are called **isochrones**, and their form will depend on the initial distribution of excess pore water pressure and the drainage conditions at the boundaries of the soil layer. A layer for which both the upper and lower boundaries are free-draining is described as an **open layer** (sometimes called double drainage); a layer for which only one boundary is free-draining is a **half-closed layer** (sometimes called single drainage). Examples of isochrones are shown in Figure 4.13. In part (a) of the figure the initial distribution of u_i is constant and for an open layer of thickness $2d$ the isochrones are symmetrical about the centre line. The upper half of this diagram also represents the case of a half-closed layer of thickness d. The slope of an isochrone at any depth gives the hydraulic gradient and also indicates the direction of flow. In parts (b) and (c) of the figure, with a triangular distribution of u_i, the direction of flow changes over certain parts of the layer. In part (c) the lower boundary is impermeable, and for a time swelling takes place in the lower part of the layer.

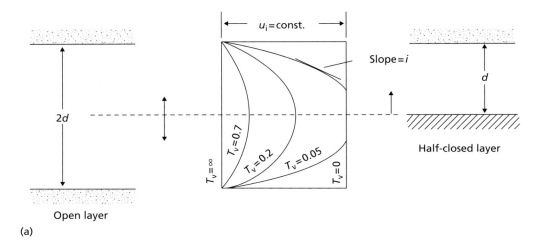

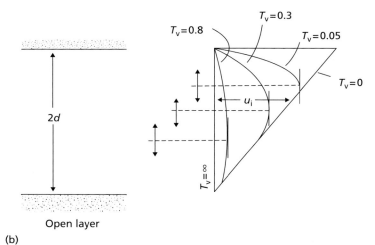

Figure 4.13 Isochrones.

Consolidation

The degree of consolidation at depth z and time t can be obtained by substituting the value of u_e (Equation 4.20) in Equation 4.12, giving

$$U_v = 1 - \sum_{m=0}^{m=\infty} \frac{2}{M}\left(\sin\frac{Mz}{d}\right)\exp(-M^2 T_v) \tag{4.21}$$

In practical problems it is the average degree of consolidation (U_v) over the depth of the layer as a whole that is of interest, the consolidation settlement at time t being given by the product of U_v and the final settlement. The average degree of consolidation at time t for constant u_i is given by

$$U_v = 1 - \frac{(1/2d)\int_0^{2d} u_e \, dz}{u_i}$$
$$= 1 - \sum_{m=0}^{m=\infty} \frac{2}{M^2}\exp(M^2 T_v) \tag{4.22}$$

The relationship between U_v and T_v given by Equation 4.22 is represented by curve 1 in Figure 4.14. Equation 4.22 can be represented almost exactly by the following empirical equations:

$$T_v = \begin{cases} \dfrac{\pi}{4}U_v^2 & U_v < 0.60 \\ -0.933\log(1-U_v) - 0.085 & U_v > 0.60 \end{cases} \tag{4.23}$$

If u_i is not constant, the average degree of consolidation is given by

$$U_v = 1 - \frac{\int_0^{2d} u_e \, dz}{\int_0^{2d} u_i \, dz} \tag{4.24}$$

where

$$\int_0^{2d} u_e \, dz = \text{area under isochrone at the time in question}$$

and

$$\int_0^{2d} u_i \, dz = \text{area under initial isochrone}$$

(For a half-closed layer, the limits of integration are 0 and d in the above equations.)

The initial variation of excess pore water pressure in a clay layer can usually be approximated in practice to a linear distribution. Curves 1, 2 and 3 in Figure 4.14 represent the solution of the consolidation equation for the cases shown in Figure 4.15, where

i Represents the initial conditions u_i in the oedometer test, and also the field case where the height of the water table has been changed (water table raised = positive u_i; water table lowered = negative u_i);
ii Represents virgin (normal) consolidation;
iii Represents approximately the field condition when a surface loading is applied (e.g. placement of surcharge or construction of foundation).

Development of a mechanical model for soil

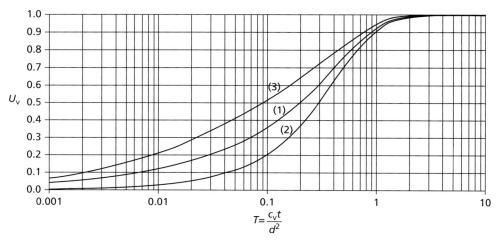

Figure 4.14 Relationships between average degree of consolidation and time factor.

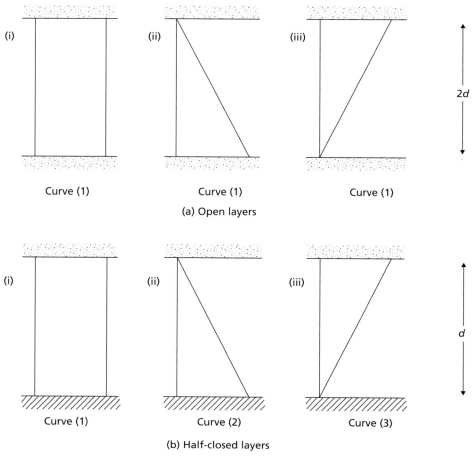

Figure 4.15 Initial variations of excess pore water pressure.

4.6 Determination of coefficient of consolidation

In order to apply consolidation theory in practice, it is necessary to determine the value of the coefficient of consolidation for use in either Equation 4.23 or Figure 4.14. The value of c_v for a particular pressure increment in the oedometer test can be determined by comparing the characteristics of the experimental and theoretical consolidation curves, the procedure being referred to as curve-fitting. The characteristics of the curves are brought out clearly if time is plotted to a square root or a logarithmic scale. It should be noted that once the value of c_v has been determined, the coefficient of permeability can be calculated from Equation 4.16, the oedometer test being a useful method for obtaining the permeability of fine-grained soils.

The log time method (due to Casagrande)

The forms of the experimental and theoretical curves are shown in Figure 4.16. The experimental curve is obtained by plotting the dial gauge readings in the oedometer test against the time in minutes, plotted on a logarithmic axis. The theoretical curve (inset) is given as the plot of the average degree of consolidation against the logarithm of the time factor. The theoretical curve consists of three parts: an initial curve which approximates closely to a parabolic relationship, a part which is linear and a final curve to which the horizontal axis is an asymptote at $U_v = 1.0$ (or 100%). In the experimental curve, the point corresponding to $U_v = 0$ can be determined by using the fact that the initial part of the curve represents an approximately parabolic relationship between compression and time. Two points on the curve are selected (A and B in Figure 4.16) for which the values of t are in the ratio of 1:4, and the vertical distance between them, ζ, is measured. In Figure 4.16, point A is shown at one minute and point B at four minutes. An equal distance of ζ above point A fixes the dial gauge reading (a_s) corresponding to $U_v = 0$.

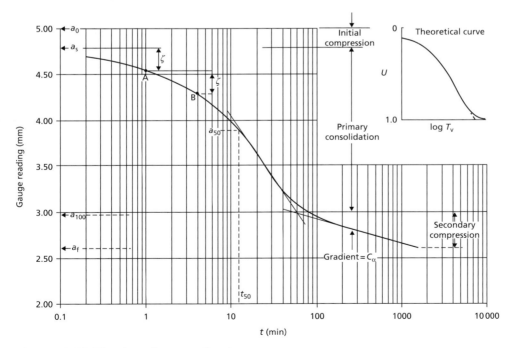

Figure 4.16 The log time method.

As a check, the procedure should be repeated using different pairs of points. The point corresponding to $U_v=0$ will not generally correspond to the point (a_0) representing the initial dial gauge reading, the difference being due mainly to the compression of small quantities of air in the soil, the degree of saturation being marginally below 100%: this compression is called **initial compression**. The final part of the experimental curve is linear but not horizontal, and the point (a_{100}) corresponding to $U_v=100\%$ is taken as the intersection of the two linear parts of the curve. The compression between the a_s and a_{100} points is called **primary consolidation**, and represents that part of the process accounted for by Terzaghi's theory. Beyond the point of intersection, compression of the soil continues at a very slow rate for an indefinite period of time and is called **secondary compression** (see Section 4.7). The point a_f in Figure 4.16 represents the final dial gauge reading before a subsequent total stress increment is applied.

The point corresponding to $U_v=50\%$ can be located midway between the a_s and a_{100} points, and the corresponding time t_{50} obtained. The value of T_v corresponding to $U_v=50\%$ is 0.196 (Equation 4.22 or Figure 4.14, curve 1), and the coefficient of consolidation is given by

$$c_v = \frac{0.196d^2}{t_{50}} \tag{4.25}$$

the value of d being taken as half the average thickness of the specimen for the particular pressure increment, due to the two-way drainage in the oedometer cell (to the top and bottom). If the average temperature of the soil in-situ is known and differs from the average test temperature, a correction should be applied to the value of c_v, correction factors being given in test standards (see Section 4.2).

The root time method (due to Taylor)

Figure 4.17 shows the forms of the experimental and theoretical curves, the dial gauge readings being plotted against the square root of time in minutes and the average degree of consolidation against the square root of time factor respectively. The theoretical curve is linear up to approximately 60% consolidation, and at 90% consolidation the abscissa (AC) is 1.15 times the abscissa (AB) of the extrapolation of the linear part of the curve. This characteristic is used to determine the point on the experimental curve corresponding to $U_v=90\%$.

The experimental curve usually consists of a short curved section representing initial compression, a linear part and a second curve. The point (D) corresponding to $U_v=0$ is obtained by extrapolating the linear part of the curve to the ordinate at zero time. A straight line (DE) is then drawn having abscissae 1.15 times the corresponding abscissae on the linear part of the experimental curve. The intersection of the line DE with the experimental curve locates the point (a_{90}) corresponding to $U_v=90\%$ and the corresponding value $\sqrt{t_{90}}$ can be obtained. The value of T_v corresponding to $U_v=90\%$ is 0.848 (Equation 4.22 or Figure 4.14, curve 1), and the coefficient of consolidation is given by

$$c_v = \frac{0.848d^2}{t_{90}} \tag{4.25}$$

If required, the point (a_{100}) on the experimental curve corresponding to $U_v=100\%$, the limit of primary consolidation, can be obtained by proportion. As in the log time plot, the curve extends beyond the 100% point into the secondary compression range. The root time method requires compression readings covering a much shorter period of time compared with the log time method, which requires the accurate definition of the second linear part of the curve well into the secondary compression range. On the other hand, a straight-line portion is not always obtained on the root time plot, and in such cases the log time method should be used.

Other methods of determining c_v have been proposed by Naylor and Doran (1948), Scott (1961) and Cour (1971).

Consolidation

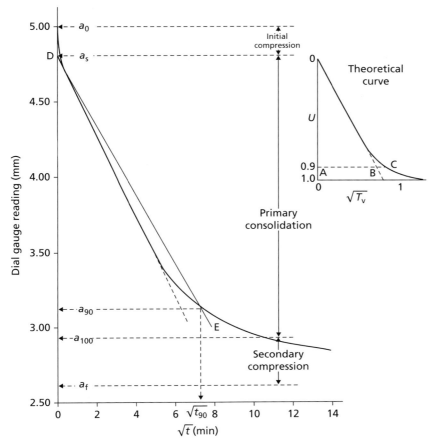

Figure 4.17 The root time method.

Compression ratios

The relative magnitudes of the initial compression, the compression due to primary consolidation and the secondary compression can be expressed by the following ratios (refer to Figures 4.16 and 4.17).

$$\text{Initial compression ratio}: r_0 = \frac{a_0 - a_s}{a_0 - a_f} \tag{4.27}$$

$$\text{Primary compression ratio (log time)}: r_p = \frac{a_s - a_{100}}{a_0 - a_f} \tag{4.28}$$

$$\text{Primary compression ratio (root time)}: r_p = \frac{10(a_s - a_{90})}{9(a_0 - a_f)} \tag{4.29}$$

$$\text{Secondary compression ratio}: r_s = 1 - (r_0 + r_p) \tag{4.30}$$

123

Development of a mechanical model for soil

In-situ value of c_v

Settlement observations have indicated that the rates of settlement of full-scale structures are generally much greater than those predicted using values of c_v obtained from oedometer tests on small specimens (e.g. 75 mm diameter × 20 mm thick). Rowe (1968) has shown that such discrepancies are due to the influence of the soil macro-fabric on drainage behaviour. Features such as laminations, layers of silt and fine sand, silt-filled fissures, organic inclusions and root-holes, if they reach a major permeable stratum, have the effect of increasing the overall permeability of the soil mass. In general, the macro-fabric of a field soil is not represented accurately in a small oedometer specimen, and the permeability of such a specimen will be lower than the mass permeability in the field.

In cases where fabric effects are significant, more realistic values of c_v can be obtained by means of the hydraulic oedometer developed by Rowe and Barden (1966) and manufactured for a range of specimen sizes. Specimens 250 mm in diameter by 100 mm in thick are considered large enough to represent the natural macro-fabric of most clays: values of c_v obtained from tests on specimens of this size have been shown to be consistent with observed settlement rates.

Details of a hydraulic oedometer are shown in Figure 4.18. Vertical pressure is applied to the specimen by means of water pressure acting across a convoluted rubber jack. The system used to apply the pressure must be capable of compensating for pressure changes due to leakage and specimen volume change. Compression of the specimen is measured by means of a central spindle passing through a sealed housing in the top plate of the oedometer. Drainage from the specimen can be either vertical or radial, and pore water pressure can be measured during the test The apparatus can also be used for flow tests, from which the coefficient of permeability can be determined directly (see Section 2.2).

Piezometers installed into the ground (see Chapter 6) can be used for the in-situ determination of c_v, but the method requires the use of three-dimensional consolidation theory. The most satisfactory procedure is to maintain a constant head at the piezometer tip (above or below the ambient pore water pressure in the soil) and measure the rate of flow into or out of the system. If the rate of flow is measured at various times, the value of c_v (and of the coefficient of permeability k) can be deduced. Details have been given by Gibson (1966, 1970) and Wilkinson (1968).

Another method of determining c_v is to combine laboratory values of m_v (which from experience are known to be more reliable than laboratory values of c_v) with in-situ measurements of k, using Equation 4.16.

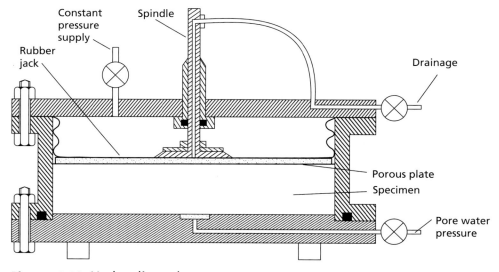

Figure 4.18 Hydraulic oedometer.

Example 4.3

The following compression readings were taken during an oedometer test on a saturated clay specimen ($G_s = 2.73$) when the applied pressure was increased from 214 to 429 kPa:

TABLE G

Time (min)	0	$\frac{1}{4}$	$\frac{1}{2}$	1	$2\frac{1}{4}$	4	9	16	25
Gauge (mm)	5.00	4.67	4.62	4.53	4.41	4.28	4.01	3.75	3.49
Time (min)	36	49	64	81	100	200	400	1440	
Gauge (mm)	3.28	3.15	3.06	3.00	2.96	2.84	2.76	2.61	

After 1440 min, the thickness of the specimen was 13.60 mm and the water content was 35.9%. Determine the coefficient of consolidation from both the log time and the root time methods and the values of the three compression ratios. Determine also the value of the coefficient of permeability.

Solution

Total change in thickness during increment $= 5.00 - 2.61 = 2.39$ mm

Average thickness during increment $= 13.60 + \dfrac{2.39}{2} = 14.80$ mm

Length of drainage path $d = \dfrac{14.80}{2} = 7.40$ mm

From the log time plot (data shown in Figure 4.16),

$t_{50} = 12.5$ min

$$c_v = \frac{0.196 d^2}{t_{50}} = \frac{0.196 \times 7.40^2}{12.5} \times \frac{1440 \times 365}{10^6} = 0.45 \, \text{m}^2/\text{year}$$

$$r_0 = \frac{5.00 - 4.79}{5.00 - 2.61} = 0.088$$

$$r_p = \frac{4.79 - 2.98}{5.00 - 2.61} = 0.757$$

$$r_s = 1 - (0.088 + 0.757) = 0.155$$

From the root time plot (data shown in Figure 4.17) $\sqrt{t_{90}} = 7.30$, and therefore

$t_{90} = 53.3$ min

$$c_v = \frac{0.848 d^2}{t_{90}} = \frac{0.848 \times 7.40^2}{53.3} \times \frac{1440 \times 365}{10^6} = 0.46 \, \text{m}^2/\text{year}$$

$$r_0 = \frac{5.00 - 4.81}{5.00 - 2.61} = 0.080$$

$$r_p = \frac{10(4.81 - 3.12)}{9(5.00 - 2.61)} = 0.785$$

$$r_s = 1 - (0.080 + 0.785) = 0.135$$

Development of a mechanical model for soil

In order to determine the permeability, the value of m_v must be calculated.

Final void ratio, $e_1 = w_1 G_s = 0.359 \times 2.73 = 0.98$

Initial void ratio, $e_0 = e_1 + \Delta e$

Now

$$\frac{\Delta e}{\Delta H} = \frac{1+e_0}{H_0}$$

i.e.

$$\frac{\Delta e}{2.39} = \frac{1.98 + \Delta e}{15.99}$$

Therefore

$$\Delta e = 0.35 \text{ and } e_0 = 1.33$$

Now

$$m_v = \frac{1}{1+e_0} \cdot \frac{e_0 - e_1}{\sigma_1' - \sigma_0'}$$

$$= \frac{1}{2.33} \times \frac{0.35}{215} = 7.0 \times 10^{-4} \text{ m}^2/\text{kN}$$

$$= 0.70 \text{ m}^2/\text{MN}$$

Coefficient of permeability:

$$k = c_v m_v \gamma_w$$

$$= \frac{0.45 \times 0.70 \times 9.8}{60 \times 1440 \times 365 \times 10^3}$$

$$= 1.0 \times 10^{-10} \text{ m/s}$$

4.7 Secondary compression

In the Terzaghi theory, it is implied by assumption 8 that a change in void ratio is due entirely to a change in effective stress brought about by the dissipation of excess pore water pressure, with permeability alone governing the time dependency of the process. However, experimental results show that compression does not cease when the excess pore water pressure has dissipated to zero but continues at a gradually decreasing rate under constant effective stress (see Figure 4.16). Secondary compression is thought to be due to the gradual readjustment of fine-grained particles into a more stable configuration following the structural disturbance caused by the decrease in void ratio, especially if the soil is laterally confined. An additional factor is the gradual lateral displacements which take place in thick layers subjected to shear stresses. The rate of secondary compression is thought to be controlled by the highly viscous film of adsorbed water surrounding clay mineral particles in the soil. A very slow viscous flow of adsorbed water takes place from the zones of film contact, allowing the solid particles to move closer together. The viscosity of the film increases as the particles move closer, resulting in a decrease in the rate of compression of the soil. It is presumed that primary consolidation and secondary compression proceed simultaneously from the time of loading.

Table 4.3 Secondary compression characteristics of natural soils (after Mitchell and Soga, 2005)

Soil type	C_α/C_c
Sands (low fines content)	0.01–0.03
Clays and silts	0.03–0.08
Organic soils	0.05–0.10

The rate of secondary compression in the oedometer test can be defined by the slope (C_α) of the final part of the compression–log time curve (Figure 4.16). Mitchell and Soga (2005) collated data on C_α for a range of natural soils, normalising the data by C_c. These data are summarised in Table 4.3, from which it can be seen that the secondary compression is typically between 1% and 10% of the primary compression, depending on soil type.

The magnitude of secondary compression occurring over a given time is generally greater in normally consolidated clays than in overconsolidated clays. In overconsolidated clays strains are mainly elastic, but in normally consolidated clays significant plastic strains occur. For certain highly plastic clays and organic clays, the secondary compression part of the compression–log time curve may completely mask the primary consolidation part. For a particular soil the magnitude of secondary compression over a given time, as a percentage of the total compression, increases as the ratio of pressure increment to initial pressure decreases; the magnitude of secondary compression also increases as the thickness of the oedometer specimen decreases and as temperature increases. Thus, the secondary compression characteristics of an oedometer specimen cannot normally be extrapolated to the full-scale situation.

In a small number of normally consolidated clays it has been found that secondary compression forms the greater part of the total compression under applied pressure. Bjerrum (1967) showed that such clays have gradually developed a reserve resistance against further compression as a result of the considerable decrease in void ratio which has occurred, under constant effective stress, over the hundreds or thousands of years since deposition. These clays, although normally consolidated, exhibit a quasi-preconsolidation pressure. It has been shown that, provided any additional applied pressure is less than approximately 50% of the difference between the quasi-preconsolidation pressure and the effective overburden pressure, the resultant settlement will be relatively small.

4.8 Numerical solution using the Finite Difference Method

The one-dimensional consolidation equation can be solved numerically by the method of finite differences. The method has the advantage that any pattern of initial excess pore water pressure can be adopted, and it is possible to consider problems in which the load is applied gradually over a period of time. The errors associated with the method are negligible, and the solution is easily programmed for the computer. A spreadsheet tool, Consolidation_CSM8.xls, which implements the FDM for solution of consolidation problems, as outlined in this section, is provided on the Companion Website.

The method is based on a depth–time grid (or mesh) as shown in Figure 4.19. This differs from the FDM described in Chapter 2, where the two-dimensional geometry meant that only the steady-state pore water pressures could be determined straightforwardly. Due to the simpler one-dimensional nature of the consolidation process, the second dimension of the finite difference grid can be used to determine the evolution of pore water pressure with time.

Development of a mechanical model for soil

The depth of the consolidating soil layer is divided into m equal parts of thickness Δz, and any specified period of time is divided into n equal intervals Δt. Any point on the grid can be identified by the subscripts i and j, the depth position of the point being denoted by i ($0 \le i \le m$) and the elapsed time by j ($0 \le j \le n$). The value of excess pore water pressure at any depth after any time is therefore denoted by $u_{i,j}$. (In this section the subscript e is dropped from the symbol for excess pore water pressure, i.e. u represents u_e as defined in Section 3.4). The following finite difference approximations can be derived from Taylor's theorem:

$$\frac{\partial u}{\partial t} = \frac{1}{\Delta t}\left(u_{i,j+1} - u_{i,j}\right)$$

$$\frac{\partial^2 u}{\partial z^2} = \frac{1}{(\Delta z)^2}\left(u_{i-1,j} + u_{i+1,j} - 2u_{i,j}\right)$$

Substituting these values in Equation 4.15 yields the finite difference approximation of the one-dimensional consolidation equation:

$$u_{i,j+1} = u_{i,j} + \frac{c_v \Delta t}{(\Delta z)^2}\left(u_{i-1,j} + u_{i+1,j} - 2u_{i,j}\right) \tag{4.31}$$

It is convenient to write

$$\beta = \frac{c_v \Delta t}{(\Delta z)^2} \tag{4.32}$$

this term being called the operator of Equation 4.31. It has been shown that for convergence the value of the operator must not exceed 1/2. The errors due to neglecting higher-order derivatives in Taylor's theorem are reduced to a minimum when the value of the operator is 1/6.

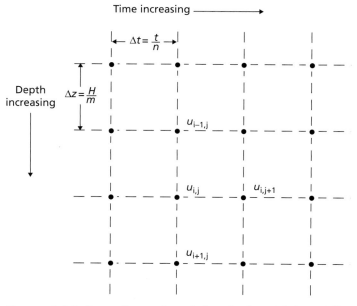

Figure 4.19 One-dimensional depth-time Finite Difference mesh.

It is usual to specify the number of equal parts m into which the depth of the layer is to be divided, and as the value of β is limited a restriction is thus placed on the value of Δt. For any specified period of time t in the case of an open layer:

$$T_v = \frac{c_v (n\Delta t)}{\left(\frac{1}{2} m\Delta z\right)^2} \qquad (4.33)$$

$$= 4\frac{n}{m^2}\beta$$

In the case of a half-closed layer the denominator becomes $(m\Delta z)^2$ and

$$T_v = \frac{n}{m^2}\beta \qquad (4.34)$$

A value of n must therefore be chosen such that the value of β in Equation 4.33 or 4.34 does not exceed 1/2.

Equation 4.31 does not apply to points on an impermeable boundary. There can be no flow across an impermeable boundary, a condition represented by the equation:

$$\frac{\partial u}{\partial z} = 0$$

which can be represented by the finite difference approximation:

$$\frac{1}{2\Delta z}\left(u_{i-1,j} - u_{i+1,j}\right) = 0$$

the impermeable boundary being at a depth position denoted by subscript i, i.e.

$$u_{i-1,j} = u_{i+1,j}$$

For all points on an impermeable boundary, Equation 4.31 therefore becomes

$$u_{i+1,j} = u_{i,j} + \frac{c_v \Delta t}{(\Delta z)^2}\left(2u_{i-1,j} - 2u_{i,j}\right) \qquad (4.35)$$

The degree of consolidation at any time t can be obtained by determining the areas under the initial isochrone and the isochrone at time t as in Equation 4.24. An implementation of this methodology may be found within Consolidation_CSM8.xls on the Companion Website.

Example 4.4

A half-closed clay layer (free-draining at the upper boundary) is 10 m thick and the value of c_v is 7.9 m²/year. The initial distribution of excess pore water pressure is as follows:

TABLE H

Depth (m)	0	2	4	6	8	10
Pressure (kPa)	60	54	41	29	19	15

Obtain the values of excess pore water pressure after consolidation has been in progress for 1 year.

Solution

The layer is half-closed, and therefore $d = 10$ m. For $t = 1$ year,

$$T_v = \frac{c_v t}{d^2} = \frac{7.9 \times 1}{10^2} = 0.079$$

Table 4.4 Example 4.4

i	j										
	0	1	2	3	4	5	6	7	8	9	10
0	0	0	0	0	0	0	0	0	0	0	0
1	54.0	40.6	32.6	27.3	23.5	20.7	18.5	16.7	15.3	14.1	13.1
2	41.0	41.2	38.7	35.7	32.9	30.4	28.2	26.3	24.6	23.2	21.9
3	29.0	29.4	29.9	30.0	29.6	29.0	28.3	27.5	26.7	26.0	25.3
4	19.0	20.2	21.3	22.4	23.3	24.0	24.5	24.9	25.1	25.2	25.2
5	15.0	16.6	18.0	19.4	20.6	21.7	22.6	23.4	24.0	24.4	24.7

The layer is divided into five equal parts, i.e. $m = 5$. Now

$$T_v = \frac{n}{m^2} \beta$$

Therefore

$$n\beta = 0.079 \times 5^5 = 1.98 \quad \text{(say 2.0)}$$

(This makes the actual value of $T_v = 0.080$ and $t = 1.01$ years.) The value of n will be taken as 10 (i.e. $\Delta t = 1/10$ year), making $\beta = 0.2$. The finite difference equation then becomes

$$u_{i,j+1} = u_{i,j} + 0.2\left(u_{i-1,j} + u_{i+1,j} - 2u_{i,j}\right)$$

but on the impermeable boundary:

$$u_{i,j+1} = u_{i,j} + 0.2\left(2u_{i-1,j} - 2u_{i,j}\right)$$

On the permeable boundary, $u = 0$ for all values of t, assuming the initial pressure of 60 kPa instantaneously becomes zero.

The calculations are set out in Table 4.4. The use of the spreadsheet analysis tool Consolidation_CSM8.xls to analyse this example, and the other worked examples in this chapter, may be found on the Companion Website.

4.9 Correction for construction period

In practice, structural loads are applied to the soil not instantaneously but over a period of time. Initially there is usually a reduction in net load due to excavation, resulting in swelling: settlement will not begin until the applied load exceeds the weight of the excavated soil. Terzaghi (1943) proposed an empirical method of correcting the instantaneous time–settlement curve to allow for the construction period.

The net load (P') is the gross load less the weight of soil excavated, and the effective construction period (t_c) is measured from the time when P' is zero. It is assumed that the net load is applied uniformly over the time t_c (Figure 4.20) and that the degree of consolidation at time t_c is the same as if the load P' had been acting as a constant load for the period $1/2t_c$. Thus the settlement at any time during the construction period is equal to that occurring for instantaneous loading at half that time; however, since the load then acting is not the total load, the value of settlement so obtained must be reduced in the proportion of that load to the total load. This procedure is shown graphically in Figure 4.20.

For the period subsequent to the completion of construction, the settlement curve will be the instantaneous curve offset by half the effective construction period. Thus at any time after the end of construction, the corrected time corresponding to any value of settlement is equal to the time from the start of loading less half the effective construction period. After a long period of time, the magnitude of settlement is not appreciably affected by the construction time.

Alternatively, a numerical solution (Section 4.8) can be used, successive increments of excess pore water pressure being applied over the construction period.

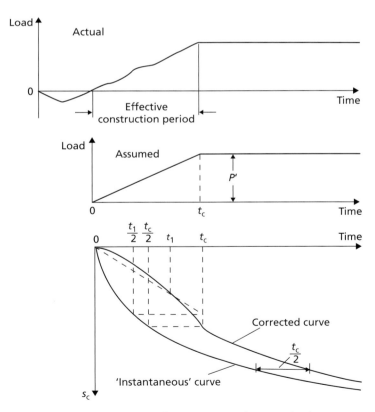

Figure 4.20 Correction for construction period.

Example 4.5

A layer of clay 8 m thick lies between two layers of sand as shown in Figure 4.21. The upper sand layer extends from ground level to a depth of 4 m, the water table being at a depth of 2 m. The lower sand layer is under artesian pressure, the piezometric level being 6 m above ground level. For the clay, $m_v = 0.94\,\text{m}^2/\text{MN}$ and $c_v = 1.4\,\text{m}^2/\text{year}$. As a result of pumping from the artesian layer, the piezometric level falls by 3 m over a period of two years. Draw the time–settlement curve due to consolidation of the clay for a period of five years from the start of pumping.

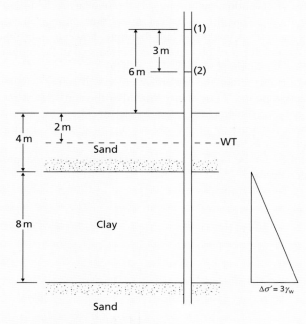

Figure 4.21 Example 4.5.

Solution

In this case, consolidation is due only to the change in pore water pressure at the lower boundary of the clay: there is no change in total vertical stress. The effective vertical stress remains unchanged at the top of the clay layer, but will be increased by $3\gamma_w$ at the bottom of the layer due to the decrease in pore water pressure in the adjacent artesian layer. The distribution of $\Delta\sigma'$ is shown in Figure 4.21. The problem is one-dimensional, since the increase in effective vertical stress is the same over the entire area in question. In calculating the consolidation settlement, it is necessary to consider only the value of $\Delta\sigma'$ at the centre of the layer. Note that in order to obtain the value of m_v it would have been necessary to calculate the initial and final values of effective vertical stress in the clay.

At the centre of the clay layer, $\Delta\sigma' = 1.5\gamma_w = 14.7\,\text{kPa}$. The final consolidation settlement is given by

$$s_{oed} = m_v \Delta\sigma' H$$
$$= 0.94 \times 14.7 \times 8$$
$$= 110\,\text{mm}$$

The clay layer is open and two-way drainage can occur due to the high permeability of the sand above and below the clay: therefore $d=4$ m. For $t=5$ years,

$$T_v = \frac{c_v t}{d^2}$$
$$= \frac{1.4 \times 5}{4^2}$$
$$= 0.437$$

From curve 1 of Figure 4.14, the corresponding value of U_v is 0.73. To obtain the time–settlement relationship, a series of values of U_v is selected up to 0.73 and the corresponding times calculated from the time factor equation (4.19): the corresponding values of consolidation settlement (s_c) are given by the product of U_v and s_{oed} (see Table 4.5). The plot of s_c against t gives the 'instantaneous' curve. Terzaghi's method of correction for the two-year period over which pumping takes place is then carried out as shown in Figure 4.22.

Table 4.5 **Example 4.5**

U_v	T_v	t (years)	s_c (mm)
0.10	0.008	0.09	11
0.20	0.031	0.35	22
0.30	0.070	0.79	33
0.40	0.126	1.42	44
0.50	0.196	2.21	55
0.60	0.285	3.22	66
0.73	0.437	5.00	80

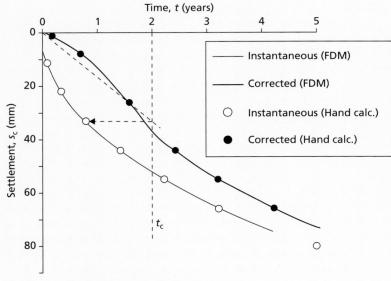

Figure 4.22 Example 4.5 (contd).

Development of a mechanical model for soil

Example 4.6

An 8 m depth of sand overlies a 6 m layer of clay, below which is an impermeable stratum (Figure 4.23); the water table is 2 m below the surface of the sand. Over a period of one year, a 3-m depth of fill (unit weight 20 kN/m³) is to be placed on the surface over an extensive area. The saturated unit weight of the sand is 19 kN/m³ and that of the clay is 20 kN/m³; above the water table the unit weight of the sand is 17 kN/m³. For the clay, the relationship between void ratio and effective stress (units kPa) can be represented by the equation

$$e = 0.88 - 0.32 \log\left(\frac{\sigma'}{100}\right)$$

and the coefficient of consolidation is 1.26 m²/year.

a   Calculate the final settlement of the area due to consolidation of the clay and the settlement after a period of three years from the start of fill placement.
b   If a very thin layer of sand, freely draining, existed 1.5 m above the bottom of the clay layer (Figure 4.23(b)), what would be the values of the final and three-year settlements?

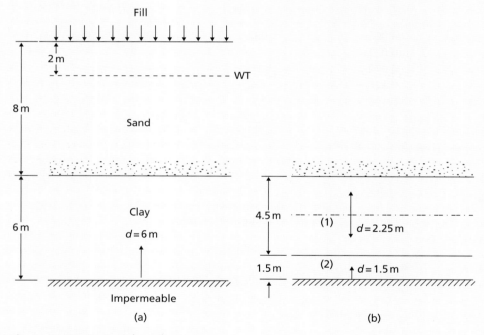

Figure 4.23 Example 4.6.

Solution

a   Since the fill covers a wide area, the problem can be considered to be one-dimensional. The consolidation settlement will be calculated in terms of C_c, considering the clay layer as a whole, and therefore the initial and final values of effective vertical stress at the centre of the clay layer are required.

$$\sigma'_0 = (17 \times 2) + (9.2 \times 6) + (10.2 \times 3) = 119.8 \, \text{kPa}$$
$$e_0 = 0.88 - 0.32 \log 1.198 = 0.88 - 0.025 = 0.855$$
$$\sigma'_1 = 119.8 + (3 \times 20) = 179.8 \, \text{kPa}$$
$$\log\left(\frac{179.8}{119.8}\right) = 0.176$$

The final settlement is calculated from Equation 4.11:

$$s_{\text{oed}} = \frac{0.32 \times 0.176 \times 6000}{1.855} = 182 \, \text{mm}$$

In the calculation of the degree of consolidation three years after the start of fill placement, the corrected value of time to allow for the one-year filling period is

$$t = 3 - \frac{1}{2} = 2.5 \, \text{years}$$

The layer is half-closed, and therefore $d = 6 \, \text{m}$. Then

$$T_v = \frac{c_v t}{d^2} = \frac{1.26 \times 2.5}{6^2}$$
$$= 0.0875$$

From curve 1 of Figure 4.14, $U_v = 0.335$. Settlement after three years:

$$s_c = 0.335 \times 182 = 61 \, \text{mm}$$

b The final settlement will still be 182 mm (ignoring the thickness of the drainage layer); only the rate of settlement will be affected. From the point of view of drainage there is now an open layer of thickness 4.5 m ($d = 2.25 \, \text{m}$) above a half-closed layer of thickness 1.5 m ($d = 1.5 \, \text{m}$): these layers are numbered 1 and 2, respectively.

By proportion

$$T_{v1} = 0.0875 \times \frac{6^2}{2.25^2} = 0.622$$
$$\therefore U_1 = 0.825$$

and

$$T_{v2} = 0.0875 \times \frac{6^2}{1.5^2} = 1.40$$
$$\therefore U_2 = 0.97$$

Now for each layer, $s_c = U_v s_{\text{oed}}$, which is proportional to $U_v H$. Hence if $\overline{U}$ is the overall degree of consolidation for the two layers combined:

$$4.5 U_1 + 1.5 U_2 = 6.0 \overline{U}$$

i.e. $(4.5 \times 0.825) + (1.5 \times 0.97) = 6.0 \overline{U}$.

Hence $\overline{U} = 0.86$ and the 3-year settlement is

$$s_c = 0.86 \times 182 = 157 \, \text{mm}$$

Development of a mechanical model for soil

4.10 Vertical drains

The slow rate of consolidation in saturated clays of low permeability may be accelerated by means of vertical drains which shorten the drainage path within the clay. Consolidation is then due mainly to horizontal radial drainage, resulting in the faster dissipation of excess pore water pressure and consequently more rapid settlement; vertical drainage becomes of minor importance. In theory the final magnitude of consolidation settlement is the same, only the rate of settlement being affected.

In the case of an embankment constructed over a highly compressible clay layer (Figure 4.24), vertical drains installed in the clay would enable the embankment to be brought into service much sooner. A degree of consolidation of the order of 80% would be desirable at the end of construction, to keep in-service settlements to an acceptable level. Any advantages, of course, must be set against the additional cost of the installation.

The traditional method of installing vertical drains was by driving boreholes through the clay layer and backfilling with a suitably graded sand, typical diameters being 200–400 mm to depths of over 30 m. Prefabricated drains are now generally used, and tend to be more economic than backfilled drains for a given area of treatment. One type of drain (often referred to as a '**sandwick**') consists of a filter stocking, usually of woven polypropylene, filled with sand. Compressed air is used to ensure that the stocking is completely filled with sand. This type of drain, a typical diameter being 65 mm, is very flexible and is generally unaffected by lateral soil displacement, the possibility of necking being virtually eliminated. The drains are installed either by insertion into pre-bored holes or, more commonly, by placing them inside a mandrel or casing which is then driven or vibrated into the ground.

Another type of prefabricated drain is the **band drain**, consisting of a flat plastic core indented with drainage channels, surrounded by a layer of filter fabric. The fabric must have sufficient strength to prevent it from being squeezed into the channels, and the mesh size must be small enough to prevent the passage of soil particles which could clog the channels. Typical dimensions of a band drain are 100×4 mm, and in design the equivalent diameter is assumed to be the perimeter divided by π. Band drains are installed by placing them inside a steel mandrel which is either pushed, driven or vibrated into the ground. An anchor is attached to the lower end of the drain to keep it in position as the mandrel is withdrawn. The anchor also prevents soil from entering the mandrel during installation.

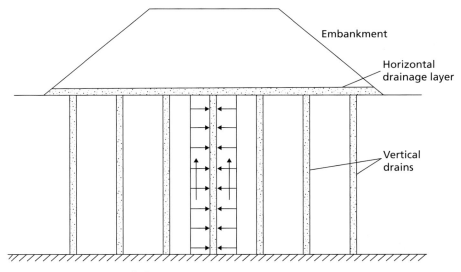

Figure 4.24 Vertical drains.

Consolidation

Drains are normally installed in either a square or a triangular pattern. As the object is to reduce the length of the drainage path, the spacing of the drains is the most important design consideration. The spacing must obviously be less than the thickness of the clay layer for the consolidation rate to be improved; there is therefore no point in using vertical drains in relatively thin layers. It is essential for a successful design that the coefficients of consolidation in both the horizontal and the vertical directions (c_h and c_v, respectively) are known as accurately as possible. In particular, the accuracy of c_h is the most crucial factor in design, being more important than the effect of simplifying assumptions in the theory used. The ratio c_h/c_v is normally between 1 and 2; the higher the ratio, the more advantageous a drain installation will be. A design complication in the case of large-diameter sand drains is that the column of sand tends to act as a weak pile (see Chapter 9), reducing the vertical stress increment imposed on the clay layer by an unknown degree, resulting in lower excess pore water pressure and therefore reduced consolidation settlement. This effect is minimal in the case of prefabricated drains because of their flexibility.

Vertical drains may not be effective in overconsolidated clays if the vertical stress after consolidation remains less than the preconsolidation pressure. Indeed, disturbance of overconsolidated clay during drain installation might even result in increased final consolidation settlement. It should be realised that the rate of secondary compression cannot be controlled by vertical drains.

In polar coordinates, the three-dimensional form of the consolidation equation, with different soil properties in the horizontal and vertical directions, is

$$\frac{\partial u_e}{\partial t} = c_h \left(\frac{\partial^2 u_e}{\partial r^2} + \frac{1}{r}\frac{\partial u_e}{\partial r} \right) + c_v \frac{\partial^2 u_e}{\partial z^2} \tag{4.36}$$

The vertical prismatic blocks of soil which are drained by and surround each drain are replaced by cylindrical blocks, of radius R, having the same cross-sectional area (Figure 4.25). The solution to Equation 4.36 can be written in two parts:

$$U_v = f(T_v)$$

and

$$U_r = f(T_r)$$

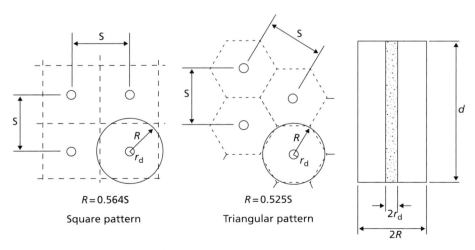

Figure 4.25 Cylindrical blocks.

Development of a mechanical model for soil

where U_v = average degree of consolidation due to vertical drainage only; U_r = average degree of consolidation due to horizontal (radial) drainage only;

$$T_v = \frac{c_v t}{d^2} \qquad (4.37)$$

= time factor for consolidation due to vertical drainage only

$$T_r = \frac{c_h t}{4R^2} \qquad (4.38)$$

= time factor for consolidation due to radial drainage only

The expression for T_r confirms the fact that the closer the spacing of the drains, the quicker consolidation due to radial drainage proceeds. The solution of the consolidation equation for radial drainage only may be found in Barron (1948); a simplified version which is appropriate for design is given by Hansbo (1981) as

$$U_r = 1 - e^{-\frac{8T_r}{\mu}} \qquad (4.39)$$

where

$$\mu = \frac{n^2}{n^2 - 1}\left(\ln n - \frac{3}{4} + \frac{1}{n^2} + \frac{1}{n^4}\right) \approx \ln n - \frac{3}{4} \qquad (4.40)$$

In Equation 4.40, $n = R/r_d$, R is the radius of the equivalent cylindrical block and r_d the radius of the drain. The consolidation curves given by Equation 4.39 for various values of n are plotted in Figure 4.26. Some vertical drainage will continue to occur even if vertical drains have been installed and it can also be shown that

$$(1-U) = (1-U_v)(1-U_r) \qquad (4.41)$$

where U is the average degree of consolidation under combined vertical and radial drainage.

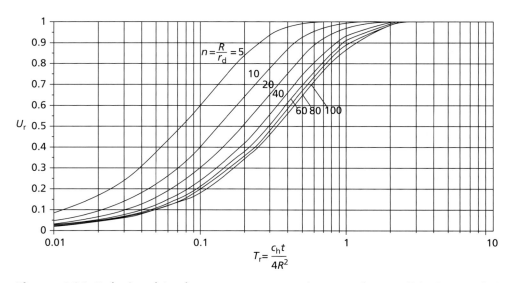

Figure 4.26 Relationships between average degree of consolidation and time factor for radial drainage.

Consolidation

Installation effects

The values of the soil properties for the soil immediately surrounding the drains may be significantly reduced due to remoulding during installation, especially if boring is used, an effect known as **smear**. The smear effect can be taken into account either by assuming a reduced value of c_h or by using a reduced drain diameter in Equations 4.39–4.41. Alternatively, if the extent and permeability (k_s) of the smeared material are known or can be estimated, the expression for μ in Equation 4.40 can be modified after Hansbo (1979) as follows:

$$\mu \approx \ln\frac{n}{S} + \frac{k}{k_s}\ln S - \frac{3}{4} \tag{4.42}$$

Example 4.7

An embankment is to be constructed over a layer of clay 10 m thick, with an impermeable lower boundary. Construction of the embankment will increase the total vertical stress in the clay layer by 65 kPa. For the clay, $c_v = 4.7\,\text{m}^2$/year, $c_h = 7.9\,\text{m}^2$/year and $m_v = 0.25\,\text{m}^2$/MN. The design requirement is that all but 25 mm of the settlement due to consolidation of the clay layer will have taken place after 6 months. Determine the spacing, in a square pattern, of 400-mm diameter sand drains to achieve the above requirement.

Solution

$$\text{Final settlement} = m_v \Delta\sigma' H = 0.25 \times 65 \times 10$$
$$= 162\,\text{mm}$$

For $t = 6$ months,

$$U = \frac{162 - 25}{162} = 0.85$$

For vertical drainage only, the layer is half-closed, and therefore $d = 10\,\text{m}$.

$$T_v = \frac{c_v t}{d^2} = \frac{4.7 \times 0.5}{10^2} = 0.0235$$

From curve 1 of Figure 4.14, or using the spreadsheet tool on the Companion Website $U_v = 0.17$.
For radial drainage the diameter of the sand drains is 0.4 m, i.e. $r_d = 0.2\,\text{m}$.
Radius of cylindrical block:

$$R = nr_d = 0.2n$$

$$T_r = \frac{c_h t}{4R^2} = \frac{7.9 \times 0.5}{4 \times 0.2^2 \times n^2} = \frac{24.7}{n^2}$$

i.e.

$$U_r = 1 - e^{-\left[\frac{8 \times 24.7}{n^2(\ln n - 0.75)}\right]}$$

Development of a mechanical model for soil

Now $(1-U) = (1-U_v)(1-U_r)$, and therefore

$$0.15 = 0.83(1-U_r)$$
$$U_r = 0.82$$

The value of n for $U_r=0.82$ may then be found by evaluating U_r at different values of n using Equations 4.39 and 4.40 and interpolating the value of n at which $U_r=0.82$. Alternatively a simple 'Goal seek' or optimisation routine in a standard spreadsheet may be used to solve the equations iteratively. For the first of these methods, Figure 4.27 plots the value of U_r against n, from which it may be seen that $n=9$. It should be noted that this process is greatly shortened by the use of a spreadsheet to perform the calculations. Therefore

$$R = 0.2 \times 9 = 1.8\,\text{m}$$

Spacing of drains in a square pattern is given by

$$S = \frac{R}{0.564} = \frac{1.8}{0.564} = 3.2\,\text{m}$$

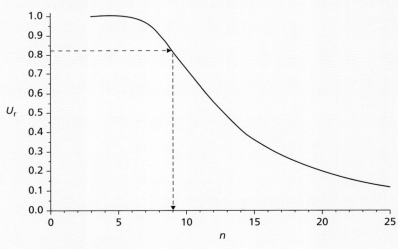

Figure 4.27 Example 4.7.

4.11 Pre-loading

If construction is to be undertaken on very compressible ground (i.e. soil with high m_v), it is likely that the consolidation settlements which will occur will be too large to be tolerable by the construction. If the compressible material is close to the surface and of limited depth, one solution would be to excavate the material and replace it with properly compacted fill (see Section 1.7). If the compressible material is of significant thickness, this may prove prohibitively expensive. An alternative solution in such a case would be to pre-load the ground for an appropriate period by applying a surcharge, normally in the form of a temporary embankment, and allowing most of the settlement to take place before the final construction is built.

Consolidation

An example for a normally consolidated soil is shown in Figure 4.28. In Figure 4.28(a), a foundation, applying a pressure of q_f is built without applying any pre-load; consolidation will proceed along the virgin compression line and the settlement will be large. In Figure 4.28(b) a pre-load is first applied by constructing a temporary embankment, applying a pressure q_p, which induces significant volumetric change (and hence settlement) along the virgin compression line. Once consolidation is complete, the pre-load is removed and the soil swells along an unload–reload line. The foundation is then built and, provided the pressure applied is less than the pre-load pressure ($q_f < q_p$), the soil will remain on the unload–reload line and the settlement in this phase (which the foundation sees) will be small.

The pre-load may have to be applied in increments to avoid undrained shear failure of the supporting soil (see Chapters 5 and 8). As consolidation proceeds, the shear strength of the soil increases, allowing further load increments to be applied. The rate of dissipation of excess pore water pressure can be monitored by installing piezometers (Chapter 6), thus providing a means of controlling the rate of loading. It should be appreciated that differential settlement will occur if non-uniform soil conditions exist and the ground surface may need re-levelling at the end of the pre-loading period. One of the principal disadvantages of pre-loading is the need to wait for consolidation to be complete under the pre-load before construction can begin on the foundation. One solution to this is to combine pre-loading with vertical drains (Section 4.10) to speed up the pre-load stage.

The principle of pre-loading can also be applied to accelerate the settlement of embankments by surcharging with an additional depth of fill. At the end of an appropriate period, surcharge is removed down to the final (formation) level.

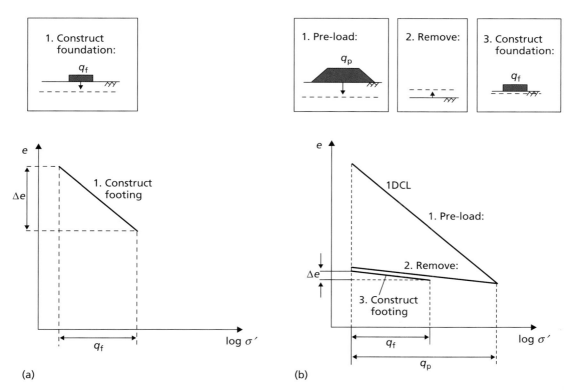

Figure 4.28 Application of pre-loading: (a) foundation construction on highly compressible soil, (b) foundation constructed following pre-loading.

Development of a mechanical model for soil

> **Summary**
>
> 1. Due to the finite permeability of soils, changes in applied (total) stress or fluctuations in groundwater level do not immediately lead to corresponding increments in effective stress. Under such conditions, the volume of the soil will change with time. Dissipation of positive excess pore pressures (due to loading or lowering of the water table) leads to compression which is largely irrecoverable. Dissipation of negative excess pore pressures (unloading or raising of the water table) leads to swelling which is recoverable and of much lower magnitude than initial compression. The compression behaviour is quantified mathematically by the virgin compression and unload/re-load lines in e–$\log\sigma'$ space. The compressibility (m_v), coefficient of consolidation (c_v) and (indirectly) the permeability (k) may be determined within the oedometer.
> 2. The final amount of settlement due to one-dimensional consolidation can be straightforwardly determined using the compressibility (m_v) and the known changes in total stress or pore pressure conditions induced by construction processes. A more detailed description of settlement (or heave) with time may be found either analytically or using the Finite Difference Method. A spreadsheet implementation of this method is available on the Companion Website.
> 3. For soils of low permeability (e.g. fine-grained soils) it may be necessary in construction to speed up the process of consolidation. This may be achieved by adding vertical drains. An appropriate drain specification and layout can be determined using standard solutions for radial pore water drainage to meet a specified level of performance (e.g. U_r % consolidation by time t).

Problems

4.1 In an oedometer test on a specimen of saturated clay ($G_s = 2.72$), the applied pressure was increased from 107 to 214 kPa and the following compression readings recorded:

TABLE I

Time (min)	0	$\frac{1}{4}$	$\frac{1}{2}$	1	$2\frac{1}{4}$	4	$6\frac{1}{4}$	9	16
Gauge (mm)	7.82	7.42	7.32	7.21	6.99	6.78	6.61	6.49	6.37
Time (min)	25	36	49	64	81	100	300	1440	
Gauge (mm)	6.29	6.24	6.21	6.18	6.16	6.15	6.10	6.02	

After 1440 min, the thickness of the specimen was 15.30 mm and the water content was 23.2%. Determine the values of the coefficient of consolidation and the compression ratios from (a) the root time plot and (b) the log time plot. Determine also the values of the coefficient of volume compressibility and the coefficient of permeability.

4.2 In an oedometer test, a specimen of saturated clay 19 mm thick reaches 50% consolidation in 20 min. How long would it take a layer of this clay 5 m thick to reach the same degree of consolidation under the same stress and drainage conditions? How long would it take the layer to reach 30% consolidation?

4.3 The following results were obtained from an oedometer test on a specimen of saturated clay:

TABLE J

Pressure (kPa)	27	54	107	214	429	214	107	54
Void ratio	1.243	1.217	1.144	1.068	0.994	1.001	1.012	1.024

A layer of this clay 8 m thick lies below a 4 m depth of sand, the water table being at the surface. The saturated unit weight for both soils is 19 kN/m³. A 4-m depth of fill of unit weight 21 kN/m³ is placed on the sand over an extensive area. Determine the final settlement due to consolidation of the clay. If the fill were to be removed some time after the completion of consolidation, what heave would eventually take place due to swelling of the clay?

4.4 Assuming the fill in Problem 4.3 is dumped very rapidly, what would be the value of excess pore water pressure at the centre of the clay layer after a period of three years? The layer is open and the value of c_v is 2.4 m²/year.

4.5 A 10-m depth of sand overlies an 8-m layer of clay, below which is a further depth of sand. For the clay, $m_v = 0.83$ m²/MN and $c_v = 4.4$ m²/year. The water table is at surface level but is to be lowered permanently by 4 m, the initial lowering taking place over a period of 40 weeks. Calculate the final settlement due to consolidation of the clay, assuming no change in the weight of the sand, and the settlement two years after the start of lowering.

4.6 An open clay layer is 6 m thick, the value of c_v being 1.0 m²/year. The initial distribution of excess pore water pressure varies linearly from 60 kPa at the top of the layer to zero at the bottom. Using the finite difference approximation of the one-dimensional consolidation equation, plot the isochrone after consolidation has been in progress for a period of three years, and from the isochrone determine the average degree of consolidation in the layer.

4.7 A half-closed clay layer is 8 m thick and it can be assumed that $c_v = c_h$. Vertical sand drains 300 mm in diameter, spaced at 3 m centres in a square pattern, are to be used to increase the rate of consolidation of the clay under the increased vertical stress due to the construction of an embankment. Without sand drains, the degree of consolidation at the time the embankment is due to come into use has been calculated as 25%. What degree of consolidation would be reached with the sand drains at the same time?

4.8 A layer of saturated clay is 10 m thick, the lower boundary being impermeable; an embankment is to be constructed above the clay. Determine the time required for 90% consolidation of the clay layer. If 300-mm diameter sand drains at 4 m centres in a square pattern were installed in the clay, in what time would the same overall degree of consolidation be reached? The coefficients of consolidation in the vertical and horizontal directions, respectively, are 9.6 and 14.0 m²/year.

References

Al-Tabbaa, A. and Wood, D.M. (1987) Some measurements of the permeability of kaolin, *Géotechnique*, **37**(4), 499–503.

ASTM D2435 (2011) *Standard Test Methods for One-Dimensional Consolidation Properties of Soils Using Incremental Loading*, American Society for Testing and Materials, West Conshohocken, PA.

Barron, R.A. (1948) Consolidation of fine grained soils by drain wells, *Transactions of the ASCE*, **113**, 718–742.

Bjerrum, L. (1967) Engineering geology of Norwegian normally-consolidated marine clays as related to settlement of buildings, *Géotechnique*, **17**(2), 83–118.

British Standard 1377 (1990) *Methods of Test for Soils for Civil Engineering Purposes*, British Standards Institution, London.

Casagrande, A. (1936) Determination of the preconsolidation load and its practical significance, in *Proceedings of the International Conference on SMFE, Harvard University, Cambridge, MA*, Vol. III, pp. 60–64.

CEN ISO/TS 17892 (2004) *Geotechnical Investigation and Testing – Laboratory Testing of Soil*, International Organisation for Standardisation, Geneva.

Cour, F.R. (1971) Inflection point method for computing c_v, Technical Note, *Journal of the ASCE*, **97**(SM5), 827–831.

EC7–2 (2007) *Eurocode 7: Geotechnical design – Part 2: Ground Investigation and Testing, BS EN 1997–2:2007*, British Standards Institution, London.

Gibson, R.E. (1966) A note on the constant head test to measure soil permeability *in-situ*, *Géotechnique*, **16**(3), 256–259.

Gibson, R.E. (1970) An extension to the theory of the constant head *in-situ* permeability test, *Géotechnique*, **20**(2), 193–197.

Hansbo, S. (1979) Consolidation of clay by band-shaped prefabricated drains, *Ground Engineering*, **12**(5), 16–25.

Hansbo, S. (1981) Consolidation of fine-grained soils by prefabricated drains, in *Proceedings of the 10th International Conference on SMFE, Stockholm*, Vol. III, pp. 677–682.

Mitchell, J.K. and Soga, K. (2005) *Fundamentals of Soil Behaviour* (3rd edn), John Wiley & Sons, New York, NY.

Naylor, A.H. and Doran, I.G. (1948) Precise determination of primary consolidation, in *Proceedings of the 2nd International Conference on SMFE, Rotterdam*, Vol. 1, pp. 34–40.

Rowe, P.W. (1968) The influence of geological features of clay deposits on the design and performance of sand drains, in *Proceedings ICE* (Suppl. Vol.), Paper 70585.

Rowe, P.W. and Barden, L. (1966) A new consolidation cell, *Géotechnique*, **16**(4), 162–170.

Schmertmann, J.H. (1953) Estimating the true consolidation behaviour of clay from laboratory test results, *Proceedings ASCE*, **79**, 1–26.

Scott, R.F. (1961) New method of consolidation coefficient evaluation, *Journal of the ASCE*, **87**, No. SM1.

Taylor, D.W. (1948) *Fundamentals of Soil Mechanics*, John Wiley & Sons, New York, NY.

Terzaghi, K. (1943) *Theoretical Soil Mechanics*, John Wiley & Sons, New York, NY.

Wilkinson, W.B. (1968) Constant head *in-situ* permeability tests in clay strata, *Géotechnique*, **18**(2), 172–194.

Further reading

Burland, J.B. (1990) On the compressibility and shear strength of natural clays, *Géotechnique*, **40**(3), 329–378.

This paper describes in detail the role of depositional structure on the initial compressibility of natural clays (rather than those reconstituted in the laboratory). It contains a large amount of experimental data, and is therefore useful for reference.

McGown, A. and Hughes, F.H. (1981) Practical aspects of the design and installation of deep vertical drains, *Géotechnique*, **31**(1), 3–17.

This paper discusses practical aspects related to the use of vertical drains, which were introduced in Section 4.10.

For further student and instructor resources for this chapter, please visit the Companion Website at www.routledge.com/cw/craig

Chapter 5

Soil behaviour in shear

> **Learning outcomes**
>
> After working through the material in this chapter, you should be able to:
>
> 1 Understand how soil may be modelled as a continuum, and how its mechanical behaviour (strength and stiffness) may be adequately described using elastic and plastic material (constitutive) models (Section 5.1–5.3);
> 2 Understand the method of operation of standard laboratory testing apparatus and derive strength and stiffness properties of soil from these tests for use in subsequent geotechnical analyses (Section 5.4);
> 3 Appreciate the different strength characteristics of coarse and fine-grained soils and derive material parameters to model these (Sections 5.5–5.6 and Section 5.8);
> 4 Understand the critical state concept and its important role in coupling strength and volumetric behaviour in soil (Section 5.7);
> 5 Use simple empirical correlations to estimate strength properties of soil based on the results of index tests (see Chapter 1) and appreciate how these may be used to support the results from laboratory tests (Section 5.9).

5.1 An introduction to continuum mechanics

This chapter is concerned with the resistance of soil to failure in shear, a knowledge of which is required in the analysis of the stability of soil masses and, therefore, for the design of geotechnical structures. Many problems can be treated by analysis in two dimensions, i.e. where only the stresses and displacements in a single plane need to be considered. This simplification will be used initially in this chapter while the framework for the constitutive behaviour of soil is described.

An element of soil in the field will typically by subjected to total normal stresses in the vertical (z) and horizontal (x) directions due to the self-weight of the soil and any applied external loading (e.g. from a foundation). The latter may also induce an applied shear stress which additionally acts on the element. The total normal stresses and shear stresses in the x- and z-directions on an element of soil are shown in Figure 5.1(a), the stresses being positive in magnitude as shown; the stresses vary across the element. The rates of change of the normal stresses in the x- and z-directions are $\partial\sigma_x/\partial x$ and $\partial\sigma_z/\partial z$ respectively;

Development of a mechanical model for soil

the rates of change of the shear stresses are $\partial \tau_{xz}/\partial x$ and $\partial \tau_{xz}/\partial z$. Every such element within a soil mass must be in static equilibrium. By equating moments about the centre point of the element, and neglecting higher-order differentials, it is apparent that $\tau_{xz}=\tau_{zx}$. By equating forces in the *x*- and *z*-directions the following equations are obtained:

$$\frac{\partial \sigma_x}{\partial x} + \frac{\partial \tau_{xz}}{\partial z} = 0 \tag{5.1a}$$

$$\frac{\partial \tau_{xz}}{\partial x} + \frac{\partial \sigma_z}{\partial z} - \gamma = 0 \tag{5.1b}$$

These are the **equations of equilibrium** in two dimensions in terms of total stresses; for dry soils, the body force (or unit weight) $\gamma = \gamma_{dry}$, while for saturated soil, $\gamma = \gamma_{sat}$. Equation 5.1 can also be written in terms of effective stress. From Terzaghi's Principle (Equation 3.1) the effective body forces will be 0 and $\gamma' = \gamma - \gamma_w$ in the *x*- and *z*-directions respectively. Furthermore, if seepage is taking place with hydraulic gradients of i_x and i_z in the *x*- and *z*-directions, then there will be additional body forces due to seepage (see Section 3.6) of $i_x \gamma_w$ and $i_z \gamma_w$ in the *x*- and *z*-directions, i.e.:

$$\frac{\partial \sigma'_x}{\partial x} + \frac{\partial \tau'_{zx}}{\partial z} - i_x \gamma_w = 0 \tag{5.2a}$$

$$\frac{\partial \tau'_{xz}}{\partial x} + \frac{\partial \sigma'_z}{\partial z} - (\gamma' + i_z \gamma_w) = 0 \tag{5.2b}$$

The effective stress components are shown in Figure 5.1(b).

Due to the applied loading, points within the soil mass will be displaced relative to the axes and to one another, as shown in Figure 5.2. If the components of displacement in the *x*- and *z*-directions are denoted by *u* and *w*, respectively, then the normal strains in these directions (ε_x and ε_z, respectively) are given by

$$\varepsilon_x = \frac{\partial u}{\partial x}, \; \varepsilon_z = \frac{\partial w}{\partial z}$$

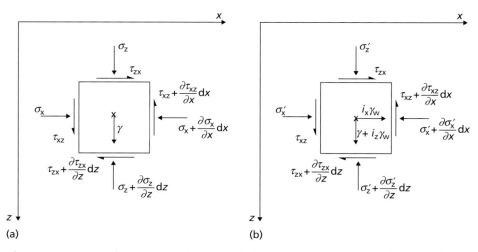

Figure 5.1 Two-dimensional state of stress in an element of soil: (a) total stresses, (b) effective stresses.

and the shear strain by

$$\gamma_{xz} = \frac{\partial u}{\partial z} + \frac{\partial w}{\partial x}$$

However, these strains are not independent; they must be compatible with each other if the soil mass as a whole is to remain continuous. This requirement leads to the following relationship, known as the **equation of compatibility** in two dimensions:

$$\frac{\partial^2 \varepsilon_x}{\partial z^2} + \frac{\partial^2 \varepsilon_z}{\partial x^2} - \frac{\partial \gamma_{xz}}{\partial x \partial z} = 0 \tag{5.3}$$

The rigorous solution of a particular problem requires that the equations of equilibrium and compatibility are satisfied for the given boundary conditions (i.e. applied loads and known displacement conditions) at all points within a soil mass; an appropriate stress–strain relationship is also required to link the two equations. Equations 5.1–5.3, being independent of material properties, can be applied to soil with any stress–strain relationship (also termed a **constitutive model**). In general, soils are non-homogeneous, exhibit **anisotropy** (i.e. have different values of a given property in different directions) and have non-linear stress–strain relationships which are dependent on stress history (see Section 4.2) and the particular stress path followed. This can make solution difficult.

In analysis therefore, an appropriate idealisation of the stress–strain relationship is employed to simplify computation. One such idealisation is shown by the dotted lines in Figure 5.3(a), linearly elastic behaviour (i.e. Hooke's Law) being assumed between O and Y′ (the assumed yield point) followed by unrestricted plastic strain (or flow) Y′P at constant stress. This idealisation, which is shown separately in Figure 5.3(b), is known as the **elastic–perfectly plastic model** of material behaviour. If only the collapse condition (soil failure) in a practical problem is of interest, then the elastic phase can be omitted and the **rigid–perfectly plastic model**, shown in Figure 5.3(c), may be used. A third idealisation is the **elastic–strain hardening plastic model**, shown in Figure 5.3(d), in which plastic strain beyond the yield point necessitates further stress increase, i.e. the soil hardens or strengthens as it strains. If unloading and reloading were to take place subsequent to yielding in the strain hardening model, as shown by the dotted

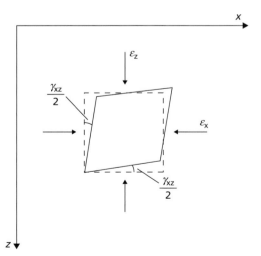

Figure 5.2 Two-dimensional induced state of strain in an element of soil, due to stresses shown in Figure 5.1.

Development of a mechanical model for soil

line Y″U in Figure 5.3(d), there would be a new yield point Y″ at a stress level higher than that at the first yield point Y′. An increase in yield stress is a characteristic of strain hardening. No such increase takes place in the case of perfectly plastic (i.e. non-hardening) behaviour, the stress at Y″ being equal to that at Y′ as shown in Figures 5.3(b) and (c). A further idealisation is the **elastic–strain softening plastic model**, represented by OY′P′in Figure 5.3(d), in which the plastic strain beyond the yield point is accompanied by stress decrease or softening of the material.

In plasticity theory (Hill, 1950; Calladine, 2000) the characteristics of yielding, hardening and flow are considered; these are described by a yield function, a hardening law and a flow rule, respectively. The yield function is written in terms of stress components or principal stresses, and defines the yield point as a function of current effective stresses and stress history. The **Mohr–Coulomb criterion**, which will be described later in Section 5.3, is one possible (simple) yield function if perfectly plastic behaviour is assumed. The hardening law represents the relationship between the increase in yield stress and the corresponding plastic strain components, i.e. defining the gradient of Y′P or Y′P′ in Figure 5.3(d). The flow rule specifies the relative (i.e. not absolute) magnitudes of the plastic strain components during yielding under a particular state of stress. The remainder of this book will consider simple elastic–perfectly plastic material models for soil, as shown in Figure 5.3(b), in which elastic behaviour is isotropic (Section 5.2) and plastic behaviour is defined by the Mohr–Coulomb criterion (Section 5.3).

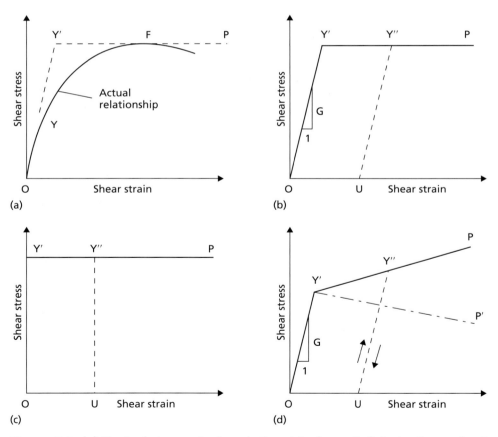

Figure 5.3 (a) Typical stress–strain relationship for soil, (b) elastic–perfectly plastic model, (c) rigid–perfectly plastic model, and (d) strain hardening and strain softening elastic–plastic models.

5.2 Simple models of soil elasticity

Linear elasticity

The initial region of soil behaviour, prior to plastic collapse (yield) of the soil, may be modeled using an elastic constitutive model. The simplest model is (isotropic) linear elasticity in which shear strain is directly proportional to the applied shear stress. Such a model is shown by the initial parts of the stress–strain relationships in Figure 5.3(a, b, d), in which the relationship is a straight line, the gradient of which is the **shear modulus**, G, i.e.

$$\tau_{xz} = \tau_{zx} = G\gamma_{xz} \tag{5.4}$$

For a general 2-D element of soil, as shown in Figures 5.1 and 5.2, loading may not just be due to an applied shear stress τ_{xz}, but also by normal stress components σ_x and σ_z. For a linearly elastic constitutive model, the relationship between stress and strain is given by Hooke's Law, in which

$$\varepsilon_x = \frac{1}{E}\left(\sigma'_x - v\sigma'_z\right) \tag{5.5a}$$

$$\varepsilon_z = \frac{1}{E}\left(\sigma'_z - v\sigma'_x\right) \tag{5.5b}$$

where E is the **Young's Modulus** (=normal stress/normal strain) and v is the **Poisson's ratio** of the soil. While the soil remains elastic, determination of the soil response (strain) to applied stresses requires knowledge only of the elastic properties of the soil, defined by G, E and v. For an **isotropically** elastic material (i.e. uniform behaviour in all directions), it can further be shown that the three elastic material constants are related by

$$G = \frac{E}{2(1+v)} \tag{5.6}$$

It is therefore only necessary to know any two of the elastic properties; the third can always be found using Equation 5.6. It is preferable in soil mechanics to use v and G as the two properties. From Equation 5.4 it can be seen that the behaviour of soil under pure shear is independent of the normal stresses and therefore is not influenced by the pore water (water cannot carry shearing stresses). G may therefore be measured for soil which is either fully **drained** (e.g. after consolidation is completed) or under an **undrained** condition (before consolidation has begun), with both values being the same. E, on the other hand, is dependent on the normal stresses in the soil (Equation 5.5), and therefore is influenced by pore water. To determine response under immediate and long-term loading it would be necessary to know two values of E, but only one of G.

Poisson's ratio, which is defined as the ratio of strains in the two perpendicular directions under uniaxial loading ($v = \varepsilon_x/\varepsilon_z$ under applied loads σ'_z, $\sigma'_x = 0$), is also dependent on the drainage conditions. For fully or partially drained conditions, $v < 0.5$, and is normally between 0.2 and 0.4 for most soils under fully drained conditions. Under undrained conditions, the soil is incompressible (no pore-water drainage has yet occurred). The volumetric strain of an element of linearly elastic material under normal stresses σ_x and σ_z is given by

$$\frac{\Delta V}{V} = \varepsilon_x + \varepsilon_z = \frac{1-2v}{E}\left(\sigma'_x + \sigma'_z\right)$$

where V is the volume of the soil element. Therefore, for undrained conditions $\Delta V/V = 0$ (no change in volume), hence $v = v_u = 0.5$. This is true for all soils provided conditions are completely undrained.

Development of a mechanical model for soil

For given applied stress components σ_x, σ_z and τ_{xz}, the resulting strains may be found by solving Equations 5.4 and 5.5 simultaneously. Alternatively, the equations may be written in matrix form

$$\begin{bmatrix} \varepsilon_x \\ \varepsilon_z \\ \gamma_{xz} \end{bmatrix} = \frac{1}{2G(1+v)} \begin{bmatrix} 1 & -v & 0 \\ -v & 1 & 0 \\ 0 & 0 & 2(1+v) \end{bmatrix} \begin{bmatrix} \sigma'_x \\ \sigma'_z \\ \tau_{xz} \end{bmatrix} \quad (5.7)$$

The elastic constants G, E and v can further be related to the constrained modulus (E'_{oed}) described in Section 4.2. In the oedometer test, soil strains in the z-direction under drained conditions, but the lateral strains are zero. From the 3-D version of Equation 5.7 which will be introduced later (Equation 5.29), the zero lateral strain condition gives:

$$\varepsilon_z = \frac{1}{E'}\left(\frac{1-v'-2v'^2}{1-v'}\right)\sigma'_z$$

where E' and v' are the Young's modulus and Poisson's ratio for fully drained conditions. Then, from the definition of E'_{oed} (Equations 4.3 and 4.5):

$$E'_{oed} = \frac{\sigma'_z}{\varepsilon_z} = \frac{E'(1-v')}{(1-v'-2v'^2)}$$

$$\therefore E' = E'_{oed}\frac{(1+v')(1-2v')}{1-v'} \quad (5.8)$$

Substituting Equation 5.6

$$2G(1+v') = E'_{oed}\frac{(1+v')(1-2v')}{1-v'}$$

$$\therefore G = E'_{oed}\frac{(1-2v')}{2(1-v')} \quad (5.9)$$

Therefore, the results of oedometer tests can be used to define the shear modulus.

Non-linear elasticity

In reality, the shear modulus of soil is not a material constant, but is a highly non-linear function of shear strain and effective confining stress, as shown in Figure 5.4(a). At very small values of strain, the shear modulus is a maximum (defined as G_0). The value of G_0 is independent of strain, but increases with increasing effective stress. As a result, G_0 generally increases with depth within soil masses. If the shear modulus is normalised by G_0 to remove the stress dependence, a single non-linear curve of G/G_0 versus shear strain is obtained (Figure 5.4(b)). Atkinson (2000) has suggested that this relationship may be approximated by Equation 5.10:

$$\frac{G}{G_0} = \frac{1-(\gamma_p/\gamma)^B}{1-(\gamma_p/\gamma_0)^B} \leq 1.0 \quad (5.10)$$

where γ is the shear strain, γ_0 defines the maximum strain at which the small-strain stiffness G_0 is still applicable (typically around 0.001% strain), γ_p defines the strain at which the soil becomes plastic

Soil behaviour in shear

(typically around 1% strain) and B defines the shape of the curve between $G/G_0=0$ and 1, typically being between 0.1–0.5 depending on the soil type. This relationship is shown in Figure 5.4(b).

For most common geotechnical structures, the operative levels of strain will mean that the shear modulus $G<G_0$. Common strain ranges are shown in Figure 5.4(c), and these may be used to estimate an appropriate linearised value of G for a given problem from the non-linear relationship. The full non-linear G–γ relationship may be determined by:

1. undertaking triaxial testing (described later in Section 5.4) on modern machines with small-strain sample measurements; this equipment is now available in most soil testing laboratories;
2. determining the value of G_0 using seismic wave techniques (either in a triaxial cell using specialist bender elements, or in-situ as described in Chapters 6 and 7) and combining this with a normalised G/G_0 versus γ relationship (e.g. Equation 5.10; see Atkinson, 2000 for further details).

Of these methods, the second is usually the cheapest and quickest to implement in practice. In principle, the value of G can also be estimated from the curve relating principal stress difference and axial strain in an undrained triaxial test (this will be described in Section 5.4). Without small-strain sample measurements, however, the data are only likely to be available for $\gamma>0.1\%$ (see Figure 5.4(d)). Because of the effects of sampling disturbance (see Chapter 6), it can be preferable to determine G (or E) from the results of in-situ rather than laboratory tests.

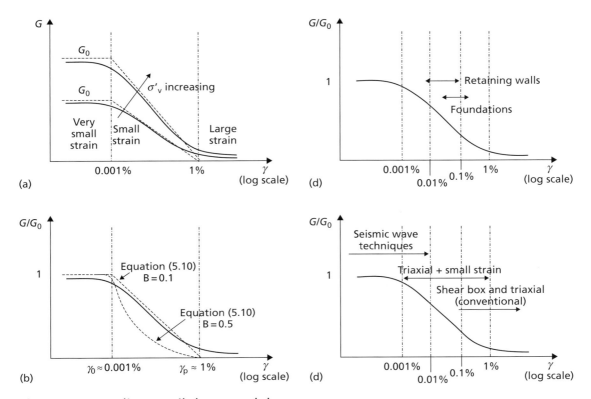

Figure 5.4 Non-linear soil shear modulus.

Development of a mechanical model for soil

5.3 Simple models of soil plasticity

Soil as a frictional material

The use of elastic models alone (as described previously) implies that the soil is infinitely strong. If at a point on any plane within a soil mass the shear stress becomes equal to the shear strength of the soil, then failure will occur at that point. Coulomb originally proposed that the limiting strength of soils was frictional, imagining that if slip (plastic failure) occurred along any plane within an element of closely packed particles (soil), then the slip plane would be rough due to all of the individual particle-to-particle contacts. Friction is commonly described by:

$$T = \mu N$$

where T is the limiting frictional force, N is the normal force acting perpendicular to the slip plane and μ is the coefficient of friction. This is shown in Figure 5.5(a). In an element of soil, it is more useful to use shear stress and normal stress instead of T and N

$$\tau_f = (\tan\phi)\sigma$$

where $\tan\phi$ is equivalent to the coefficient of friction, which is an intrinsic material property related to the roughness of the shear plane (i.e. the shape, size and angularity of the soil particles). The frictional relationship in terms of stresses is shown in Figure 5.5(b).

While Coulomb's frictional model represented loosely packed particle arrangements well, if the particles are arranged in a dense packing then additional initial interlocking between the particles can cause the frictional resistance τ_f to be higher than that predicted considering friction alone. If the normal stress is increased, it can become high enough that the contact forces between the individual particles cause particle breakage, which reduces the degree of interlocking and makes slip easier. At high normal stresses, therefore, the interlocking effect disappears and the material behaviour becomes purely frictional again. This is also shown in Figure 5.5(b).

In accordance with the principle that shear stress in a soil can be resisted only by the skeleton of solid particles and not the pore water, the shear strength (τ_f) of a soil at a point on a particular plane is normally expressed as a function of effective normal stress (σ') rather than total stress.

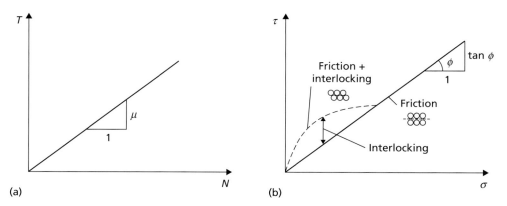

Figure 5.5 (a) Frictional strength along a plane of slip, (b) strength of an assembly of particles along a plane of slip.

Soil behaviour in shear

The Mohr–Coulomb model

As described in Section 5.1, the state of stress in an element of soil is defined in terms of the normal and shear stresses applied to the boundaries of the soil element. States of stress in two dimensions can be represented on a plot of shear stress (τ) against effective normal stress (σ'). The stress state for a 2-D element of soil can be represented either by a pair of points with coordinates (σ'_z, τ_{zx}) and (σ'_x, τ_{xz}), or by a **Mohr circle** defined by the effective **principal stresses** σ'_1 and σ'_3, as shown in Figure 5.6. The stress points at either end of a diameter through a Mohr circle at an angle of 2θ to the horizontal represent the stress conditions on a plane at an angle of θ to the minor principal stress. The circle therefore represents the stress states on all possible planes within the soil element. The principal stress components alone are enough to fully describe the position and size of the Mohr circle and so are often used to describe the stress state, as it reduces the number of stress variables from three (σ'_x, σ'_z, τ_{zx}) to two (σ'_1, σ'_3). When the

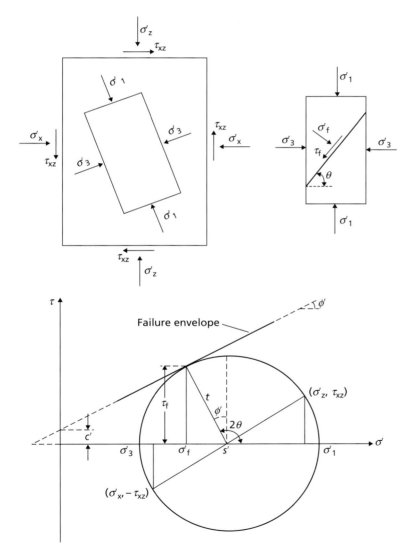

Figure 5.6 Mohr–Coulomb failure criterion.

Development of a mechanical model for soil

element of soil reaches failure, the circle will just touch the failure envelope, at a single point. The failure envelope is defined by the frictional model described above; however, it can be difficult to deal with the non-linear part of the envelope associated with interlocking, so that it is common practice to approximate the failure envelope by a straight line described by:

$$\tau_f = c' + \sigma' \tan \phi' \tag{5.11}$$

where c' and ϕ' are shear strength parameters referred to as the **cohesion intercept** and the **angle of shearing resistance**, respectively. Failure will thus occur at any point in the soil where a critical combination of shear stress and effective normal stress develops. It should be appreciated that c' and ϕ' are simply mathematical constants defining a linear relationship between shear strength and effective normal stress. Shearing resistance is developed mechanically due to inter-particle contact forces and friction, as described above; therefore, if effective normal stress is zero then shearing resistance must also be zero (unless there is cementation or some other bonding between the particles) and the value of c' would be zero. This point is crucial to the interpretation of shear strength parameters (described in Section 5.4).

A state of stress represented by a stress point that plots above the failure envelope, or by a Mohr circle, part of which lies above the envelope, is impossible.

With reference to the general case with $c'>0$ shown in Figure 5.6, the relationship between the shear strength parameters and the effective principal stresses at failure at a particular point can be deduced, compressive stress being taken as positive. The coordinates of the tangent point are τ_f and σ'_f where

$$\tau_f = \frac{1}{2}(\sigma'_1 - \sigma'_3)\sin 2\theta \tag{5.12}$$

$$\sigma'_f = \frac{1}{2}(\sigma'_1 + \sigma'_3) + \frac{1}{2}(\sigma'_1 - \sigma'_3)\cos 2\theta \tag{5.13}$$

and θ is the theoretical angle between the minor principal plane and the plane of failure. It is apparent that $2\theta = 90° + \phi'$, such that

$$\theta = 45° + \frac{\phi'}{2} \tag{5.14}$$

Now

$$\sin \phi' = \frac{\frac{1}{2}(\sigma'_1 - \sigma'_3)}{c' \cot \phi' + \frac{1}{2}(\sigma'_1 + \sigma'_3)}$$

Therefore

$$(\sigma'_1 - \sigma'_3) = (\sigma'_1 + \sigma'_3)\sin \phi' + 2c' \cos \phi' \tag{5.15a}$$

or

$$\sigma'_1 = \sigma'_3 \tan^2\left(45° + \frac{\phi'}{2}\right) + 2c' \tan\left(45° + \frac{\phi'}{2}\right) \tag{5.15b}$$

Equation 5.15 is referred to as the Mohr–Coulomb failure criterion, defining the relationship between principal stresses at failure for given material properties c' and ϕ'.

For a given state of stress it is apparent that, because $\sigma'_1 = \sigma_1 - u$ and $\sigma'_3 = \sigma_3 - u$, the Mohr circles for total and effective stresses have the same diameter but their centres are separated by the corresponding pore water pressure u, as shown in Figure 5.7. Similarly, total and effective stress points are separated by the value of u.

Soil behaviour in shear

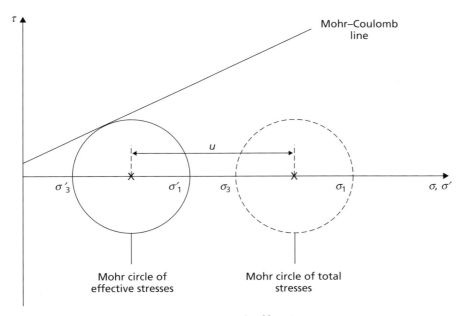

Figure 5.7 Mohr circles for total and effective stresses.

Effect of drainage conditions on shear strength

The shear strength of a soil under undrained conditions is different from that under drained conditions. The failure envelope is defined in terms of effective stresses by ϕ' and c', and so is the same irrespective of whether the soil is under drained or undrained conditions; the difference is that under a given set of applied total stresses, in undrained loading excess pore pressures are generated which change the effective stresses within the soil (under drained conditions excess pore pressures are zero as consolidation is complete). Therefore, two identical samples of soil which are subjected to the same changes in total stress but under different drainage conditions will have different internal effective stresses and therefore different strengths according to the Mohr–Coulomb criterion. Rather than have to determine the pore pressures and effective stresses under undrained conditions, the **undrained strength** can be expressed in terms of total stress. The failure envelope will still be linear, but will have a different gradient and intercept; a Mohr–Coulomb model can therefore still be used, but the shear strength parameters are different and denoted by c_u and ϕ_u (=0, see Section 5.6), with the subscripts denoting undrained behaviour. The drained strength is expressed directly in terms of the effective stress parameters c' and ϕ' described previously.

When using these strength parameters to subsequently analyse geotechnical constructions in practice, the principal consideration is the rate at which the changes in total stress (due to construction operations) are applied in relation to the rate of dissipation of excess pore water pressure (consolidation), which in turn is related to the permeability of the soil as described in Chapter 4. In fine-grained soils of low permeability (e.g. clay, silt), loading in the short term (e.g. of the order of weeks or less) will be undrained, while in the long-term, conditions will be drained. In coarse grained soils (e.g. sand, gravel) both short and long-term loading will result in drained conditions due to the higher permeability, allowing consolidation to take place rapidly. Under dynamic loading (e.g. earthquakes), loading may be fast enough to generate an undrained response in coarse-grained material. 'Short-term' is taken to be synonymous with 'during construction', while 'long-term' usually relates to the design life of the construction (usually many tens of years).

5.4 Laboratory shear tests

The shear stiffness (G) and strength parameters (c', ϕ', c_u) for a particular soil can be determined by means of laboratory tests on specimens taken from representative samples of the in-situ soil. Great care and judgement are required in the sampling operation and in the storage and handling of samples prior to testing, especially in the case of **undisturbed samples** where the object is to preserve the in-situ structure and water content of the soil. In the case of clays, test specimens may be obtained from tube or block samples, the latter normally being subjected to the least disturbance. Swelling of a clay specimen will occur due to the release of the in-situ total stresses. Sampling techniques are described in more detail in Chapter 6.

The direct shear test

Test procedures are detailed in BS 1377, Parts 8 (UK), CEN ISO/TS 17892–10 (Europe) and ASTM D3080 (US). The specimen, of cross-sectional area A, is confined in a metal box (known as the **shearbox** or direct shear apparatus) of square or circular cross-section split horizontally at mid-height, a small clearance being maintained between the two halves of the box. At failure within an element of soil under principal stresses σ'_1 and σ'_3, a slip plane will form within the element at an angle θ as shown in Figure 5.6. The shearbox is designed to represent the stress conditions along this slip plane. Porous plates are placed below and on top of the specimen if it is fully or partially saturated to allow free drainage: if the specimen is dry, solid metal plates may be used. The essential features of the apparatus are shown in Figure 5.8. In Figure 5.8(a), a vertical force (N) is applied to the specimen through a loading plate, under which the sample is allowed to consolidate. Shear displacement is then gradually applied on a horizontal plane by causing the two halves of the box to move relative to each other, the shear force required (T) being measured together with the corresponding shear displacement (Δl). The induced shear stress within the sample on the slip plane is equal to that required to shear the two halves of the box. Suggested rates of shearing for fully drained conditions to be achieved are approximately 1 mm/min for sand, 0.01 mm/min for silt and 0.001 mm/min for clay (Bolton, 1991). Normally, the change in thickness (Δh) of the specimen is also measured. If the initial thickness of the specimen is h_0, then the shear strain (γ) can be approximated by $\Delta l/h_0$ and the volumetric strain (ε_v) by $\Delta h/h_0$.

Interpretation of direct shear test data

A number of specimens of the soil are tested, each under a different vertical force, and the value of shear stress at failure ($\tau_f = T/A$) is plotted against the normal effective stress ($\sigma'_f = N/A$) for each test. The Mohr–Coulomb shear strength parameters c' and ϕ' are then obtained from the straight line which best fits the plotted points. The shear stress throughout the test may also be plotted against the shear strain; the gradient of the initial part of the curve before failure (peak shear stress) gives a crude approximation of the shear modulus G.

The test suffers from several disadvantages, the main one being that drainage conditions cannot be controlled. As pore water pressure cannot be measured, only the total normal stress can be determined, although this is equal to the effective normal stress if the pore water pressure is zero (i.e. by shearing slowly enough to achieve drained conditions). Only an approximation to the state of pure shear defined in Figure 5.2 is produced in the specimen and shear stress on the failure plane is not uniform, failure occurring progressively from the edges towards the centre of the specimen. Furthermore, the cross-sectional area of the sample under the shear and vertical loads does not remain constant throughout the test. The advantages of the test are its simplicity and, in the case of coarse-grained soils, the ease of specimen preparation.

Soil behaviour in shear

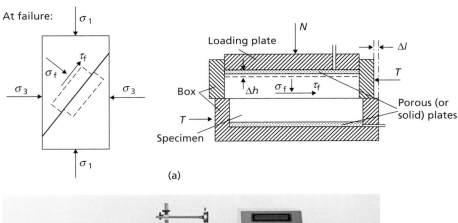

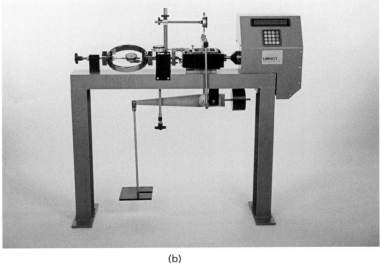

Figure 5.8 Direct shear apparatus: (a) schematic, (b) standard direct shear apparatus (image courtesy of Impact Test Equipment Ltd.).

The triaxial test

The **triaxial apparatus** is the most widely used laboratory device for measuring soil behaviour in shear, and is suitable for all types of soil. The test has the advantages that drainage conditions can be controlled, enabling saturated soils of low permeability to be consolidated, if required, as part of the test procedure, and pore water pressure measurements can be made. A cylindrical specimen, generally having a length/diameter ratio of 2, is used in the test; this sits within a chamber of pressurised water. The sample is stressed axially by a loading ram and radially by the confining fluid pressure under conditions of axial symmetry in the manner shown in Figure 5.9. The most common test, **triaxial compression**, involves applying shear to the soil by holding the confining pressure constant and applying compressive axial load through the loading ram.

The main features of the apparatus are also shown in Figure 5.9. The circular base has a central pedestal on which the specimen is placed, there being access through the pedestal for drainage and for the measurement of pore water pressure. A Perspex cylinder, sealed between a ring and the circular cell top, forms the body of the cell. The cell top has a central bush through which the loading ram passes. The cylinder and cell top clamp onto the base, a seal being made by means of an O-ring.

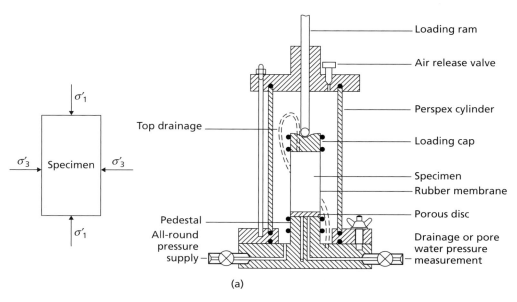

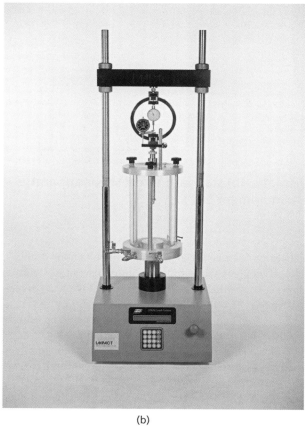

Figure 5.9 The triaxial apparatus: (a) schematic, (b) a standard triaxial cell (image courtesy of Impact Test Equipment Ltd.).

Soil behaviour in shear

The specimen is placed on either a porous or a solid disc on the pedestal of the apparatus. Typical specimen diameters (in the UK) are 38 and 100 mm. A loading cap is placed on top of the specimen, and the specimen is then sealed in a rubber membrane, O-rings under tension being used to seal the membrane to the pedestal and the loading cap to make these connections watertight. In the case of sands, the specimen must be prepared in a rubber membrane inside a rigid former which fits around the pedestal. A small negative pressure is applied to the pore water to maintain the stability of the specimen while the former is removed prior to the application of the all-round confining pressure. A connection may also be made through the loading cap to the top of the specimen, a flexible plastic tube leading from the loading cap to the base of the cell; this connection is normally used for the application of back pressure (as described later in this section). Both the top of the loading cap and the lower end of the loading ram have coned seatings, the load being transmitted through a steel ball. The specimen is subjected to an all-round fluid pressure in the cell, consolidation is allowed to take place, if appropriate, and then the axial stress is gradually increased by the application of compressive load through the ram until failure of the specimen takes place, usually on a diagonal plane through the sample (see Figure 5.6). The load is measured by means of a load ring or by a load transducer fitted to the loading ram either inside or outside the cell. The system for applying the all-round pressure must be capable of compensating for pressure changes due to cell leakage or specimen volume change.

Prior to triaxial compression, sample consolidation may be permitted under equal increments of total stress normal to the end and circumferential surfaces of the specimen, i.e. by increasing the confining fluid pressure within the triaxial cell. Lateral strain in the specimen is not equal to zero during consolidation under these conditions (unlike in the oedometer test, as described in Section 4.2). This is known as **isotropic consolidation**. Dissipation of excess pore water pressure takes place due to drainage through the porous disc at the bottom (or top, or both) of the specimen. The drainage connection leads to an external volume gauge, enabling the volume of water expelled from the specimen to be measured. Filter paper drains, in contact with the end porous disc, are sometimes placed around the circumference of the specimen; both vertical and radial drainage then take place and the rate of dissipation of excess pore water pressure is increased to reduce test time for this stage.

The pore water pressure within a triaxial specimen can usually be measured, enabling the results to be expressed in terms of effective stresses within the sample, rather than just the known applied total stresses; conditions of no flow either out of or into the specimen must be maintained, otherwise the correct pressure will be modified. Pore water pressure is normally measured by means of an electronic pressure transducer.

If the specimen is partially saturated, a fine porous ceramic disc must be sealed into the pedestal of the cell if the correct pore water pressure is to be measured. Depending on the pore size of the ceramic, only pore water can flow through the disc, provided the difference between the pore air and pore water pressures is below a certain value, known as the **air entry value** of the disc. Under undrained conditions the ceramic disc will remain fully saturated with water, provided the air entry value is high enough, enabling the correct pore water pressure to be measured. The use of a coarse porous disc, as normally used for a fully saturated soil, would result in the measurement of the pore air pressure in a partially saturated soil.

Test limitations and corrections

The average cross-sectional area (A) of the specimen does not remain constant throughout the test, and this must be taken into account when interpreting stress data from the axial ram load measurements. If the original cross-sectional area of the specimen is A_0, the original length is l_0 and the original volume is V_0, then, if the volume of the specimen decreases during the test,

$$A = A_0 \frac{1-\varepsilon_v}{1-\varepsilon_a} \tag{5.16}$$

Development of a mechanical model for soil

where ε_v is the volumetric strain ($\Delta V/V_0$) and ε_a is the axial strain ($\Delta l/l_0$). If the volume of the specimen increases during a drained test, the sign of ΔV will change and the numerator in Equation 5.16 becomes $(1+\varepsilon_v)$. If required, the radial strain (ε_r) could be obtained from the equation

$$\varepsilon_v = \varepsilon_a + 2\varepsilon_r \tag{5.17}$$

In addition, the strain conditions in the specimen are not uniform due to frictional restraint produced by the loading cap and pedestal disc; this results in dead zones at each end of the specimen, which becomes barrel-shaped as the test proceeds. Non-uniform deformation of the specimen can be largely eliminated by lubrication of the end surfaces. It has been shown, however, that non-uniform deformation has no significant effect on the measured strength of the soil, provided the length/diameter ratio of the specimen is not less than 2. The compliance of the rubber membrane must also be accounted for.

Interpretation of triaxial test data: strength

Triaxial data may be presented in the form of Mohr circles at failure; it is more straightforward, however, to present it in terms of stress invariants, such that a given set of effective stress conditions can be represented by a single point instead of a circle. Under 2-D stress conditions, the state of stress represented in Figure 5.6 could also be defined by the radius and centre of the Mohr circle. The radius is usually denoted by $t=\frac{1}{2}(\sigma'_1-\sigma'_3)$, with the centre point denoted by $s'=\frac{1}{2}(\sigma'_1-\sigma'_3)$. These quantities ($t$ and s') also represent the maximum shear stress within the element and the average principal effective stress, respectively. The stress state could also be expressed in terms of total stress. It should be noted that

$$\frac{1}{2}(\sigma'_1-\sigma'_3) = \frac{1}{2}(\sigma_1-\sigma_3)$$

i.e. parameter t, much like shear stress τ is independent of u. This parameter is known as the **deviatoric stress invariant** for 2-D stress conditions, and is analogous to shear stress τ acting on a shear plane (alternatively, τ may be thought of as the deviatoric stress invariant in direct shear). Parameter s' is also known as the **mean stress invariant**, and is analogous to the normal effective stress acting on a shear plane (i.e. causing volumetric change but no shear). Substituting Terzaghi's Principle into the definition of s', it is apparent that

$$\frac{1}{2}(\sigma'_1+\sigma'_3) = \frac{1}{2}(\sigma_1+\sigma_3)-u$$

or, re-written, $s'=s-u$, i.e. Terzaghi's Principle rewritten in terms of the 2-D stress invariants.

The stress conditions and Mohr circle for a 3-D element of soil under a general distribution of stresses is shown in Figure 5.10. Unlike 2-D conditions when there are three unique stress components (σ'_x, σ'_z, τ_{zx}), in 3-D there are six stress components (σ'_x, σ'_y, σ'_z, τ_{xy}, τ_{yz}, τ_{zx}). These can, however, be reduced to a set of three principal stresses, σ'_1, σ'_2 and σ'_3. As before, σ'_1 and σ'_3 are the major (largest) and minor (smallest) principal stresses; σ'_2 is known as the **intermediate principal stress**. For the general case where the three principal stress components are different ($\sigma'_1>\sigma'_2>\sigma'_3$, also described as **true triaxial** conditions), a set of three Mohr circles can be drawn as shown in Figure 5.10.

As, in the case of 2-D stress conditions, it was possible to describe the stress state in terms of a mean and deviatoric invariant (s' and t respectively), so it is possible in 3-D conditions. To distinguish between 2-D and 3-D cases, the triaxial mean invariant is denoted p' (effective stress) or p (total stress), and the deviatoric invariant by q. As before, the mean stress invariant causes only volumetric change (does not induce shear), and is the average of the three principal stress components:

$$p' = \frac{\sigma'_1+\sigma'_2+\sigma'_3}{3} \tag{5.18}$$

Soil behaviour in shear

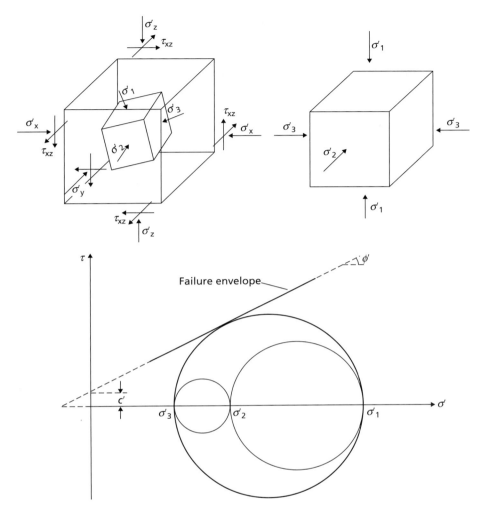

Figure 5.10 Mohr circles for triaxial stress conditions.

Equation 5.18 may also be written in terms of total stresses. Similarly, q, as the deviatoric invariant, induces shearing within the sample and is independent of the pore fluid pressure, u:

$$q = \frac{1}{\sqrt{2}}\left[(\sigma_1 - \sigma_2)^2 + (\sigma_2 - \sigma_3)^2 + (\sigma_3 - \sigma_1)^2\right]^{\frac{1}{2}} \tag{5.19}$$

A full derivation of Equation 5.19 may be found in Atkinson and Bransby (1978). In the triaxial cell, the stress conditions are simpler than the general case shown in Figure 5.10. During a standard compression test, $\sigma_1 = \sigma_a$, and due to the axial symmetry, $\sigma_2 = \sigma_3 = \sigma_r$, so that Equations 5.18 and 5.19 reduce to:

$$p = \frac{\sigma_a + 2\sigma_r}{3} \quad p' = \frac{\sigma'_a + 2\sigma'_r}{3} \tag{5.20}$$

$$q = \sigma'_a - \sigma'_r = \sigma_a - \sigma_r \tag{5.21}$$

Development of a mechanical model for soil

The confining fluid pressure within the cell (σ_r) is the minor principal stress in the standard triaxial compression test. The sum of the confining pressure and the applied axial stress from the loading ram is the major principal stress (σ_a), on the basis that there are no shear stresses applied to the surfaces of the specimen. The applied axial stress component from the loading ram is thus equal to the deviatoric stress q, also referred to as the **principal stress difference**. As the intermediate principal stress is equal to the minor principal stress, the stress conditions at failure can be represented by a single Mohr circle, as shown in Figure 5.6 with $\sigma'_1 = \sigma'_a$ and $\sigma'_3 = \sigma'_r$. If a number of specimens are tested, each under a different value of confining pressure, the failure envelope can be drawn and the shear strength parameters for the soil determined.

Because of the axial symmetry in the triaxial test, both 2-D and 3-D invariants are in common usage, with stress points represented by s',t or p',q respectively. The parameters defining the strength of the soil (ϕ' and c') are not affected by the invariants used; however, the interpretation of the data to find these properties does vary with the set of invariants used. Figure 5.11 shows the Mohr–Coulomb failure envelope for an element of soil plotted in terms of direct shear (σ', τ), 2-D (s', t) and 3-D/triaxial (p', q) stress invariants. Under direct shear conditions, it has already been described in Section 5.3 that the gradient of the failure envelope τ/σ' is equal to the tangent of the angle of shearing resistance and the intercept is equal to c' (Figure 5.11(a)). For 2-D conditions, Equation 5.15a gives:

$$(\sigma'_1 - \sigma'_3) = (\sigma'_1 + \sigma'_3)\sin\phi' + 2c'\cos\phi'$$
$$2t = 2s'\sin\phi' + 2c'\cos\phi' \tag{5.22}$$
$$t = s'\sin\phi' + c'\cos\phi'$$

Therefore, if the points of ultimate failure in triaxial tests are plotted in terms of s' and t, a straight-line failure envelope will be obtained – the gradient of this line is equal to the sine of the angle of shearing resistance and the intercept = $c'\cos\phi'$ (Figure 5.11(b)). Under triaxial conditions with $\sigma_2 = \sigma_3$, from the definitions of p' and q (Equations 5.20 and 5.21) it is clear that:

$$(\sigma'_1 + \sigma'_3) = (\sigma'_a + \sigma'_r) = \frac{6p' + q}{3} \tag{5.23}$$

and

$$(\sigma'_1 - \sigma'_3) = (\sigma'_a - \sigma'_r) = q \tag{5.24}$$

The gradient of the failure envelope in triaxial conditions is described by the parameter $M = q/p'$, (Figure 5.11(c)) so that from Equation 5.15a:

$$(\sigma'_1 - \sigma'_3) = (\sigma'_1 + \sigma'_3)\sin\phi' + 2c'\cos\phi'$$
$$q = \left(\frac{6p' + q}{3}\right)\sin\phi' + 2c'\cos\phi' \tag{5.25}$$
$$q = \left(\frac{6\sin\phi'}{3 - \sin\phi'}\right)p' + \frac{6c'\cos\phi'}{3 - \sin\phi'}$$

Equation 5.25 represents a straight line when plotted in terms of p' and q with a gradient given by

$$M = \frac{q}{p'} = \frac{6\sin\phi'}{3 - \sin\phi'}$$
$$\therefore \sin\phi' = \frac{3M}{6 + M} \tag{5.26}$$

Soil behaviour in shear

Equations 5.25 and 5.26 apply only to triaxial compression, where $\sigma_a > \sigma_r$. For samples under **triaxial extension**, $\sigma_r > \sigma_a$, the cell pressure becomes the major principal stress. This does not affect Equation 5.23, but Equation 5.24 becomes

$$(\sigma_1' - \sigma_3') = (\sigma_r' - \sigma_a') = -q \tag{5.27}$$

giving:

$$\sin\phi' = \frac{3M}{6-M} \tag{5.28}$$

As the friction angle of the soil must be the same whether it is measured in triaxial compression or extension, this implies that different values of M will be observed in compression and extension.

Interpretation of triaxial test data: stiffness

Triaxial test data can also be used to determine the stiffness properties (principally the shear modulus, G) of a soil. Assuming that the soil is isotropically linear elastic, for 3-D stress conditions Equation 5.7 can be extended to give:

$$\begin{bmatrix} \varepsilon_x \\ \varepsilon_y \\ \varepsilon_z \\ \gamma_{xy} \\ \gamma_{yz} \\ \gamma_{zx} \end{bmatrix} = \frac{1}{2G(1+v)} \begin{bmatrix} 1 & -v & -v & 0 & 0 & 0 \\ -v & 1 & -v & 0 & 0 & 0 \\ -v & -v & 1 & 0 & 0 & 0 \\ 0 & 0 & 0 & 2(1+v) & 0 & 0 \\ 0 & 0 & 0 & 0 & 2(1+v) & 0 \\ 0 & 0 & 0 & 0 & 0 & 2(1+v) \end{bmatrix} \begin{bmatrix} \sigma_x' \\ \sigma_y' \\ \sigma_z' \\ \tau_{xy} \\ \tau_{yz} \\ \tau_{zx} \end{bmatrix} \tag{5.29}$$

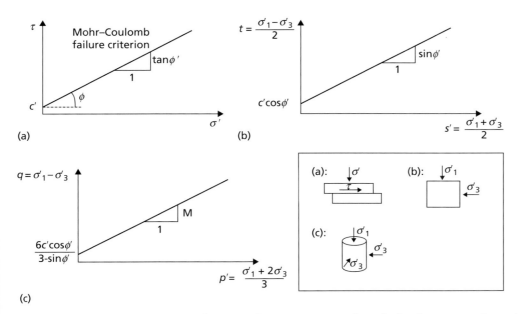

Figure 5.11 Interpretation of strength parameters c' and ϕ' using stress invariants.

Development of a mechanical model for soil

This may alternatively be simplified in terms of principal stresses and strains to give:

$$\begin{bmatrix} \varepsilon_1 \\ \varepsilon_2 \\ \varepsilon_3 \end{bmatrix} = \frac{1}{2G(1+v)} \begin{bmatrix} 1 & -v & -v \\ -v & 1 & -v \\ -v & -v & 1 \end{bmatrix} \begin{bmatrix} \sigma'_1 \\ \sigma'_2 \\ \sigma'_3 \end{bmatrix} \quad (5.30)$$

The deviatoric shear strain ε_s within a triaxial cell (i.e. the strain induced by the application of the deviatoric stress, q, also known as the **triaxial shear strain**) is given by:

$$\varepsilon_s = \frac{2}{3}(\varepsilon_1 - \varepsilon_3) \quad (5.31)$$

Substituting for ε_1 and ε_3 from Equation 5.30 and using Equation 5.6, Equation 5.31 reduces to:

$$\varepsilon_s = \frac{1}{3G} q \quad (5.32)$$

This implies that on a plot of q versus ε_s (i.e. deviatoric stress versus deviatoric strain), the value of the gradient of the curve prior to failure is equal to three times the shear modulus. Generally, in order to determine ε_s within a triaxial test, it is necessary to measure both the axial strain ε_a ($=\varepsilon_1$ for triaxial compression) and the radial strain ε_r ($=\varepsilon_3$). While the former is routinely measured, direct measurement of the latter parameter requires sophisticated sensors to be attached directly to the sample, though the volume change during drained shearing may also be used to infer ε_r using Equation 5.17. If, however, the test is conducted under undrained conditions, then there will be no volume change ($\varepsilon_v=0$) and hence from Equation 5.17:

$$\varepsilon_r = -\frac{1}{2}\varepsilon_a \quad (5.33)$$

From Equation 5.31 it is then clear that for undrained conditions, $\varepsilon_s=\varepsilon_a$. A plot of q versus ε_a for an undrained test will thus have a gradient equal to $3G$. Undrained triaxial testing is therefore extremely useful for determining shear modulus, using measurements which can be made on even the most basic triaxial cells. As G is independent of the drainage conditions within the soil, the value obtained applies equally well for subsequent analysis of the soil under drained loadings. In addition to reducing instrumentation requirements, undrained testing is also much faster than drained testing, particularly for saturated clays of low permeability.

If drained triaxial tests are conducted in which volume change is permitted, radial strain measurements should be made such that G can be determined from a plot of q versus ε_s. Under these test conditions the drained Poisson's ratio (v') can also be determined, being:

$$v' = -\frac{\varepsilon_r}{\varepsilon_a} \quad (5.34)$$

Under undrained conditions it is not necessary to measure Poisson's ratio (v_u), as, comparing Equations 5.33 and 5.34, it is clear that $v_u=0.5$ for there to be no volume change.

Testing under back pressure

Testing under back pressure involves raising the pore water pressure within the sample artificially by connecting a source of constant fluid pressure through a porous disc to one end of a triaxial specimen. In a drained test this connection remains open throughout the test, drainage taking place against the back pressure; the back pressure is then the datum for excess pore water pressure measurement. In a consolidated-undrained test (described later) the connection to the back pressure source is closed at the end of the consolidation stage, before the application of the principal stress difference is commenced.

Soil behaviour in shear

The object of applying a back pressure is to ensure full saturation of the specimen or to simulate in-situ pore water pressure conditions. During sampling, the degree of saturation of a fine-grained soil may fall below 100% owing to swelling on the release of in-situ stresses. Compacted specimens will also have a degree of saturation below 100%. In both cases, a back pressure is applied which is high enough to drive the pore air into solution in the pore water.

It is essential to ensure that the back pressure does not by itself change the effective stresses in the specimen. It is necessary, therefore, to raise the cell pressure simultaneously with the application of the back pressure and by an equal increment. Consider an element of soil, of volume V and porosity n, in equilibrium under total principal stresses σ_1, σ_2 and σ_3, as shown in Figure 5.12, the pore pressure being u_0. The element is subjected to equal increases in confining pressure $\Delta\sigma_3$ in each direction i.e. an isotropic increase in stress, accompanied by an increase Δu_3 in pore pressure.

The increase in effective stress in each direction $= \Delta\sigma_3 - \Delta u_3$

Reduction in volume of the soil skeleton $= C_s V(\Delta\sigma_3 - \Delta u_3)$

where C_s is the compressibility of the soil skeleton under an isotropic effective stress increment.

Reduction in volume of the pore space $= C_v n V \Delta u_3$

where C_v is the compressibility of pore fluid under an isotropic pressure increment.

If the soil particles are assumed to be incompressible and if no drainage of pore fluid takes place, then the reduction in volume of the soil skeleton must equal the reduction in volume of the pore space, i.e.

$$C_s V (\Delta\sigma_3 - \Delta u_3) = C_v n V \Delta u_3$$

Therefore,

$$\Delta u_3 = \Delta\sigma_3 \left(\frac{1}{1 + n(C_v / C_s)} \right)$$

Writing $1/[1 + n(C_v/C_s)] = B$, defined as a **pore pressure coefficient**,

$$\Delta u_3 = B \Delta\sigma_3 \tag{5.35}$$

In fully saturated soils the compressibility of the pore fluid (water only) is considered negligible compared with that of the soil skeleton, and therefore $C_v/C_s \to 0$ and $B \to 1$. In partially saturated soils the compressibility of the pore fluid is high due to the presence of pore air, and therefore $C_v/C_s > 0$ and $B < 1$. The variation of B with degree of saturation for a particular soil is shown in Figure 5.13.

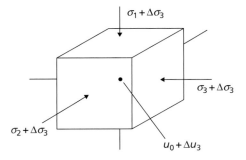

Figure 5.12 Soil element under isotropic stress increment.

Development of a mechanical model for soil

The value of B can be measured in the triaxial apparatus (Skempton, 1954). A specimen is set up under any value of all-round pressure and the pore water pressure measured. Under undrained conditions the all-round pressure is then increased (or reduced) by an amount $\Delta\sigma_3$ and the change in pore water pressure (Δu) from the initial value is measured, enabling the value of B to be calculated from Equation 5.35. A specimen is normally considered to be saturated if the pore pressure coefficient B has a value of at least 0.95.

Types of triaxial test

Many variations of test procedure are possible with the triaxial apparatus, but the three principal types of test are as follows:

1. *Unconsolidated–Undrained (UU)*. The specimen is subjected to a specified confining pressure and then the principal stress difference is applied immediately, with no drainage/consolidation being permitted at any stage of the test. The test procedure is standardised in BS1377, Part 7 (UK), CEN ISO/TS 17892–8 (Europe) and ASTM D2850 (US).
2. *Consolidated–Undrained (CU)*. Drainage of the specimen is permitted under a specified confining pressure until consolidation is complete; the principal stress difference is then applied with no further drainage being permitted. Pore water pressure measurements may be made during the undrained part of the test to determine strength parameters in terms of effective stresses. The consolidation phase is isotropic in most standard testing, denoted by CIU. Modern computer-controlled triaxial machines (also known as **stress path cells**) use hydraulic pressure control units to control the cell (confining) pressure, back pressure and ram load (axial stress) independently (Figure 5.14). Such an apparatus can therefore apply a 'no-lateral strain' condition where stresses are anisotropic, mimicking the one-dimensional compression that occurs in an oedometer test. These tests are often denoted by CAU (the 'A' standing for anisotropic). The test procedure is standardised in BS1377, Part 8 (UK), CEN ISO/TS 17892–9 (Europe) and ASTM D4767 (US).
3. *Consolidated–Drained (CD)*. Drainage of the specimen is permitted under a specified confining pressure until consolidation is complete; with drainage still being permitted, the principal stress difference is then applied at a rate slow enough to ensure that the excess pore water pressure is maintained at zero. The test procedure is standardised in BS1377, Part 8 (UK), CEN ISO/TS 17892–9 (Europe) and ASTM D7181 (US).

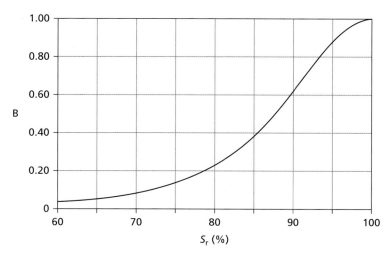

Figure 5.13 Typical relationship between B and degree of saturation.

Soil behaviour in shear

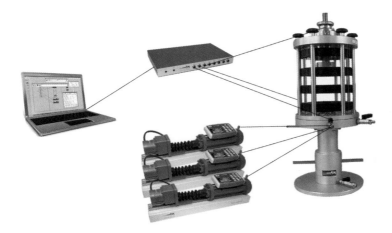

Figure 5.14 Stress path triaxial cell (image courtesy of GDS Instruments).

The use of these test procedures for determining the strength and stiffness properties of both coarse and fine grained soils will be discussed in the following sections (5.5 and 5.6).

Other tests

Although direct shear and triaxial tests are the most commonly used laboratory tests for quantifying the constitutive behaviour of soil, there are other tests which are used routinely. An **unconfined compression test** is essentially a triaxial test in which the confining pressure $\sigma_3 = 0$. The reported result from such a test is the **unconfined compressive strength** (UCS), which is the major principal (axial) stress at failure (which, because $\sigma_3' = 0$, is also the deviatoric stress at failure). As with the triaxial test, a Mohr circle may be plotted for the test as shown in Figure 5.15; however, as only one test is conducted it is not

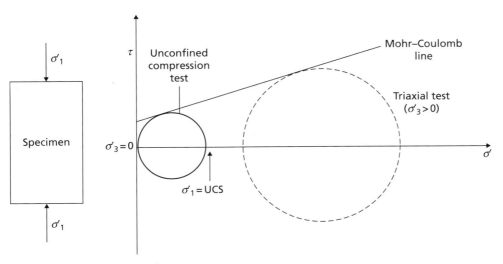

Figure 5.15 Unconfined compression test interpretation.

possible to define the Mohr–Coulomb shear strength envelope without conducting further triaxial tests. The test is not suitable for cohesionless soils ($c' \approx 0$), which would fail immediately without the application of confining pressure. It is usually used with fine-grained soils, and is particularly popular for testing rock.

A **laboratory vane** is sometimes used in fine-grained soils for measuring the undrained shear strength parameter c_u. It is much quicker and simpler for this purpose than conducting a UU triaxial test. It is not discussed further in this chapter, as its principle of operation is similar to the field vane, which is described in detail in Section 7.3.

The **simple shear apparatus** (SSA) is an alternative to the direct shear apparatus (DSA) described earlier in this section. Rather than using a split shear box, the side walls of the simple shear apparatus are usually formed of either plates which can rotate, or a series of laminations which can move relative to each other or are flexible, such arrangements allowing rotation of the sides of the sample which imposes a state of pure simple shear as defined in Figures 5.1 and 5.2. Analysis of the test data is identical to the DSA, although the approximations and assumptions made for the conditions within the DSA are truer for the SSA. While the SSA is better for testing soil, it cannot conduct interface shear tests and is therefore not as versatile or popular as the DSA.

5.5 Shear strength of coarse-grained soils

The shear strength characteristics of coarse-grained soils such as sands and gravels can be determined from the results of either direct shear tests or drained triaxial tests, only the drained strength of such soils normally being relevant in practice. The characteristics of dry and saturated sands or gravels are the same, provided there is zero excess pore water pressure generated in the case of saturated soils, as strength and stiffness are dependent on effective stress. Typical curves relating shear stress and shear strain for initially dense and loose sand specimens in direct shear tests are shown in Figure 5.16(a). Similar curves are obtained relating principal stress difference and axial strain in drained triaxial compression tests.

In dense deposits (high relative density, I_D, see Chapter 1) there is a considerable degree of interlocking between particles. Before shear failure can take place, this interlocking must be overcome in addition to the frictional resistance at the points of inter-granular contact. In general, the degree of interlocking is greatest in the case of very dense, well-graded soils consisting of angular particles. The characteristic stress–strain curve for initially dense sand shows a peak stress at a relatively low strain, and thereafter, as interlocking is progressively overcome, the stress decreases with increasing strain. The reduction in the degree of interlocking produces an increase in the volume of the specimen during shearing as characterised by the relationship between volumetric strain and shear strain in the direct shear test, shown in Figure 5.16(c). In the drained triaxial test, a similar relationship would be obtained between volumetric strain and axial strain. The change in volume is also shown in terms of void ratio (e) in Figure 5.16(d). Eventually the specimen becomes loose enough to allow particles to move over and around their neighbours without any further net volume change, and the shear stress reduces to an ultimate value. However, in the triaxial test non-uniform deformation of the specimen becomes excessive as strain is progressively increased, and it is unlikely that the ultimate value of principal stress difference can be reached.

The term **dilatancy** is used to describe the increase in volume of a dense coarse-grained soil during shearing, and the rate of dilation can be represented by the gradient $d\varepsilon_v/d\gamma$, the maximum rate corresponding to the peak stress (Figure 5.16(c)). The **angle of dilation** (ψ) is defined as $\tan^{-1}(d\varepsilon_v/d\gamma)$. The concept of dilatancy can be illustrated in the context of the direct shear test by considering the shearing of dense and loosely packed spheres (idealised soil particles) as shown in Figure 5.17. During shearing of a dense soil (Figure 5.17(a)), the macroscopic shear plane is horizontal but sliding between individual particles takes place on numerous microscopic planes inclined at various angles above the horizontal, as

Soil behaviour in shear

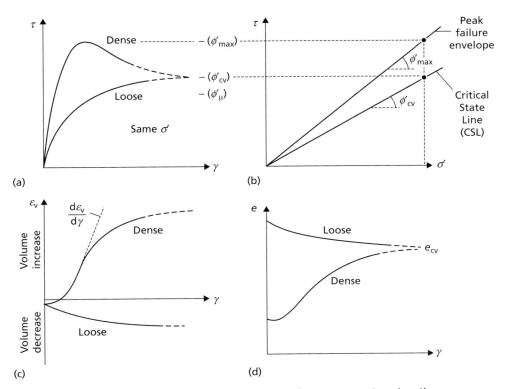

Figure 5.16 Shear strength characteristics of coarse-grained soils.

the particles move up and over their neighbours. The angle of dilation represents an average value of these angles for the specimen as a whole. The loading plate of the apparatus is thus forced upwards, work being done against the normal stress on the shear plane. For a dense soil, the maximum (or peak) angle of shearing resistance (ϕ'_{max}) determined from peak stresses (Figure 5.16(b)) is significantly greater than the true angle of friction (ϕ_μ) between the surfaces of individual particles, the difference representing the work required to overcome interlocking and rearrange the particles.

In the case of initially loose soil (Figure 5.17(b)), there is no significant particle interlocking to be overcome and the shear stress increases gradually to an ultimate value without a prior peak, accompanied by a decrease in volume. The ultimate values of shear stress and void ratio for dense and loose specimens of the same soil under the same values of normal stress in the direct shear test are essentially equal, as indicated in Figures 5.16(a) and (d). The ultimate resistance occurs when there is no further change in volume or shear stress (Figures 5.16(a) and (c)), which is known as the **critical state**. Stresses at the critical state define a straight line (Mohr–Coulomb) failure envelope intersecting the origin, known as the **critical state line (CSL)**, the slope of which is tan ϕ_{cv}' (Figure 5.16(b)). The corresponding angle of shearing resistance at critical state (also called the **critical state angle of shearing resistance**) is usually denoted ϕ'_{cv} or ϕ'_{crit}. The difference between ϕ'_μ and ϕ_{cv}' represents the work required to rearrange the particles. The friction angles ϕ'_{cv} and ϕ'_{max} are related to ψ after the relationship given by Bolton (1986):

$$\phi'_{max} = \phi'_{cv} + 0.8\psi \tag{5.36}$$

Equation 5.36 applies for conditions of plane strain within soil, such as those induced within the DSA or

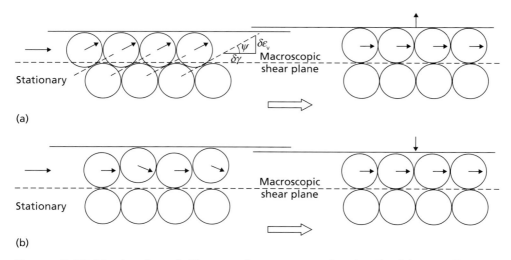

Figure 5.17 Mechanics of dilatancy in coarse-grained soils: (a) initially dense soil, exhibiting dilation, (b) initially loose soil, showing contraction.

SSA. Under triaxial conditions, the final term becomes approximately 0.5ψ.

It can be difficult to determine the value of the parameter ϕ'_{cv} from laboratory tests because of the relatively high strain required to reach the critical state. In general, the critical state is identified by extrapolation of the stress–strain curve to the point of constant stress, which should also correspond to the point of zero rate of dilation ($d\varepsilon_v/d\gamma = 0$) on the volumetric strain–shear strain curve.

An alternative method of representing the results from laboratory shear tests is to plot the **stress ratio** (τ/σ' in direct shear) against shear strain. Plots of stress ratio against shear strain representing tests on three specimens of sand in a direct shear test, each having the same initial void ratio, are shown in Figure 5.18(a), the values of effective normal stress (σ') being different in each test. The plots are labelled A, B and C, the effective normal stress being lowest in test A and highest in test C. Corresponding plots of void ratio against shear strain are also shown in Figure 5.18(b). Such results indicate that both the maximum stress ratio and the ultimate (or critical) void ratio decrease with increasing effective normal stress. Thus, dilation is suppressed by increasing mean stress (normal stress σ' in direct shear). This is descried in greater detail by Bolton (1986). The ultimate values of stress ratio ($= \tan \phi'_{cv}$), however, are the same. From Figure 5.18(a) it is apparent that the difference between peak and ultimate stress decreases with increasing effective normal stress; therefore, if the maximum shear stress is plotted against effective normal stress for each individual test, the plotted points will lie on an envelope which is slightly curved, as shown in Figure 5.18(c). Figure 5.18(c) also shows the **stress paths** for each of the three specimens leading up to failure. For any type of shear test, two stress paths may be plotted: the **total stress path (TSP)** plots the variation of σ and τ through the test; the **effective stress path (ESP)** plots the variation of σ' and τ. If both stress paths are plotted on the same axis, the horizontal distance between the two paths at a given value of τ (i.e. $\sigma - \sigma'$) represents the pore water pressure in the sample from Terzaghi's Principle (Equation 3.1). In direct shear tests, the pore water pressure is approximately zero such that the TSP and ESP lie on the same line, as shown in Figure 5.18(c). Remembering that it is the effective (not total) stresses that govern soil shear strength (Equation 5.11), failure occurs when the ESP reaches the failure envelope.

The value of ϕ'_{max} for each test can then be represented by a secant parameter: in the shearbox test, $\phi'_{max} = \tan^{-1}(\tau_{max}/\sigma')$. The value of ϕ'_{max} decreases with increasing effective normal stress until it becomes equal to ϕ'_{cv}. The reduction in the difference between peak and ultimate shear stress with

Soil behaviour in shear

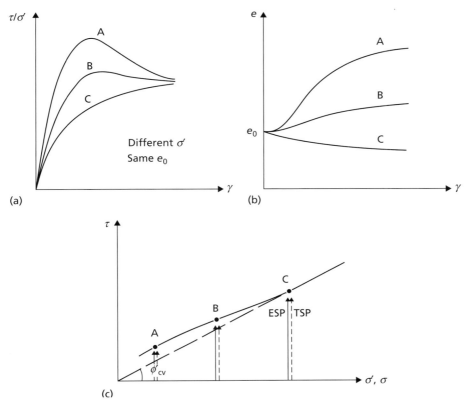

Figure 5.18 Determination of peak strengths from direct shear test data.

increasing normal stress is mainly due to the corresponding decrease in ultimate void ratio. The lower the ultimate void ratio, the less scope there is for dilation. In addition, at high stress levels some fracturing or crushing of particles may occur, with the consequence that there will be less particle interlocking to be overcome. Crushing thus causes the suppression of dilatancy and contributes to the reduced value of ϕ'_{max}.

In the absence of any cementation or bonding between particles, the curved peak failure envelopes for coarse-grained soils would show zero shear strength at zero normal effective stress. Mathematical representations of the curved envelopes may be expressed in terms of power laws, i.e. of the form $\tau_f = A\gamma^B$. These are not compatible with many standard analyses for geotechnical structures which require soil strength to be defined in terms of a straight line (Mohr–Coulomb model). It is common, therefore, in practice to fit a straight line to the peak failure points to define the peak strength in terms of an angle of shearing resistance ϕ' and a cohesion intercept c'. It should be noted that the parameter c' is only a mathematical line-fitting constant used to model the peak states, and should not be used to imply that the soil has shear strength at zero normal effective stress. This parameter is therefore also commonly referred to as the **apparent cohesion** of the soil. In soils which do have natural cementation/bonding, the cohesion intercept will represent the combined effects of any apparent cohesion and the **true cohesion** due to the interparticle bonding.

Once soil has been sheared to the critical state (ultimate conditions), the effects of any true or apparent cohesion are destroyed. This is important when selecting strength properties for use in design, particularly where soil has been tested under its in-situ condition (where line-fitting may suggest $c' > 0$), then

Development of a mechanical model for soil

sheared during excavation and subsequently placed to support a foundation or used to backfill behind a retaining structure. In such circumstances the excavation/placement imposes large shear strains within the soil such that critical state conditions (with $c'=0$) should be assumed in design.

Figure 5.19 shows the behaviour of soils A, B and C as would be observed in a drained triaxial test. The main differences compared to the behaviour in direct shear (Figure 5.18) lie in the stress paths and failure envelope shown in Figure 5.19(c). In standard triaxial compression, radial stress is held constant ($\Delta\sigma_r = 0$) while axial stress is increased (by $\Delta\sigma_a$). From Equations 5.20 and 5.21, this gives $\Delta p = \Delta\sigma_a/3$ and $\Delta q = \Delta\sigma_a$. The gradient of the TSP is therefore $\Delta q/\Delta p = 3$. In a drained test there is no change in pore water pressure, so the ESP is parallel to the TSP. If the sample is dry, the TSP and ESP lie along the same line; if the sample is saturated and a back pressure of u_0 applied, the TSP and ESP are parallel, maintaining a constant horizontal separation of u_0 throughout the test, as shown in Figure 5.19(c). As before, failure occurs when the ESP meets the failure envelope. The value of ϕ'_{max} for each test is determined by finding $M (=q/p')$ at failure and using Equation 5.26, with the resulting value of $\phi' = \phi'_{max}$.

In practice, the routine laboratory testing of sands is difficult because of the problem of obtaining undisturbed specimens and setting them up, still undisturbed, in the test apparatus. If required, tests can be undertaken on specimens reconstituted in the apparatus at appropriate densities, but the in-situ structure is then unlikely to be reproduced.

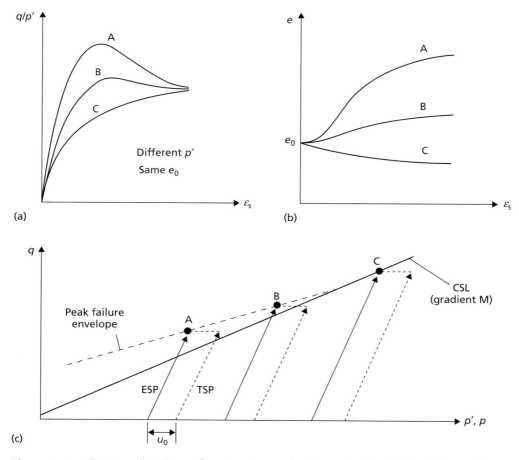

Figure 5.19 Determination of peak strengths from drained triaxial test data.

Soil behaviour in shear

> ### Example 5.1
>
> The results shown in Figure 5.20 were obtained from direct shear tests on reconstituted specimens of sand taken from loose and dense deposits and compacted to the in-situ density in each case. The raw data from the tests and the use of a spreadsheet to process the test data may be found on the Companion Website. Plot the failure envelopes of each sand for both peak and ultimate states, and hence determine the critical state friction angle ϕ'_{cv}.
>
>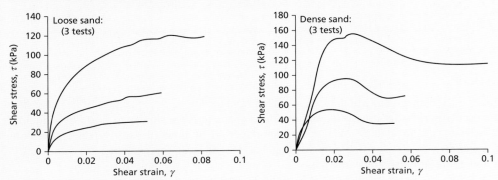
>
> **Figure 5.20** Example 5.1.
>
> ### Solution
>
> The values of shear stress at peak and ultimate states are read from the curves in Figure 5.20 and are plotted against the corresponding values of normal stress, as shown in Figure 5.21. The failure envelope is the line having the best fit to the plotted points; for ultimate conditions a straight line through the origin is appropriate (CSL). From the gradients of the failure envelopes, $\phi'_{cv} = 33.4°$ for the dense sand and 32.6° for the loose sand. These are within 1° of each other, and confirm that the critical state friction angle is an intrinsic soil property which is independent of state (i.e. density). The loose sand does not exhibit peak behaviour, while the peak failure envelope for the dense sand may be characterised by $c' = 15.4$ kPa and $\phi' = 38.0°$ (tangent value) or by secant values as given in Table 5.1.
>
>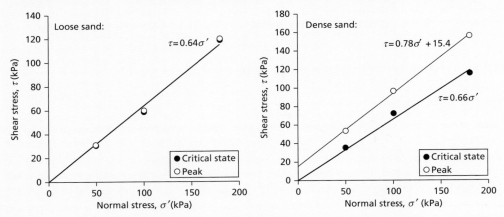
>
> **Figure 5.21** Example 5.1: Failure envelopes for (a) loose, and (b) dense sand samples.

Table 5.1 Example 5.1

Normal stress (kPa)	Secant friction angle (°)
50	46.8
100	43.7
181	40.8

Liquefaction

Liquefaction is a phenomenon in which loose saturated sand loses a large percentage of its shear strength due to high excess pore water pressures, and develops characteristics similar to those of a liquid. It is usually induced by cyclic loading over a very short period of time (usually seconds), resulting in undrained conditions in the sand. Cyclic loading may be caused, for example, by vibrations from machinery and, more seriously, by earthquakes.

Loose sand tends to compact under cyclic loading. The decrease in volume causes an increase in pore water pressure which cannot dissipate under undrained conditions. Indeed, there may be a cumulative increase in pore water pressure under successive cycles of loading. If the pore water pressure becomes equal to the maximum total stress component, normally the overburden pressure, σ_v, the value of effective stress will be zero by Terzaghi's Principle, as described in Section 3.7 – i.e. interparticle forces will be zero, and the sand will exist in a liquid state with negligible shear strength. Even if the effective stress does not fall to zero, the reduction in shear strength may be sufficient to cause failure.

Liquefaction may develop at any depth in a sand deposit where a critical combination of in-situ density and cyclic deformation occurs. The higher the void ratio of the sand and the lower the confining pressure, the more readily liquefaction will occur. The larger the strains produced by the cyclic loading, the lower the number of cycles required for liquefaction.

Liquefaction may also be induced under static conditions where pore pressures are increased as a result of seepage. The techniques described in Chapter 2 and Section 3.7 may be used to determine the pore water pressures and, by Terzaghi's Principle, the effective stresses in the soil for a given seepage event. The shear strength at these low effective stresses is then approximated by the Mohr–Coulomb criterion.

5.6 Shear strength of saturated fine-grained soils

Isotropic consolidation

If a saturated clay specimen is allowed to consolidate in the triaxial apparatus under a sequence of equal confining cell pressures (σ_3), sufficient time being allowed between successive increments to ensure that consolidation is complete, the relationship between void ratio and effective stress can be obtained. This is similar to an oedometer test, though triaxial data are conventionally expressed in terms of specific volume (v) rather than void ratio (e). Consolidation in the triaxial apparatus under equal cell pressure is referred to as isotropic consolidation. Under these conditions, $\sigma_a = \sigma_r = \sigma_3$, such that $p = p' = \sigma_3$, from Equation 5.20 and $q = 0$. As the deviatoric stress is zero, there is no shear induced in the specimen ($\varepsilon_s = 0$), though volumetric strain (ε_v) does occur under the inreasing p'. Unlike one-dimensional (anisotropic) consolidation (discussed in Chapter 4), the soil element will strain both axially and radially by equal amounts (Equation 5.31).

Soil behaviour in shear

The relationship between void ratio and effective stress during isotropic consolidation depends on the stress history of the clay, defined by the overconsolidation ratio (Equation 4.6), as described in Section 4.2. Overconsolidation is usually the result of geological factors, as described in Chapter 4; overconsolidation can also be due to higher stresses previously applied to a specimen in the triaxial apparatus.

One-dimensional and isotropic consolidation characteristics are compared in Figure 5.22. The key difference between the two relationships is the use of the triaxial mean stress invariant p' for isotropic conditions and the one-dimensional normal stress σ' for 1-D consolidation. This has an effect on the gradients of the virgin compression line (here defined as the **isotropic compression line, ICL**) and the unload–reload lines, denoted by λ and κ respectively, which are different to the values of C_c and C_e. It should be realised that a state represented by a point to the right of the ICL is impossible.

As a result of the similarity between the two processes, it is possible to use values of λ and κ determined from the consolidation stage of a triaxial test to directly estimate the 1-D compression relationship as $C_c \approx 2.3\lambda$ and $C_e \approx 2.3\kappa$.

Strength in terms of effective stress

The strength of a fine-grained soil in terms of effective stress, i.e. for drained or long-term loading, can be determined by either the consolidated–undrained triaxial test with pore water pressure measurement, or the drained triaxial test. In a drained test, the stress path concept can be used to determine the failure envelope as described earlier for sands, by determining where the ESP reaches its peak and ultimate (critical state) values, and using these to determine ϕ'_{max} and ϕ'_{cv} respectively. However, because of the low permeability of most fine-grained soils, drained tests are rarely used for such materials because consolidated undrained (CU) tests can provide the same information in less time as consolidation does not need to occur during the shearing phase. The undrained shearing stage of the CU test must, however, be run at a rate of strain slow enough to allow equalisation of pore water pressure throughout the specimen, this rate being a function of the permeability of the clay.

The key principle in using undrained tests (CU) to determine drained properties is that the ultimate state will always occur when the ESP reaches the critical state line. Unlike in the drained test, significant excess pore pressures will develop in an undrained test, resulting in divergence of the TSP and ESP. Therefore, in order to determine the ESP for undrained conditions from the known TSP applied to the sample, pore water pressure in the specimen must be measured.

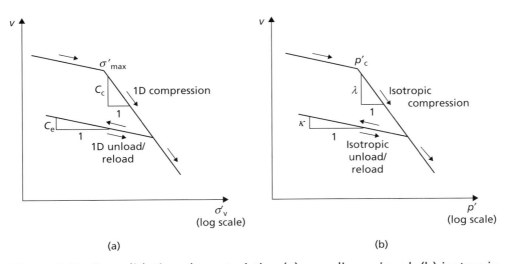

Figure 5.22 Consolidation characteristics: (a) one-dimensional, (b) isotropic.

Development of a mechanical model for soil

Typical test results for specimens of normally consolidated and overconsolidated clays are shown in Figure 5.23. In CU tests, axial stress and pore water pressure are plotted against axial strain. For normally consolidated clays, axial stress reaches an ultimate value at relatively large strain, accompanied by an increase in pore water pressure to a steady value. For overconsolidated clays, axial stress increases to a peak value and then decreases with subsequent increase in strain. However, it is not usually possible to reach the ultimate stress due to excessive specimen deformation. Pore water pressure increases initially and then decreases; the higher the overconsolidation ratio, the greater the decrease. Pore water pressure may become negative in the case of heavily overconsolidated clays, as shown by the dotted line in Figure 5.23(b).

In drained tests, axial stress and volume change are plotted against axial strain. For normally consolidated clays, an ultimate value of stress is again reached at relatively high strain. A decrease in volume takes place during shearing, and the clay hardens. For overconsolidated clays, a peak value of axial stress is reached at relatively low strain. Subsequently, axial stress decreases with increasing strain but, again, it is not usually possible to reach the ultimate stress in the triaxial apparatus. After an initial decrease, the volume of an overconsolidated clay increases prior to and after peak stress and the clay softens. For overconsolidated clays, the decrease from peak stress towards the ultimate value becomes less pronounced as the overconsolidation ratio decreases.

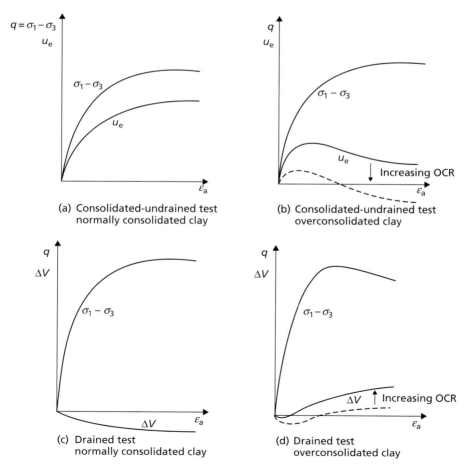

Figure 5.23 Typical results from consolidated–undrained and drained triaxial tests.

Soil behaviour in shear

Failure envelopes for normally consolidated and overconsolidated clays are of the forms shown in Figure 5.24. This figure also shows typical stress paths for samples of soil which are initially consolidated isotropically to the same initial mean stress p'_c; the overconsolidated sample is then partially unloaded isotropically prior to shearing to induce overconsolidation. For a normally consolidated or lightly overconsolidated clay with negligible cementation (Figure 5.24(a)), the failure envelope should pass through the origin (i.e. $c' \approx 0$). The envelope for a heavily overconsolidated clay is likely to exhibit curvature over the stress range up to approximately $p'_c/2$ (Figure 5.24(b)). The corresponding Mohr–Coulomb failure envelopes in terms of σ' and τ for use in subsequent geotechnical analyses are shown in Figure 5.24(c). The gradient of the straight part of the failure envelope is approximately $\tan \phi'_{cv}$. The value of ϕ'_{cv} can be found from the gradient of the critical state line M using Equation 5.26. If the critical state value of ϕ'_{cv} is required for a heavily overconsolidated clay, then, if possible, tests should be performed at stress levels that are high enough to define the critical state envelope, i.e. specimens should be consolidated at all-round pressures in excess of the preconsolidation value. Alternatively, an estimated value of ϕ'_{cv} can be obtained from tests on normally consolidated specimens reconsolidated from a slurry.

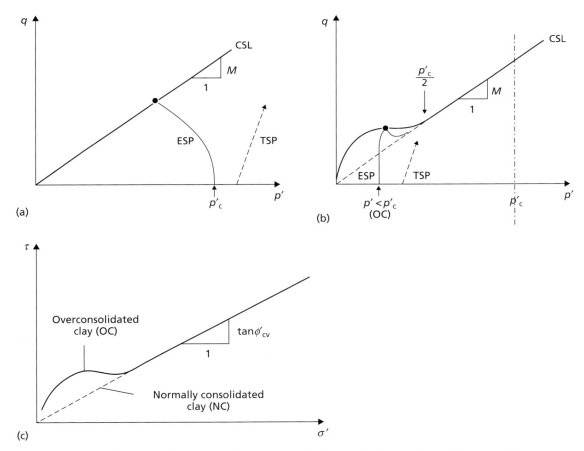

Figure 5.24 Failure envelopes and stress paths in triaxial tests for: (a) normally consolidated (NC) clays, (b) overconsolidated (OC) clays, (c) corresponding Mohr–Coulomb failure envelope.

Example 5.2

The results shown in Table 5.2 were obtained for peak failure in a series of consolidated–undrained triaxial tests, with pore water pressure measurement, on specimens of a saturated clay. Determine the values of the effective stress strength parameters defining the peak failure envelope.

Table 5.2 Example 5.2

All-round pressure (kPa)	Principal stress difference (kPa)	Pore water pressure (kPa)
150	192	80
300	341	154
450	472	222

Solution

Values of effective principal stresses σ'_3 and σ'_1 at failure are calculated by subtracting pore water pressure at failure from the total principal stresses as shown in Table 5.3 (all stresses in kPa). The Mohr circles in terms of effective stress are drawn in Figure 5.25. In this case, the failure envelope is slightly curved and a different value of the secant parameter ϕ' applies to each circle. For circle (a) the value of ϕ' is the slope of the line OA, i.e. 35°. For circles (b) and (c) the values are 33° (OB) and 31° (OC), respectively.

Table 5.3 Example 5.2 (contd.)

σ_3 (kPa)	σ_1 (kPa)	σ'_3 (kPa)	σ'_1 (kPa)
150	342	70	262
300	641	146	487
450	922	228	700

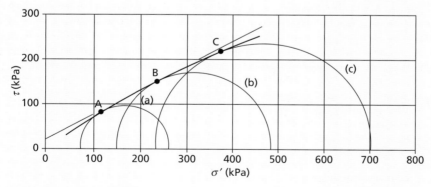

Figure 5.25 Example 5.2.

Tangent parameters can be obtained by approximating the curved envelope to a straight line over the stress range relevant to the problem. In Figure 5.25 a linear approximation has been drawn for the range of effective normal stress 200–300 kPa, giving parameters $c' = 20$ kPa and $\phi' = 29°$.

Undrained strength

In principle, the unconsolidated–undrained (UU) triaxial test enables the undrained strength of a fine-grained soil in its in-situ condition to be determined, the void ratio of the specimen at the start of the test being unchanged from the in-situ value at the depth of sampling. In practice, however, the effects of sampling and preparation result in a small increase in void ratio, particularly due to swelling when the in-situ stresses are removed. Experimental evidence (e.g. Duncan and Seed, 1966) has shown that the in-situ undrained strength of saturated clays is significantly anisotropic, the strength depending on the direction of the major principal stress relative to the in-situ orientation of the specimen. Thus, undrained strength is not a unique parameter, unlike the critical state angle of shearing resistance.

When a specimen of saturated fine-grained soil is placed on the pedestal of the triaxial cell the initial pore water pressure is negative due to capillary tension, total stresses being zero and effective stresses positive. After the application of confining pressure the effective stresses in the specimen remain unchanged because, for a fully saturated soil under undrained conditions, any increase in all-round pressure results in an equal increase in pore water pressure (see Figure 4.11). Assuming all specimens to have the same void ratio and composition, a number of UU tests, each at a different value of confining pressure, should result, therefore, in equal values of principal stress difference at failure. The results are expressed in terms of total stress as shown in Figure 5.26, the failure envelope being horizontal, i.e. $\phi_u = 0$, and the shear strength is given by $\tau_f = c_u$, where c_u is the undrained shear strength. Undrained strength may also be determined without the use of Mohr circles; the principal stress difference at failure (q_f) is the diameter of the Mohr circle, while τ_f is its radius, therefore, in an undrained triaxial test:

$$c_u = \frac{q_f}{2} \tag{5.37}$$

It should be noted that if the values of pore water pressure at failure were measured in the series of tests, then in principle only one effective stress circle, shown dotted in Figure 5.26, would be obtained. The circle representing an unconfined compression test (i.e. with a cell pressure of zero) would lie to the left of the effective stress circle in Figure 5.26 because of negative pore water pressure (suction) in the specimen. The unconfined strength of a soil is due to a combination of friction and pore water suction.

If the best common tangent to the Mohr circles obtained from a series of UU tests is not horizontal, then the inference is that there has been a reduction in void ratio during each test due to the presence of air in the voids – i.e. the specimen had not been fully saturated at the outset. It should never be inferred that $\phi_u > 0$. It could also be that an initially saturated specimen has partially dried prior to testing, or has been repaired. Another reason could be the entrapment of air between the specimen and the membrane.

In the case of fissured clays the failure envelope at low values of confining pressure is curved, as shown in Figure 5.26. This is because the fissures open to some extent on sampling, resulting in a lower strength, and only when the confining pressure becomes high enough to close the fissures again does the strength become constant. Therefore, the unconfined compression test is not appropriate in the case of fissured clays. The size of a fissured clay specimen must also be large enough to represent the mass

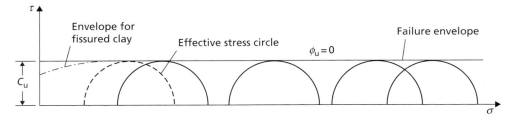

Figure 5.26 Unconsolidated–undrained triaxial test results for saturated clay.

structure, i.e. to contain fissures representative of those in-situ, otherwise the measured strength will be greater than the in-situ strength. Large specimens are also required for clays exhibiting other features of macro-fabric. Curvature of the undrained failure envelope at low values of confining pressure may also be exhibited in heavily overconsolidated clays due to relatively high negative pore water pressure at failure causing cavitation (pore air coming out of solution in the pore water).

The results of unconsolidated–undrained tests are usually presented as a plot of c_u against the corresponding depth from which the specimen originated. Considerable scatter can be expected on such a plot as the result of sampling disturbance and macro-fabric features (if present). For normally consolidated fine-grained soils, the undrained strength will generally increase linearly with increasing effective vertical stress σ'_v (i.e. with depth if the water table is at the surface). If the water table is below the surface of the clay the undrained strength between the surface and the water table will be significantly higher than that immediately below the water table, due to drying.

The consolidated–undrained (CU) triaxial test can be used to determine the undrained strength of the clay after the void ratio has been changed from the initial value by consolidation. If this is undertaken, it should be realised that clays in-situ have generally been consolidated under conditions of zero lateral strain, the effective vertical and horizontal stresses being unequal – i.e. the clay has been consolidated one-dimensionally (see Chapter 4). A stress release then occurs on sampling. In the standard CU triaxial test the specimen is consolidated again, though usually under isotropic conditions, to the value of the effective vertical stress in-situ. Isotropic consolidation in the triaxial test under a pressure equal to the in-situ effective vertical stress results in a void ratio lower than the in-situ (one-dimensional) value, and therefore an undrained strength higher than the (actual) in-situ value.

The unconsolidated–undrained test and the undrained part of the consolidated–undrained test can be carried out rapidly (provided no pore water pressure measurements are to be made), failure normally being produced within a period of 10–15 minutes. However, a slight decrease in strength can be expected if the time to failure is significantly increased, and there is evidence that this decrease is more pronounced the greater the plasticity index of the clay. Each test should be continued until the maximum value of principal stress difference has been passed or until an axial strain of 20% has been attained.

Sensitivity

Some fine-grained soils are very sensitive to remoulding, suffering considerable loss of strength due to their natural structure being damaged or destroyed. The **sensitivity** of a soil is defined as the ratio of the undrained strength in the undisturbed state to the undrained strength, at the same water content, in the remoulded state, and is denoted by S_t. Remoulding for test purposes is normally brought about by the process of kneading. The sensitivity of most clays is between 1 and 4. Clays with sensitivities between 4 and 8 are referred to as **sensitive**, and those with sensitivities between 8 and 16 as **extrasensitive**. **Quick clays** are those having sensitivities greater than 16; the sensitivities of some quick clays may be of the order of 100. Typical values of S_t are given in Section 5.9 (Figure 5.37).

Sensitivity can have important implications for geotechnical structures and soil masses. In 1978, a landslide occurred in a deposit of quick clay at Rissa in Norway. Spoil from excavation works was deposited on the gentle slope forming the shore of Lake Botnen. This soil already had a significant in-situ shear stress applied due to the ground slope, and the additional load caused the undisturbed undrained shear strength to be exceeded, resulting in a small slide. As the soil strained, it remoulded itself, breaking down its natural structure and reducing the amount of shear stress which it could carry. The surplus was transferred to the adjacent undisturbed soil, which then exceeded its undisturbed undrained shear strength. This process continued progressively until, after a period of 45 minutes, 5–6 million cubic metres of soil had flowed out into the lake, at some points reaching a flow velocity of 30–40 km/h, destroying seven farms and five homes (Figure 5.27). It is fortunate that the slide occurred beneath a sparsely populated rural area. Further information regarding the Rissa landslide can be found on the Companion Website.

Soil behaviour in shear

Figure 5.27 Damage observed following the Rissa quick clay flow-slide (photo: Norwegian Geotechnical Institute – NGI).

Example 5.3

The results shown in Figure 5.28 were obtained at failure in a series of UU triaxial tests on specimens taken from the same approximate depth within a layer of soft saturated clay. The raw data from the tests and the use of a spreadsheet to interpret the test data may be found on the Companion Website. Determine the undrained shear strength at this depth within the soil.

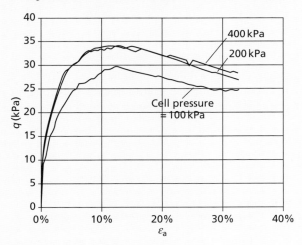

Figure 5.28 Example 5.3.

Solution

The maximum value of q ($=q_f$) is read from Figure 5.28 for each test, and Equation 5.37 used to determine c_u for each sample. As the samples are all from the same depth but tested at different confining pressures, c_u should theoretically be the same for all samples, so an average is taken as $c_u = 16.3$ kPa.

Development of a mechanical model for soil

Example 5.4

The results shown in Table 5.4 were obtained at failure in a series of triaxial tests on specimens of a saturated clay initially 38 mm in diameter by 76 mm long. Determine the values of the shear strength parameters with respect to (a) total stress and (b) effective stress.

Table 5.4 Example 5.4

Type of test		Confining pressure (kPa)	Axial load (N)	Axial deformation (mm)	Volume change (ml)
(a)	Undrained (UU)	200	222	9.83	–
		400	215	10.06	–
		600	226	10.28	–
(b)	Drained (D)	200	403	10.81	6.6
		400	848	12.26	8.2
		600	1265	14.17	9.5

Solution

The principal stress difference at failure in each test is obtained by dividing the axial load by the cross-sectional area of the specimen at failure (Table 5.5). The corrected cross-sectional area is calculated from Equation 5.16. There is, of course, no volume change during an undrained test on a saturated clay. The initial values of length, area and volume for each specimen are:

$$l_0 = 76 \text{ mm}, \quad A_0 = 1135 \text{ mm}^2, \quad V_0 = 86 \times 10^3 \text{ mm}^3$$

The Mohr circles at failure and the corresponding failure envelopes for both series of tests are shown in Figure 5.29. In both cases the failure envelope is the line nearest to a common tangent to the Mohr circles. The total stress parameters, representing the undrained strength of the clay, are

$$c_u = 85 \text{ kPa}, \quad \phi_u = 0$$

Table 5.5 Example 5.4 (contd.)

	σ_3 (kPa)	$\Delta l/l_0$	$\Delta V/V_0$	Area (mm²)	$\sigma_1 - \sigma_3$ (kPa)	σ_1 (kPa)
(a)	200	0.129	–	1304	170	370
	400	0.132	–	1309	164	564
	600	0.135	–	1312	172	772
(b)	200	0.142	0.077	1222	330	530
	400	0.161	0.095	1225	691	1091
	600	0.186	0.110	1240	1020	1620

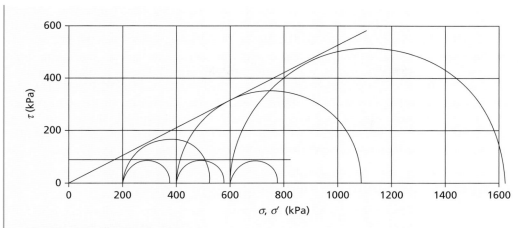

Figure 5.29 Example 5.4.

The effective stress parameters, representing the drained strength of the clay, are

$$c' = 0, \quad \phi' = 27°$$

5.7 The critical state framework

The critical state concept, originally presented by Roscoe *et al.* (1958), represents an idealisation of the observed patterns of behaviour of saturated clays under applied principal stresses. However, the critical state concept applies to all soils, both coarse- (as described in Section 5.5) and fine-grained. The concept relates the effective stresses and the corresponding specific volume ($v = 1 + e$) of a clay during shearing under drained or undrained conditions, thus unifying the characteristics of shear strength and deformation (volumetric change). It was demonstrated that a characteristic surface exists which limits all possible states of the clay, and that all effective stress paths reach or approach a line on that surface when yielding occurs at constant volume under constant effective stress. This line represents the critical states of the soil, and links p', q and v at these states. The model was originally derived based on observations of behaviour from triaxial tests, and in this section the critical state concept will be 'rediscovered' in a similar way by considering the triaxial behaviour that has previously been discussed in this chapter.

Volumetric behaviour during undrained shear

The behaviour of a saturated fine-grained soil during undrained shearing in the triaxial cell is first considered, as presented in Figure 5.23. As the pore pressure within the sample change during undrained shearing, the ESP cannot be determined only from the TSP as for the drained test (Figure 5.19(c)). However, during undrained shearing it is known that there must be no change in volume ($\Delta v = 0$), so use can be made of the volumetric behaviour shown in Figure 5.22(b). As q is independent of u, any excess pore water pressure generated during undrained shearing must result from a change in mean stress ($\Delta p'$), i.e. $u_e = \Delta p'$. The normally consolidated and overconsolidated clays from Figure 5.23 are plotted in Figure 5.30(a), assuming that the samples are all consolidated under the same cell pressure prior to undrained shearing. The normally consolidated (NC) clay has an initial state on the ICL and shows a large positive ultimate value of u_e at critical state (ultimate strength), corresponding to a reduction in p'. The lightly overconsolidated (LOC) clay has

Development of a mechanical model for soil

a higher initial volume, having swelled a little, with its initial state lying on the unload–reload line. At critical state, the excess pore water pressure is lower than the NC clay, so the change in p' is less. The heavily overconsolidated (HOC) clay starts at a higher initial volume, having swelled more than the LOC sample. The sample exhibits negative excess pore water pressure at critical state, i.e. an increase in p' due to shearing. It will be seen from Figure 5.30(a) that the points representing the critical states of the three samples all lie on a line parallel to the ICL and slightly below it. This is the critical state line (CSL) that has been described earlier, but represented in v–p' space, and represents the specific volume of the soil at critical state, i.e. when q is at its ultimate value. This line is a projection of the set of critical states of a soil in the v–p' plane, just as the critical state line on the q–p' plane ($q=Mp'$, Figure 5.11(c)) is a projection of the same set of critical states in terms of the stress parameters q and p'.

Volumetric behaviour during drained shear

Considering the behaviour of the coarse-grained soil under triaxial compression described in Figure 5.19, the volumetric behaviour is plotted in Figure 5.30(b). Samples A and B both exhibit dilation (increase in v) due to shearing; as the confining stress increases, the amount of dilation reduces. As the confining stress is increased still further (sample C), compression occurs in place of dilation. Again, the final points at critical state lie on a straight line (the CSL) parallel to the ICL; however, the CSL for this coarse-grained soil will have different numerical values of the intercept and gradient compared to the fine-grained soil considered in Figure 5.30(a), the CSL being an intrinsic soil property.

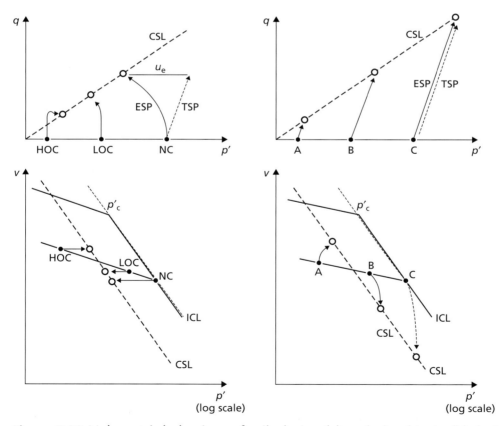

Figure 5.30 Volumetric behaviour of soils during (a) undrained tests, (b) drained tests.

Defining the critical state line

Figure 5.31 shows the CSL plotted in three dimensions, along with the projections in the q–p' and v–p' planes. The projection in the q–p' plane is a straight line with gradient M. As described previously in both this and the previous chapters, it is common to plot v–p' data with a logarithmic axis for the mean stress p'. The CSL in this plane is defined by

$$v_{cv} = \Gamma - \lambda \ln p' \tag{5.38}$$

where Γ is the value of v on the critical state line at $p' = 1\,\text{kPa}$ and is an intrinsic soil property; λ is the gradient of the ICL as defined previously. In order to use the critical state framework, it is also necessary to know the initial state of the soil, which depends on its stress history; in a triaxial test this is usually governed by previous consolidation and swelling under isotropic conditions, prior to shearing. The equation of the normal consolidation line (ICL) is

$$v_{icl} = N - \lambda \ln p' \tag{5.39}$$

where N is the value of v at $p' = 1\,\text{kPa}$. The swelling and recompression (unload–reload) relationships can be approximated to a single straight line of slope $-\kappa$, represented by the equation

$$v = v_k - \kappa \ln p' \tag{5.40}$$

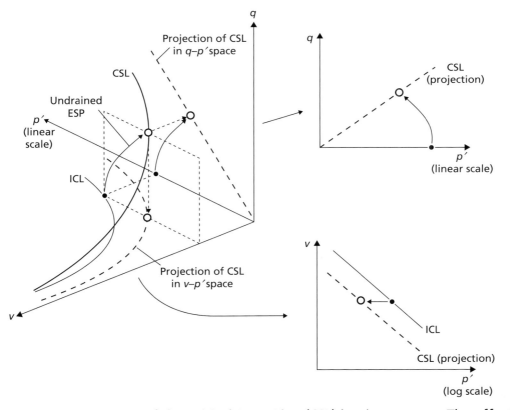

Figure 5.31 Position of the Critical State Line (CSL) in p'–q–v space. The effective stress path in an undrained triaxial test is also shown.

Development of a mechanical model for soil

where v_k is the value of v at $p'=1$ kPa, and will depend on the preconsolidation pressure (i.e. it is not a material constant). The initial volume prior to shearing (v_0) can alternatively be defined by a single equation:

$$v_0 = N - \lambda \ln p'_c + \kappa \ln\left(\frac{p'_c}{p'}\right) \tag{5.41}$$

In order to use the critical state framework to predict the strength of soil, it is therefore necessary to define five constants: N, λ, κ, Γ, M. These will allow for the determination of the initial state and the CSL in both the q–p' and v–$\ln(p')$ planes. These parameters may theoretically all be determined from a single CU test, though, typically, multiple tests will be carried out at different values of p'_c.

Example 5.5

The data shown in Figure 5.32 were obtained from a series of CU triaxial compression tests on a soft saturated clay in a modern computer-controlled stress-path cell. The samples were isotropically consolidated to confining pressures of 250, 500 and 750 kPa prior to undrained shearing. Determine the critical state parameters N, λ, Γ and M. The raw data from these tests and their interpretation using a spreadsheet is provided on the Companion Website.

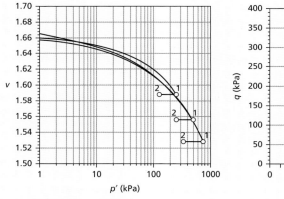
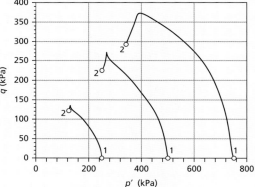

Figure 5.32 Example 5.5.

Solution

The values of p', q and v at the end of both the consolidation stage (1) and the undrained shearing stage (2) are read from the figures and summarised in Table 5.6. Assuming that all of the consolidation pressures are high enough to exceed any pre-existing preconsolidation stress, at the end of consolidation the samples should all lie on the ICL. By plotting v_1 versus p'_1, N and λ may be found by fitting a straight line as shown in Figure 5.33, the intercept being $N = 1.886$ and the gradient $\lambda = 0.054$. If the points at the end of shearing (p'_2, v_2) are plotted in a similar way, the straight line fit lies almost exactly parallel to the ICL – this is the CSL (assuming that the strain induced in the shearing stage has been sufficient to reach the critical state in each case). The intercept is $\Gamma = 1.867$ and the gradient $\lambda = 0.057$. To find M, the points (p'_2, q_2) at the end of shearing (critical state) are plotted as shown in Figure 5.33. The best-fit straight line goes through the origin with gradient $M = 0.88$.

Soil behaviour in shear

Table 5.6 Example 5.5

Confining pressure, σ_3 (kPa)	After consolidation:			After shearing:		
	p'_1	q_1	v_1	p'_2	q_2	v_2
250	250	0	1.588	125	121	1.588
500	500	0	1.556	250	225	1.556
750	750	0	1.528	340	292	1.528

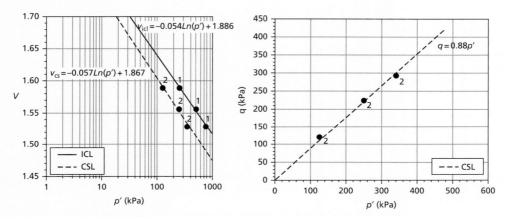

Figure 5.33 Example 5.5 – determination of critical state parameters from line-fitting.

Example 5.6

Estimate the values of principal stress difference and void ratio at failure in undrained and drained triaxial tests on specimens of the clay described in Example 5.5, isotropically consolidated under a confining pressure of 300 kPa. What would be the expected value of ϕ'_{cv}?

Solution

After normal consolidation to $p'_c = 300$ kPa the sample will be on the ICL with specific volume (v_0) given by

$$v_0 = N - \lambda \ln p'_c = 1.886 - 0.054 \ln 300 = 1.578$$

In an undrained test the volume change is zero, and therefore the specific volume at critical state (v_{cs}) will also be 1.58, i.e. the corresponding void ratio will be $e_{cs(U)} = 0.58$.
Assuming failure to take place on the critical state line,

$$q'_f = Mp'_f$$

and the value of p'_f can be obtained from Equation 5.38. Therefore

$$q'_{f(U)} = M \exp\left(\frac{\Gamma - v_0}{\lambda}\right)$$

$$= 0.88 \exp\left(\frac{1.867 - 1.578}{0.054}\right)$$

$$= 186 \text{ kPa}$$

For a drained test the slope of the stress path on a q–p' plot is 3, i.e.

$$q'_f = 3(p'_f - p'_c) = 3\left(\frac{q'_f}{M} - p'_c\right)$$

Therefore,

$$q'_{f(D)} = \frac{3Mp'_c}{3 - M}$$

$$= \frac{3 \times 0.88 \times 300}{3 - 0.88}$$

$$= 374 \text{ kPa}$$

Then

$$p'_f = \frac{q_f}{M} = \frac{374}{0.88} = 425 \text{ kPa}$$

$$\therefore v_{cs} = \Gamma - \lambda \ln p'_f = 1.867 - 0.054 \ln 425 = 1.540$$

Therefore, $e_{cs(D)} = 0.54$ and

$$\phi'_{cv} = \sin^{-1}\left(\frac{3M}{6 + M}\right)$$

$$= \sin^{-1}\left(\frac{3 \times 0.88}{6.88}\right)$$

$$= 23°$$

5.8 Residual strength

In the drained triaxial test, most clays would eventually show a decrease in shear strength with increasing strain after the peak strength has been reached. However, in the triaxial test there is a limit to the strain which can be applied to the specimen. The most satisfactory method of investigating the shear strength of clays at large strains is by means of the ring shear apparatus (Bishop et al., 1971; Bromhead, 1979), an annular direct shear apparatus. The annular specimen (Figure 5.34(a)) is sheared, under a given normal stress, on a horizontal plane by the rotation of one half of the apparatus relative to the other; there is no restriction to the magnitude of shear displacement between the two halves of the specimen. The rate of rotation must be slow enough to ensure that the specimen remains in a drained condition. Shear stress, which is calculated from the applied torque, is plotted against shear displacement as shown in Figure 5.34(b).

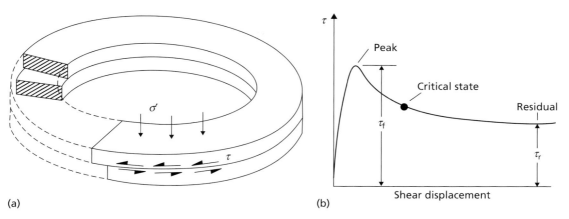

Figure 5.34 (a) Ring shear test, and (b) residual strength.

The shear strength falls below the peak value, and the clay in a narrow zone adjacent to the failure plane will soften and reach the critical state. However, because of non-uniform strain in the specimen, the exact point on the curve corresponding to the critical state is uncertain. With continuing shear displacement the shear strength continues to decrease, below the critical state value, and eventually reaches a residual value at a relatively large displacement. If the soil contains a relatively high proportion of plate-like particles, a reorientation of these particles parallel to the failure plane will occur (in the narrow zone adjacent to the failure plane) as the strength decreases towards the residual value. However, reorientation may not occur if the plate-like particles exhibit high interparticle friction. In this case, and in the case of soils containing a relatively high proportion of bulky particles, rolling and translation of particles takes place as the residual strength is approached. It should be appreciated that the critical state concept envisages continuous deformation of the specimen as a whole, whereas in the residual condition there is preferred orientation or translation of particles in a narrow shear zone. The original soil structure in this narrow shear zone is completely destroyed as a result of particle reorientation. A remoulded specimen can therefore be used in the ring shear apparatus if only the residual strength (and not the peak strength) is required.

The results from a series of tests, under a range of values of normal stress, enable the failure envelope for both peak and residual strength to be obtained, the residual strength parameters in terms of effective stress being denoted c'_r and ϕ'_r. Residual strength data from ring shear testing for a large range of soils have been published (e.g. Lupini et al., 1981; Mesri and Cepeda-Diaz, 1986; Tiwari and Marui, 2005), which indicate that the value of c'_r can be taken to be zero. Thus, the residual strength can be expressed as

$$\tau_r = \sigma'_f \tan \phi'_r \tag{5.42}$$

Typical values of ϕ'_r are given in Section 5.9 (Figure 5.39).

5.9 Estimating strength parameters from index tests

In order to obtain reliable values of soil strength parameters from triaxial and shear box testing, undisturbed samples of soil are required. Methods of sampling are discussed in Section 6.3. No sample will be fully undisturbed, and obtaining high quality samples is almost always difficult and often expensive. As a result, the principle strength properties of soils (ϕ'_{cv}, c_u, S_t, ϕ'_r) are here correlated to the basic index

Development of a mechanical model for soil

properties described in Chapter 1 (I_P and I_L for fine-grained soils and e_{max} and e_{min} for coarse-grained soils). These simple tests can be undertaken using disturbed samples. The spoil generated from drilling a borehole provides essentially a continuous disturbed sample for testing so a large amount of information describing the strength of the ground can be gathered without performing detailed laboratory or field testing. The use of correlations can therefore be very useful during the preliminary stages of a Ground Investigation (GI). Ground investigation can also be cheaper/more efficient using mainly disturbed samples, supplemented by fewer undisturbed samples to verify the properties.

It should be noted that the use of any of the approximate correlations presented here should be regarded as estimates only; they should be used to support, and never replace, laboratory tests on undisturbed samples, particularly when a given strength property is critical to the design or analysis of a given geotechnical construction. They are most useful for checking the results of laboratory tests, for increasing the amount of data from which the determination of strength properties is made and for estimating parameters before the results of laboratory tests are known (e.g. for feasibility studies and when planning a ground investigation).

Critical state angle of shearing resistance (ϕ'_{cv})

Figure 5.35(a) shows data of ϕ'_{cv} for 65 coarse-grained soils determined from shearbox and triaxial testing collected by Bolton (1986), Miura *et al.* (1998) and Hanna (2001). The data are correlated against $e_{max} - e_{min}$. This correlation is appropriate as both e_{max} and e_{min} are independent of the soil state/density (defined by void ratio e), just as ϕ'_{cv} is an intrinsic property and independent of e. A correlation line, representing a best fit to the data is also shown in Figure 5.35(a), which suggests that ϕ'_{cv} increases as the potential for volumetric change ($e_{max} - e_{min}$) increases. The large scatter in the data may be attributed to the influence of particle shape (angularity), which is only partially captured by $e_{max} - e_{min}$, and grain roughness (which is a function of the parent material(s) from which the soil was produced).

Figure 5.35(b) shows data of ϕ'_{cv} for 32 undisturbed and 32 remoulded fine-grained soils from triaxial testing collected by Kenney (1959), Parry (1960) and Zhu and Yin (2000). The data are correlated against plasticity index (I_P) which is also independent of the current soil state (defined by the current water content w). A correlation line, representing a best fit to the undisturbed data is also shown in Figure 5.35(b), which suggests that ϕ'_{cv} reduces as the plasticity index increases. This may be attributed to the increase in clay fraction (i.e. fine platy particles) as I_P increases, with these particles tending to

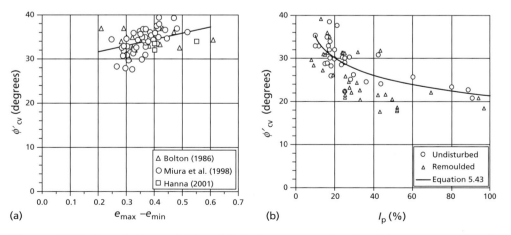

Figure 5.35 Correlation of ϕ'_{cv} with index properties for (a) coarse-grained, and (b) fine-grained soils.

have lower friction than the larger particles within the soil matrix The scatter in the data may be attributed to the minerality of these clay-sized particles – as outlined in Chapter 1, the main clay minerals (kaolinite, smectite, illite, montmorillonite) have very different particle shapes, specific surface and frictional properties. The equation of the line is given by:

$$\phi'_{cv} = 57(I_P)^{-0.21} \tag{5.43}$$

Undrained shear strength and sensitivity (c_u, S_t)

As described in Section 5.6, the undrained strength of a fine-grained soil is not an intrinsic material property, but is dependent on the state of the soil (as defined by w) and the stress level. The undrained shear strength should therefore be correlated to a state-dependent index parameter, namely liquidity index I_L. Figure 5.36 shows data of c_u plotted against I_L for 62 remoulded fine-grained soils collected by Skempton and Northey (1953), Parry (1960), Leroueil et al. (1983) and Jardine et al. (1984). At the liquid limit, $c_{ur} \approx 1.7$ kPa. The undrained shear strength at the plastic limit is defined as 100 times that at the liquid limit, which suggests that the relationship between c_u (in kPa) and I_L should be of the form

$$c_{ur} \approx 1.7 \times 100^{(1-I_L)} \tag{5.44}$$

It should be noted that the undrained strength in Equation 5.44 is that appropriate to the soil in a remoulded (fully disturbed) state. Sensitive clays may therefore exhibit much higher apparent undrained strengths in their undisturbed condition than would be predicted by Equation 5.44.

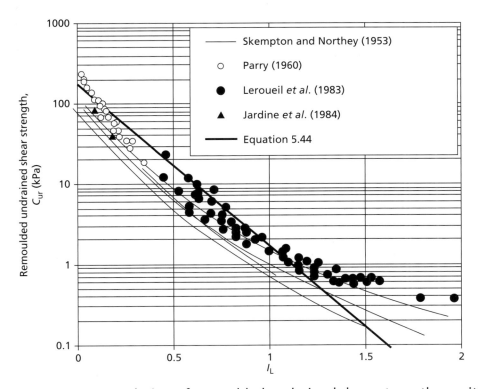

Figure 5.36 Correlation of remoulded undrained shear strength c_{ur} with index properties.

Figure 5.37 shows data for 49 sensitive clays collated from Skempton and Northey (1953), Bjerrum (1954) and Bjerrum and Simons (1960), in which sensitivity is also correlated to I_L. The data suggest a linear correlation described by:

$$S_t \approx 100^{(0.43 I_L)} \tag{5.45}$$

A correlation for the undisturbed (in-situ) undrained shear strength may also be estimated from the definition of sensitivity as the ratio of undisturbed to remoulded strength (i.e. by combining Equations 5.44 and 5.45):

$$c_u = S_t c_{ur} \approx 1.7 \times 100^{(1-0.57 I_L)} \tag{5.46}$$

To demonstrate how useful Equations 5.44–5.46 can be, Figure 5.38 shows examples of two real soil profiles. In Figure 5.38(a), the soil is a heavily overconsolidated insensitive clay (Gault clay, near Cambridge). Applying Equation 5.45, it can be seen that $S_t \approx 1.0$ everywhere, such that Equations 5.44 and 5.46 give almost identical values (solid markers). This is compared with the results of UU triaxial tests which measure the undisturbed undrained shear strength. It will be seen that although there is significant scatter (in both datasets), both the magnitude and variation with depth is captured well using the approximate correlations. Figure 5.38(b) shows data for a normally consolidated clay with $S_t \approx 5$ (Bothkennar clay, near Edinburgh). For sensitive clay such as this, the undisturbed strength is likely to be much higher than the remoulded strength, and it will be seen that the predictions of Equations 5.44 and 5.46 are different. To validate these predictions, Field Vane Testing (FVT), which is further described in Section 7.3, was used as this can obtain both the undisturbed and remoulded strengths (and therefore also the sensitivity) in soft fine-grained soils. It can be seen from Figure 5.38(b) that the correlations presented here do reasonably predict the values measured by the more reliable in-situ testing.

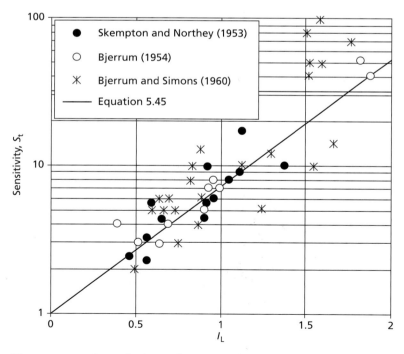

Figure 5.37 Correlation of sensitivity S_t with index properties.

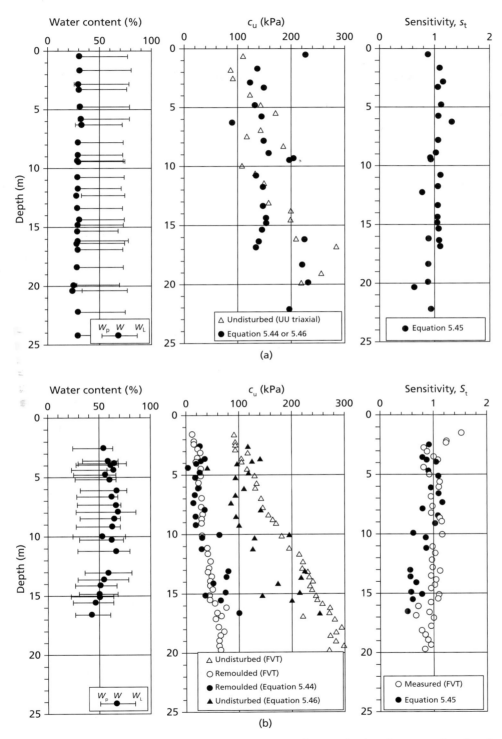

Figure 5.38 Use of correlations to estimate the undrained strength of cohesive soils: (a) Gault clay, (b) Bothkennar clay.

Development of a mechanical model for soil

Residual friction angle (ϕ'_r)

As noted in Section 5.8, the residual strength of a soil reduces with an increasing proportion of platy particles (e.g. clays) and is therefore correlated with clay fraction and plasticity index. Figure 5.39 shows data of ϕ'_r determined from ring shear tests plotted against I_P for 89 clays and tills and 23 shales, collected by Lupini *et al.* (1981), Mesri and Cepeda-Diaz (1986) and Tiwari and Marui (2005). As observed for ϕ'_{cv} (Figure 5.35(b)), there is considerable scatter around the correlation line which may be attributed to the different characteristics of the clay minerals present within these soils. The data appears to follow a power law where:

$$\phi'_r = 93 (I_P)^{-0.56} \qquad (5.47)$$

with ϕ'_r given in degrees.

In most conventional geotechnical applications, the strains are normally small enough that residual conditions will never be reached. Residual strength is particularly important in the study of slope stability, however, where historical slip may have been of sufficient magnitude to align the particles along the slip plane (slope stability is discussed further in Chapter 12). Under these conditions, a slip which is reactivated (e.g. due to reduction of effective stress on the shear plane due to an increase in the water table (rain) or seepage) will be at residual conditions and ϕ'_r is the most appropriate measure of strength to use in analysis. As an example of the use of this correlation, Figure 5.39 also shows a series of data points for a range of slopes around the UK with historical slips, as reported by Skempton (1985). It will be seen that using I_P alone, the predicted values of ϕ'_r are within 2° of the values measured by large direct shear tests on the slip surfaces in the field.

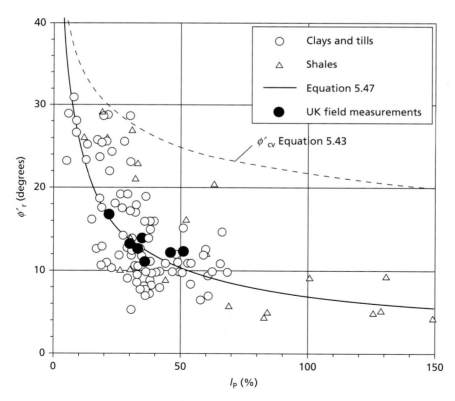

Figure 5.39 Correlation of ϕ'_r with index properties for fine-grained soils, showing application to UK slope case studies.

Summary

1. Knowledge of the strength and stiffness of soil (its constitutive behaviour) is fundamental for assessing the stability and performance of geotechnical constructions. It relates ground stresses (which are in equilibrium with the applied loads) to ground strains (giving compatible deformations). The constitutive behaviour of soil is highly non-linear and dependent on the level of confining stress, but for most practical problems, it can be modelled/idealised using isotropic linear elasticity coupled with Mohr–Coulomb (stress dependent) plasticity.

2. The strength and stiffness of soil may be directly measured in the laboratory by direct shear tests, triaxial tests or residual (ring) shear tests (amongst others). The method of operation, set-up and strengths/weaknesses of each test have been described. These tests may be used to derive strength properties for the Mohr–Coulomb model (c and ϕ) and the shear stiffness of the soil (G). Digital examplars provided on the Companion Website have demonstrated how digital data from modern computerised test apparatus may be efficiently processed.

3. For most common rates of loading or geotechnical processes, coarse-grained soils will behave in a drained way. The peak shear strength of such soils is governed by dilatancy (volumetric change) which is density (state) dependent. This behaviour may be modeled by secant peak friction angles or using a linearised Mohr–Coulomb model ($\tau = c'\tau + \sigma' \tan \phi'$). If such soils are sheared to large enough strains, volume change will cease and the soil will reach a critical state (ultimate strength). Soil properties may typically be obtained from drained direct shear or drained triaxial tests. Fine-grained soils will behave in a similar way if they are allowed to drain (i.e. for slow processes or long-term conditions). If loaded rapidly, they will respond in an undrained way, and the strength is defined in terms of total, rather than effective stresses using the Mohr–Coulomb model ($\tau = c_u$). Drained and undrained strength properties and the shear modulus G (which is independent of drainage conditions) are typically quantified using triaxial tests (CD, CU, UU) in such soils. At very large strains, the strength of fine-grained soils may reduce below the critical state value to a residual value defined by a friction angle ϕ'_r, which may be measured using a ring shear apparatus.

4. With increasing strain with any drainage conditions, all saturated soils will move towards a critical state where they achieve their ultimate shear strength. The Critical State Line defines the critical states for any initial state of the soil, and is therefore an intrinsic soil property. By linking volume change to shear strength, the critical state concept shows that drained and undrained strength both represent the effective stress path arriving at the critical state line (for the two extreme amounts of drainage).

5. Simple index tests (described in Chapter 1) may be used with empirical correlations to estimate values of a range of strength properties (ϕ'_{cv}, c_u and ϕ'_r). Such data can be useful when high quality laboratory test data are unavailable and for providing additional data to support the results of such tests. While useful, empirical correlations should never be used to replace a comprehensive programme of laboratory testing.

Development of a mechanical model for soil

Problems

5.1 What is the shear strength in terms of effective stress on a plane within a saturated soil mass at a point where the total normal stress is 295 kPa and the pore water pressure 120 kPa? The strength parameters of the soil for the appropriate stress range are $\phi'=30°$ and $c'=12$ kPa.

5.2 Three separate direct shear tests are carried out on identical samples of dry sand. The shearbox is 60×60 mm in plan. The hanger loads, peak forces and ultimate forces measured during the tests are summarised below:

TABLE K

Test	Hanger load (N)	Peak shear force (N)	Ultimate shear force (N)
1	180	162	108
2	360	297	216
3	540	423	324

Determine the Mohr coulomb parameters (ϕ', c') for modeling the peak strength of the soil, the critical state angle of shearing resistance, and the dilation angles in the three tests.

5.3 A series of drained triaxial tests with zero back pressure were carried out on specimens of a sand prepared at the same porosity, and the following results were obtained at failure.

TABLE L

Confining pressure (kPa)	100	200	400	800
Principal stress difference (kPa)	452	908	1810	3624

Determine the value of the angle of shearing resistance ϕ'.

5.4 In a series of unconsolidated–undrained (UU) triaxial tests on specimens of a fully saturated clay, the following results were obtained at failure. Determine the values of the shear strength parameters c_u and ϕ_u.

TABLE M

Confining pressure (kPa)	200	400	600
Principal stress difference (kPa)	222	218	220

5.5 The results below were obtained at failure in a series of consolidated–undrained (CU) triaxial tests, with pore water pressure measurement, on specimens of a fully saturated clay. Determine the values of the shear strength parameters c' and ϕ'. If a specimen of the same soil were consolidated under an all-round pressure of 250 kPa and the principal stress difference applied with the all-round pressure changed to 350 kPa, what would be the expected value of principal stress difference at failure?

TABLE N

σ_3 (kPa)	150	300	450	600
$\sigma_1 - \sigma_3$ (kPa)	103	202	305	410
u (kPa)	82	169	252	331

5.6 A consolidated–undrained (CU) triaxial test on a specimen of saturated clay was carried out under an all-round pressure of 600 kPa. Consolidation took place against a back pressure of 200 kPa. The following results were recorded during the test:

TABLE O

$\sigma_1 - \sigma_3$ (kPa)	0	80	158	214	279	319
u (kPa)	200	229	277	318	388	433

Draw the stress paths (total and effective). If the clay has reached its critical state at the end of the test, estimate the critical state friction angle.

5.7 The following results were obtained at failure in a series of consolidated-drained (CD) triaxial tests on fully saturated clay specimens originally 38 mm in diameter by 76 mm long, with a back pressure of zero. Determine the secant value of ϕ' for each test and the values of tangent parameters c' and ϕ' for the stress range 300–500 kPa.

TABLE P

All-round pressure (kPa)	200	400	600
Axial compression (mm)	7.22	8.36	9.41
Axial load (N)	565	1015	1321
Volume change (ml)	5.25	7.40	9.30

5.8 In a triaxial test, a soil specimen is allowed to consolidate fully under an all-round pressure of 200 kPa. Under undrained conditions the all-round pressure is increased to 350 kPa, the pore water pressure then being measured as 144 kPa. Axial load is then applied under undrained conditions until failure takes place, the following results being obtained.

TABLE Q

Axial strain (%)	0	2	4	6	8	10
Principal stress difference (kPa)	0	201	252	275	282	283
Pore water pressure (kPa)	144	211	228	222	212	209

Determine the value of the pore pressure coefficient B and determine whether the test can be considered as saturated. Plot the stress–strain (q–ε_s) curve for the test, and hence determine the shear modulus (stiffness) of the soil at each stage of loading. Draw also the stress paths (total and effective) for the test and estimate the critical state friction angle.

References

ASTM D2850 (2007) *Standard Test Method for Unconsolidated-Undrained Triaxial Compression Test on Cohesive Soils*, American Society for Testing and Materials, West Conshohocken, PA.

ASTM D3080 (2004) *Standard Test Method for Direct Shear Test of Soils Under Consolidated Drained Conditions*, American Society for Testing and Materials, West Conshohocken, PA.

ASTM D4767 (2011) *Standard Test Method for Consolidated Undrained Triaxial Compression Test for Cohesive Soils*, American Society for Testing and Materials, West Conshohocken, PA.

ASTM D7181 (2011) *New Test Method for Consolidated Drained Triaxial Compression Test for Soils*, American Society for Testing and Materials, West Conshohocken, PA.

Atkinson, J.H. (2000) Non-linear soil stiffness in routine design, *Géotechnique*, **50**(5), 487–508.

Atkinson, J.H. and Bransby, P.L. (1978) *The Mechanics of Soils: An Introduction to Critical State Soil Mechanics*, McGraw-Hill Book Company (UK) Ltd, Maidenhead, Berkshire.

Bishop, A.W., Green, G.E., Garga, V.K., Andresen, A. and Brown, J.D. (1971) A new ring shear apparatus and its application to the measurement of residual strength, *Géotechnique*, **21**(4), 273–328.

Bjerrum, L. (1954) Geotechnical properties of Norwegian marine clays, *Géotechnique*, **4**(2), 49–69.

Bjerrum, L. and Simons, N.E. (1960) Comparison of shear strength characteristics of normally consolidated clays, in *Proceedings of Research Conference on Shear Strength of Cohesive Soils, Boulder, Colorado*, pp. 711–726.

Bolton, M.D. (1986) The strength and dilatancy of sands, *Géotechnique*, **36**(1), 65–78.

Bolton, M.D. (1991) *A Guide to Soil Mechanics*, Macmillan Press, London.

British Standard 1377 (1990) *Methods of Test for Soils for Civil Engineering Purposes*, British Standards Institution, London.

Bromhead, E.N. (1979) A simple ring shear apparatus, *Ground Engineering*, **12**(5), 40–44.

Calladine, C.R. (2000) *Plasticity for Engineers: Theory and Applications*, Horwood Publishing, Chichester, W. Sussex.

CEN ISO/TS 17892 (2004) *Geotechnical Investigation and Testing – Laboratory Testing of Soil*, International Organisation for Standardisation, Geneva.

Duncan, J.M. and Seed, H.B. (1966) Strength variation along failure surfaces in clay, *Journal of the Soil Mechanics & Foundations Division, ASCE*, **92**, 81–104.

Hanna, A. (2001) Determination of plane-strain shear strength of sand from the results of triaxial tests, *Canadian Geotechnical Journal*, **38**, 1231–1240.

Hill, R. (1950) *Mathematical Theory of Plasticity*, Oxford University Press, New York, NY.

Jardine, R.J., Symes, M.J. and Burland J.B. (1984) The measurement of soil stiffness in the triaxial apparatus, *Géotechnique* **34**(3), 323–340.

Kenney, T.C. (1959) Discussion of the geotechnical properties of glacial clays, *Journal of the Soil Mechanics and Foundations Division, ASCE*, **85**, 67–79.

Leroueil, S. Tavenas, F. and Le Bihan, J.P. (1983) Propriétés caractéristiques des argiles de l'est du Canada, *Canadian Geotechnical Journal*, **20**, 681–705 (in French).

Lupini, J.F., Skinner, A.E. and Vaughan, P.R. (1981) The drained residual strength of cohesive soils, *Géotechnique*, **31**(2), 181–213.

Mesri, G., and Cepeda-Diaz, A.F. (1986) Residual shear strength of clays and shales, *Géotechnique*, **36**(2), 269–274.

Miura, K., Maeda, K., Furukawa, M. and Toki, S. (1998) Mechanical characteristics of sands with different primary properties, *Soils and Foundations*, **38**(4), 159–172.

Parry, R.H.G. (1960) Triaxial compression and extension tests on remoulded saturated clay, *Géotechnique*, **10**(4), 166–180.

Roscoe, K.H., Schofield, A.N. and Wroth, C.P. (1958) On the yielding of soils, *Géotechnique*, **8**(1), 22–53.

Skempton, A.W. (1954) The pore pressure coefficients A and B, *Géotechnique*, **4**(4), 143–147.

Skempton, A.W. (1985) Residual strength of clays in landslides, folded strata and the laboratory, *Géotechnique*, **35**(1), 1–18.

Skempton, A.W. and Northey, R.D. (1953) The sensitivity of clays, *Géotechnique*, **3**(1), 30–53.

Tiwari, B., and Marui, H. (2005) A new method for the correlation of residual shear strength of the soil with mineralogical composition, *Journal of Geotechnical and Geoenvironmental Engineering*, ASCE, **131**(9), 1139–1150.

Zhu, J. and Yin, J. (2000) Strain-rate-dependent stress–strain behaviour of overconsolidated Hong Kong marine clay, *Canadian Geotechnical Journal*, **37**(6), 1272–1282.

Further reading

Atkinson, J.H. (2000) Non-linear soil stiffness in routine design, *Géotechnique*, **50**(5), 487–508.
This paper provides a state-of-the-art review of soil stiffness (including non-linear behaviour) and the selection of an appropriate shear modulus in geotechnical design.

Head, K.H. (1986) *Manual of Soil Laboratory Testing*, three volumes, Pentech, London.
This book contains comprehensive descriptions of the set-up of the major laboratory tests for soil testing and practical guidance on test procedures, to accompany the various test standards.

Muir Wood, D. (1991) *Soil Behaviour and Critical State Soil Mechanics*, Cambridge University Press, Cambridge.
This book covers the behaviour of soil entirely within the critical state framework, building substantially on the material presented herein and also forming a useful reference volume.

For further student and instructor resources for this chapter, please visit the Companion Website at www.routledge.com/cw/craig

Chapter 6

Ground investigation

> **Learning outcomes**
>
> After working through the material in this chapter, you should be able to:
>
> 1 Specify a basic site investigation strategy to identify soil deposits and determine the depth, thickness and areal extent of such deposits within the ground;
> 2 Understand the applications and limitations of a wide range of methods available for profiling the ground, and interpret their findings (Sections 6.2, 6.5–6.7);
> 3 Appreciate the effects of sampling on the quality of soil samples taken for laboratory testing, and the implications of these effects for the interpretation of such test data (Section 6.3).

6.1 Introduction

An adequate ground investigation is an essential preliminary to the execution of a civil engineering project. Sufficient information must be obtained to enable a safe and economic design to be made and to avoid any difficulties during construction. The principal objects of the investigation are: (1) to determine the sequence, thicknesses and lateral extent of the soil strata and, where appropriate, the level of bedrock; (2) to obtain representative samples of the soils (and rock) for identification and classification, and, if necessary, for use in laboratory tests to determine relevant soil parameters; (3) to identify the groundwater conditions. The investigation may also include the performance of **in-situ** tests to assess appropriate soil characteristics. In-situ testing will be discussed in Chapter 7. Additional considerations arise if it is suspected that the ground may be contaminated. The results of a ground investigation should provide adequate information, for example, to enable the most suitable type of foundation for a proposed structure to be selected, and to indicate if special problems are likely to arise during construction.

Before any ground investigation work is started on-site, a **desk study** should be conducted. This involves collating available relevant information about the site to assist in planning the subsequent fieldwork. A study of geological maps and memoirs, if available, should give an indication of the probable soil conditions of the site in question. If the site is extensive and if no existing information is available, the use of aerial photographs, topographical maps or satellite imagery can be useful in identifying existing features of geological significance. Existing borehole or other site investigation data may have been collected for previous uses of the site; in the UK, for example, the National Geological Records Centre may be a useful source of such information. Links to online sources of desk study materials are provided on the Companion Website. Particular care must be taken for sites that have been used previously where additional ground hazards may exist, including buried foundations, services, mine workings etc. Such previous uses may be obtained by examining historical mapping data.

Before the start of fieldwork an inspection of the site and the surrounding area should be made on foot. River banks, existing excavations, quarries and road or railway cuttings, for example, can yield valuable information regarding the nature of the strata and groundwater conditions; existing structures should be examined for signs of settlement damage. Previous experience of conditions in the area may have been obtained by adjacent owners or local authorities. Consideration of all of the information obtained in the desk study enables the most suitable type of investigation to be selected, and allows the fieldwork to be targeted to best characterise the site. This will ultimately result in a more effective site investigation.

The actual investigation procedure depends on the nature of the strata and the type of project, but will normally involve the excavation of boreholes or trial pits. The number and location of **boreholes**, **trial pits** and **CPT soundings** (Section 6.5) should be planned to enable the basic geological structure of the site to be determined and significant irregularities in the subsurface conditions to be detected. Approximate guidance on the spacing of these **investigation points** is given in Table 6.1. The greater the degree of variability of the ground conditions, the greater the number of boreholes or pits required. The locations should be offset from areas on which it is known that foundations are to be sited. A preliminary investigation on a modest scale may be carried out to obtain the general characteristics of the strata, followed by a more extensive and carefully planned investigation including sampling and in-situ testing.

It is essential that the investigation is taken to an adequate depth. This depth depends on the type and size of the project, but must include all strata liable to be significantly affected by the structure and its construction. The investigation must extend below all strata which might have inadequate shear strength for the support of foundations, or which would give rise to significant settlement. If the use of deep foundations (Chapter 9) is anticipated the investigation will thus have to extend to a considerable depth below the surface. If rock is encountered, it should be penetrated by at least 3 m in more than one location to confirm that bedrock (and not a large boulder) has been reached, unless geological knowledge indicates otherwise. The investigation may have to be taken to depths greater than normal in areas of old mine workings or other underground cavities. Boreholes and trial pits should be backfilled after use. Backfilling with compacted soil may be adequate in many cases, but if the groundwater conditions are altered by a borehole and the resultant flow could produce adverse effects then it is necessary to use a cement-based grout to seal the hole.

Table 6.1 Guidance on spacing of ground investigation points (Eurocode 7, Part 2: 2007)

Type of construction	Spacing of investigation points
High-rise and industrial structures	Grid pattern, at spacing of 15–40 m
Large-area structures	Grid pattern, spacing ≤60 m
Linear structures (e.g. roads, railways, retaining walls etc.)	Along route, at spacing of 20–200 m
Special structures (e.g. bridges, chimneys/stacks, machine foundations)	2–6 investigation points per foundation
Dams and weirs	25– to 75-m spacing, along relevant sections

The cost of an investigation depends on the location and extent of the site, the nature of the strata and the type of project under consideration. In general, the larger the project, and the less critical the ground conditions are to the design and construction of the project, the lower the cost of the ground investigation as a percentage of the total cost. The cost of a ground investigation is generally within the range 0.1–2% of the project cost. To reduce the scope of an investigation for financial reasons alone is never justified. Chapman (2008) provides an interesting example of this, considering the development of a six-storey office building in central London with a construction cost of £30 million. It is demonstrated that in halving the cost of site investigation (from £45 000 to £22 500), poorer foundation design due to less available ground information carries a cost of around £210 000; If project completion is delayed by more than one month due to unforeseen ground conditions (this happens in approximately 20% of projects), the associated cost could exceed £800 000 in lost building rents and re-design. Clients' costs are therefore very vulnerable to the unexpected. The aim of a good ground investigation should be to ensure that unforeseen ground conditions do not increase this vulnerability.

6.2 Methods of intrusive investigation

Trial pits

The excavation of trial pits is a simple and reliable method of investigation, but is limited to a maximum depth of 4–5 m. The soil is generally removed by means of the back-shovel of a mechanical excavator. Before any person enters the pit, the sides must always be supported unless they are sloped at a safe angle or are stepped; the excavated soil should be placed at least 1 m from the edge of the pit (see Section 12.2 for a discussion of stability of trial pits and trenches). If the pit is to extend below the water table, some form of dewatering is necessary in the more permeable soils, resulting in increased costs. The use of trial pits enables the in-situ soil conditions to be examined visually, and thus the boundaries between strata and the nature of any macro-fabric can be accurately determined. It is relatively easy to obtain disturbed or undisturbed soil samples: in fine-grained soils block samples can be cut by hand from the sides or bottom of the pit, and tube samples can be obtained below the bottom of the pit. Trial pits are suitable for investigations in all types of soil, including those containing cobbles or boulders.

Shafts and headings

Deep pits or shafts are usually advanced by hand excavation, the sides being supported by timbering. **Headings** or adits are excavated laterally from the bottom of shafts or from the surface into hillsides, both the sides and roof being supported. It is unlikely that shafts or headings would be excavated below the water table. Shafts and headings are very costly, and their use would be justified only in investigations for very large structures, such as dams, if the ground conditions could not be ascertained adequately by other means.

Percussion boring

The boring rig (Figure 6.1) consists of a derrick, a power unit, and a winch carrying a light steel cable which passes through a pulley on top of the derrick. Most rigs are fitted with road wheels, and when folded down can be towed behind a vehicle. Various boring tools can be attached to the cable. The borehole is advanced by the percussive action of the tool which is alternately raised and dropped (usually over a distance of 1–2 m) by means of the winch unit. The two most widely used tools are the **shell** (Figure 6.1(b)) and the **clay cutter** (Figure 6.1(c)). If necessary, a heavy steel element called a sinker bar can be fitted immediately above the tool to increase the impact energy.

Development of a mechanical model for soil

The shell, which is used in sands and other coarse-grained soils, is a heavy steel tube fitted with a flap or clack valve at the lower end. Below the water table, the percussive action of the shell loosens the soil and produces a slurry in the borehole. Above the water table, a slurry is produced by introducing water into the borehole. The slurry passes through the clack valve during the downward movement of the shell and is retained by the valve during the upward movement. When full, the shell is raised to the surface to be emptied. In cohesionless soils (e.g. sands and gravels), the borehole must be cased to prevent collapse. The casing, which consists of lengths of steel tubing screwed together, is lowered into the borehole and will normally slide down under its own weight; however, if necessary, installation of the casing can be aided by driving. On completion of the investigation the casing is recovered by means of the winch or by the use of jacks: excessive driving during installation may make recovery of the casing difficult.

The clay cutter, which is used in fine-grained soils (e.g. clays, silts and tills), is an open steel tube with a cutting shoe and a retaining ring at the lower end; the tool is used in a dry borehole. The percussive action of the tool cuts a plug of soil which eventually fractures near its base due to the presence of the retaining ring. The ring also ensures that the soil is retained inside the cutter when it is raised to the surface to be emptied.

Small boulders, cobbles and hard strata can be broken up by means of a **chisel**, aided by the additional weight of a sinker bar if necessary.

Borehole diameters can range from 150 to 300 mm. The maximum borehole depth is generally between 50 and 60 m. Percussion boring can be employed in most types of soil, including those containing cobbles and boulders. However, there is generally some disturbance of the soil below the bottom of the borehole, from which samples are taken, and it is extremely difficult to detect thin soil layers and minor geological features with this method. The rig is extremely versatile, and can normally be fitted with a hydraulic power unit and attachments for mechanical augering, rotary core drilling and in-situ testing (Chapter 7).

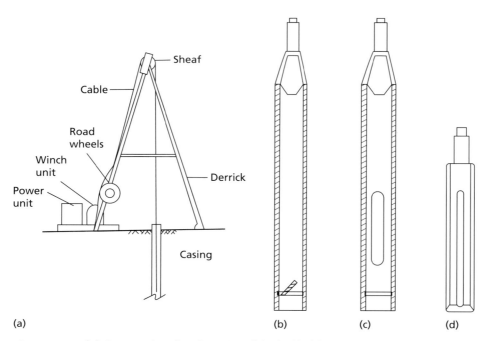

Figure 6.1 (a) Percussion boring rig, (b) shell, (c) clay cutter, and (d) chisel.

Ground investigation

Mechanical augers

Power-operated augers are generally mounted on vehicles or in the form of attachments to the derrick used for percussion boring. The power required to rotate the auger depends on the type and size of the auger itself, and the type of soil to be penetrated. Downward pressure on the auger can be applied hydraulically, mechanically or by dead weight. The types of tool generally used are the **flight auger** and the **bucket auger**. The diameter of a flight auger is usually between 75 and 300 mm, although diameters as large as 1 m are available; the diameter of a bucket auger can range from 300 mm to 2 m. However, the larger sizes are used principally for excavating shafts for bored piles. Augers are used mainly in soils in which the borehole requires no support and remains dry, i.e. mainly in stiffer, overconsolidated clays. The use of casing would be inconvenient because of the necessity of removing the auger before driving the casing; however, it is possible to use bentonite slurry to support the sides of unstable holes (Section 12.2). The presence of cobbles or boulders creates difficulties with the smaller-sized augers.

Short-flight augers (Figure 6.2(a)) consist of a helix of limited length, with cutters below the helix. The auger is attached to a steel shaft, known as a kelly bar, which passes through the rotary head of the rig. The auger is advanced until it is full of soil, then it is raised to the surface where the soil is ejected by rotating the auger in the reverse direction. Clearly, the shorter the helix, the more often the auger must be raised and lowered for a given borehole depth. The depth of the hole is limited by the length of the kelly bar.

Continuous-flight augers (Figure 6.2(b)) consist of rods with a helix covering the entire length. The soil rises to the surface along the helix, obviating the necessity for withdrawal; additional lengths of auger are added as the hole is advanced. Borehole depths up to 50 m are possible with continuous-flight augers, but there is a possibility that different soil types may become mixed as they rise to the surface, and it may be difficult to determine the depths at which changes of strata occur.

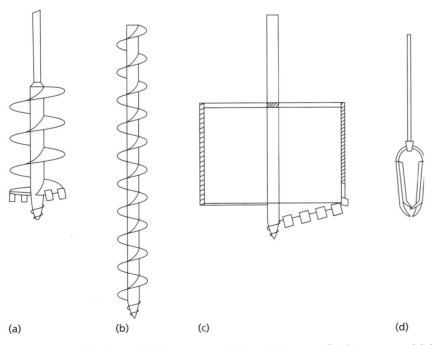

Figure 6.2 (a) Short-flight auger, (b) continuous-flight auger, (c) bucket auger, and (d) Iwan (hand) auger.

Continuous-fight augers with hollow stems are also used. When boring is in progress, the hollow stem is closed at the lower end by a plug fitted to a rod running inside the stem. Additional lengths of auger (and internal rod) are again added as the hole is advanced. At any depth the rod and plug may be withdrawn from the hollow stem to allow undisturbed samples to be taken, a sample tube mounted on rods being lowered down the stem and driven into the soil below the auger. If bedrock is reached, drilling can also take place through the hollow stem. The internal diameter of the stem can range from 75 to 150 mm. As the auger performs the function of a casing it can be used in permeable soils (e.g. sands) below the water table, although difficulty may be experienced with soil being forced upwards into the stem by hydrostatic pressure; this can be avoided by filling the stem with water up to water table level.

Bucket augers (Figure 6.2(c)) consist of a steel cylinder, open at the top but fitted with a base plate on which cutters are mounted, adjacent to slots in the plate: the auger is attached to a kelly bar. When the auger is rotated and pressed downwards, the soil removed by the cutters passes through the slots into the bucket. When the bucket is full, it is raised to the surface to be emptied by releasing the hinged base plate.

Augered holes of 1-m diameter and larger can be used for the examination of the soil strata in-situ, by lowering a remote CCTV camera into the borehole.

Hand and portable augers

Hand augers can be used to excavate boreholes to depths of around 5 m using a set of extension rods. The auger is rotated and pressed down into the soil by means of a T-handle on the upper rod. The two common types are the Iwan or post-hole auger (Figure 6.2(d)), with diameters up to 200 mm, and the small helical auger, with diameters of about 50 mm. Hand augers are generally used only if the sides of the hole require no support and if particles of coarse gravel size and above are absent. The auger must be withdrawn at frequent intervals for the removal of soil. Undisturbed samples can be obtained by driving small-diameter tubes below the bottom of the borehole.

Small portable power augers, generally transported and operated by two persons, are suitable for boring to depths of 10–15 m; the hole diameter may range from 75 to 300 mm. The borehole may be cased if necessary, and therefore the auger can be used in most soil types provided the larger particle sizes are absent.

Wash boring

In this method, water is pumped through a string of hollow boring rods and is released under pressure through narrow holes in a chisel attached to the lower end of the rods (Figure 6.3). The soil is loosened and broken up by the water jets and the up-and-down movement of the chisel. There is also provision for the manual rotation of the chisel by means of a tiller attached to the boring rods above the surface. The soil particles are washed to the surface between the rods and the side of the borehole, and are allowed to settle out in a sump. The rig consists of a derrick with a power unit, a winch and a water pump. The winch carries a light steel cable which passes through the sheaf of the derrick and is attached to the top of the boring rods. The string of rods is raised and dropped by means of the winch unit, producing the chopping action of the chisel. The borehole is generally cased, but the method can be used in uncased holes. Drilling fluid may be used as an alternative to water in the method, eliminating the need for casing.

Wash boring can be used in most types of soil, but progress becomes slow if particles of coarse-gravel size and larger are present. The accurate identification of soil types is difficult, due to particles being broken up by the chisel and to mixing as the material is washed to the surface; in addition, segregation of particles takes place as they settle out in the sump. However, a change in the feel of the boring tool can sometimes be detected, and there may be a change in the colour of the water rising to the

Ground investigation

surface, when the boundaries between different strata are reached. The method is unacceptable as a means of obtaining soil samples. It is used only as a means of advancing a borehole to enable tube samples to be taken or in-situ tests to be carried out below the bottom of the hole. An advantage of the method is that the soil immediately below the hole remains relatively undisturbed.

Rotary drilling

Although primarily intended for investigations in rock, the method is also used in soils. The drilling tool, which is attached to the lower end of a string of hollow drilling rods (Figure 6.4), may be either a cutting bit or a coring bit; the coring bit is fixed to the lower end of a core barrel, which in turn is carried by the drilling rods. Water or drilling fluid is pumped down the hollow rods and passes under pressure through narrow holes in the bit or barrel; this is the same principle as used in wash boring. The drilling fluid cools and lubricates the drilling tool, and carries the loose debris to the surface between the rods and the side of the hole. The fluid also provides some support to the sides of the hole if no casing is used.

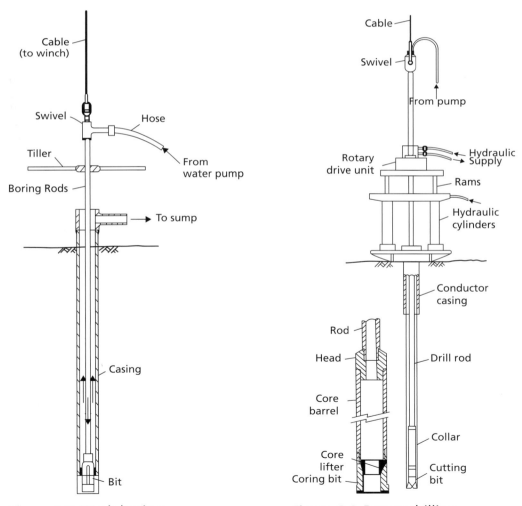

Figure 6.3 Wash boring.

Figure 6.4 Rotary drilling.

The rig consists of a derrick, power unit, winch, pump and a drill head to apply high-speed rotary drive and downward thrust to the drilling rods. A rotary head attachment can be supplied as an accessory to a percussion boring rig.

There are two forms of rotary drilling: open-hole drilling and core drilling. Open-hole drilling, which is generally used in soils and weak rock, uses a cutting bit to break down all the material within the diameter of the hole. Open-hole drilling can thus be used only as a means of advancing the hole; the drilling rods can then be removed to allow tube samples to be taken or in-situ tests to be carried out. In core drilling, which is used in rocks and hard clays, the diamond or tungsten carbide bit cuts an annular hole in the material and an intact core enters the barrel, to be removed as a sample. However, the natural water content of the material is liable to be increased due to contact with the drilling fluid. Typical core diameters are 41, 54 and 76 mm, but can range up to 165 mm.

The advantage of rotary drilling in soils is that progress is much faster than with other investigation methods and disturbance of the soil below the borehole is slight. The method is not suitable if the soil contains a high percentage of gravel-sized (or larger) particles, as they tend to rotate beneath the bit and are not broken up.

Groundwater observations

An important part of any ground investigation is the determination of water table level and of any artesian pressure in particular strata. The variation of level or pore water pressure over a given period of time may also require determination. Groundwater observations are of particular importance if deep excavations are to be carried out.

Water table level can be determined by measuring the depth to the water surface in a borehole. Water levels in boreholes may take a considerable time to stabilise; this time, known as the **response time**, depends on the permeability of the soil. Measurements, therefore, should be taken at regular intervals until the water level becomes constant. It is preferable that the level should be determined as soon as the borehole has reached water table level. If the borehole is further advanced it may penetrate a stratum under artesian pressure (Section 2.1), resulting in the water level in the hole being above water table level. It is important that a stratum of low permeability below a perched water table (Section 2.1) should not be penetrated before the water level has been established. If a perched water table exists, the borehole must be cased in order that the main water table level is correctly determined; if the perched aquifer is not sealed, the water level in the borehole will be above the main water table level.

When it is required to determine the pore water pressure in a particular stratum, a **piezometer** should be used. A piezometer consists of an element filled with de-aired water and incorporating a porous tip which provides continuity between the pore water in the soil and the water within the element. The element is connected to a pressure-measuring system. A high air-entry ceramic tip is essential for the measurement of pore water pressure in partially saturated soils (e.g. compacted fills), the air-entry value being the pressure difference at which air would bubble through a saturated filter. Therefore the air-entry value must exceed the difference between the pore air and pore water pressures, otherwise pore air pressure will be recorded. A coarse porous tip can only be used if it is known that the soil is fully saturated. If the pore water pressure is different from the pressure of the water in the measuring system, a flow of water into or out of the element will take place. This, in turn, results in a change in pressure adjacent to the tip, and consequent seepage of pore water towards or away from the tip. Measurement involves balancing the pressure in the measuring system with the pore water pressure in the vicinity of the tip. However the response time taken for the pressures to equalise depends on the permeability of the soil and the flexibility of the measuring system. The response time of a piezometer should be as short as possible. Factors governing flexibility are the volume change required to actuate the measuring device, the expansion of the connections, and the presence of entrapped air. A de-airing unit forms an essential part of the equipment: efficient de-airing during installation is essential if errors in pressure measurement are

Ground investigation

to be avoided. To achieve a rapid response time in soils of low permeability the measuring system must be as stiff as possible, requiring the use of a closed hydraulic system in which virtually no flow of water is required to operate the measuring device.

The simplest type is the open standpipe piezometer (Figure 6.5), which is used in a cased borehole and is suitable if the soil is fully saturated and the permeability is relatively high. The water level is normally determined by means of an electrical dipper, a probe with two conductors on the end of a measuring tape; the battery-operated circuit closes, triggering an indicator, when the conductors come into contact with the water. The standpipe is normally a plastic tube of 50-mm diameter or smaller, the lower end of which is either perforated or fitted with a porous element. A relatively large volume of water must pass through the porous element to change the standpipe level, therefore a short response time will only be obtained in soils of relatively high permeability. Sand or fine gravel is packed around the lower end, and the standpipe is sealed in the borehole with clay (generally by the use of bentonite pellets) immediately above the level at which pore pressure is to be measured. The remainder of the borehole is backfilled with sand except near the surface, where a second seal is placed to prevent the inflow of surface water. The top of the standpipe is fitted with a cap, again to prevent ingress of water. Open standpipe piezometers which can be pushed or driven into the ground have also been developed. Piezometers may also be used to collect water samples for further chemical analysis in the laboratory on sites which may be contaminated (Section 6.8).

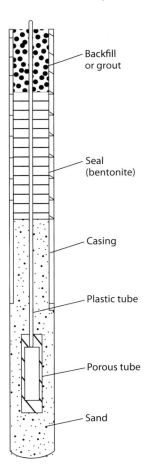

Figure 6.5 Open standpipe piezometer.

Development of a mechanical model for soil

The open standpipe piezometer has a long response time in soils of low permeability, and in such cases it is preferable to install a hydraulic piezometer having a relatively short response time. To achieve a rapid response time in soils of low permeability the measuring system must be as stiff as possible, requiring the use of a closed hydraulic system in which no flow of water is required to operate the measuring device. Three types of piezometer for use with a closed hydraulic system are illustrated in Figure 6.6. The piezometers consist of a brass or plastic body into which a porous tip of ceramic, bronze or stone is sealed. Two tubes lead from the device, one of which is connected to a transducer, allowing results to be recorded automatically. The tubes are of nylon coated with polythene, nylon being impermeable to air and polythene to water.

The two tubes enable the system to be kept air-free by the periodic circulation of de-aired water. Allowance must be made for the difference in level between the tip and the measuring instrument, which should be sited below the tip whenever possible.

6.3 Sampling

Once an opening (trial pit or borehole) has been made in the ground, it is often desirable to recover samples of the soil to the surface for laboratory testing. Soil samples are divided into two main categories, **undisturbed** and **disturbed**. Undisturbed samples, which are required mainly for shear strength and consolidation tests (Chapters 4 and 5), are obtained by techniques which aim at preserving the in-situ structure and water content of the soil as far as is practically possible. In boreholes, undisturbed samples can be obtained by withdrawing the boring tools (except when hollow-stem continuous-flight augers are used) and driving or pushing a sample tube into the soil at the bottom of the hole. The sampler is normally attached to a length of boring rod which can be lowered and raised by the cable of the percussion rig. When the tube is brought to the surface, some soil is removed from each end and molten wax is applied, in thin layers, to form a water-tight seal approximately 25 mm thick; the ends of the tube

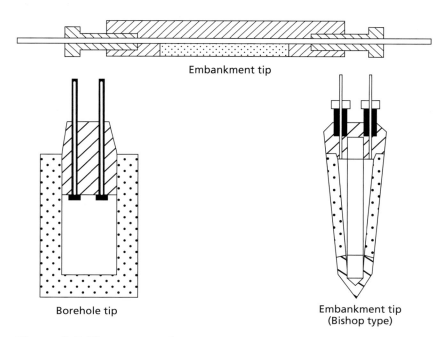

Figure 6.6 Piezometer tips.

are then covered by protective caps. Undisturbed block samples can be cut by hand from the bottom or sides of a trial pit. During cutting, the samples must be protected from water, wind and sun to avoid any change in water content; the samples should be covered with molten wax immediately they have been brought to the surface. It is impossible to obtain a sample that is completely undisturbed, no matter how elaborate or careful the ground investigation and sampling technique might be. In the case of clays, for example, swelling will take place adjacent to the bottom of a borehole due to the reduction in total stresses when soil is removed, and structural disturbance may be caused by the action of the boring tools; subsequently, when a sample is removed from the ground the total stresses are reduced to zero.

Soft clays are extremely sensitive to sampling disturbance, the effects being more pronounced in clays of low plasticity than in those of high plasticity. The central core of a soft clay sample will be relatively less disturbed than the outer zone adjacent to the sampling tube. Immediately after sampling, the pore water pressure in the relatively undisturbed core will be negative due to the release of the in-situ total stresses. Swelling of the relatively undisturbed core will gradually take place due to water being drawn from the more disturbed outer zone and resulting in the dissipation of the negative excess pore water pressure; the outer zone of soil will consolidate due to the redistribution of water within the sample. The dissipation of the negative excess pore water pressure is accompanied by a corresponding reduction in effective stresses. The soil structure of the sample will thus offer less resistance to shear, and will be less stiff than the in-situ soil.

A disturbed sample is one having the same particle size distribution as the in-situ soil but in which the soil structure has been significantly damaged or completely destroyed; in addition, the water content may be different from that of the in-situ soil. Disturbed samples, which are used mainly for soil classification tests (Chapter 1), visual classification and compaction tests, can be excavated from trial pits or obtained from the tools used to advance boreholes (e.g. from augers and the clay cutter). The soil recovered from the shell in percussion boring will be deficient in fines and will be unsuitable for use as a disturbed sample. Samples in which the natural water content has been preserved should be placed in airtight, non-corrosive containers; all containers should be completely filled so that there is negligible air space above the sample.

Samples should be taken at changes of stratum (as observed from the soil recovered by augering/drilling) and at a specified spacing within strata of not more than 3 m. All samples should be clearly labelled to show the project name, date, location, borehole number, depth and method of sampling; in addition, each sample should be given a unique serial number. Special care is required in the handling, transportation and storage of samples (particularly undisturbed samples) prior to testing.

The sampling method used should be related to the quality of sample required. Quality can be classified as shown in Table 6.2, with Class 1 being most useful and of the highest quality, and Class 5 being useful only for basic visual identification of soil type. For Classes 1 and 2, the sample must be undisturbed. Samples of Classes 3, 4 and 5 may be disturbed. The principal types of tube samplers are described below.

Table 6.2 Sample quality related to end use (after EC7–2: 2007)

Soil property	Class 1	Class 2	Class 3	Class 4	Class 5
Sequence of layers	•	•	•	•	•
Strata boundaries	•	•	•	•	
Particle size distribution	•	•	•	•	
Atterberg limits, organic content	•	•	•	•	
Water content	•	•	•		
(Relative) Density, porosity	•	•			
Permeability	•	•			
Compressibility, shear strength	•				

Development of a mechanical model for soil

Open drive sampler

An open drive sampler (Figure 6.7(a)) consists of a long steel tube with a screw thread at each end. A cutting shoe is attached to one end of the tube. The other end of the tube screws into a sampler head, to which, in turn, the boring rods are connected. The sampler head also incorporates a non-return valve to allow air and water to escape as the soil fills the tube, and to help retain the sample as the tube is withdrawn. The inside of the tube should have a smooth surface, and must be maintained in a clean condition.

The internal diameter of the cutting edge (d_c) should be approximately 1% smaller than that of the tube to reduce frictional resistance between the tube and the sample. This size difference also allows for slight elastic expansion of the sample on entering the tube, and assists in sample retention. The external diameter of the cutting shoe (d_w) should be slightly greater than that of the tube to reduce the force required to withdraw the tube. The volume of soil displaced by the sampler as a proportion of the sample volume is represented by the area ratio (C_a) of the sampler, where

$$C_a = \frac{d_w^2 - d_c^2}{d_c^2} \tag{6.1}$$

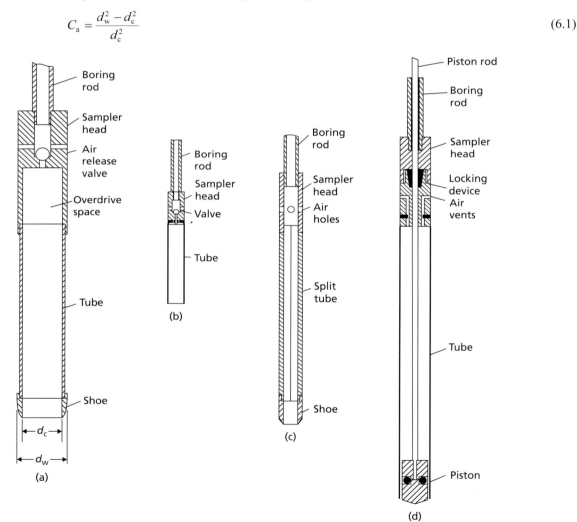

Figure 6.7 Types of sampling tools: (a) open drive sampler, (b) thin-walled sampler, (c) split-barrel sampler, and (d) stationary piston sampler.

The area ratio is generally expressed as a percentage. Other factors being equal, the lower the area ratio, the lower is the degree of sample disturbance.

The sampler can be driven dynamically by means of a drop weight or sliding hammer, or statically by hydraulic or mechanical jacking. Prior to sampling, all loose soil should be removed from the bottom of the borehole. Care should be taken to ensure that the sampler is not driven beyond its capacity, otherwise the sample will be compressed against the sampler head. Some types of sampler head have an overdrive space below the valve to reduce the risk of sample damage. After withdrawal, the cutting shoe and sampler head are detached and the ends of the sample are sealed.

The most widely used sample tube has an internal diameter of 100 mm and a length of 450 mm; the area ratio is approximately 30%. This sampler is suitable for all clay soils. When used to obtain samples of sand, a **core-catcher**, a short length of tube with spring-loaded flaps, should be fitted between the tube and cutting shoe to prevent loss of soil. The class of sample obtained depends on soil type.

Thin-walled sampler

Thin-walled samplers (Figure 6.7(b)) are used in soils which are sensitive to disturbance, such as soft to firm clays and plastic silts. The sampler does not employ a separate cutting shoe, the lower end of the tube itself being machined to form a cutting edge. The internal diameter may range from 35 to 100 mm. The area ratio is approximately 10%, and samples of first-class quality can be obtained provided the soil has not been disturbed in advancing the borehole. In trial pits and shallow boreholes, the tube can often be driven manually.

Split-barrel sampler

Split-barrel samplers (Figure 6.7(c)) consist of a tube which is split longitudinally into two halves; a shoe and a sampler head incorporating air-release holes are screwed onto the ends. The two halves of the tube can be separated when the shoe and head are detached to allow the sample to be removed. The internal and external diameters are 35 and 50 mm, respectively, the area ratio being approximately 100%, with the result that there is considerable disturbance of the sample (Class 3 or 4). This sampler is used mainly in sands, being the tool specified in the **Standard Penetration Test** (SPT, see Chapter 7).

Stationary piston sampler

Stationary piston samplers (Figure 6.7(d)) consist of a thin-walled tube fitted with a piston. The piston is attached to a long rod which passes through the sampler head and runs inside the hollow boring rods. The sampler is lowered into the borehole with the piston located at the lower end of the tube, the tube and piston being locked together by means of a clamping device at the top of the rods. The piston prevents water or loose soil from entering the tube. In soft soils the sampler can be pushed below the bottom of the borehole, bypassing any disturbed soil. The piston is held against the soil (generally by clamping the piston rod to the casing) and the tube is pushed past the piston (until the sampler head meets the top of the piston) to obtain the sample. The sampler is then withdrawn, a locking device in the sampler head holding the piston at the top of the tube as this takes place. The vacuum between the piston and the sample helps to retain the soil in the tube; the piston thus serves as a non-return valve.

Piston samplers should always be pushed down by hydraulic or mechanical jacking; they should never be driven. The diameter of the sampler is usually between 35 and 100 mm, but can be as large as 250 mm. The samplers are generally used for soft clays, and can produce samples of first-class quality up to 1 m in length.

Continuous sampler

The continuous sampler is a highly specialised type of sampler which is capable of obtaining undisturbed samples up to 25 m in length; the sampler is used mainly in soft clays. Details of the soil fabric can be determined more easily if a continuous sample is available. An essential requirement of continuous samplers is the elimination of frictional resistance between the sample and the inside of the sampler tube. In one type of sampler, developed by Kjellman *et al.* (1950), this is achieved by superimposing thin strips of metal foil between the sample and the tube. The lower end of the sampler (Figure 6.8(a)) has a sharp cutting edge above which the external diameter is enlarged to enable 16 rolls of foil to be housed in recesses within the wall of the sampler. The ends of the foil are attached to a piston, which fits loosely inside the sampler; the piston is supported on a cable which is fixed at the surface. Lengths of sample tube (68 mm in diameter) are attached as required to the upper end of the sampler.

As the sampler is pushed into the soil the foil unrolls and encases the sample, the piston being held at a constant level by means of the cable. As the sampler is withdrawn the lengths of tube are uncoupled and a cut is made, between adjacent tubes, through the foil and sample. Sample quality is generally Class 1 or 2.

Another type is the Delft continuous sampler, of either 29 or 66 mm in diameter. The sample feeds into an impervious nylon stockinette sleeve. The sleeved sample, in turn, is fed into a fluid-supported thin-walled plastic tube.

Compressed air sampler

The compressed air sampler (Figure 6.8(b)) is used to obtain undisturbed samples of sand (generally Class 2) below the water table. The sample tube, usually 60 mm in diameter, is attached to a sampler head having a relief valve which can be closed by a rubber diaphragm. Attached to the sampler head is a hollow guide rod surmounted by a guide head. An outer tube, or bell, surrounds the sample tube, the bell being attached to a weight which slides on the guide rod. The boring rods fit loosely into a plain socket in the top of the guide head, the weight of the bell and sampler being supported by means of a shackle which hooks over a peg in the lower length of boring rod; a light cable, leading to the surface, is fixed to the shackle. Compressed air, produced by a foot pump, is supplied through a tube leading to the guide head, the air passing down the hollow guide rod to the bell.

The sampler is lowered on the boring rods to the bottom of the borehole, which will contain water below the level of the water table. When the sampler comes to rest at the bottom of the borehole the shackle springs off the peg, removing the connection between the sampler and the boring rods. The tube is pushed into the soil by means of the boring rods, a stop on the guide rod preventing overdriving; the boring rods are then withdrawn. Compressed air is now introduced to expel the water from the bell and to close the valve in the sampler head by pressing the diaphragm downwards. The tube is withdrawn into the bell by means of the cable, and then the tube and bell together are raised to the surface. The sand sample remains in the tube by virtue of arching and the slight negative pore water pressure in the soil. A plug is placed at the bottom of the tube before the suction is released and the tube is removed from the sampler head.

Window sampler

This sampler, which is most suited to dry fine-grained soils, employs a series of tubes, usually 1 m in length and of different diameters (typically 80, 60, 50 and 36 mm). Tubes of the same diameter can be coupled together. A cutting shoe is attached to the end of the bottom tube. The tubes are driven into the soil by percussion using either a manual or rig-supported device, and are extracted either manually or by means of the rig. The tube of largest diameter is the first to be driven and extracted with its sample inside. A tube of lesser diameter is then driven below the bottom of the open hole left by extraction of

Ground investigation

the larger tube. The operation is repeated using tubes of successively lower diameter, and depths of up to 8 m can be reached. There are longitudinal slots or 'windows' in the walls at one side of the tubes to allow the soil to be examined and enable disturbed samples of Class 3 or 4 to be taken.

6.4 Selection of laboratory test method(s)

The ultimate aim of careful sampling is to obtain the mechanical characteristics of soil for use in subsequent geotechnical analyses and design. These basic characteristics and the laboratory tests for their determination have been described in detail in Chapters 1–5. Table 6.3 summarises the mechanical characteristics which can be obtained from each type of laboratory test discussed previously using undisturbed samples (i.e. Class 1/2). The table demonstrates why the triaxial test is so popular in soil

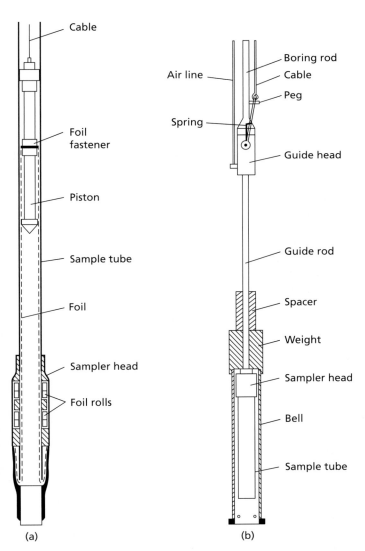

Figure 6.8 (a) Continuous sampler, (b) compressed air sampler.

Development of a mechanical model for soil

mechanics, with modern computer-controlled machines being able to conduct various different test stages on a single sample of soil, thereby making best use of the material recovered from site.

Disturbed samples may be effectively used to support the tests described in Table 6.3 by determining the index properties (w, w_L, w_P, I_P and I_L for fine-grained soils; e_{max}, e_{min} and e for coarse-grained soils) and using the empirical correlations presented in Section 5.9 (i.e. Equations 5.43–5.47 inclusive).

6.5 Borehole logs

After an investigation has been completed and the results of any laboratory tests are available, the ground conditions discovered in each borehole (or trial pit) are summarised in the form of a borehole (or trial pit) log. An example of such a log appears in Table 6.4, but details of the layout can vary. The last few columns are originally left without headings to allow for variations in the data presented. The method of investigation and details of the equipment used should be stated on each log. The location, ground level and diameter of the hole should be specified, together with details of any casing used. The names of the client and the project should be stated.

The log should enable a rapid appraisal of the soil profile to be made. The log is prepared with reference to a vertical depth scale. A detailed description of each stratum is given, and the levels of strata boundaries clearly shown; the level at which boring was terminated should also be indicated. The different soil (and rock) types are represented by means of a legend using standard symbols. The depths (or ranges of depth) at which samples were taken or at which in-situ tests were performed are recorded; the type of sample is also specified. The results of certain laboratory or in-situ tests may be given in the log – Table 6.4 shows N values which are the result of Standard Penetration tests (SPT), which are described in Chapter 7. The depths at which groundwater was encountered and subsequent changes in levels, with times, should be detailed.

Table 6.3 Derivation of key soil properties from undisturbed samples tested in the laboratory

Parameter	Oedometer (Chapter 4)	Shearbox (Chapter 5)	Triaxial cell (Chapter 5)	Permeameter (Chapter 2)	Particle size distribution (PSD) (Chapter 1)
Consolidation characteristics: m_v, C_c	YES		C_c from λ		
Stiffness properties: G, G_0			YES		
Drained strength properties: ϕ', c'		YES	CD/CU tests		
Undrained strength properties: c_u (in-situ)			UU tests		
Permeability: k			FH* – sands & gravels CH* – finer soils	FH – sands & gravels CH – finer soils	sands & gravels (Eq. 2.4)

Notes: * FH = Falling head test; CH = Constant head test. These tests can be conducted in a modern stress-path cell by controlling the back pressure and maintaining a zero lateral strain condition.

Table 6.4 Sample borehole log

BOREHOLE LOG

Location: Barnhill
Client: RFG Consultants
Boring method: Shell and auger to 14.4 m
Rotary core drilling to 17.8 m
Diameter: 150 mm
NX
Casing: 150 mm to 5 m

Borehole No.1
Sheet 1 of 1
Ground level: 36.30
Date: 30.7.77
Scale: 1:100

Description of strata	Level	Legend	Depth	Samples	N	C_u (kN/m²)
TOPSOIL	35.6		0.7			
Loose, light brown SAND	33.7		2.6	D	6	
Medium dense, brown gravelly SAND ▽	32.5			D	15	
	31.9		4.4			
				U		80
Firm, yellowish-brown, closely fissured CLAY				U		86
				U		97
				U		105
	24.1		12.2			
Very dense, red, silty SAND with decomposed SANDSTONE				D	50 for 210 mm	
	21.9		14.4			
Red, medium-grained, granular, fresh SANDSTONE, moderately weak, thickly bedded	18.5		17.8			

U: Undisturbed sample
D: Disturbed sample
B: Bulk disturbed sample
W: Water sample
▽: Water table

REMARKS

Water level (0930 h)
29.7.77   32.2 m
30.7.77   32.5 m
31.7.77   32.5 m

Development of a mechanical model for soil

The soil description should be based on particle size distribution and plasticity, generally using the rapid procedure in which these characteristics are assessed by means of visual inspection and feel; disturbed samples are generally used for this purpose. The description should include details of soil colour, particle shape and composition; if possible, the geological formation and type of deposit should be given. The structural characteristics of the soil mass should also be described, but this requires an examination of undisturbed samples or of the soil in-situ (e.g. in a trial pit). Details should be given of the presence and spacing of bedding features, fissures and other relevant characteristics. The density index of sands and the consistency of clays (Table 1.3) should be indicated.

6.6 Cone penetration testing (CPT)

The **Cone Penetrometer** is one of the most versatile tools available for soil exploration (Lunne et al., 1997). In this section, the use of the CPT to identify stratigraphy and the materials which are present in the ground will be presented. However, the technique can also be used to determine a wide range of standard geotechnical parameters for these materials instead of or in addition to the laboratory tests summarised in Section 6.4. This will be further discussed in Chapter 7. Furthermore, because of the close analogy between a CPT and a pile under vertical loading, CPT data may also be used directly in the design of deep foundations (Chapter 9).

The penetrometer consists of a short cylindrical element, at the end of which is a cone-shaped tip. The cone has an apex angle of 60° and a cross-sectional area of 1000 mm². This is pushed vertically into the ground using a **thrust machine** at a constant rate of penetration of 20 mm/s (ISO, 2006). For onshore applications, the thrust machine is commonly a CPT truck which provides a reaction mass due to its self-weight (typically 15–20 tonnes). A 20-tonne (200-kN) thrust will normally permit penetration to around 30 m in dense sands or stiff clays (Lunne et al., 1997). For deeper investigations where higher resistances to penetration will be encountered, this may be further ballasted or temporarily anchored to the ground (Figure 6.9). As the instrument penetrates, additional **push rods** of the same diameter as the instrument are attached to extend the string. Cables passing up through the centre of the push rods carry data from instruments within the penetrometer to the surface. In a standard electrical cone (CPT), a load cell between the cone and the body of the instrument continuously records the resistance to penetration of the cone (cone tip resistance q_c), and a friction sleeve is used to measure the interface shearing resistance (f_s) along the cylindrical body of the instrument. Different types of soil will exhibit different proportions of sleeve friction to end resistance: for example, gravels generally have low f_s and high q_c, while clays have high f_s and low q_c. By examining an extensive database of CPT test data, Robertson (1990) proposed a chart which may be used for identifying soil types based on normalised versions of these parameters (Q_t and F_r), which is shown in Figure 6.10.

For a standard CPT cone, q_t is approximated by q_c. During penetration, however, excess pore water pressures around the cone will increase, particularly in fine-grained soils, which artificially reduce q_c. More sophisticated **piezocones** (CPTU) include localised measurement of the excess pore water pressures around the cone which are induced by penetration. These are most commonly measured immediately behind the cone (u_2), though measurements may also be made on the cone itself (u_1) and/or at the other end of the friction sleeve (u_3), as shown in Figure 6.11. When such measurements are made, the corrected cone resistance q_t is determined using

$$q_t = q_c + u_2(1-a) \tag{6.2}$$

The parameter a is an area correction factor depending on the penetrometer, and typically varies between 0.5–0.9. As before, different soils will experience different changes in pore water pressure during penetration: coarse-grained soils (sands and gravels) will exhibit little excess pore water pressure generation due to their high permeability, while fine-grained soils of lower permeability typically exhibit larger values of u_2.

Ground investigation

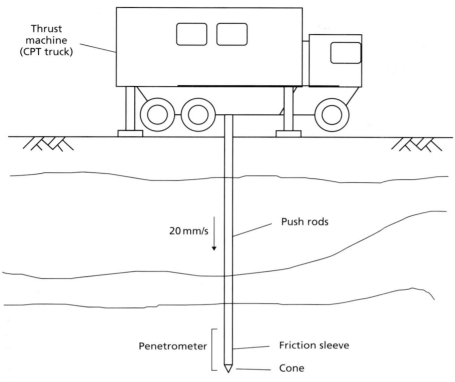

Figure 6.9 Schematic of Cone Penetrometer Test (CPT) showing standard terminology.

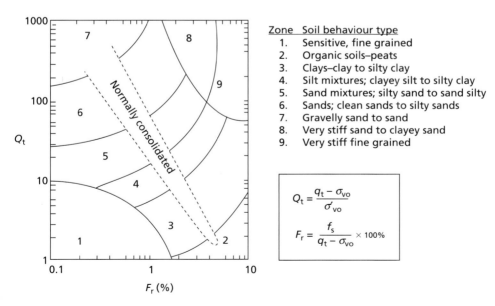

Zone   Soil behaviour type
1.   Sensitive, fine grained
2.   Organic soils–peats
3.   Clays–clay to silty clay
4.   Silt mixtures; clayey silt to silty clay
5.   Sand mixtures; silty sand to sand silty
6.   Sands; clean sands to silty sands
7.   Gravelly sand to sand
8.   Very stiff sand to clayey sand
9.   Very stiff fine grained

$$Q_t = \frac{q_t - \sigma_{vo}}{\sigma'_{vo}}$$

$$F_r = \frac{f_s}{q_t - \sigma_{vo}} \times 100\%$$

Figure 6.10 Soil behaviour type classification chart based on normalised CPT data (reproduced after Robertson, 1990).

Development of a mechanical model for soil

If the soil is heavily overconsolidated, u_2 may be negative. The pore water pressure measurement u_2 therefore provides a third continuous parameter which may be used to identify soil types, using the chart shown in Figure 6.12 (Robertson, 1990). If CPTU data are available, both Figures 6.10 and 6.12 should be used to determine soil type. In some instances, the two charts may give different interpretations of the ground conditions. In these cases, judgement is required to correctly identify the ground conditions. CPT soundings may also struggle to identify inter-bedded soil layers (Lunne *et al.*, 1997).

As data are recorded continuously with increasing depth during a CPT sounding, the technique can be used to produce a ground profile showing soil stratigraphy and classification, similar to a borehole log. An example of the use of the foregoing identification charts is shown in Figure 6.13. The CPT test is quick, relatively cheap, and has the advantage of not leaving a large void in the ground as in the case of a borehole or trial pit. However, due to the difficulties in interpretation which have previously been discussed, CPT data are most effective at 'filling in the gaps' between widely spaced boreholes. Under these conditions, use of the CPT(U) soil identification charts can be informed by the observations from the boreholes. The CPT(U) data will then provide useful information as to how the levels of different soil strata vary across a site, and may identify localised hard inclusions or voids which may have been missed by the boreholes.

Given the large amount of data that is generated from a continuous CPT sounding, and that soil behaviour types (zones 1–9 in Figures 6.10 and 6.12) fall within ranges defined by the magnitude of the measured parameters, interpretation of soil stratigraphy from CPT data benefits greatly from automation/computerisation. It may be seen from Figure 6.10 that the boundaries between zones 2–7 are close to being circular arcs with an origin around the top left corner of the plot. Robertson and Wride (1998) quantified the radius of these arcs by a parameter I_c:

$$I_c = \sqrt{(3.47 - \log Q_t)^2 + (\log F_r + 1.22)^2} \tag{6.3}$$

where Q_t and F_r are the normalised tip resistance and friction ratio as defined in Figure 6.10. Figure 6.14 shows curves plotted using Equation 6.3 for values of I_c which most closely represent the soil boundaries in Figure 6.10. Using a single equation to define the soil behaviour type makes the processing of CPT

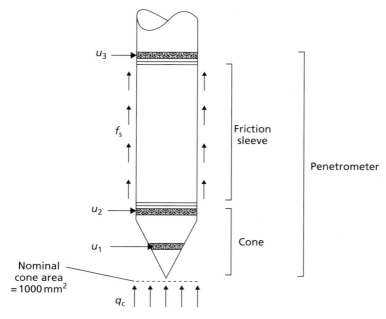

Figure 6.11 Schematic of piezocone (CPTU).

Ground investigation

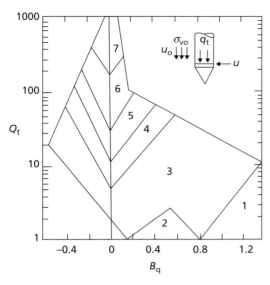

Figure 6.12 Soil behaviour type classification chart based on normalised CPTU data.

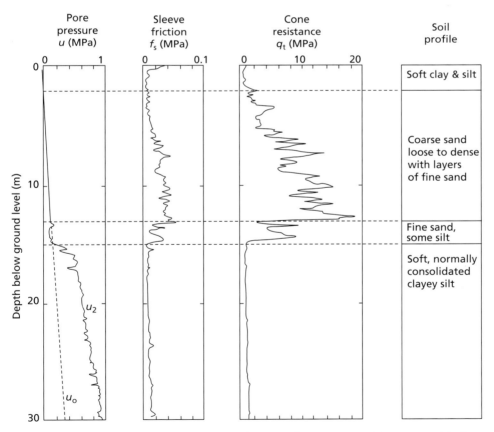

Figure 6.13 Example showing use of CPTU data to provide ground information.

Development of a mechanical model for soil

data amenable to automated analysis using a spreadsheet, by applying Equation 6.3 to each test datapoint. A spreadsheet tool, CPTic_CSM8.xls, which implements the I_c method may be found on the Companion Website. Care must be taken when using the method, as it will not correctly identify soil types 1, 8 and 9 (Figure 6.10); however, for most routine use, the I_c method provides a valuable tool for the interpretation of stratigraphic information from CPT soundings.

6.7 Geophysical methods

Under certain conditions geophysical methods may be useful in ground investigation, especially at the reconnaissance stage. However, the methods are not suitable for all ground conditions and there are limitations to the information that can be obtained; thus they must be considered mainly as supplementary methods. It is possible to locate strata boundaries only if the physical properties of the adjacent materials are significantly different. It is always necessary to check the results against data obtained by direct methods such as boring or CPT soundings. Geophysical methods can produce rapid and economic results, making them useful for the filling in of detail between widely spaced boreholes or to indicate where additional boreholes may be required. The methods can also be useful in estimating the depth to bedrock or to the water table, or for locating buried metallic objects (e.g. unexploded ordnance) and voids. They can be particularly useful in investigating sensitive contaminated sites, as the methods are non-intrusive – unlike boring or CPT. There are several geophysical techniques, based on different physical principles. Three of these techniques are described below.

Seismic refraction

The seismic refraction method depends on the fact that seismic waves have different velocities in different types of soil (or rock), as shown in Table 6.5; in addition, the waves are refracted when they cross the boundary between different types of soil. The method enables the general soil types and the approximate depths to strata boundaries, or to bedrock, to be determined.

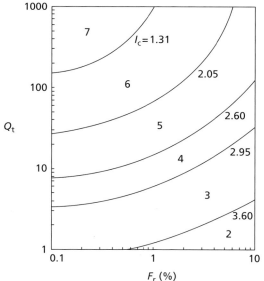

Figure 6.14 Soil behaviour type classification using the I_c method.

Ground investigation

Table 6.5 Shear wave velocities of common geotechnical materials (after Borcherdt, 1994)

Soil/rock type	Shear wave velocity, V_s (m/s)
Hard rocks (e.g. metamorphic)	1400+
Firm to hard rocks (e.g. igneous, conglomerates, competent sedimentary)	700–1400
Gravelly soils and soft rocks (e.g. sandstone, shale, soils with >20% gravel)	375–700
Stiff clays and sandy soils	200–375
Soft soils (e.g. loose submerged fills and soft clays)	100–200
Very soft soils (e.g. marshland, reclaimed soil)	50–100

Waves are generated either by the detonation of explosives or by striking a metal plate with a large hammer. The equipment consists of one or more sensitive vibration transducers, called geophones, and an extremely accurate time-measuring device called a seismograph. A circuit between the detonator or hammer and the seismograph starts the timing mechanism at the instant of detonation or impact. The geophone is also connected electrically to the seismograph: when the first wave reaches the geophone the timing mechanism stops and the time interval is recorded in milliseconds.

When detonation or impact takes place, waves are emitted in every direction. One particular wave, called the direct (or surface) wave, will travel parallel to the surface in the direction of the geophone. Other waves travel in a downward direction, at various angles to the horizontal, and will be refracted if they pass into a stratum of different **shear wave** (or seismic) **velocity** (V_s). If the shear wave velocity of the lower stratum is higher than that of the upper stratum, one particular wave will travel along the top of the lower stratum, parallel to the boundary, as shown in Figure 6.15(a): this wave continually 'leaks' energy back to the surface. Energy from this refracted wave can be detected by geophones at the surface.

The test procedure consists of installing either a single geophone in turn at a number of points in a straight line, at increasing distances from the source of wave generation, or, alternatively, using a single geophone position and producing a series of vibration sources at increasing distances from the geophone (as shown in Figure 6.15(a)). The length of the line of points should be three to five times the required depth of investigation, and the spacing between measurement/shot points is approximately 3 m. For each shot (detonation or impact), the arrival time of the first wave at the geophone position is recorded. When the distance between the source and the geophone is short, the arrival time will be that of the direct wave. When the distance between the source and the geophone exceeds a certain value (depending on the thickness of the upper stratum), the refracted wave will be the first to be detected by the geophone. This is because the path of the refracted wave, although longer than that of the direct wave, is partly through a stratum of higher shear wave velocity. The use of explosives is generally necessary if the source–geophone distance exceeds 30–50 m or if the upper soil stratum is loose.

Arrival time is plotted against the distance between the source and the geophone, a typical plot being shown in Figure 6.15(b). If the source–geophone spacing is less than d_1, the direct wave reaches the geophone in advance of the refracted wave and the time–distance relationship is represented by a straight line through the origin. On the other hand, if the source–geophone distance is greater than d_1 but less than d_2, the refracted wave arrives in advance of the direct wave and the time–distance relationship is represented by a straight line at a different slope. For still larger spacings, a third straight line may be observed representing the third layer, and so on. The slopes of the lines which are read from the graph are the reciprocals of the shear wave velocities (V_{s1}, V_{s2} and V_{s3}) of the upper, middle and lower strata,

Development of a mechanical model for soil

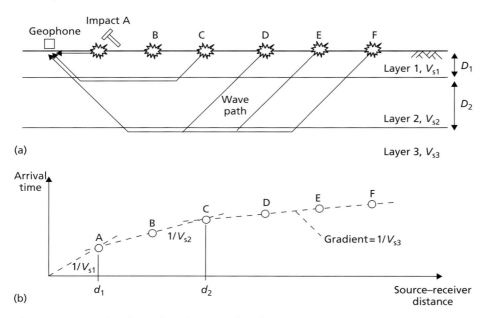

Figure 6.15 Seismic refraction method.

respectively. The general types of soil or rock can be determined from a knowledge of these velocities, for example by using Table 6.5. The depths (D_1 and D_2) of the boundaries between the soil strata (provided the thickness of the strata is constant) can be estimated from the formulae

$$D_1 = \frac{d_1}{2}\sqrt{\left(\frac{V_{s2} - V_{s1}}{V_{s2} + V_{s1}}\right)} \tag{6.4}$$

$$D_2 = 0.8 D_1 + \frac{d_2}{2}\sqrt{\left(\frac{V_{s3} - V_{s2}}{V_{s3} + V_{s2}}\right)} \tag{6.5}$$

The method can also be used where there are more than three strata and procedures exist for the identification of inclined strata boundaries and vertical discontinuities.

The formulae used to estimate the depths of strata boundaries are based on the assumptions that each stratum is homogeneous and isotropic, the boundaries are plane, each stratum is thick enough to produce a change in slope on the time–distance plot, and the shear wave velocity increases in each successive stratum from the surface downwards. Softer, lower-velocity material (e.g. soft clay) can therefore be masked by overlying stronger, higher-velocity material (e.g. gravel). Other difficulties arise if the velocity ranges of adjacent strata overlap, making it difficult to distinguish between them, and if the velocity increases with depth in a particular stratum. It is important that the results are correlated with data from borings.

Knowledge of V_s can also be used to directly determine the stiffness of the soil. As shear waves induce only very small strains in the materials as they pass, the material behaviour is elastic. In an elastic medium, the small strain shear modulus (G_0) is related to shear wave velocity by:

$$G_0 = \rho V_s^2 \tag{6.6}$$

Once G_0 is known, the full non-linear G–γ relationship may be inferred using normalised relationships such as Equation 5.10 (Section 5.2).

Spectral analysis of surface waves (SASW)

This is a relatively new method which in a practical sense is very similar to seismic refraction; indeed, the test procedure is essentially identical. However, instead of measuring arrival times, a detailed analysis of the frequency content of the signal received at the geophone is conducted, and only the lower-frequency surface wave is considered. The analysis is automated using a spectrum analyser to produce a dispersion curve, and a computer is used to develop a ground model (a series of layers of different shear wave velocity) which would lead to the measured signal, a process known as **inversion**. In the SASW procedure only a single geophone is used, which can make it unreliable in urban areas where there is substantial low-frequency background noise which is difficult to distinguish from the signal of interest. A modified technique which overcomes these difficulties uses a set of geophones in a linear array. This is known as **multi-channel analysis of surface waves** (MASW), and has the additional advantage that analysis can be reliably automated.

Both techniques are quick to conduct, and by conducting tests in different locations a two-dimensional map of the ground layering (based on shear wave velocities) can be determined – a process known as tomography. The main advantages of MASW over the refraction method is that the test can infer layers of low shear wave velocity beneath material of high shear wave velocity, which would be masked in the refraction method. As with the previous method, the MASW data should always be correlated with data from borings. It may similarly be used to determine G_0 from V_s using Equation 6.6.

Electrical resistivity

This method depends on differences in the electrical resistance of different soil (and rock) types. The flow of current through a soil is mainly due to electrolytic action, and therefore depends on the concentration of dissolved salts in the pore water: the mineral particles of a soil are poor conductors of current. The resistivity of a soil therefore decreases as both the water content and the concentration of salts increase. A dense, clean sand above the water table, for example, would exhibit a high resistivity due to its low degree of saturation and the virtual absence of dissolved salts. A saturated clay of high void ratio, on the other hand, would exhibit a low resistivity due to the relative abundance of pore water and the free ions in that water.

In its usual form (Figure 6.16(a)), the method involves driving four electrodes into the ground at equal distances (L) apart in a straight line. Current (I), from a battery, flows through the soil between the two outer electrodes, producing an electric field within the soil. The potential drop (E) is then measured between the two inner electrodes. The apparent resistivity (R_Ω) is given by the equation

$$R_\Omega = \frac{2\pi LE}{I} \tag{6.7}$$

The apparent resistivity represents a weighted average of true resistivity in a large volume of soil, the soil close to the surface being more heavily weighted than the soil at depth. The presence of a stratum of soil of high resistivity lying below a stratum of low resistivity forces the current to flow closer to the surface, resulting in a higher voltage drop and hence a higher value of apparent resistivity. The opposite is true if a stratum of low resistivity lies below a stratum of high resistivity.

When the variation of resistivity with depth is required, the following method can be used to make rough estimates of the types and depths of strata. A series of readings are taken, the (equal) spacing of the electrodes being increased for each successive reading; however, the centre of the four electrodes remains at a fixed point. As the spacing is increased, the apparent resistivity is influenced by a greater depth of soil. If the resistivity increases with increasing electrode spacing, it can be concluded that an underlying stratum of higher resistivity is beginning to influence the readings. If increased separation

Development of a mechanical model for soil

produces decreasing resistivity, on the other hand, a stratum of lower resistivity is beginning to influence the readings. The greater the thickness of a layer, the greater the electrode spacing over which its influence will be observed, and vice versa.

Apparent resistivity is plotted against electrode spacing, preferably on log–log paper. Characteristic curves for a two-layer structure are illustrated in Figure 6.16(b). For curve A, the resistivity of layer 1 is lower than that of layer 2; for curve B, layer 1 has a higher resistivity than layer 2. The curves become asymptotic to lines representing the true resistivities R_1 and R_2 of the respective layers. Approximate layer thicknesses can be obtained by comparing the observed curve of resistivity versus electrode spacing with a set of standard curves. Other methods of interpretation have also been developed for two-layer and three-layer systems.

The procedure known as profiling is used in the investigation of lateral variation of soil types. A series of readings is taken, the four electrodes being moved laterally as a unit for each successive reading; the electrode spacing remains constant for each reading. Apparent resistivity is plotted against the centre position of the four electrodes, to natural scales; such plots can be used to locate the positions of soil of high or low resistivity. Contours of resistivity can be plotted over a given area.

The apparent resistivity for a particular soil or rock type can vary over a wide range of values, as shown in Table 6.6; in addition, overlap occurs between the ranges for different types. This makes the identification of soil or rock type and the location of strata boundaries extremely uncertain. The presence of irregular features near the surface and of stray potentials can also cause difficulties in interpretation. It is essential, therefore, that the results obtained are correlated with borehole data. The method is not considered to be as reliable as the seismic methods described previously.

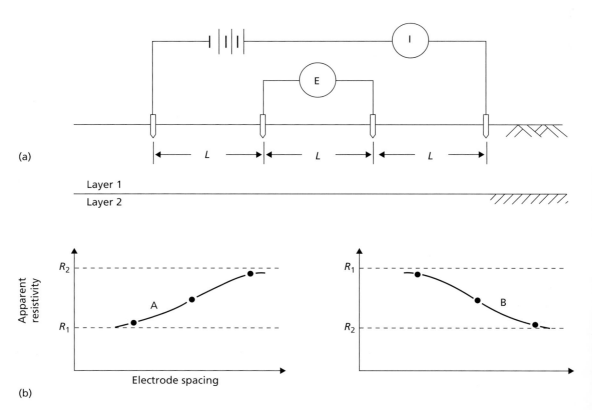

Figure 6.16 (a) Electrical resistivity method, (b) identification of soil layers by sounding.

Table 6.6 Typical resistivities of common geotechnical materials (collated after Campanella and Weemes, 1990; McDowell et al., 2002; Hunt, 2005)

Material	Resistivity, R_Ω (Ωm)
Bedrock (massive)	>2400
Bedrock (fractured), dry deposits of sand and gravel	300–2400
Mudstone	20–60
Sandy soils (saturated)	15–300
Clayey/silty soils (saturated)	1.5–15
Glacial till	20–30
Landfill leachate	0.5–10
Freshwater	20–60
Sea water	0.18–0.24

6.8 Contaminated ground

The scope of an investigation must be extended if it is known or suspected that the ground in question has been contaminated. In such cases the soil and groundwater may contain potentially harmful substances, such as organic or inorganic chemicals, fibrous materials such as asbestos, toxic or explosive gases, biological agents, and/or radioactive elements. The contaminant may be in solid, liquid or gaseous form. Chemical contaminants may be adsorbed on the surfaces of fine soil particles. The presence of contamination influences all other aspects of ground investigation, and may have consequences for foundation design and the general suitability of the site for the project under consideration. Adequate precautions must be taken to ensure the safety from health hazards of all personnel working on the site and when dealing with samples. During the investigation, precautions must be taken to prevent the spread of contaminants by personnel, by surface or groundwater flow and by wind.

At the outset, possible contamination may be predicted from information on previous uses of the site or adjacent areas, such as by certain types of industry, mining operations, or reported leakage of hazardous liquids on the surface or from underground pipelines. This information should be collected during the initial desk study. The visual presence of contaminants and the presence of odours give direct evidence of potential problems. Remote sensing and geophysical techniques (e.g. infra-red photography and electrical resistivity testing, respectively) can be useful in assessing possible contamination.

Soil and groundwater samples are normally obtained from shallow trial pits or boreholes, as contamination from incidents of pollution is often highest towards the ground surface. The depths at which samples are taken depend on the probable source of contamination, and details of the types and structures of the strata. Experience and judgement are thus required in formulating the sampling programme. Solid samples, which would normally be taken at depth intervals of 100–150 mm, are obtained by means of stainless steel tools, which are easily cleaned and are not contaminated, or in driven steel tubes. Samples should be sealed in water-tight containers made of material that will not react with the sample. Care must be taken to avoid the loss of volatile contaminants to the atmosphere. Groundwater samples may be required for chemical analysis to determine if they contain agents that would chemically attack steel and concrete which may be used in the subsurface works (e.g. sulphates, chlorides, magnesium and ammonium), or substances which may be harmful to potential users of the site and the environment. It is important to ensure that samples are not contaminated or diluted. Samples can be taken directly from trial pits or by means of specially designed sampling probes; however, it is preferable to obtain samples from standpipe piezometers if these have been installed to measure the pore water pressure regime. In cases where groundwater contamination is present, care must be taken to ensure that boring does not lead to preferential flow paths being created within the ground which allow the contaminants to spread. A sample should be taken immediately after the water-bearing stratum is reached in boring, and subsequent samples should be taken over a suitable period to determine if properties are constant or variable. This may be used to imply whether pollution or migration of the contaminants is ongoing. Gas samples can be obtained from tubes suspended in a perforated standpipe in a borehole, or from special probes. There are several types of receptacle suitable for collecting gases. Details of the sampling process should be based on the advice of specialist analysts who will undertake the testing programme and report on the results.

When designing a site investigation in contaminated ground, the locations of the investigation points should be related to the position of the contaminants. If the position of a polluting source is known, or can be inferred from a desk study (e.g. the position of a factory waste outfall or spoil heap), **targeted sampling** should be undertaken to determine the spread of contaminant from the source. This typically involves radial lines of sampling out from the contaminant source. An initially wide spacing can be used initially to determine the extent of the contaminant, with subsequent investigation 'filling in the gaps' as necessary. Some **non-targeted sampling** should always be additionally conducted, where the site is split into smaller areas within each of which a single investigation point is made. The aim of this type of sampling is to identify potentially unexpected contaminants within the ground. The spacing of ground investigation points for non-targeted sampling should typically be between 18 and 24 m (BSI, 2001).

Summary

1. Ground investigation using intrusive methods (trial pits, boreholes, CPT) or non-intrusive geophysical methods can be used to determine the location, extent and identity of soil deposits within the ground.
2. Boreholes and trial pits are essential for visually confirming the geotechnical materials within the ground, and for obtaining samples for laboratory testing (using the methods outlined in Chapters 2, 4 and 5). To be most effective, these measurements should ideally be used in combination with CPT and geophysical methods to minimise the chances of encountering unforeseen ground conditions once construction has started, which may significantly increase project duration and cost.
3. The quality (and often also the cost) of the sampling that will be required will depend on the geotechnical characteristics/properties that are required from laboratory tests on such samples.

References

Borcherdt, R. (1994) Estimates of site-dependent response spectra for design (methodology and justification), *Earthquake Spectra*, **10**, 617–653.

BSI (2001) *Investigation of Potentially Contaminated Sites – Code of Practice BS 10175:2001*, British Standards Institution, London.

Campanella, R.G. and Weemes, I.A. (1990) Development and use of an electrical resistivity cone for groundwater contamination studies, in *Geotechnical Aspects of Contaminated Sites, 5th Annual Symposium of the Canadian Geotechnical Society, Vancouver*.

Chapman, T.J.P. (2008) The relevance of developer costs in geotechnical risk management, in *Proceedings of the 2nd BGA International Conference on Foundations, Dundee*.

EC7-2 (2007) *Eurocode 7: Geotechnical Design – Part 2: Ground Investigation and Testing, BS EN 1997–2:2007*, British Standards Institution, London.

Hunt, R.E. (2005) *Geotechnical Engineering Investigation Handbook* (2nd edn), Taylor & Francis Group, Boca Raton, FL.

ISO (2006) *Geotechnical Investigation and Testing – Field Testing – Part 1: Electrical Cone and Piezocone Penetration Tests, ISO TC 182/SC 1*, International Standards Organisation, Geneva.

Kjellman, W., Kallstenius, T. and Wager, O. (1950) Soil sampler with metal foils, in *Proceedings of the Royal Swedish Geotechnical Institute*, No. 1.

Lunne, T., Robertson, P.K. and Powell, J.J.M. (1997) *Cone Penetration Testing in Geotechnical Practice*, E & FN Spon, London.

McDowell, P.W., Barker, R.D., Butcher, A.P., Culshaw, M.G., Jackson, P.D., McCann, D.M., Skipp, B.O., Matthews, S.L. and Arthur, J.C.R. (2002) *Geophysics in Engineering Investigations*, CIRIA Publication C562, CIRIA, London.

Robertson, P.K. (1990) Soil classification using the cone penetration test, *Canadian Geotechnical Journal*, **27**(1), 151–158.

Robertson, P.K. and Wride, C.E. (1998). Evaluating cyclic liquefaction potential using the cone penetration test, *Canadian Geotechnical Journal*, **35**(3): 442–459.

Further reading

BSI (2002) *Geotechnical Investigation and Testing – Identification and Classification of Soil. Identification and Description (AMD Corrigendum 14181 & 16930) BS EN ISO 14688–1:2002*, British Standards Institution, London.

Standard (in UK and Europe) by which soil is identified and described when in the field and for producing borehole logs/ground investigation reports. ASTM D2488 and D5434 are the equivalent standards used in the US and elsewhere.

BSI (2006) *Geotechnical Investigation and Testing – Sampling Methods and Groundwater Measurements. Technical Principles for Execution BS EN ISO 22475–1:2006*, British Standards Institution, London.

Standard (in UK and Europe) describing the detailed procedures and practical requirements for collecting soil samples of high quality for laboratory testing. In the US and elsewhere, a range of standards cover a similar area, including ASTM D5730, D1452, D6151, D1587 and D6519.

Clayton, C.R.I., Matthews, M.C. and Simons, N.E. (1995) *Site Investigation* (2nd edn), Blackwell, London.

A comprehensive book on site investigation which contains much useful practical advice and greater detail than can be provided in this single chapter.

Lerner, D.N. and Walter, R.G. (eds) (1998) Contaminated land and groundwater: future directions, *Geological Society, London, Engineering Geology Special Publication*, **14**, 37–43.

This article provides a more comprehensive introduction to the additional ground investigation issues associated with contaminated land.

Rowe, P.W. (1972) The relevance of soil fabric to site investigation practice, *Géotechnique*, **22**(2), 195–300.

Presents 35 case studies demonstrating how the depositional/geological history of the soil deposits influences the selection of laboratory tests and the associated requirements of sampling techniques.

For further student and instructor resources for this chapter, please visit the Companion Website at www.routledge.com/cw/craig

Chapter 7

In-situ testing

> ### Learning outcomes
>
> After working through the material in this chapter, you should be able to:
>
> 1 Understand the rationale behind testing soils in-situ to obtain their constitutive properties, and appreciate the part it plays with laboratory testing and the use of empirical correlations in establishing a reliable ground model (Section 7.1);
> 2 Understand the principle of operation of four common in-situ testing devices, their applicability and the constitutive properties that can be reliably obtained from them (Sections 7.2–7.5);
> 3 Process the test data from these methods with the help of a computer and use this to derive key strength and stiffness properties (Sections 7.2–7.5).

7.1 Introduction

In Chapter 5, laboratory tests for determining the constitutive behaviour of soil (strength and stiffness properties) were described. While such tests are invaluable in quantifying the mechanical behaviour of an element of soil, there remain a number of disadvantages. First, to obtain high quality data through tri-axial testing, undisturbed samples must be obtained, which can be difficult and expensive in some deposits (e.g. sands and sensitive clays, see Chapter 6). Second, in deposits where there are significant features within the macro-fabric (e.g. fissuring in stiff clays) the response of a small element of soil may not represent the behaviour of the complete soil mass, if the sample happens to be taken such that it does not contain any of these features. As a result of such limitations, in-situ testing methods have been developed which can overcome these limitations and provide a rapid assessment of key parameters which can be conducted during the ground investigation phase.

In this chapter the four principal in-situ testing techniques will be considered, namely:

- the Standard Penetration Test (SPT);
- the Field Vane Test (FVT);
- the Pressuremeter Test (PMT);
- the Cone Penetration Test (CPT).

In each case the testing methodology will be briefly described, but the focus will be on the parameters which can be measured/estimated from each test and the theoretical/empirical models for achieving this, the range of application to different soils, interpretation of constitutive properties (e.g. ϕ', c_u, G), stress history (OCR) and stress state (K_0) from the test data, and limitations of the data collected. The worked examples in the main text and problems at the end of the chapter are all based on actual test data from real sites which have been collated from the literature. In these examples/problems extensive use will be made of spreadsheets to perform the required calculations, and digital data for this purpose are provided on the Companion Website

The four techniques listed above are not the only in-situ testing techniques; a Dilatometer Test (DMT), for example, is similar in principle of operation and in the properties which can be measured to the PMT, involving expanding a cavity within the soil to determine mechanical properties. This common test will not be discussed herein, but references are provided at the end of the chapter for further reading on this topic. Plate Loading Tests (PLT) are also in common usage – these involve performing a load test on a small plate which is essentially a model shallow foundation, and are most commonly used to derive soil data for foundation works due to the close similarity of the test procedure to the ultimate construction. Soil parameters are then back-calculated using standard techniques for analysing shallow foundations, which are described in detail in Chapter 8. It should also be noted that geophysical methods for profiling which use seismic methods (e.g. SASW/MASW, seismic refraction) measure the shear wave velocity (V_s) in-situ from which G_0 can be determined and are therefore in-situ tests in their own right (these methods were previously described in Chapter 6).

The data collected from in-situ tests should always be considered as complementing rather than replacing sampling and laboratory testing. Indeed, three of the tests that will be discussed (SPT, FVT, PMT) require the prior drilling of a borehole, so a single borehole may be used very efficiently to gain visual identification of materials from spoil (Section 6.4), disturbed samples for index testing and use of subsequent empirical correlations (Section 5.9), undisturbed samples for laboratory testing (Chapters 5 and 6) and in-situ measurements of soil properties (this chapter). These independent observations should be used to support each other in identifying and characterising the deposits of soil in the ground and producing a detailed and accurate ground model for subsequent geotechnical analyses (Part 2 of this book).

7.2 Standard Penetration Test (SPT)

The SPT is one of the oldest and most widely used in-situ tests worldwide. The technical standards governing its use are EN ISO 22476, Part 3 (UK and Europe) and ASTM D1586 (US). Its popularity is largely due to its low cost and simplicity, and the fact that testing may be conducted rapidly as a borehole is drilled. A borehole is first drilled (using casing where appropriate) to just above the test depth. A **split-barrel sampler** (Figure 7.1) with a smaller diameter than the borehole is then attached to a string of rods and driven into the soil at the base of the borehole by a **drop hammer** (a known mass falling under gravity from a known height). An initial seating drive to 150 mm penetration is first performed to embed the sampler into the soil. This is followed by the test itself, in which the sampler is driven further into the soil by 300 mm (this is usually marked off on the rod string at the surface). The number of blows of the hammer to achieve this penetration is recorded; this is the (uncorrected) **SPT blowcount**, N.

A wide range of equipment is used worldwide to undertake testing which influences the amount of energy transferred to the sampler with each blow of the drop hammer. The constitutive properties of a given soil deposit should not vary with the equipment used, and so N is conventionally corrected to a value N_{60}, representing a standardised energy ratio of 60%. The blowcount also needs to be corrected for the size of the borehole and for tests done at shallow depths (<10 m). These corrections are achieved using

In-situ testing

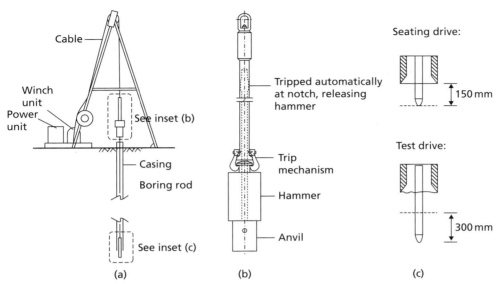

Figure 7.1 The SPT test: (a) general arrangement, (b) UK standard hammer system, (c) test procedure.

$$N_{60} = N\zeta \left(\frac{ER}{60} \right) \qquad (7.1)$$

where ζ is the correction factor for rod length (i.e. test depth) and borehole size from Table 7.1, and ER is the Energy Ratio of the equipment used. BS EN ISO 22476 (2005) describes how ER may be measured for any SPT apparatus; for most purposes, however, it is sufficient to use the values given in Table 7.2 (after Skempton, 1986).

Interpretation of SPT data in coarse-grained soils (I_D, ϕ'_{max})

The SPT is most suited to the investigation of coarse-grained soils. In sands and gravels, the corrected blowcounts are further normalised to account for the overburden pressure (σ'_{v0}) at the test depth, as the penetration resistance will naturally increase with stress level and this may mask smaller changes in constitutive properties. The **normalised blowcount** $(N_1)_{60}$ is obtained from

Table 7.1 SPT correction factor ζ (after Skempton, 1986)

Rod length/depth (m)	Borehole diameter (mm)		
	65–115	150	200
3–4	0.75	0.79	0.86
4–6	0.85	0.89	0.98
6–10	0.95	1.00	1.09
>10	1.00	1.05	1.15

Development of a mechanical model for soil

Table 7.2 Common Energy ratios in use worldwide (after Skempton, 1986)

Country	ER (%)
UK	60
USA	45–55
China	55–60
Japan	65–78

$$(N_1)_{60} = C_N N_{60} \qquad (7.2)$$

where C_N is the overburden correction factor and is given by:

$$C_N = \frac{A}{B + \sigma'_{v0}} \qquad (7.3)$$

In Equation 7.3, σ'_{v0} should be entered in kPa, and A and B vary with density, coarseness and OCR. For normally consolidated (NC) fine sands ($D_{50} < 0.5$ mm) of medium relative density ($I_D \approx 40$–60%), $A = 200$ and $B = 100$; for overconsolidated (OC) fine sands, $A = 170$ and $B = 70$; for dense coarse sands ($D_{50} > 0.5$ mm, $I_D \approx 60$–80%) $A = 300$ and $B = 200$ (Skempton, 1986). The differences between the three classes of soil are most pronounced at lower values of σ'_{v0} (i.e. at shallower depths), as shown in Figure 7.2.

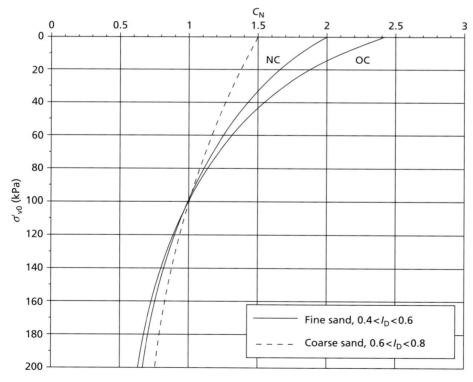

Figure 7.2 Overburden correction factors for coarse-grained soils (after Skempton, 1986).

In-situ testing

The density of a coarse-grained deposit may then be found from $(N_1)_{60}$. Skempton (1986) proposed that for most natural deposits $(N_1)_{60}/I_D^2 \approx 60$ would provide a reasonable estimate of I_D if nothing else is known about the deposit. As a cautionary note however, Skempton's own case studies showed $35 < (N_1)_{60}/I_D^2 < 85$, so the SPT can only provide an estimate at best of the relative density. However as determination of a more precise density in coarse-grained soils requires estimation of the in-situ void ratio e and the use of Equation 1.23 (requiring high quality, usually frozen or resin-injected samples, which is expensive), SPT data are extremely valuable. Skempton further showed that for more recent depositions, $(N_1)_{60}/I_D^2$ reduces as shown in Figure 7.3. This is important when analysing test data from hydraulic fills, such as those used to build artificial islands (e.g. The World and Palm developments in Dubai; Chek Lap Kok airport, Hong Kong), or when testing in river sediments.

In Chapter 5 it was demonstrated that density (i.e. I_D) is one of the key parameters governing the peak strength of coarse-grained soil. As the normalised SPT blowcount can be correlated against density, it follows that it may further be correlated to the peak angle of shearing resistance, ϕ'_{max}. This is also mechanically reasonable, as the SPT involves penetration of the soil (i.e. continuously exceeding its capacity) such that the soil resistance is expected to be governed by the peak strength. Figure 7.4 shows correlations for silica sands and gravels from Stroud (1989) and based on the suggestions from Eurocode 7: Part 2 (2007), assuming $(N_1)_{60}/I_D^2 = 60$. The former data shows the effect of overconsolidation on the interpretation of ϕ'_{max}, where increasing OCR is shown to reduce the peak friction angle, as expected from Chapter 5. The ranges for the latter represent bounds on the uniformity of the soil from uniform (lower bound) to well-graded (upper bound).

Interpretation of SPT data in fine-grained soils (c_u)

The SPT may, in principle, be used in fine-grained soils to determine an estimate of the in-situ undrained shear strength (the test is rapid so undrained conditions can be assumed). Correlations between c_u and blowcount depend on a number of factors in such soils, including OCR and any resultant fissuring, soil plasticity (I_P) and sensitivity (S_t). This high level of dependency means that SPT data should normally by used qualitatively in such soils, to support other in-situ and laboratory test data. However, if enough experience can be gained in a certain type of soil, reliable soil-specific correlations can be developed to provide quantitative data (EC7–2, 2007).

As an example, Stroud (1989) demonstrated that for overconsolidated UK clays, $c_u/N_{60} \approx 5$ for $I_P > 30\%$. For clays at lower plasticity index, this value increases to approximately $c_u/N_{60} = 7$ at $I_P = 15\%$.

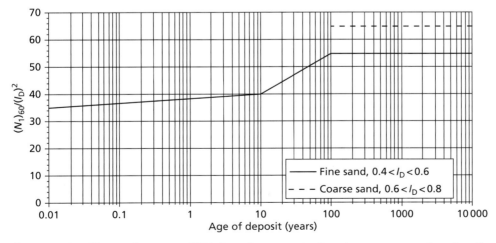

Figure 7.3 Effect of age on SPT data interpretation in coarse-grained soils.

Development of a mechanical model for soil

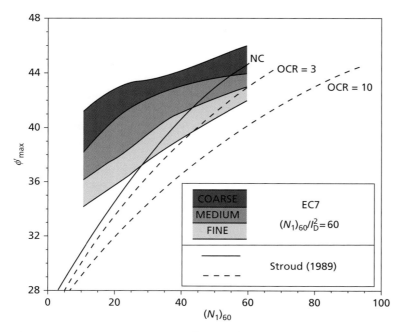

Figure 7.4 Determination of ϕ'_{max} from SPT data in coarse-grained soils.

This grouping of clays has a number of similarities, generally being overconsolidated, fissured and insensitive. The scatter in the correlation is consequently low and the SPT can be used in such soils with some confidence, which explains its popularity in UK practice. It should be noted that the non-normalised blow-count (N_{60}) is used as c_u is a total stress parameter, and is therefore independent of the effective stress in the ground. Clayton (1995) showed, by using London Clay (which was part of Stroud's database), that if the fissuring is removed by remoulding such soils, $c_u/N_{60} \approx 11$ – i.e. in unfissured soils, a higher value of c_u/N_{60} should be used when interpreting SPT data in fine-grained soils. This is consistent with US practice, where $c_u/N_{60} = 10$ is routinely used (Terzaghi and Peck, 1967). In sensitive soils, Schmertman (1979) suggested that the sides of the SPT sampler, which contribute approximately 70% of the penetration resistance in clays, are generally governed by the remoulded strength, while the base is influenced by the undisturbed undrained shear strength in the soil beneath the sampler. This suggests that in sensitive soils,

$$\frac{c_u}{N_{60}} = \frac{CS_t}{0.7 + 0.3S_t} \tag{7.4}$$

where C is the value of c_u/N_{60} for an insensitive clay ($S_t = 1$). To use Equation 7.4, the sensitivity may be estimated using Equation 5.45. Combining Equation 7.4 and the other recommendations outlined in this section, a tentative interpretation for SPT data in fine-grained soils is shown in Figure 7.5.

7.3 Field Vane Test (FVT)

In contrast to the SPT test, which is predominantly used for coarse-grained soils, the FVT test is principally used for the in-situ determination of the undrained strength characteristics of intact, fully saturated clays. Silts and glacial tills may also be characterised using this method, though the reliability of such data is more questionable and should be supported with other test data where possible. The test is not suitable for coarse-grained soils. In particular, the FVT is very suitable for soft clays, the shear strength

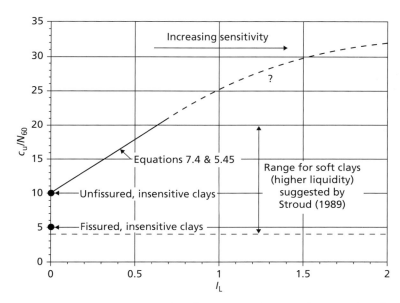

Figure 7.5 Estimation of c_u from SPT data in fine-grained soils.

of which, if measured in the laboratory, may be significantly altered by the sampling process and subsequent handling. Generally, this test is only used in clays with $c_u < 100$ kPa. The test may not give reliable results if the clay contains sand or silt laminations.

The technical standards governing its use are EN ISO 22476, Part 9 (UK and Europe) and ASTM D2573 (US). The equipment consists of a stainless steel **vane** (Figure 7.6) of four thin rectangular blades at 90° to each other, carried on the end of a high-tensile steel rod; the rod is enclosed by a sleeve packed with grease. The length of the vane is equal to twice its overall width, typical dimensions being 150 mm by 75 mm and 100 mm by 50 mm. Preferably, the diameter of the rod should not exceed 12.5 mm.

The vane and rod are pushed into the soil below the bottom of a borehole to a depth of at least three times the borehole diameter; if care is taken this can be done without appreciable disturbance of the clay. Steady bearings are used to keep the rod and sleeve central in the borehole casing. In soft clays, tests may be conducted without a borehole by direct penetration of the vane from ground level; in this case a shoe is required to protect the vane during penetration. Small, hand-operated vanes are also available for use in exposed clay strata.

Torque is applied gradually to the upper end of the rod until the clay fails in shear due to rotation of the vane. Shear failure takes place over the surface and ends of a cylinder having a diameter equal to the overall width of the vane. The rate of rotation of the vane should be within the range of 6–12° per minute. The shear strength is calculated from the expression

$$T = \pi c_{uFV} \left(\frac{d^2 h}{2} + \frac{d^3}{6} \right) \tag{7.5}$$

where T is the torque at failure, d the overall vane width and h the vane length (see Figure 7.6). However, the shear strength over the cylindrical vertical surface may be different from that over the two horizontal end surfaces, as a result of anisotropy. The shear strength is normally determined at intervals over the depth of interest. If, after the initial test, the vane is rotated rapidly through several revolutions, the soil will become remoulded; the shear strength in this condition can then be determined if required. The ratio of the in-situ and fully remoulded values of c_u determined in this way gives the sensitivity of the soil.

Development of a mechanical model for soil

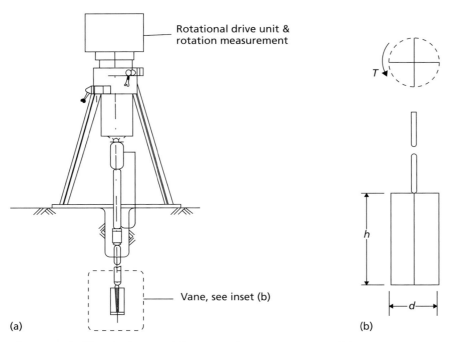

Figure 7.6 The FVT test: (a) general arrangement, (b) vane geometry.

The undrained strength as measured by the vane test is generally greater than the average strength mobilised along a failure surface in a field situation (Bjerrum, 1973). The discrepancy was found to be greater the higher the plasticity index of the clay, and is attributed primarily to differences in the rate of loading between the two cases. In the vane test shear failure occurs within a few minutes, whereas in a field situation the stresses are usually applied over a period of a few weeks or months. A secondary factor may be anisotropy. Bjerrum and, later, Azzouz et al. (1983) presented correction factors (μ), correlated empirically with I_P, as shown in Figure 7.7. The probable field strength (c_u) is then determined from the measured FVT strength (c_{uFV}) using

$$c_u = \mu \cdot c_{uFV} \tag{7.6}$$

The FVT can further be used to estimate the overconsolidation ratio (OCR) of the soil, as demonstrated by Mayne and Mitchell (1988). This is achieved using a second empirical factor, α_{FV}, where:

$$\text{OCR} = \alpha_{FV} \cdot \left(\frac{c_{uFV}}{\sigma'_{v0}} \right) \tag{7.7}$$

By considering a large database of test results from 96 different sites, it has been shown that

$$\alpha_{FV} \approx 22 (I_P)^{-0.48} \tag{7.8}$$

where I_P is entered in percent. The relationship between α_{FV} and I_P in Equation 7.8 is similar in shape to that between μ and I_P (Figure 7.7), such that $\alpha_{FV} \approx 4\mu$. Mayne and Mitchell (1988) further demonstrated good agreement between this method and the results of conventional oedometer tests for determining OCR (as previously described in Chapter 4) over a range of $I_P = 8$–100%.

In-situ testing

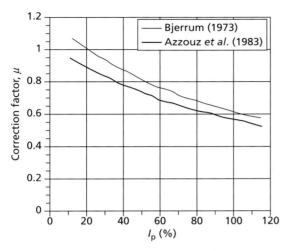

Figure 7.7 Correction factor μ for undrained strength as measured by the FVT.

Example 7.1

The FVT test data for the Bothkennar clay of Figure 5.38(b) has been provided in electronic form on the Companion Website (both peak and remoulded strengths are given). The index test data (w, w_P and w_L) from Figure 5.38(b) are also provided in electronic form. The water table is 0.8 m below ground level (BGL); the soil above the water table has a bulk unit weight of $\gamma = 18.7$ kN/m^3, while the soil below the water table has $\gamma = 16$ kN/m^3. Using these data, estimate the variation of S_t and OCR with depth. The latter should be compared with oedometer test data on the same soil shown in Figure 7.8(a).

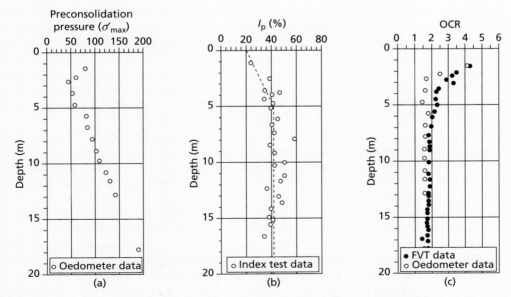

Figure 7.8 Example 7.1 (a) Oedometer test data, (b) I_p calculated from index test data, (c) OCR from FVT and oedometer data.

Development of a mechanical model for soil

> **Solution**
>
> The detailed calculations were conducted using the spreadsheet provided on the Companion Website. The sensitivity can be directly found for each test by dividing the peak strength by the remoulded value. The results of these calculations are shown in Figure 5.38(b), where they agree well with values estimated using the empirical correlations of Section 5.9. To determine OCR from the FVT data, the in-situ stress conditions (σ_{v0}, u_0 and σ'_{v0}) are found at each test depth. As the index test data were not conducted at the same depths as the FVT tests (this is common in practice), I_P is determined at each depth and an average trend determined as shown in Figure 7.8(b). This is defined by $I_P = 4.4z + 20\%$ for $0 \leq z \leq 5$ m and $I_P = 42\%$ for $z \geq 5$ m. Using these values of I_P, α_{FV} is calculated for each FVT depth from Equation 7.8. The OCR at each depth is then found using Equation 7.7. The resulting data are compared to the oedometer test data in Figure 7.8(c), where the two sets of data show similar trends, though the FVT data slightly overpredicts OCR compared to the oedometer test data.

7.4 Pressuremeter Test (PMT)

The **pressuremeter** was developed in the 1950s by Ménard to provide a high quality in-situ test which could be used to derive both strength and stiffness parameters for soil as an alternative to triaxial testing. As the test is in-situ, it overcame the problem of sampling disturbance associated with the latter; as the pressuremeter influences a much larger volume of soil than normal laboratory tests, it also ensures that the macro-fabric of the soil is adequately represented. Ménard's original design, illustrated in Figure 7.9(a), consists of three cylindrical rubber cells of equal diameter arranged coaxially. The device is lowered into a (slightly oversize) borehole to the required depth and the central measuring cell is expanded against the borehole wall by means of water pressure, measurements of the applied pressure and the corresponding increase in cell volume being recorded. Pressure is applied to the water by compressed gas (usually nitrogen) in a control cylinder at the surface. The increase in volume of the measuring cell is determined from the movement of the gas–water interface in the control cylinder. The pressure is corrected for (a) the head difference between the water level in the cylinder and the test level in the borehole, (b) the pressure required to stretch the rubber cell and (c) the expansion of the control cylinder and tubing under pressure. The two outer guard cells are expanded under the same pressure as in the measuring cell but using compressed gas; the increase in volume of the guard cells is not measured. The function of the guard cells is to eliminate end effects, ensuring a state of plane strain adjacent to the measuring cell.

In modern developments of the pressuremeter, the measuring cell is expanded directly by gas pressure. This pressure and the **cavity strain** (radial expansion of the rubber membrane) are recorded by means of electrical transducers within the cell. In addition, a pore water pressure transducer is fitted into the cell wall such that it is in contact with the soil during the test. A considerable increase in accuracy is obtained with these pressuremeters compared with the original Ménard device. It is also possible to adjust the cell pressure continuously, using electronic control equipment, to achieve a constant rate of increase in circumferential strain (i.e. a strain-controlled test), rather than applying the pressure in increments (a stress-controlled test). The technical standards governing the use of pressuremeters in pre-bored holes are EN ISO 22476, Part 5 (UK and Europe) and ASTM D4719 (US). The Ménard device is still popular in some parts of Europe; this is governed by EN ISO 22476, Part 4.

Some soil disturbance adjacent to a borehole is inevitable due to the boring process, and the results of pressuremeter tests in pre-formed holes can be sensitive to the method of boring. The **self-boring pressuremeter** (SBPM) was developed to overcome this problem, and is suitable for use in most types of soil; however, special insertion techniques are required in the case of sands. This device, illustrated in

Figure 7.9(b), is jacked slowly into the ground and the soil is broken up by a rotating cutter fitted inside a cutting head at the lower end, the optimum position of the cutter being a function of the shear strength of the soil. Water or drilling fluid is pumped down the hollow shaft to which the cutter is attached, and the resulting slurry is carried to the surface through the annular space adjacent to the shaft; the device is thus inserted with minimal disturbance of the soil. The only correction required is for the pressure required to stretch the membrane. If a self-boring pressuremeter is used, EN ISO 22476, Part 6 is the relevant standard (an ASTM standard has not yet been released).

The membrane of a pressuremeter may be protected against possible damage (particularly in coarse soils) by a thin stainless steel sheath with longitudinal cuts, designed to cause only negligible resistance to the expansion of the cell.

Like the FVT described in Section 7.3, soil parameters are derived from the cell pressure and cell volume change (or cavity strain) using a theoretical model (see Equation 7.5) rather than empirical correlations. These analyses will be described in the following sections.

Interpretation of PMT data in fine-grained soils (G, c_u)

In fine-grained soils, the following analysis is derived from Gibson and Anderson (1961). During the pressuremeter test, the cavity (borehole) is expanded radially from its initial radius of r_c to a new radius $r_c + y_c$ by an amount y_c (the displacement at the cavity wall). There is no displacement or strain in the

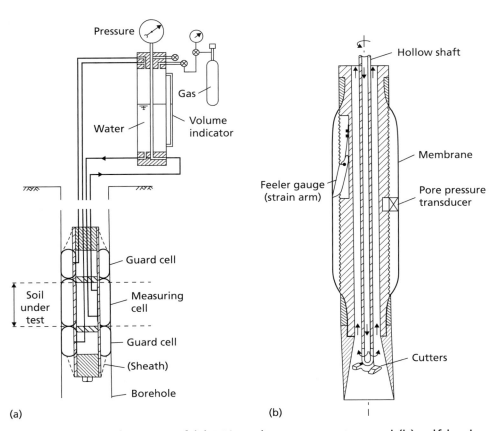

Figure 7.9 Basic features of (a) Ménard pressuremeter, and (b) self-boring pressuremeter.

Development of a mechanical model for soil

vertical direction (along the axis of the borehole) – these conditions are known as **plane strain**, as the soil may only strain in a single plane (in this case, horizontal). During this expansion, the volume of the cavity increases by an amount dV. The soil at any radius r away from the centre of the cavity similarly expands from its initial radius r to a new radius $r+y$ by an amount y. In an undrained test there must be no overall change in the volume of the soil, so the change in volume of the soil expanding from r to $r+y$ must be equal to dV (i.e. the two shaded annuli in Figure 7.10(a) must have equal areas), giving

$$2\pi r y dz = dV$$
$$y = \frac{dV}{2\pi r dz} \tag{7.9}$$

The soil may strain radially by an amount ε_r and circumferentially by an amount ε_θ. The strain in the axial direction $\varepsilon_a = 0$ (plane strain). If the soil is isotropic, ε_r, ε_θ and ε_a are principal strains. The shear strain in the soil is then given by

$$\gamma = \varepsilon_r - \varepsilon_\theta \tag{7.10}$$

And the volumetric strain ε_v by

$$\varepsilon_v = \varepsilon_r + \varepsilon_\theta + \varepsilon_a \tag{7.11}$$

In an undrained test $\varepsilon_v = 0$, so from Equation 7.11, $\varepsilon_r = -\varepsilon_\phi$ and from Equation 7.10

$$\gamma = -2\varepsilon_\theta \tag{7.12}$$

The circumference of the soil annulus is initially $2\pi r$ (Figure 7.10(a)) and increases to $2\pi(r+y)$ giving an extension of $2\pi y$. Choosing compression of the soil as positive, the circumferential strain is then:

$$\varepsilon_\theta = -\frac{2\pi y}{2\pi r} = -\frac{y}{r} \tag{7.13}$$

Substituting Equations 7.13 and 7.9 into Equation 7.12 gives the equation of compatibility for the pressuremeter test (cylindrical cavity expansion):

$$\gamma = \frac{2y}{r} = \frac{dV}{\pi r^2 dz} \tag{7.14}$$

Equation 7.14 was derived by considering compatibility of the soil displacements from the cavity wall outwards. As described in Section 5.1, equilibrium must also be satisfied within the soil mass. The stresses acting on a segment of the soil along the annulus defined by r is shown in Figure 7.10(b). For there to be radial equilibrium:

$$(\sigma_r + d\sigma_r)(r + dr)d\theta = \sigma_r r d\theta + \sigma_\theta dr d\theta$$
$$\therefore r\frac{d\sigma_r}{dr} + (\sigma_r - \sigma_\theta) = 0 \tag{7.15}$$

The stresses σ_r and σ_θ are principal stresses, being associated with the principal strains ε_r and ε_ϕ respectively. Considering the Mohr Circle represented by these stresses, the associated maximum shear stress is $\tau = (\sigma_r - \sigma_\theta)/2$, which when substituted into Equation 7.15 gives the equation of equilibrium for the pressuremeter test (cylindrical cavity expansion):

$$r\frac{d\sigma_r}{dr} + 2\tau = 0 \tag{7.16}$$

In-situ testing

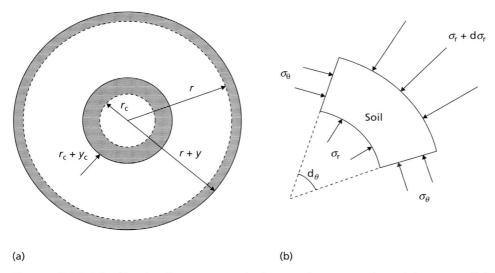

Figure 7.10 Idealised soil response during cavity expansion: (a) compatible displacement field, (b) equilibrium stress field.

Having considered both compatibility and equilibrium, it only remains to consider the constitutive model to link the shear stresses (in Equation 7.16) with the shear strains (Equation 7.14).

Linear elastic soil behaviour
While the soil is behaving elastically (Figure 7.11(a)), the constitutive relationship is given by

$$\tau = G\gamma \tag{7.17}$$

Substituting Equations 7.14 and 7.17 into Equation 7.16 gives

$$r\frac{d\sigma_r}{dr} + 2G\frac{dV}{\pi r^2 dz} = 0 \tag{7.18}$$

This is the governing differential equation governing the response of all of the soil around the pressuremeter.

At the cavity wall, $r=r_c$ and $\sigma_r=p$ (where p is the pressure within the pressuremeter); far from the cavity the soil is unaffected by the pressuremeter expansion so that at $r=\infty$, $\sigma_r=\sigma_{h0}$ where σ_{h0} is the in-situ horizontal total stress in the ground (also termed the **lift-off pressure**). Equation 7.18 may then be integrated using these limits:

$$\int_{\sigma_{h0}}^{p} d\sigma_r = -\int_{\infty}^{r_c} \frac{2G dV}{\pi r^3 dz} dr$$

$$\left[\sigma_r\right]_{\sigma_{h0}}^{p} = -2G\frac{dV}{\pi}\left[-\frac{1}{2r^2}\right]_{\infty}^{r_c}$$

$$p - \sigma_{h0} = G\frac{dV}{\pi r_c^2}$$

Development of a mechanical model for soil

Recognising that πr_c^2 is the volume of the cavity (V), the following relationship is obtained:

$$p = G\frac{dV}{V} + \sigma_{h0} \tag{7.19}$$

Equation 7.19 suggests that, for elastic soil behaviour, if the cavity pressure p is plotted against the volumetric strain in the cavity dV/V, a straight line will be given, the gradient of which gives the soil stiffness G and the intercept gives the initial total horizontal stress at the test depth, σ_{h0}. This is shown in Figure 7.11(b).

Elasto-plastic soil behaviour

In reality, the soil cannot remain elastic forever and will yield when the shear stress reaches τ_{max}. Pressuremeter tests are usually conducted rapidly compared to the consolidation time required in most fine-grained soils so the behaviour is undrained and $\tau_{max} = c_u$ (Figure 7.12(a)). Yield will occur when the cavity pressure reaches p_y defined by

$$p_y = \sigma_{h0} + c_u \tag{7.20}$$

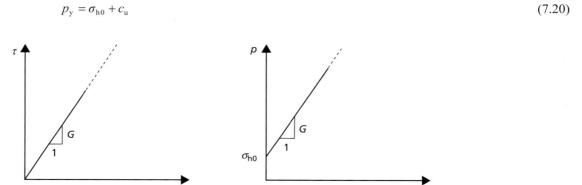

Figure 7.11 Pressuremeter interpretation during elastic soil behaviour: (a) constitutive model (linear elasticity), (b) derivation of G and σ_{h0} from measured p and dV/V.

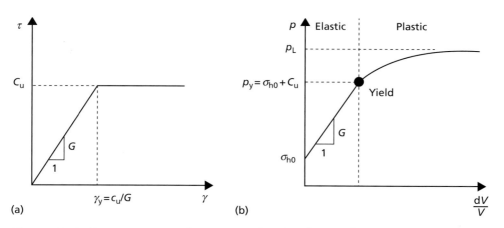

Figure 7.12 Pressuremeter interpretation in elasto-plastic soil: (a) constitutive model (linear elasticity, Mohr–Coulomb plasticity), (b) non-linear characteristics of measured p and dV/V.

In-situ testing

This is shown in Figure 7.12(b). Yielding will not occur in all of the soil (to $r=\infty$) simultaneously. There will be a zone of yielding soil immediately around the cavity out to a radius $r=r_y$ where the stresses are highest – in this zone the radial stresses will be $\sigma_r=p_y$ everywhere. Outside of this zone (i.e. from $r_y<r<\infty$), the soil will be elastic. In the plastic zone, $\tau=c_u$ everywhere. Substituting this into Equation 7.16 gives

$$r\frac{d\sigma_r}{dr}+2c_u = 0 \tag{7.21}$$

Equation 7.21 is only valid within the plastic zone, so that the equation should be integrated between the limits of $\sigma_r=p_y$ at $r=r_y$ to $\sigma_r=p$ at $r=r_c$ (as before):

$$\int_{p_y}^{p} d\sigma_r = -\int_{r_y}^{r_c} \frac{2c_u}{r} dr$$

$$p-p_y = 2c_u \ln\left(\frac{r_y}{r_c}\right) \tag{7.22}$$

In order to make use of Equation 7.22, the parameter r_y needs to be expressed in terms of familiar parameters. Irrespective of whether the soil is elastic or plastic, there must be no overall change in volume in an undrained test. Therefore, from Equation 7.14

$$\gamma_y = \frac{2y_y}{r_y} \tag{7.23}$$

and from Equation 7.9

$$y_y r_y = y_c r_c \tag{7.24}$$

Equations 7.23 and 7.24 are combined to eliminate the unknown y_y, giving

$$\gamma_y = \frac{2y_y}{r_y}$$

$$\left(\frac{r_y}{r_c}\right)^2 = \frac{1}{\gamma_y}\left(\frac{2y_c}{r_c}\right) \tag{7.25}$$

From Figure 7.12(a) $\gamma_y=c_u/G$, and from Equation 7.14 $2y_c/r_c=dV/V$. Substituting these relationships into Equation 7.25 gives

$$\left(\frac{r_y}{r_c}\right)^2 = \frac{G}{c_u}\left(\frac{dV}{V}\right) \tag{7.26}$$

Substituting Equations 7.20 and 7.26 into Equation 7.22 gives

$$p = c_u \ln\left(\frac{dV}{V}\right)+\sigma_{h0}+c_u+c_u \ln\left(\frac{G}{c_u}\right). \tag{7.27}$$

Equation 7.27 suggests that, for linearly elastic–perfectly plastic soil behaviour, if the cavity pressure p is plotted against the logarithm of the volumetric strain in the cavity $\ln(dV/V)$, the data will approach a

Development of a mechanical model for soil

straight line asymptote as dV/V gets large (i.e. as $\ln(dV/V)$ tends towards zero). The gradient of this asymptote will be the undrained shear strength c_u, as shown in Figure 7.13. Furthermore, $\ln(dV/V) = 0$ represents the limiting case of infinite expansion in the soil. This corresponds to an ultimate or **limit pressure** p_L, which is shown in Figures 7.12(b) and 7.13. It should be noted that this pressure is impossible to achieve in practice.

Practical derivation of soil parameters

In practice, pressuremeter test data are usually plotted as a graph of cavity pressure versus cavity strain ε_c (Figure 7.14). By considering the strains at the cavity wall due to a change in cavity volume from V to $V + dV$, it can be shown that

$$\frac{dV}{V} = 1 - \left(\frac{1}{1+\varepsilon_c}\right)^2 \tag{7.28}$$

At most stages of a pressuremeter test, the strains are small enough that Equation 7.28 can be approximated by

$$\frac{dV}{V} \approx 2\varepsilon_c \tag{7.29}$$

From Equation 7.29, the volumetric change is directly proportional to the cavity strain at small strains. Therefore the graph of p versus ε_c will have the same shape as Figure 7.12(b), so that σ_{h0} may be directly read from the graph, and G determined from the gradient of the curve at any point. It should be noted that unlike in Figure 7.11(b) where the gradient was G, on a plot of p versus dV/V, the gradient will be $2G$ from Equation 7.29, i.e.

$$G = \frac{1}{2}\left(\frac{dp}{d\varepsilon_c}\right) \tag{7.30}$$

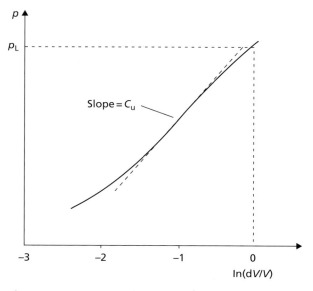

Figure 7.13 Determination of undrained shear strength from pressuremeter test data.

In-situ testing

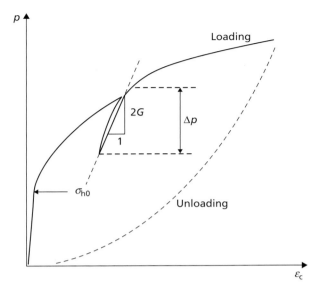

Figure 7.14 Direct determination of G and σ_{h0} in fine-grained soils from pressuremeter test data.

after Palmer (1972). Rather than find G from the slope of the initial elastic portion of the p–ε_c curve as suggested by Figure 7.11(b), in practice the modulus is obtained from the slope of an unloading–reloading cycle as shown in Figure 7.14, ensuring that the soil remains in the 'elastic' state during unloading. Wroth (1984) has shown that, in the case of a clay, this requirement will be satisfied if the reduction in pressure during the unloading stage is less than $2c_u$, i.e.

$$\Delta p < 2c_u \tag{7.31}$$

Most modern pressuremeters have multiple strain arms arranged in diametrically opposite pairs around the cylindrical body of the pressuremeter. The lift-off pressure (σ_{h0}) is normally an average value inferred from all of the gauges. However the data from individual pairs of gauges can be used to infer differences in in-situ σ_{h0} within the ground, i.e. stress anisotropy (Dalton and Hawkins, 1982).

Example 7.2

Figure 7.15 shows data from a self-boring pressuremter test undertaken at a depth of 8 m below ground level in the Gault clay shown in Figure 5.38(a). The raw data for this test is provided in electronic form on the Companion Website. Determine the following parameters: σ_{h0}, c_u, G (for both unload–reload loops conducted during the test).

Solution

The in-situ horizontal total stress can be directly estimated from inspection of the p–ε_c curve in Figure 7.15, where the lift-off pressure $\sigma_{h0} \approx 395$ kPa. The values of G for each of the unload–reload loops is found by plotting the p–ε_c data over the range of the loop only and then fitting a straight line to the data. The spreadsheet on the Companion Website shows this procedure using the trendline fit in MS Excel. Note that it was necessary to vary the intercept of the trendline to force the fitting line along the major axis of the loop. From Equation 7.30 the gradients of a trendline = $2G$, giving values of $G = 32.3$ and 27.0 MPa for the first and second loops, respectively. To obtain the undrained shear strength, the cavity strains are converted into volumetric strain (dV/V) using Equation 7.28 and the data are replotted in the form of Figure 7.13. This is also demonstrated within the spreadsheet on the Companion Website. A straight line is fitted to the data, the gradient of which gives $c_u = 111$ kPa. This compares favourably with c_u measured from triaxial tests at this depth from Figure 5.38(a).

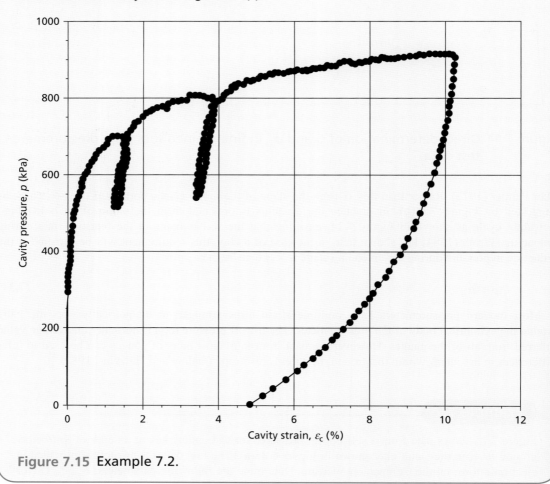

Figure 7.15 Example 7.2.

Interpretation of PMT data in coarse-grained soils (G, ϕ', ψ)

The analysis of the pressuremeter test in a drained soil is similar to the analysis described in the previous section for undrained soil, involving the formation of equations of compatibility, equilibrium and constitutive behaviour, but with stresses now expressed in terms of effective rather than total components. However, use can no longer be made of the 'no volume change' criterion such that the dilatancy of the

In-situ testing

soil must be considered. This makes the constitutive law more complex. A complete analysis is given by Hughes *et al.* (1977). The analysis enables values for the angle of shearing resistance (ϕ') and the angle of dilation (ψ) to be determined, and the derivation of these parameters from pressuremeter test data is described below.

It is typical, as for fine-grained soils, to plot cavity pressure versus cavity strain. A typical test curve is shown in Figure 7.16. If the total cavity pressure is plotted (Figure 7.16(a)), the lift-off pressure defines σ_{h0} as before. If following cavity expansion the pressuremeter is completely unloaded, the cavity pressure at which ε_c returns to zero represents the initial pore pressure within the ground (u_0). This pore pressure is constant throughout the test, as no excess pore water pressures are generated during drained shearing of soil (see Chapter 5). Unload–reload cycles are usually conducted to determine G. The gradient of these loops is $2G$ as before (Equation 7.30). The soil will remain completely elastic during these stages as long as the reduction in pressure during the unloading stage satisfies

$$\Delta p < \frac{2\sin\phi'}{1+\sin\phi'}(p-u_0) \tag{7.32}$$

It is common to subsequently correct all of the cavity pressures by subtracting the value of u_0 to give the effective cavity pressure $p-u_0$ (Figure 7.16(b)). The lift-off pressure identified from this graph then represents the in-situ effective horizontal stress σ'_{h0}.

In order to determine the strength parameters (ϕ' and ψ), the data are replotted on different axes (see Figure 7.13). In the case of drained analysis, the data are replotted as $\log(p-u_0)$ versus $\log(\varepsilon_c)$; alternatively, the corrected data may be replotted on log–log axes. The data should then lie approximately on a straight line, the gradient of which is defined as s (Figure 7.17). Once the value of s has been determined, ϕ' and ψ can be estimated using

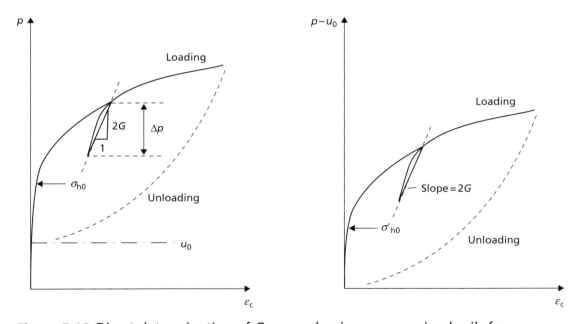

Figure 7.16 Direct determination of G, σ_{h0} and u_0 in coarse-grained soils from pressuremeter test data: (a) uncorrected curve, (b) corrected for pore pressure u_0.

Development of a mechanical model for soil

$$\sin \phi' = \frac{s}{1+(s-1)\sin \phi'_{cv}} \tag{7.33}$$

$$\sin \psi = s + (s-1)\sin \phi'_{cv} \tag{7.34}$$

The angle of shearing resistance ϕ' from Equation 7.33 represents the peak value (ϕ'_{max}). Equations 7.33 and 7.34 have been plotted in Figure 7.18 for graphical solution. Interpretation of the strength data relies on knowing the critical state angle of shearing resistance for the soil (ϕ'_{cv}). It is recommended that this is found from drained triaxial tests on loose samples of the soil such that the peak and critical state strengths are coincident (ϕ'_{cv} is independent of density, Chapter 5). If these data are not available, a value may be estimated from index test data using Figure 5.35.

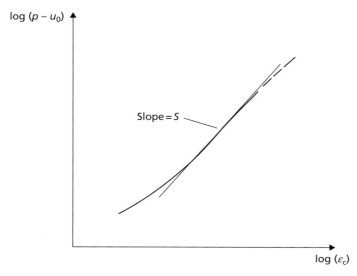

Figure 7.17 Determination of parameter s from pressuremeter test data.

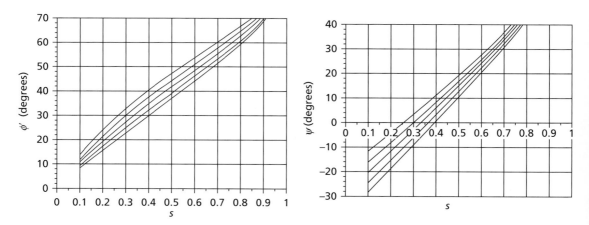

Figure 7.18 Determination of ϕ' and ψ from parameter s.

In-situ testing

> **Example 7.3**
>
> Figure 7.19 shows data from a self-boring pressuremter test undertaken in a deposit of sand. The raw data for this test are provided in electronic form on the Companion Website. Determine the following parameters: σ_{h0}, u_0, σ'_{h0}, ϕ'_{max} and ψ. State also whether the unload–reload loops have been conducted over an appropriate (fully elastic) stress range.
>
>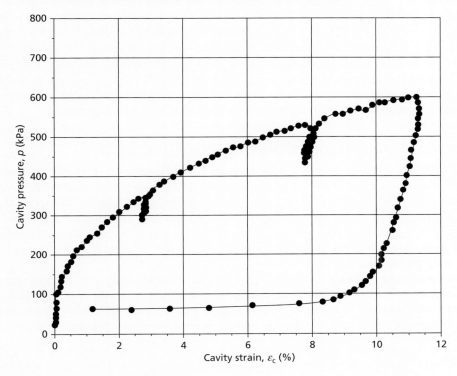
>
> **Figure 7.19** Example 7.3.
>
> ## Solution
>
> The in-situ horizontal total stress can be directly estimated from inspection of the p–ε_c curve in Figure 7.19, where the lift-off pressure $\sigma_{h0} \approx 107\,\text{kPa}$. The in-situ pore pressure is given by the intercept of the unloading portion of the curve at $\varepsilon_c = 0$ (i.e. cavity pressure at the end of the test), giving $u_0 = 63\,\text{kPa}$. The effective horizontal stress is then found using Terzaghi's Principle: $\sigma'_{h0} = 107 - 63 = 44\,\text{kPa}$. To obtain the strength parameters, the cavity pressures are corrected for u_0 and the data replotted in the form of Figure 7.17. This is shown within the spreadsheet on the Companion Website. A straight line is fitted to the data, the gradient of which gives $s = 0.50$. This is then used either in Equations 7.33 and 7.34, or in Figure 7.18, to give $\phi' = \phi'_{max} = 41.2°$ and $\psi = 14.5°$. Finally, Δp is calculated using Equation 7.32 for each data point using the corrected cavity pressure ($p' = p - u_0$) and the value of ϕ' found in the previous step. This is subtracted from the values of p (ignoring the unload–reload loops themselves) to give a curved locus, offset from the test data by an amount Δp (shown in the worksheet on the Companion Website). The unload–reload loops are not unloaded below this line, so the behaviour is expected to be fully elastic.

Development of a mechanical model for soil

7.5 Cone Penetration Test (CPT)

The CPT test was described in Chapter 6, where its use in identifying and profiling the different strata within the ground was demonstrated. The standards governing its use as an in-situ testing tool are EN ISO 22476, Part 1 (UK and Europe) and ASTM D5778 (US). The data collected by the CPT during profiling may further be used to estimate a range of soil properties, via empirical correlations. The CPT is a much more sophisticated test than either the SPT or FVT tests described previously, which only measure a single parameter (blowcount and maximum torque respectively); even the basic cone (CPT) measures two independent parameters (q_c and f_s), while a piezocone (CPTU) extends this to three independent parameters (u_2, added to the previous two) and the most sophisticated seismic cone (SCPTU) measures four parameters (q_c, f_s, u_2 and V_s). As a result, the CPT can be used to reliably estimate a wider range of soil properties, including strength, stiffness, state and consolidation parameters. Furthermore, unlike the SPT, FVT or PMT, where measurements may only be taken at discrete points, the CPT measures continuously, so that by using the interpolated soil profile the complete variation of correlated soil properties with depth may also be determined.

Interpretation of CPT data in coarse-grained soils (I_D, ϕ'_{max}, G_0)

A large database of CPT data in coarse-grained soils is available in the literature. In such soils, q_c is normally used in correlations as increases in density or soil strength will increase the resistance to penetration. Sleeve friction (f_s) is usually small and of little use in interpretation, other than for identifying the soil in question as coarse-grained. As the permeability of coarse-grained deposits is usually high (Table 2.1) it is not necessary to correct q_c for pore pressure effects (Equation 6.2) such that $q_t \approx q_c$, and a basic cone is suitable for most testing. In most correlations it is usual to correct the cone resistance for overburden stresses by using the parameter $q_c/(\sigma'_{v0})^{0.5}$.

Figure 7.20 shows correlations between I_D and $q_c/(\sigma'_{v0})^{0.5}$ for a database of nearly 300 tests in a range of normally consolidated (NC) silica and carbonate sands, collated by Jamiolkowski et al. (2001) and Mayne (2007). There is a considerable amount of scatter shown, which is predominantly a function of the compressibility of the soil. The best-fit lines for use in the interpretation of new data are given by

$$I_D = D + E \log\left(\frac{q_c}{\sigma'^{0.5}_{v0}}\right) \quad (7.35)$$

where q_c and σ'_{v0} are in kPa. For silica sands of average compressibility, $D = -1.21$ and $E = 0.584$ (best-fit line shown in Figure 7.20). For highly compressible silica sands D may be as high as -1.06 (upper-bound envelope to test data), while for very low compressibility soils D may be as low as -1.36 (lower bound envelope to test data); the value of E (the gradient of the line) is insensitive to soil compressibility. Carbonate sands are more highly compressible than silica sands due to their highly crushable grains, and the data for such soils therefore lie above the silica data (i.e. on the highly compressible side of the graph). The relationship between I_D and q_c for these soils may still be characterised using Equation 7.35, but with $D = -1.97$ and $E = 0.907$ (line also shown in Figure 7.20).

As with the SPT test, CPT data in coarse-grained soils may further be correlated against ϕ'_{max} as shown in Figure 7.21. The data used in determining this correlation are from Mayne (2007), and show very low scatter for soils with low fines content. The best-fit lines for use in the interpretation of new data are given by

$$\phi'_{max} = 6.6 + 11 \log\left(\frac{q_c}{\sigma'^{0.5}_{v0}}\right) \quad (7.36)$$

where ϕ'_{max} is in degrees and q_c and σ'_{v0} are in kPa as before.

In-situ testing

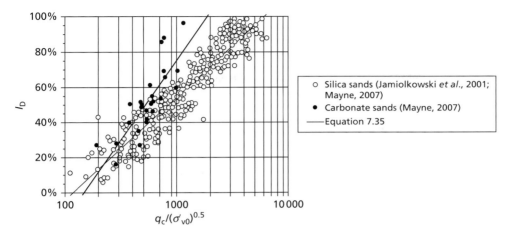

Figure 7.20 Determination of I_D from CPT/CPTU data.

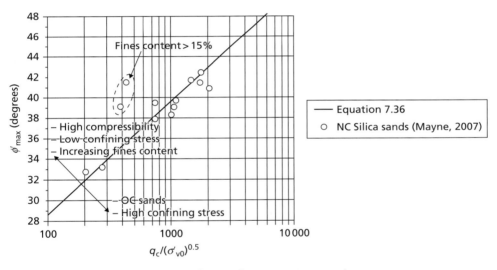

Figure 7.21 Determination of ϕ'_{max} from CPT/CPTU data.

The use of a seismic cone (SCPTU) allows for discrete seismic soundings to be made during a CPT test such that the shear wave velocity is additionally determined. The small strain shear modulus (G_0) can then be determined as for the geophysical methods described in Section 6.7 using Equation 6.6:

$$G_0 = \rho V_s^2$$

Interpretation of CPT data in fine-grained soils (c_u, *OCR*, K_0, ϕ'_{max}, G_0)

In fine-grained soils, CPT data are most commonly used to assess the in-situ undrained shear strength of the soil. As mentioned previously, the CPT provides these data continuously over the full depth of such a layer, unlike triaxial tests on undisturbed samples which can only give a limited number of discrete values. CPT data should always be calibrated against another form of testing (e.g. UU triaxial

Development of a mechanical model for soil

compression tests or FVT data) in a given material, but once this has been done the CPT can then be used directly to determine c_u at other locations within the same geological unit.

The calibration process described above varies slightly depending on the type of cone used, though the principle is identical in each case. If only basic CPT data are available, c_u is determined using:

$$c_u = \frac{q_c - \sigma_{v0}}{N_k} \tag{7.37}$$

where N_k is the 'calibration factor'. This is determined by using the results of a series of laboratory tests (e.g. UU triaxial test), from which c_u is known, and interpolating the value of q_c and σ_{v0} from the CPT log at the depths for which the laboratory tests were sampled (see Example 7.4). Once an appropriate average value of N_k has been determined for a given unit of soil, Equation 7.37 is then applied to the full CPT log to determine the variation of c_u with depth. Figure 7.22(a) shows reported values of N_k as a function of plasticity index for different fine-grained soils for general guidance and checking of N_k. It will be seen that $N_k = 15$ is normally a good first approximation, though in fissured clays (those shown are from the UK) the value can be significantly higher (i.e. using a value of $N_k = 15$ would overestimate c_u in a fissured clay).

If CPTU data are available, the process is the same; however q_t replaces q_c so that excess pore pressures generated during penetration are corrected for. With this modification, calibration factors will not be the same as described above and it is conventional to then modify Equation 7.37 to read

$$c_u = \frac{q_t - \sigma_{v0}}{N_{kt}} \tag{7.38}$$

where N_{kt} is the calibration factor for CPTU data. Figure 7.22(b) shows reported values of N_{kt} which can be seen to be a function of the pore pressure parameter B_q (defined in Figure 6.12). The scatter in the data here is much lower as, between them, q_t and B_q implicitly include the effects of overconsolidation (see following discussion). The best fit line to the data is given by

$$N_{kt} = 7.2 (B_q)^{-0.77} \tag{7.39}$$

It should be noted that in Figure 7.22(a), the reference value of c_u is that from UU triaxial compression tests; for fissured clays these were conducted on large (100-mm diameter) samples to account for the effects of the fissures. In non-fissured clays, there should be little difference in the values of N_k or N_{kt} for different sizes of triaxial sample. If there are different units of clay indicated in a single log (e.g. soft marine deposited clay overlying fissured clay), it may be necessary to use different values of N_k or N_{kt} in the different strata.

In addition to determining the undrained strength properties of fine-grained soils, CPTU data can also be used to estimate the effective stress strength parameter ϕ'_{max}. Mayne and Campanella (2005) suggested that this parameter can be correlated to the normalised cone resistance $Q_t = (q_t - \sigma_{v0})/\sigma'_{v0}$ and the pore pressure parameter B_q by

$$\phi'_{max} \approx 29.5 (B_q)^{0.121} \left[0.256 + 0.336 B_q + \log\left(\frac{q_t - \sigma_{v0}}{\sigma'_{v0}}\right) \right] \tag{7.40}$$

Equation 7.40 is applicable for $0.1 < B_q < 1.0$. For soils with $B_q < 0.1$ (i.e. sands), Equation 7.36 should be used instead.

CPT data can also be reliably used in most fine-grained soils for detailed determination of OCR with depth, and thereby quantifying the stress history of the soil. Based on a large database of test data for non-fissured clays, Mayne (2007) suggests that OCR may be estimated using

$$\text{OCR} = 0.33\left(\frac{q_t - \sigma_{v0}}{\sigma'_{v0}}\right) \quad (7.41)$$

Equation 7.41 is compared in Figure 7.23 with data for marine deposited clays from Lunne *et al.* (1989), and it can be seen that the method is reliable in non-fissured soils. Also shown in Figure 7.23 is a zone for fissured clays based on an additional smaller database from Mayne (2007), where the coefficient in Equation 7.41 should be increased to between 0.66 and 1.65. Given the wide extent of this zone, the CPT should be considered less reliable for determining OCR in fissured soils, and should always be supported by data from other tests (e.g. oedometer test data).

The CPT may also be used to estimate the in-situ horizontal stresses in the ground. The ratio of in-situ horizontal effective stress (σ'_{h0}) to in-situ vertical effective stress (σ'_{v0}) is expressed by

$$K_0 \approx \frac{\sigma'_{h0}}{\sigma'_{v0}} \quad (7.42)$$

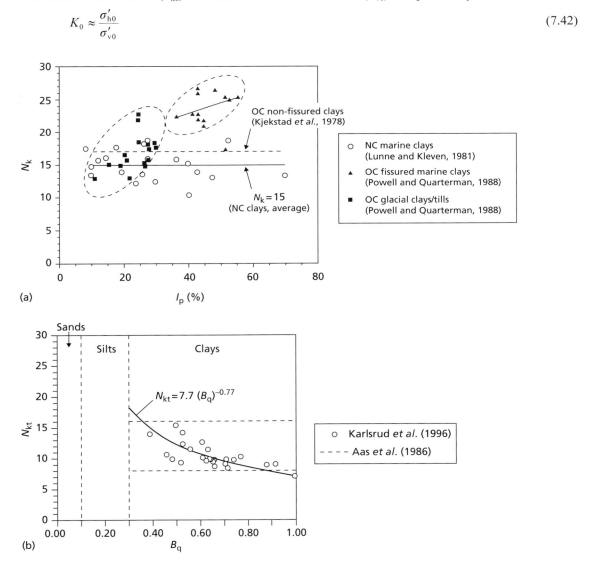

Figure 7.22 Database of calibration factors for determination of c_u: (a) N_k, (b) N_{kt}.

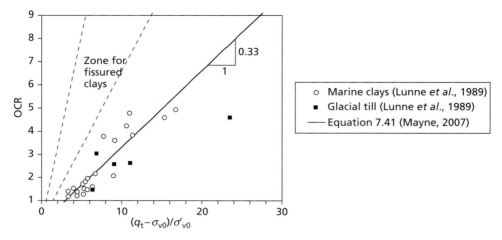

Figure 7.23 Determination of OCR from CPTU data.

where K_0 is the **coefficient of lateral earth pressure (at rest)**. Kulhawy and Mayne (1990) presented an empirical correlation for K_0 from CPTU data, meaning that the CPT can also provide information on the in-situ stress state within the ground:

$$K_0 \approx 0.1 \left(\frac{q_t - \sigma_{v0}}{\sigma'_{v0}} \right) \qquad (7.43)$$

Once σ'_{h0} has been determined using Equation 7.42 and 7.43, the total horizontal stress may then be found by adding the in-situ pore pressure (Terzaghi's Principle). The correlation represented by Equation 7.43 is shown in Figure 7.24. There is considerable scatter in the data, such that the CPT should only be used to interpret K_0 if no other data are available. If a more accurate value is required, the PMT should be used to directly measure the in-situ horizontal stresses from which K_0 may be determined using Equation 7.42.

As in coarse-grained soils, the use of a seismic cone (SCPTU) allows for measurements of G_0 from shear wave velocity to be made, using Equation 6.6.

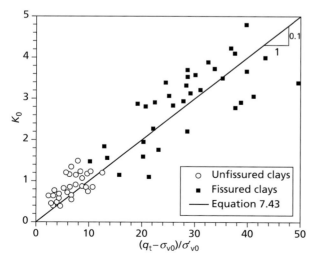

Figure 7.24 Estimation of K_0 from CPTU data.

In-situ testing

Example 7.4

CPTU data are shown in Figure 7.25 for the Bothkennar clay of Example 7.1. Figure 7.26 shows laboratory test data consisting of UU triaxial test data and oedometer test data for the same soil. Both sets of data are provided electronically on the Companion Website. Using the two sets of data: (a) determine the value of N_{kt} appropriate for the CPT in this geological unit; (b) determine the variation of OCR with depth from the CPTU data and compare this to the oedometer test data.

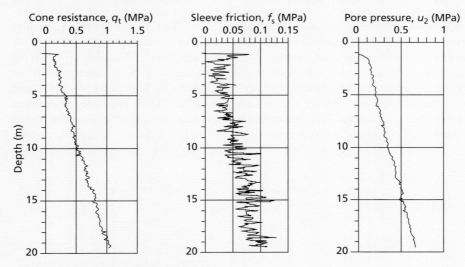

Figure 7.25 Example 7.4: CPTU data.

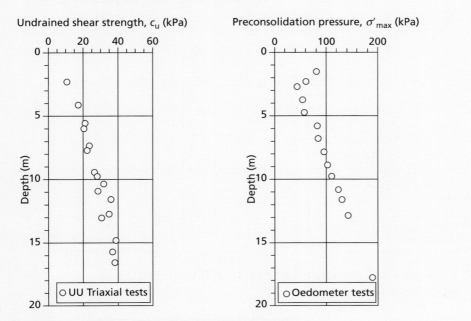

Figure 7.26 Example 7.4: Laboratory test data.

Solution

Values of σ_{v0}, u_0 and σ'_{v0} are first found at each depth sampled with the CPTU using the soil unit weight and water table information (Example 7.1). The normalised CPT parameters (Q_t, F_r and B_q) can then be found. An initial guess for N_{kt} is entered and used with Equation 7.38 to determine the value of c_u at each sampled depth during the test. The undrained shear strength is then plotted against depth for both the CPT and triaxial test data on the same graph. The value of N_{kt} can be manually adjusted until a good match between the two datasets is achieved. As an alternative to this trial-and-error manual approach, values of c_u from the CPT can be interpolated at each of the triaxial test depths. The difference between these and the triaxial values can then be found at each depth and the sum of the squares of the differences found. The value of N_{kt} giving the best fit can then be found by minimising the sum of the squares of the differences using an optimisation routine (subject to the constraint of N_{kt} being positive). This gives $N_{kt} = 14.4$, and the two sets of undrained strength data are compared in Figure 7.27. OCR is directly determined from the CPT data using Equation 7.41. The oedometer test data processing is identical to that described in Example 7.1, and is compared to the CPTU data in Figure 7.27, showing good agreement.

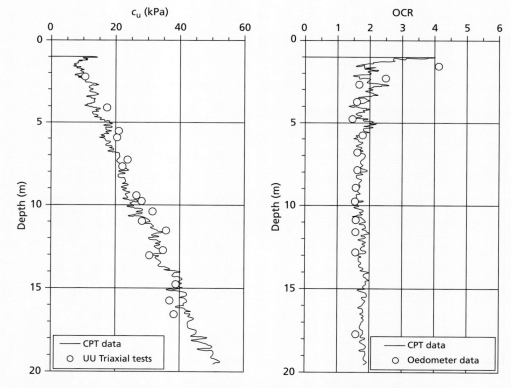

Figure 7.27 Example 7.4: Comparison of c_u and OCR from CPTU and laboratory tests.

In-situ testing

Example 7.5

Figure 7.28 shows CPTU data for a site in Canada. SPT test data at the same site are presented in Table 7.3. Both sets of data are provided electronically on the Companion Website. The site has been identified, both from borehole logs and the CPTU record, as consisting of 15 m of silica sand, overlying soft clay. The unit weight of both soils has been estimated at $\gamma \approx 17\,kN/m^3$ and the water table is at 2 m BGL. Determine the relative density and peak friction angle of the sand layer and the undrained shear strength of the clay layer using both datasets.

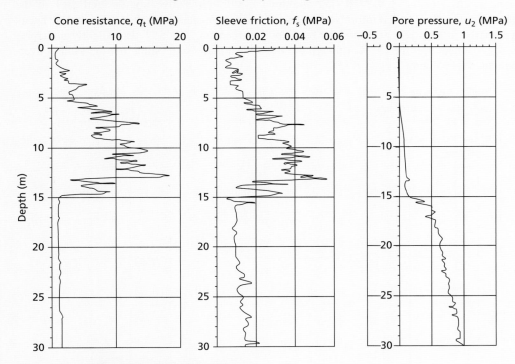

Figure 7.28 Example 7.5: CPTU data.

Table 7.3 Example 7.5: SPT data

Depth (m)	1.3	3.4	5.1	6.7	8.2	9.8	11.3	12.8	14.3	15.2	17.3	19.0	21.2	22.6	24.0
N_{60}	2	3	6	17	15	24	19	17	6	4	4	4	6	6	5

Solution

The CPTU data are processed initially as in Example 7.4 (σ_{v0}, u_0, σ'_{v0}, Q_t, F_r and B_q are found). For the data down to 15 m depth (sand), the relative density (I_D) is found at each sampled depth using Equation 7.35 with $D=-1.21$ and $E=0.584$ (best-fit parameters are used as there is no information regarding the compressibility of the sand). The peak friction angle is similarly found using Equation 7.36. Below 15-m depth (in the clay) c_u is determined at each sampled depth,

Development of a mechanical model for soil

where N_{kt} at each depth is found from B_q using Equation 7.39. For the SPT data, σ_{v0}, u_0 and σ'_{v0} are first determined at each test depth. The values of σ'_{v0} are then used to determine correction factors (C_N) at each depth using Equation 7.3, with $A=200$ and $B=100$ (in the absence of any more detailed information regarding grading). These values are used to determine corrected blow-counts $(N_1)_{60}$, from which relative density is approximated for tests down to 15-m depth using $(N_1)_{60}/I_D^2=60$, followed by determination of ϕ'_{max} values from Figure 7.4. For test points below 15 m which are in the clay, undrained shear strengths are determined directly from the normalised blowcounts (N_{60}) using $c_u/N_{60} \approx 10$ (i.e. assuming a soft, insensitive non-fissured clay in the absence of any more detailed information). The derived parameters from the CPTU and SPT data are compared in Figure 7.29, and show reasonable agreement. The SPT data slightly underpredicts ϕ'_{max} compared to the CPTU (though the trend is similar), suggesting that the sand is overconsolidated (see Figure 7.4).

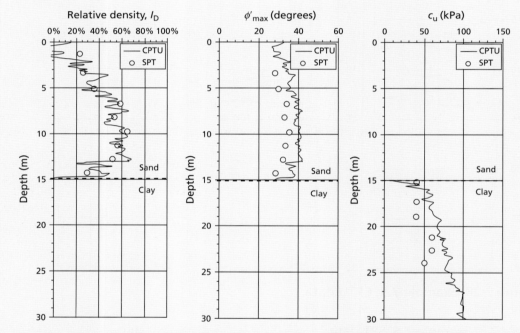

Figure 7.29 Example 7.5: Interpretation of ground properties from CPTU and SPT.

7.6 Selection of in-situ test method(s)

Section 6.4 described the parameters that could be determined from the various laboratory tests to aid in the design of a programme of ground investigation. The same can be done for the in-situ tests described in this and the previous chapters. Table 7.4 summarises the mechanical characteristics which can be obtained from each type of in-situ test, including those which were mentioned in Section 7.1, but which have not been described in detail (DMT, PLT). It will be shown in Part 2 of this book (Chapters 8–13 inclusive) that modern design approaches require both stiffness and strength parameters to verify that an appropriate level of performance will be achieved in a rigorous way. This is in contrast to older 'traditional' approaches which relied on strength parameters only and applied highly empirical global factors

Table 7.4 Derivation of key soil properties from in-situ tests

Parameter	SPT	FVT	PMT	CPT	DMT	PLT
Consolidation characteristics: m_v, C_c						
Stiffness properties: G, G_0			G	G_0* (SCPTU)	G, G_0*	YES
Drained strength properties: ϕ', c'	YES		YES	YES	YES	YES
Undrained strength properties: c_u (in-situ)	YES	YES	YES	YES	YES	YES
Soil state properties: I_D, OCR, K_0	I_D		K_0 (via σ_{h0})	I_D, OCR (K_0)	ALL	
Permeability: k				YES†	YES‡	

Notes: *Using a seismic instrument (i.e. SCPTU or SDMT). †Via a dissipation test on a piezocone (CPTU or SCPTU) – i.e. stopping penetration and measuring decay of u_2 (see Further Reading). ‡By stopping DMT expansion and measuring decay of cavity pressure (see Further Reading).

of safety to ensure adequate performance. The prevalence of 'traditional' approaches until recently explains the popularity of the SPT, as it can determine the necessary strength parameters while being simple, quick and cheap. It is expected that over the coming years CPT and PMT will become more popular in general use as they can provide reliable data on both the strength and stiffness of soil.

The end use should also be considered when determining which in-situ technique to use. For shallow foundation design (Chapter 8), PLT is useful as the test procedure is representative of the final construction (particularly in terms of defining an appropriate stiffness). For deep foundations (Chapter 9) CPT is usually preferred, due to the close analogy between a CPT probe and a jacked pile. In the case of retaining structures (Chapter 11), PMT or DMT are preferred as it is very important to accurately define the lateral earth pressures in such problems, and these tests are most reliable for this.

Summary

1. In-situ testing can be a valuable tool for evaluating the constitutive properties of the ground. Generally, a much larger body of soil is influenced during such tests, which can be advantageous over laboratory tests on small samples in certain soils (e.g. fissured clays). The tests also remove many of the issues associated with sampling, though attention must be paid instead to the disturbance of the soil that might occur during installation of the test device. The data collected from in-situ tests complements (rather than replaces) laboratory testing and the use of empirical correlations (Chapter 5). The use of in-situ testing can dramatically reduce the amount of sampling and laboratory testing required, provided it is calibrated against at least a small body of high-quality laboratory data. It can therefore be invaluable for the cost-effective ground investigation of large sites.
2. The four principle in-situ tests that are conducted in practice are (arguably) the Standard Penetration Test (SPT), Field Vane Test (FVT), Pressuremeter Test (PMT) and Cone Penetration Test (CPT). The first two of these are the simplest and

Development of a mechanical model for soil

cheapest tests, and are used to determine the strength characteristics of coarse-grained soil (SPT) and soft fine-grained soils (FVT); the SPT may also be used in stiffer fine-grained soils with greater caution. The latter two tests (PMT and CPT) represent modern devices making use of miniaturised sensors and computer control/data logging to measure multiple parameters, providing more detailed data and increasing their range of applicability. These tests are applicable to both coarse- and fine-grained soils, and can be used to reliably determine both strength and stiffness characteristics via theoretical models (PMT) or empirical correlations (CPT).

3 It has been demonstrated, through application to real soil test data, that spreadsheets are a useful tool for processing and interpreting a detailed suite of in-situ test data; moreover, they are essential for processing data from computerised PMT and CPT devices, which provide digital output. Digital exemplars have been provided for the worked examples in this chapter on the Companion Website, utilising data from all four of the tests discussed.

Problems

7.1 Table 7.5 presents corrected SPT blowcounts for a site consisting of 5 m of silt overlying a thick deposit of clean silica sand. These data are provided electronically on the Companion Website. The saturated unit weight of both soils is $\gamma \approx 16\,\text{kN/m}^3$ and the water table is 1.6 m below ground level (BGL). Determine the average relative density and peak friction angle of the sand between 10–20 m BGL.

Table 7.5 Problem 7.1

Depth (m)	N_{60} (blows)	Depth (m)	N_{60} (blows)	Depth (m)	N_{60} (blows)
1.32	4	11.27	23	21.06	35
2.50	10	12.29	22	22.21	27
3.29	7	13.39	30	23.16	28
4.30	2	14.34	29	24.32	24
5.34	8	15.20	19	25.20	30
6.44	11	16.34	9	26.08	30
7.31	10	17.33	30	27.10	30
8.41	12	18.28	30	28.12	32
9.29	18	19.23	34	29.25	11
10.40	24	20.25	30	30.22	22

7.2 Figure 7.30 presents the results of a self-boring pressuremeter test undertaken at a depth of 10.4 m BGL in the sand deposit described in Problem 7.1. The data are provided electronically on the Companion Website. Determine the following parameters: σ_{h0}, u_0, σ'_{h0}, ϕ'_{max}, ψ. Find also the value of G from the unload–reload loop.

In-situ testing

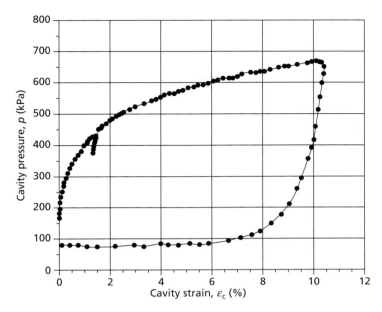

Figure 7.30 Problem 7.2.

7.3 Figure 7.31 presents the results of a CPT test undertaken in the sand deposit described in Problems 7.1 and 7.2. The data are provided electronically on the Companion Website. Determine the variation of relative density and peak friction angle with depth, and compare the results to those from the SPT tests in Problem 7.1 (the SPT data are also provided electronically).

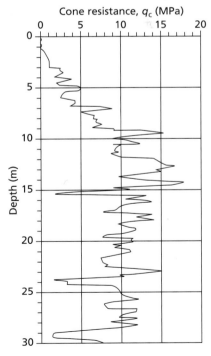

Figure 7.31 Problem 7.3.

263

Development of a mechanical model for soil

7.4 Figure 7.32 presents the results of a self-boring pressuremeter test undertaken deep within a deposit of clay. The data are provided electronically on the Companion Website. Determine the in-situ horizontal total stress and the undrained shear strength of the clay at the test depth. Is the unload–reload loop appropriately sized?

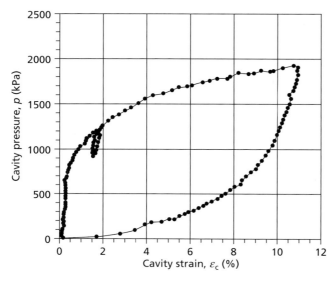

Figure 7.32 Problem 7.4.

7.5 Figure 7.33 presents the results of a CPTU test undertaken in the Gault clay shown in Figure 5.38(a). The data are provided electronically on the Companion Website, along with data from a series of SBPM tests and SPT tests conducted in the same material. The clay has a unit weight of $\gamma \approx 19\,\text{kN/m}^3$ above and below the water table, which is at a depth of 1 m BGL.
  a Using the SBPM data as a reference, determine the value of N_{kt} appropriate for obtaining c_u from the CPTU data. What does this suggest about the clay deposit?
  b Using your answer to (a) or otherwise, determine the variation of c_u with depth from the SPT data.
  c Determine the variation of K_0 with depth using both the SBPM and CPTU data.

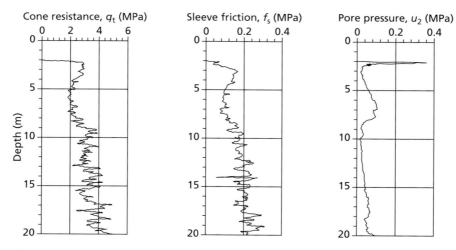

Figure 7.33 Problem 7.5.

264

References

Aas, G., Lacasse, S., Lunne, T. and Høeg, K. (1986) Use of in situ tests for foundation design on clay, in *Proceedings of the ASCE Speciality Conference In-situ '86: Use of In-situ Tests in Geotechnical Engineering, Blacksburg, VA*, pp. 1–30.

ASTM D1586 (2008) *Standard Test Method for Standard Penetration Test (SPT) and Split-Barrel Sampling of Soils*, American Society for Testing and Materials, West Conshohocken, PA.

ASTM D2573 (2008) *Standard Test Method for Field Vane Shear Test in Cohesive Soil*, American Society for Testing and Materials, West Conshohocken, PA.

ASTM D4719 (2007) *Standard Test Method for Prebored Pressuremeter Testing in Soils*, American Society for Testing and Materials, West Conshohocken, PA.

ASTM D5778 (2007) *Standard Test Method for Electronic Friction Cone and Piezocone Penetration Testing of Soils*, American Society for Testing and Materials, West Conshohocken, PA.

Azzouz, A., Baligh, M. and Ladd, C.C. (1983) Corrected field vane strength for embankment design, *Journal of Geotechnical and Geoenvironmental Engineering*, **109**(5), 730–734.

Bjerrum, L. (1973) Problems of soil mechanics and construction on soft clays, in *Proceedings of the 8th International Conference on SMFE, Moscow*, Vol. 3, pp. 111–159.

BS 1377 (1990) *Methods of test for soils for civil engineering purposes, Part 9: In-situ tests*, British Standards Institution, UK.

BS EN ISO 22476 (2005) *Geotechnical Investigation and Testing – Field Testing*, British Standards Institution, UK.

Clayton, C.R.I. (1995) The Standard Penetration Test (SPT): methods and use, *CIRIA Report 143*, CIRIA, London.

Dalton, J.C.P. and Hawkins, P.G. (1982) Fields of stress – some measurements of the in-situ stress in a meadow in the Cambridgeshire countryside. *Ground Engineering*, **15**, 15–23.

EC7–2 (2007) Eurocode 7: Geotechnical design – Part 2: Ground investigation and testing, BS EN 1997–2:2007, British Standards Institution, London.

Gibson, R.E. and Anderson, W.F. (1961) In-situ measurement of soil properties with the pressuremeter, *Civil Engineering and Public Works Review*, **56**, 615–618.

Hughes, J.M.O., Wroth, C.P. and Windle, D. (1977) Pressuremeter tests in sands, *Géotechnique*, **27**(4), 455–477.

Jamiolkowski, M., LoPresti, D.C.F. and Manassero M. (2001) Evaluation of relative density and shear strength of sands from Cone Penetration Test and Flat Dilatometer Test, *Soil Behavior and Soft Ground Construction, Geotechnical Special Publication 119*, American Society of Civil Engineers, Reston, VA, pp. 201–238.

Karlsrud, K., Lunne, T. and Brattlien, K. (1996). Improved CPTU correlations based on block samples, in *Proceedings of the Nordic Geotechnical Conference, Reykjavik*, Vol 1, pp. 195–201.

Kjekstad, O., Lunne, T. and Clausen, C.J.F. (1978) Comparison between in situ cone resistance and laboratory strength for overconsolidated North Sea clays. *Marine Geotechnology*, **3**, 23–36.

Kulhawy, F.H. and Mayne, P.W. (1990), *Manual on Estimating Soil Properties for Foundation Design*, Report EPRI EL-6800, Electric Power Research Institute, Palo Alto, CA.

Lunne, T. and Kleven, A. (1981) Role of CPT in North Sea foundation engineering. *Session at the ASCE National Convention: Cone Penetration testing and materials*, St Louis, American Society of Civil Engineers, Reston, VA, 76–107.

Lunne, T., Lacasse, S. and Rad, N.S. (1989) SPT, CPT, pressuremeter testing and recent developments on in situ testing of soils, in *General Report, Proceedings of the 12th International Conference of SMFE, Rio de Janeiro*, Vol. 4, pp. 2339–2403.

Mayne, P.W. (2007) *Cone penetration testing: a synthesis of highway practice*. NCHRP Synthesis Report 368, Transportation Research Board, Washington DC.

Mayne, P.W. and Mitchell, J.K. (1988) Profiling of overconsolidation ratio in clays by field vane. *Canadian Geotechnical Journal*, **25**(1), 150–157.

Mayne, P.W. and Campanella, R.G. (2005) Versatile site characterization by seismic piezocone, in *Proceedings of the 16th International Conference on SMFE*, Osaka, Vol. 2, pp. 721–724.

Palmer, A.C. (1972) Undrained plane strain expansion of a cylindrical cavity in clay: a simple interpretation of the pressuremeter test, *Géotechnique*, **22**(3), 451–457.

Powell, J.J.M. and Quarterman, R.S.T. (1988) The Interpretation of Cone Penetration Tests in Clays with Particular Reference to Rate Effects, *Penetration Testing 1988, Orlando*, **2**, 903–909.

Schmertmann, J.H. (1979) Statics of SPT. *Journal of the Geotechnical Division, Proceedings of the ASCE*, **105**(GT5), 655–670.

Skempton, A.W. (1986) Standard penetration test procedures and the effects in sands of overburden pressure, relative density, particle size, ageing and overconsolidation, *Géotechnique*, **36**(3), 425–447.

Stroud, M.A. (1989) The Standard Penetration Test – its application and interpretation, in *Proceedings of the ICE Conference on Penetration Testing in the UK*, Thomas Telford, London.

Terzaghi, K. and Peck, R.B. (1967) *Soil Mechanics in Engineering Practice* (2nd edn), Wiley, New York.

Wroth, C.P. (1984) The interpretation of in-situ soil tests, *Géotechnique*, **34**(4), 449–489.

Further reading

Clayton, C.R.I., Matthews, M.C. and Simons, N.E. (1995) *Site Investigation* (2nd edn), Blackwell, London.

A comprehensive book on site investigation which contains much useful practical advice and guidance on in-situ testing.

Lunne, T., Robertson, P.K. and Powell, J.J.M. (1997) *Cone Penetration Testing in Geotechnical Practice*, E & FN Spon, London.

A comprehensive reference on all aspects of the CPT, including apparatus, test set-up, test procedure, and interpretation (of a wider range of parameters than covered herein). Includes a wealth of data from real sites all over the world, and includes interpretation in more challenging ground conditions (e.g. frozen ground, volcanic soil).

Marchetti, S. (1980) In-situ tests by flat dilatometer, *Journal of the Geotechnical Engineering Division, Proceedings of the ASCE*, **106**(GT3), 299–321.

This paper (by the test developer) describes the Marchetti flat dilatometer, the device most commonly used for the DMT, and its use in deriving soil properties.

Powell, J.J.M. and Uglow, I.M. (1988) The interpretation of the Marchetti Dilatometer Test in UK clays, *Proceedings of the ICE Conference on Penetration Testing in the UK*, Thomas Telford, London.

A companion to the previous item of further reading, describing the properties that can be reliably derived from the DMT, correlations for use in practice and an idea of the scatter associated with such observations.

For further student and instructor resources for this chapter, please visit the Companion Website at www.routledge.com/cw/craig

Part 2

Applications in geotechnical engineering

Chapter 8

Shallow foundations

> **Learning outcomes**
>
> After working through the material in this chapter, you should be able to:
>
> 1 Understand the working principles behind shallow foundations;
> 2 Solve simple foundation capacity problems using Terzaghi's bearing capacity equation and/or limit analysis techniques;
> 3 Calculate the stresses induced beneath shallow foundations and the resultant foundation settlement using elastic solutions and consolidation theory;
> 4 Understand the philosophy behind limit state design codes and how this is applied in design to Eurocode 7;
> 5 Design a shallow foundation within a limit-state design framework (Eurocode 7), either analytically (based on fundamental ground properties) or directly from in-situ test data.

8.1 Introduction

A foundation is that part of a structure which transmits loads directly to the underlying soil, a process known as **soil–structure interaction**. This is shown indicatively in Figure 8.1(a). To perform in a satisfactory way, the foundation must be designed to meet two principal performance requirements (known as **limit states**), namely:

1 such that its capacity or resistance is sufficient to support the loads (actions) applied (i.e. so that it doesn't collapse);
2 to avoid excessive deformation under these applied loads, which might damage the supported structure or lead to a loss of function.

These criteria are shown schematically in Figure 8.1(b). **Ultimate limit states (ULS)** are those involving the collapse or instability of the structure as a whole, or the failure of one of its components (point 1 above). **Serviceability limit states (SLS)** are those involving excessive deformation, leading to damage or loss of function. Both ultimate and serviceability limit states must always be considered in design. The philosophy of limit states is the basis of Eurocode 7 (EC7, BSI 2004), a standard specifying all of the situations which must be considered in design in the UK and the rest of Europe. Similar standards

Applications in geotechnical engineering

are beginning to appear in other parts of the world (e.g. GeoCode 21 in Japan). In the United States, the limit state design philosophy is termed Load and Resistance Factor Design (LRFD).

This chapter will focus on the basic behaviour and design of shallow foundations under vertical loading. If a soil stratum near the surface is capable of adequately supporting the structural loads it is possible to use either **footings** or a **raft**, these being referred to in general as **shallow foundations**. A footing is a relatively small slab giving independent support to part of the structure. A footing supporting a single column is referred to as an individual footing or **pad**; one supporting a closely spaced group of columns is referred to as a combined footing, and one that supports a load-bearing wall as a **strip footing**. A raft is a relatively large single slab, usually stiffened with cross members, supporting the structure as a whole. The resistance of a shallow foundation is quantified by its **bearing resistance** (a limiting load) or **bearing capacity** (a limiting pressure), the determination of which will be addressed in Sections 8.2–8.4. Bearing capacity is directly related to the shear strength of soil and therefore requires the strength properties ϕ', c' or c_u, depending on whether drained or undrained conditions are maintained, respectively. The deformation of a foundation under a vertical load will be addressed in Sections 8.5–8.8, and depends on the stiffness of the soil (G or Young's modulus E). Limit state design of shallow

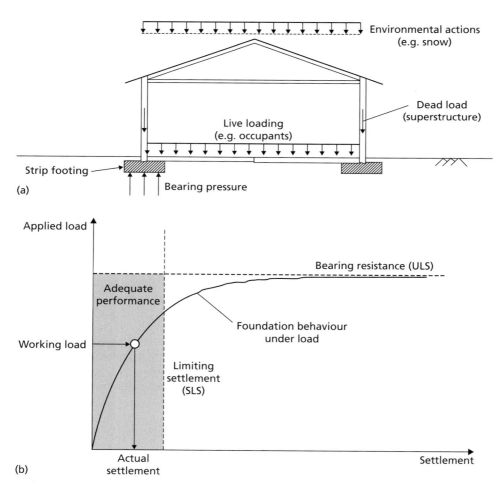

Figure 8.1 Concepts related to shallow foundation design: (a) soil–structure interaction under vertical actions, (b) foundation performance and limit state design.

foundations will be addressed in Section 8.9. When individual shallow foundation elements are used as part of a foundation system (e.g. Figure 8.1(a), consisting of two separate strips supporting the structure) they may additionally carry horizontal loads and or moments, e.g. induced by wind loading (environmental action); this will be addressed in Chapter 10.

If the soil near the surface is incapable of adequately supporting the structural loads, **piles**, or other forms of **deep foundations** such as piers or caissons, are used to transmit the applied loads to suitable soil (or rock) at greater depth where the effective stresses (and hence shear strength, as described in Chapter 5) are larger. Deep foundations are addressed in Chapter 9.

In addition to being located within an adequate bearing stratum, a shallow foundation should be below the depth which is subjected to frost action (around 0.5 m in the UK) and, where appropriate, the depth to which seasonal swelling and shrinkage of the soil takes place. Consideration must also be given to the problems arising from excavating below the water table if it is necessary to locate foundations below this level. The choice of foundation level may also be influenced by the possibility of future excavations for services close to the structure, and by the effect of construction, particularly excavation, on existing structures and services.

8.2 Bearing capacity and limit analysis

Bearing capacity (q_f) is defined as the pressure which would cause shear failure of the supporting soil immediately below and adjacent to a foundation.

Three distinct modes of failure have been identified, and these are illustrated in Figure 8.2; they will be described with reference to a strip footing. In the case of **general shear failure**, continuous failure surfaces develop between the edges of the footing and the ground surface, as shown in Figure 8.2. As the pressure is increased towards the value q_f a state of plastic equilibrium is reached initially in the soil around the edges of the footing, which subsequently spreads downwards and outwards. Ultimately, the state of plastic equilibrium is fully developed throughout the soil above the failure surfaces. Heave of the ground surface occurs on both sides of the footing, although in many cases the final slip movement occurs only on one side, accompanied by tilting of the footing, as the footing will not be perfectly level and will hence be biased to fail towards one side. This mode of failure is typical of soils of low compressibility (i.e. dense coarse-grained or stiff fine-grained soils), and the pressure–settlement curve is of the general form shown in Figure 8.2, the ultimate bearing capacity being well defined. In the mode of **local shear failure** there is significant compression of the soil under the footing, and only partial development of the state of plastic equilibrium. The failure surfaces, therefore, do not reach the ground surface and only slight heaving occurs. Tilting of the foundation would not be expected. Local shear failure is associated with soils of high compressibility and, as indicated in Figure 8.2, is characterised by the occurrence of relatively large settlements (which would be unacceptable in practice) and the fact that the ultimate bearing capacity is not clearly defined. **Punching shear failure** occurs when there is relatively high compression of the soil under the footing, accompanied by shearing in the vertical direction around the edges of the footing. There is no heaving of the ground surface away from the edges, and no tilting of the footing. Relatively large settlements are also a characteristic of this mode, and again the ultimate bearing capacity is not well defined. Punching shear failure will also occur in a soil of low compressibility if the foundation is located at considerable depth. In general, the mode of failure depends on the compressibility of the soil and the depth of the foundation relative to its breadth.

The bearing capacity problem can be considered in terms of plasticity theory. It is assumed that the stress–strain behavior of the soil can be represented by the rigid–perfectly plastic idealisation, shown in Figure 8.3, in which both yielding and shear failure occur at the same state of stress: unrestricted plastic flow takes place at this stress level. A soil mass is said to be in a state of plastic equilibrium if the shear stress at every point within the mass reaches the value represented by point Y′.

Applications in geotechnical engineering

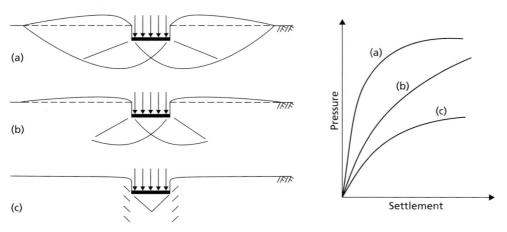

Figure 8.2 Modes of failure: (a) general shear, (b) local shear, and (c) punching shear.

Plastic collapse occurs after the state of plastic equilibrium has been reached in part of a soil mass, resulting in the formation of an unstable mechanism: that part of the soil mass slips relative to the rest of the mass. The applied load system, including body forces, for this condition is referred to as the collapse load. Determination of the collapse load is achieved using the limit theorems of plasticity (also known as limit analysis) to calculate **lower** and **upper bounds** to the true collapse load. In certain cases the theorems produce the same result, which would then be the exact value of the collapse load. The limit theorems can be stated as follows.

Lower bound (LB) theorem

If a state of stress can be found which at no point exceeds the failure criterion for the soil and is in equilibrium with a system of external loads (which includes the self-weight of the soil), then collapse cannot occur; the external load system thus constitutes a lower bound to the true collapse load (because a more efficient stress distribution may exist, which would be in equilibrium with higher external loads).

Upper bound (UB) theorem

If a **kinematically admissible** mechanism of plastic collapse is postulated and if, in an increment of displacement, the work done by a system of external loads is equal to the dissipation of energy by the

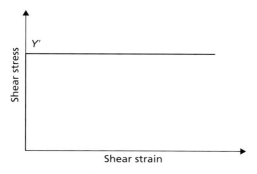

Figure 8.3 Idealised stress–strain relationship in a perfectly plastic material.

Shallow foundations

internal stresses, then collapse must occur; the external load system thus constitutes an upper bound to the true collapse load (because a more efficient mechanism may exist resulting in collapse under lower external loads).

In the upper bound approach, a mechanism of plastic collapse is formed by choosing a slip surface and the work done by the external forces is equated to the loss of energy by the stresses acting along the slip surface, without consideration of equilibrium. The chosen collapse mechanism is not necessarily the true mechanism, but it must be kinematically admissible – i.e. the motion of the sliding soil mass must remain continuous and be compatible with any boundary restrictions.

8.3 Bearing capacity in undrained materials

Analysis using the upper bound theorem

Upper bound approach, mechanism UB-1

It can be shown that for undrained conditions the failure mechanism within the soil mass should consist of slip lines which are either straight lines or circular arcs (or a combination of the two). A simple mechanism consisting of three sliding blocks of soil is shown in Figure 8.4 for a strip footing under pure vertical loading.

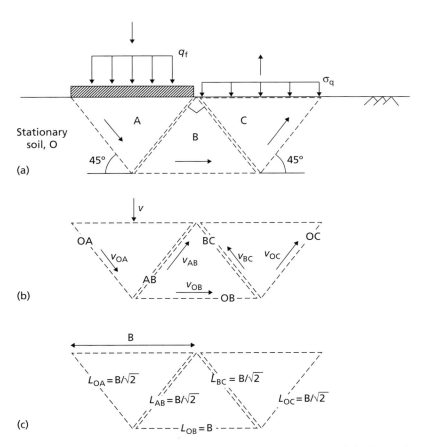

Figure 8.4 (a) Simple proposed mechanism, UB-1, (b) slip velocities, (c) dimensions.

Applications in geotechnical engineering

If the foundation inputs work to the system by moving vertically downwards with a velocity v, the blocks will have to move as shown in Figure 8.4 to form a mechanism and therefore be kinematically admissible. This movement generates relative velocities between each slipping block and its neighbours, which will result in energy dissipation along all of the slip lines shown (OA, OB, OC, AB and BC). To determine the relative velocities along the slip lines a **hodograph** (velocity diagram) is drawn, the construction procedure for which is shown in Figure 8.5. Starting with the known vertical displacement of the footing (v) gives point f (foundation) on the hodograph as shown in Figure 8.5(a). Block A must move relative to the stationary soil at an angle of 45°; the vertical component of this motion must be equal to v so the footing and soil remain in contact. Two construction lines may be added to the hodograph as shown in Figure 8.5(a) to represent these two limiting conditions. The crossing point of these two lines fixes the position of point a in the hodograph and therefore, the velocity v_{OA}. Block B moves horizontally with respect to O and at 45° relative to A. Adding two construction lines for these conditions fixes the position of b as shown in Figure 8.5(b) and therefore the velocities v_{OB} and v_{BA}. Block C moves at 45° from points o and b as shown in Figure 8.5(c) which fixes the position of c and therefore the relative velocities v_{OC} and v_{CB}. Basic trigonometry may then be used on the final hodograph to determine the lengths of the lines which represent the relative velocities in the mechanism in terms of the known foundation movement v, as shown in Figure 8.5(d).

The energy dissipated (E_i) due to shearing at relative velocity v_i along a slip line of length L_i is given by:

$$E_i = \tau_f \cdot L_i \cdot v_i \tag{8.1}$$

energy being force multiplied by velocity (stress multiplied by length is force per metre length of the slip plane into the page). The strength τ_f is used as the shear stress along the slip line, as the soil is at plastic failure along this line by definition (i.e. $\tau = \tau_f$). The total energy dissipated in the soil can then be found by summing E_i for all slip lines, as shown in Table 8.1.

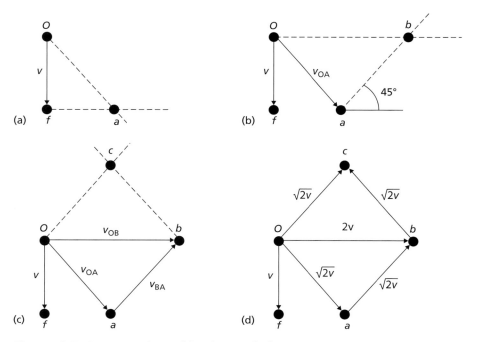

Figure 8.5 Construction of hodograph for mechanism UB-1.

Shallow foundations

Table 8.1 Energy dissipated within the soil mass in mechanism UB-1

Slip line	Stress, τ_f	Length, L_i	Relative velocity, v_i	Energy dissipated, E_i
OA	c_u	$B/\sqrt{2}$	$\sqrt{2}v$	$c_u Bv$
OB	c_u	B	$2v$	$2c_u Bv$
OC	c_u	$B/\sqrt{2}$	$\sqrt{2}v$	$c_u Bv$
AB	c_u	$B/\sqrt{2}$	$\sqrt{2}v$	$c_u Bv$
BC	c_u	$B/\sqrt{2}$	$\sqrt{2}v$	$c_u Bv$
			Total Energy, $\Sigma E_i =$	$6c_u Bv$

Work may be input to the system by the foundation moving downwards, where the work input is positive as the forcing and velocity are in the same direction. If there is a surcharge pressure σ_q acting around the foundation as shown in Figure 8.4, this will be moved upwards as a result of the vertical component of the motion of block C. As the force and velocity are in opposite directions (the surcharge is being moved upwards against gravity), this will do negative work (effectively dissipate energy). The work done W_i by a pressure q_i acting over an area per unit length B_i moving at velocity v_i is given by

$$W_i = q_i \cdot B_i \cdot v_i \tag{8.2}$$

The total work done can then be found by summing W_i for all components, as shown in Table 8.2.

By the upper bound theorem, if the system is at plastic collapse, the work done by the external loads/pressures must be equal to the energy dissipated within the soil, i.e.

$$\sum W_i = \sum E_i \tag{8.3}$$

Therefore, for mechanism UB-1, inserting the values from Tables 8.1 and 8.2 gives the bearing capacity q_f:

$$\begin{aligned}(q_f - \sigma_q)Bv &= 6c_u Bv \\ q_f &= 6c_u + \sigma_q\end{aligned} \tag{8.4}$$

Upper bound approach, mechanism UB-2

The mechanism UB-1 requires sharp changes in the direction of movement of the blocks to translate the downwards motion beneath the footing into upwards motion of the adjacent soil. A more efficient mechanism (i.e. one that dissipates less energy) replaces block B in Figure 8.4 with a number of smaller wedges as shown in Figure 8.6(a). These wedges describe a circular arc of radius R between the rigid blocks A and C, as shown in the figure, known as a **shear fan**. The hodograph for this mechanism is shown in Figure 8.6(d) – blocks A and C will move in the same direction and by the same magnitude as

Table 8.2 Work done by the external pressures, mechanism UB-1

Component	Pressure, p_i	Area, B_i	Relative velocity, v_i	Work done, W_i
Footing pressure	q_f	B	v	$q_f Bv$
Surcharge	σ_q	B	$-v$	$-\sigma_q Bv$
			Total work done, $\Sigma W_i =$	$(q_f - \sigma_q)Bv$

Applications in geotechnical engineering

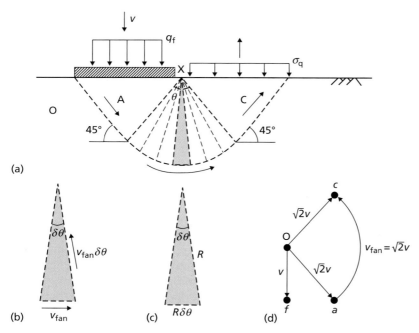

Figure 8.6 (a) Refined mechanism UB-2, (b) slip velocities on wedge *i*, (c) geometry of wedge *i*, (d) hodograph.

in Figure 8.5; the velocity around the edge of the circular arc will be constant as the soil in this region rotates about the point X.

The energy dissipated due to shearing between wedge *i* (of internal angle $\delta\theta$) and the stationary soil can then be found using Equation 8.1 where the length along the slip plane is $R\delta\theta$ as shown in Figure 8.6(b) and (c):

$$E_i = c_u \cdot (R\delta\theta) \cdot v_{fan} \tag{8.5}$$

The energy dissipated due to the shearing occurring between each wedge and the next can be found similarly, with the length along the slip line being R and the relative velocity from the hodograph $v\delta\phi$:

$$E_{i,j} = c_u \cdot R \cdot (v_{fan}\delta\theta) \tag{8.6}$$

The total amount of energy dissipated within this zone is then given by summing the components from Equations 8.5 and 8.6 across all wedges. If the wedge angle $\delta\theta$ is made vanishingly small, this summation becomes an integral over the full internal angle of the zone (θ):

$$E_{fan} = \sum_i (E_i + E_{i,j}) = \sum_i 2c_u R v_{fan} \delta\theta = \int_0^\theta 2c_u R v_{fan} \delta\theta$$
$$= 2c_u R v_{fan} \theta \tag{8.7}$$

The energy component E_{fan} replaces the terms along slip lines OB, AB and BC in mechanism UB-1. If blocks A and C still move in the same directions as before, the wedge angle $\theta = \pi/2$ (90°), $v_{fan} = v_{OA} = v_{OC} = \sqrt{2}v$ and $R = B/\sqrt{2}$. The energy dissipated in UB-2 is therefore as shown in Table 8.3.

Shallow foundations

Table 8.3 Energy dissipated within the soil mass in mechanism UB-2

Slip line	Stress, τ_f	Length, L_i	Relative velocity, v_i	Energy dissipated, E_i
OA	c_u	$B/\sqrt{2}$	$\sqrt{2}v$	$c_u Bv$
Fan zone ($\theta=\pi/2$)	c_u	$R=B/\sqrt{2}$	$v_{fan}=\sqrt{2}v$	$\pi c_u Bv$ (from Eq. 8.7)
OC	c_u	$B/\sqrt{2}$	$\sqrt{2}v$	$c_u Bv$
			Total energy, $\Sigma E_i =$	$(2+\pi)c_u Bv$

The work dissipated in mechanism UB-2 is the same as for UB-1 (values in Table 8.2) such that applying Equation 8.3 yields:

$$(q_f - \sigma_q)Bv = (2+\pi)c_u Bv$$
$$q_f = (2+\pi)c_u + \sigma_q \tag{8.8}$$

The bearing pressure in Equation 8.8 is lower than for UB-1 (Equation 8.4), so UB-2 represents a better estimate of the true collapse load by the upper bound theorem.

Analysis using the lower bound theorem

Lower bound approach, stress state LB-1

In the lower bound approach, the conditions of equilibrium and yield are satisfied without consideration of the mode of deformation. For undrained conditions the yield criterion is represented by $\tau_f = c_u$ at all points within the soil mass. The simplest possible stress field satisfying equilibrium that may be drawn for a strip footing is shown in Figure 8.7(a). Beneath the foundation (zone 1), the major principal stress (σ_1) will be vertical. In the soil to either side of this (zone 2), the major principal stress will be horizontal. The minor principal stresses (σ_3) in each zone will be perpendicular to the major principal stresses. These two distinct zones are separated by a single frictionless stress discontinuity permitting the rotation of the major principal stress direction.

Mohr circles (see Chapter 5) may be drawn for each zone of soil as shown in Figure 8.7(b). In order for the soil to be in equilibrium, σ_1 in zone 2 must be equal to σ_3 in zone 1. This requirement causes the circles to just touch, as shown in Figure 8.7(b). The major principal stress at any point within zone 1 is

$$\sigma_1 = q_f + \gamma z \tag{8.9}$$

i.e. the total vertical stress due to the weight of the soil (γz) plus the applied footing pressure (q_f). In zone two, the minor principal stress is similarly

$$\sigma_3 = \sigma_q + \gamma z \tag{8.10}$$

If the soil is undrained with shear strength c_u and the soil is everywhere in a state of plastic yielding, the diameter of each circle is $2c_u$. Therefore at the point where the circles meet

$$q_f + \gamma z - 2c_u = \sigma_q + \gamma z + 2c_u$$
$$\therefore q_f = 4c_u + \sigma_q \tag{8.11}$$

Applications in geotechnical engineering

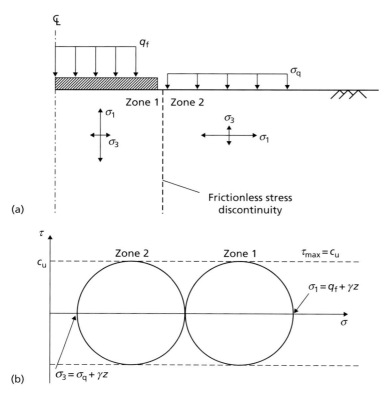

Figure 8.7 (a) Simple proposed stress state LB-1, (b) Mohr circles.

Lower bound approach, stress state LB-2

As for the upper bound UB-1, the sharp change in the stress field across the single discontinuity in stress state LB-1 is only a crude representation of the actual stress field within the ground. A more realistic stress state can be found by considering a series of frictional stress discontinuities along which a significant proportion of the soil strength can be mobilised, forming a **fan zone** which gradually rotate the major principal stress from vertical beneath the footing to horizontal outside. This is shown in Figure 8.8(a).

The change in direction of the major principal stresses across a frictional discontinuity depends on the frictional strength along the discontinuity (τ_d, Figure 8.8(b)). The Mohr circles representing the stress states in the zones either side of a discontinuity are shown in Figure 8.8(c). The mean stress in each zone is represented by s (see Chapter 5). As with mechanism LB-1 the circles will touch, but at a point where $\tau = \tau_d$ as shown in the figure. This defines the relative position of the two circles, i.e. the difference $s_A - s_B$. In crossing the discontinuity the major principal stress will rotate by an amount $\delta\theta$ (Figure 8.8(b)):

$$\delta\theta = \frac{\pi}{2} - \Delta \tag{8.12}$$

As the radius of the Mohr circles are c_u, from Figure 8.8(c)

$$s_A - s_B = 2c_u \cos\Delta \tag{8.13}$$

Shallow foundations

Equation 8.12 may then be substituted into Equation 8.13 for Δ. Then, in the limit as $s_A - s_B \to \delta s'$, $\sin\delta\theta \to \delta\theta$

$$\begin{aligned}\delta s &= 2c_u \cos\left(\frac{\pi}{2} + \delta\theta\right) \\ &= 2c_u \sin\delta\theta \\ &= 2c_u \delta\theta\end{aligned} \tag{8.14}$$

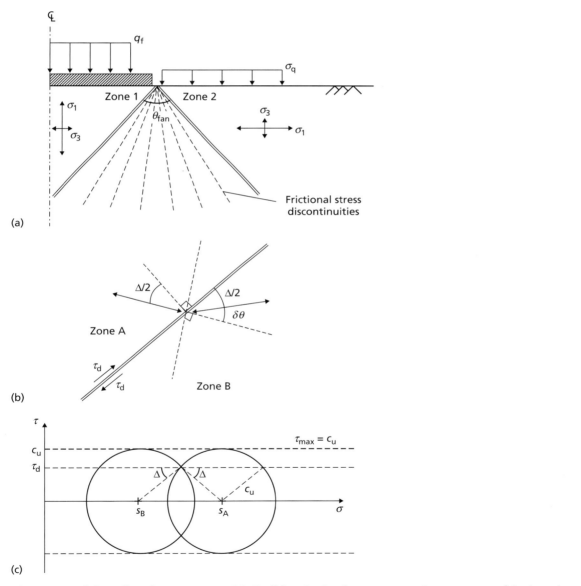

Figure 8.8 (a) Refined stress state LB-2, (b) principal stress rotation across a frictional stress discontinuity, (c) Mohr circles.

Applications in geotechnical engineering

For a fan zone of frictional stress discontinuities subtending an angle θ, Equation 8.14 may be integrated from zone 1 to zone 2 in Figure 8.8(a) across the fan angle θ_{fan}, i.e.

$$\int_{s_2}^{s_1} \delta s = \int_0^{\theta_{fan}} 2c_u \delta\theta \tag{8.15}$$
$$s_1 - s_2 = 2c_u \theta_{fan}$$

For the shallow footing problem shown in Figure 8.9, σ_1 in zone 1 is still given by Equation 8.9 and σ_3 in zone 2 is still given by Equation 8.10. Therefore the principal stress rotation required in the fan is $\theta_{fan} = \pi/2$ (90°) giving, from Equation 8.15,

$$\left(q_f + \gamma z - c_u\right) - \left(\sigma_q + \gamma z + c_u\right) = 2c_u \frac{\pi}{2} \tag{8.16}$$
$$\therefore q_f = \left(2 + \pi\right)c_u + \sigma_q$$

The bearing pressure in Equation 8.16 is higher than for LB-1 (Equation 8.11), so LB-2 represents a better estimate of the true collapse load by the lower bound theorem.

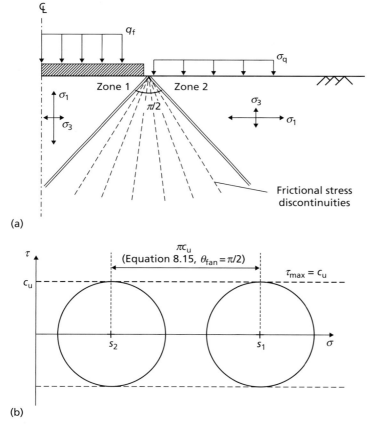

Figure 8.9 Stress state LB-2 for shallow foundation on undrained soil.

Shallow foundations

Combining upper and lower bounds to obtain the true collapse load

Ignoring the surcharge pressure σ_q (which is the same in all of the expressions derived so far), the bearing capacity from UB-1 is $6c_u$ while that from LB-1 is $4c_u$. These form upper and lower bounds to the true collapse load, so that $4c_u \leq q_f \leq 6c_u$. However comparing the refined analyses, UB-2 and LB-2 both give the same value of $q_f = (2+\pi)c_u = 5.14c_u$, so by the upper and lower bound theorems, this must be the exact solution – i.e. LB-2 represents the stress state just as mechanism UB-2 is formed.

For this problem it was possible with fairly minimal effort to determine the exact solution. In solving any generalised problem using limit analysis, it may be necessary to try many different mechanisms and stress states to determine the bearing capacity exactly (or if not exactly, with only a very narrow range between the best upper and lower bound solutions). Computers can be used to automate this optimisation process. Links to suitable software which is available for both commercial and academic use are provided on the Companion Website.

Bearing capacity factors

Comparing Equations 8.8 and 8.16, the bearing capacity of a shallow foundation on an undrained material may be written in a generalised form as

$$q_f = s_c N_c c_u + \sigma_q \tag{8.17}$$

Where, for the case of a footing surrounded by a surcharge pressure σ_q, $N_c = 5.14$. N_c is the **bearing capacity factor** for a strip foundation under undrained conditions ($\tau_f = c_u$). The parameter s_c in Equation 8.17 is a shape factor ($s_c = 1.0$ for a strip foundation). In principle, UB and LB analyses similar to those presented above may be conducted for various other cases (e.g. a footing near to a slope or on layered soils). In many cases, however, such analyses have been conducted and the results published as design charts for selecting the value of N_c for use in Equation 8.17.

Foundations are not normally located on the surface of a soil mass, but are embedded at a depth d below the surface. The soil above the **founding plane** (the level of the underside of the foundation) is considered as a surcharge imposing a uniform pressure $\sigma_q = \gamma d$ on the horizontal plane at foundation level. This assumes that the shear strength of the soil between the surface and depth d is neglected. This is a reasonable assumption provided that d is not greater than the breadth of the foundation B. The soil above foundation level is normally weaker, especially if backfilled, than the soil at greater depth.

Skempton (1951) presented values of N_c for embedded strip foundations in undrained soil as a function of d based on empirical evidence, which are given in Figure 8.10; also included are values suggested by Salgado et al. (2004), which are described by

$$N_c = (2+\pi)\left(1 + 0.27\sqrt{\frac{d}{B}}\right) \tag{8.18}$$

For a general rectangular footing of dimensions $B \times L$ (where $B < L$), Eurocode 7 recommends that the shape factor s_c in Equation 8.17 is given by:

$$s_c = 1 + 0.2\frac{B}{L} \tag{8.19}$$

Equations 8.18 and 8.19 are compared to Skempton's data for the extreme cases of strip ($B/L=0$) and square footings ($B/L=1$) in Figure 8.10. N_c for circular footings may be obtained by taking the square values. In practice, N_c is normally limited at a value of 9.0 for very deeply embedded square or circular

Applications in geotechnical engineering

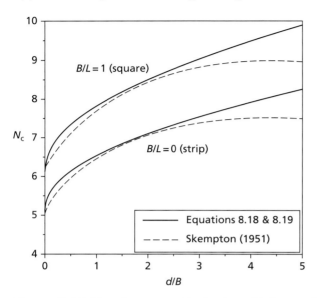

Figure 8.10 Bearing capacity factors N_c for embedded foundations in undrained soil.

foundations. Values of N_c obtained from Figure 8.10 may be used for stratified deposits, provided the value of c_u for a particular stratum is not greater than nor less than the average value for all strata within the significant depth by more than 50% of that average value.

For layered soils, Merifield *et al.* (1999) presented upper and lower bound values of N_c for strip footings resting on a two-layer cohesive soil as a function of the thickness H of the upper layer of soil of strength c_{u1} which overlies a deep deposit of material of strength c_{u2}. Proposed design values of N_c for this case are given in Figure 8.11(a) which are valid if the undrained shear strength of the upper layer is used in Equation 8.17 (i.e. $c_u = c_{u1}$). Subsequently, Merifield and Nguyen (2006) conducted further analyses for square footings with $B/L = 1.0$. The resulting shape factors they obtained are shown in Figure 8.11(b).

If a shallow foundation is built close to a slope, its bearing capacity may be dramatically reduced. This is a common case for transport infrastructure (e.g. a road or railway line) which is situated on an embankment. These types of construction are commonly very long, and so will always behave as strip foundations. Georgiadis (2010) presented charts for N_c for strip foundations set back from the crest of a slope of angle β by a multiple λ of the foundation width. These are based on upper bound analyses in which an optimal failure mechanism was found giving the lowest upper bound. For this case, it was important to include both 'local' failure mechanisms (bearing capacity failure of the foundation alone) and 'global' mechanisms (failure of the whole slope including the foundation). Slope stability is discussed in greater detail in Chapter 12. From Figure 8.12, the presence of a nearby slope reduces N_c (and hence also the bearing capacity). If the foundation is set far enough back from the crest of the slope ($\lambda > 2B$), then the slope will have no effect on the bearing capacity and $N_c = 2 + \pi$ as for level ground.

It is common for undrained strength to vary with depth, rather than be uniform (constant with depth). Davis and Booker (1973) conducted upper and lower bound plasticity analyses for soil with a linear variation of undrained shear strength with depth z below the founding plane, i.e.

$$c_u(z) = c_{u0} + Cz \tag{8.20}$$

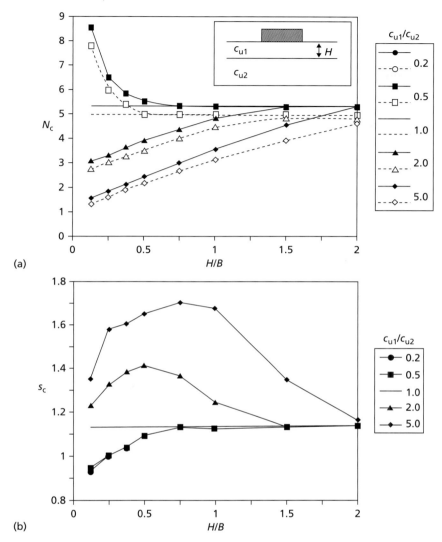

Figure 8.11 (a) Bearing capacity factors N_c for strip foundations of width B on layered undrained soils (after Merifield et al., 1999), solid lines – UB, dashed lines – LB, (b) shape factors s_c (after Merifield and Nguyen, 2006).

where c_{u0} is the undrained shear strength at the founding plane ($z=0$) and C is the gradient of the c_u-z relationship. The bearing capacity is expressed in a different form compared to Equation 8.17, as

$$q_f = \left[(2+\pi)c_{u0} + \frac{CB}{4}\right] F_z \qquad (8.21)$$

The parameter F_z is read from Figure 8.13. If the ratio of $CB/c_{u0} \leq 20$, the value of F_z may be read using the left side of the figure; if $CB/c_{u0} \geq 20$, it is more convenient to express the ratio as c_{u0}/CB and use the right side of the figure. For the special case of $C=0$ and $c_{u0}=c_u$ (uniform strength with depth), $CB/c_{u0}=0$, $F_z=1$ such that Equation 8.21 reduces to $q_f=5.14c_u$ as before.

Applications in geotechnical engineering

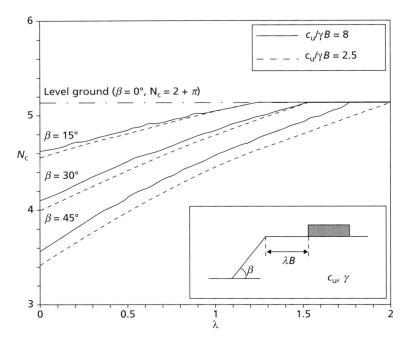

Figure 8.12 Bearing capacity factors N_c for strip foundations of width B at the crest of a slope of undrained soil (after Georgiadis, 2010).

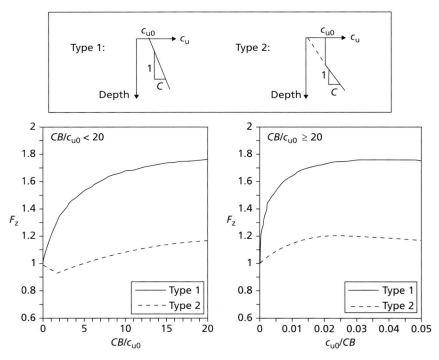

Figure 8.13 Factor F_z for strip foundations on non-uniform undrained soil (after Davis and Booker, 1973).

Shallow foundations

> **Example 8.1**
>
> A strip foundation 2.0 m wide is located at a depth of 2.0 m in a stiff clay of saturated unit weight 21 kN/m³. The undrained shear strength is uniform with depth, with $c_u = 120$ kPa. Determine the undrained bearing capacity of the foundation under the following conditions:
>
> a the foundation is constructed in level ground;
> b a cutting at a gradient of 1:2 is subsequently made adjacent to the foundation, with the crest 1.5 m from the edge of the foundation.
>
> ## Solution
>
> For case (a) $d/B = 2.0/2.0 = 1$, so, from Figure 8.10a, $N_c = 6.4$ (using Skempton's values). As the footing is a strip, $s_c = 1.0$. The surcharge pressure $\sigma_q = \gamma d = 21 \times 2.0 = 42$ kPa. Therefore:
>
> $$q_f = s_c N_c c_u + \sigma_q$$
> $$= (1.0 \cdot 6.4 \cdot 120) + 42$$
> $$= 810 \text{ kPa}$$
>
> For case (b), a 1:2 slope has an angle $\beta = \tan^{-1}(1/2) = 26.6°$. The parameter $c_u/\gamma B = 120/(21 \times 2.0) = 2.9$ and $\lambda = 1.5/2.0 = 0.75$. Interpolation between lines in Figure 8.12 is then used to find $N_c = 4.4$. The shape factor is the same as before, giving:
>
> $$q_f = s_c N_c c_u + \sigma_q$$
> $$= (1.0 \cdot 4.4 \cdot 120) + 42$$
> $$= 570 \text{ kPa}$$
>
> Construction of the slope will therefore reduce the bearing capacity of the foundation. It should be noted that the actual bearing capacity in case (b) is likely to be higher as the value of N_c is uncorrected for the depth of embedment.

8.4 Bearing capacity in drained materials

Analysis using the upper bound theorem

It can be shown that for drained conditions the slip surfaces within a kinematically admissible failure mechanism should consist of either straight lines or curves of a specific form known as logarithmic spirals (or a combination of the two). The conditions on a slip surface at any point will be analogous to a direct shear test (as described in Section 5.4). This is shown schematically in Figure 8.14 for a cohesionless soil ($c' = 0$). Referring to Figure 5.16, all drained materials will exhibit some amount of dilatancy during shear (quantified by the angle of dilation, ψ). In general, all soils will have $\psi \leq \phi'$. However, for the special case of $\psi = \phi'$ the direction of movement will be perpendicular to the resultant force (R_s) on the shear plane, so, by Equation 8.1, there is no energy dissipated in shearing along the slip line (i.e. there is no movement in the direction of the resultant force). This condition is known as the **normality principle**. This special case of $\psi = \phi'$ represents an **associative flow rule**, and the use of this flow rule considerably simplifies limit analysis in drained materials.

Applications in geotechnical engineering

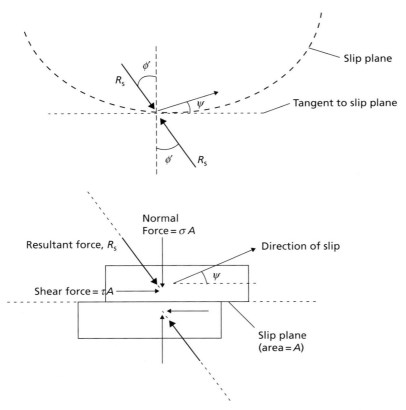

Figure 8.14 Conditions along a slip plane in drained material.

Figure 8.15(a) shows a failure mechanism in a weightless cohesionless soil ($\gamma=c'=0$) with a friction angle of ϕ'. This is similar to UB-2 (Figure 8.6), but with a logarithmic spiral replacing the circular arc for the fan shear zone B. Beneath the footing, a rigid block will form with slip surfaces at an angle of $\pi/4+\phi'/2$ to the horizontal as shown. It should be noted that this is a general value for the internal wedge angles in any soil; the internal wedge angles of 45° ($\pi/4$) used earlier for the undrained case arise from the fact that $\phi'=\phi_u=0$ in undrained materials. These angles fix the lengths of the slip planes OA and AB. To determine the geometry of the rest of the mechanism, an equation describing the logarithmic spiral must first be found. Figure 8.15(b) shows the change in geometry between two points on the slip line relative to the centre of rotation of the shear zone, from which it can be seen that if $d\theta$ is small

$$\tan \psi = \frac{dr}{r d\theta} \tag{8.22}$$

Rearranging Equation 8.22, it may be integrated from an initial radius r_0 at $\theta=0$ to a general radius r at θ:

$$\int_{r_0}^{r} \frac{dr}{r} = \int_{0}^{\theta} \tan \psi \, d\theta$$

$$\ln\left(\frac{r}{r_0}\right) = \theta \tan \psi \tag{8.23}$$

$$r = r_0 e^{\theta \tan \psi}$$

Equation 8.23 can then be applied to the mechanism in Figure 8.15(a), where $r_0 = L_{AB}$, $r = L_{BC}$ and $\theta = \pi/2$. For associative flow, $\psi = \phi'$. Length L_{AB} is found from trigonometry with the known foundation width B and wedge angles $\pi/4 + \phi'/2$, i.e.

$$L_{AB} = \frac{B}{2\cos\left(\dfrac{\pi}{4} + \dfrac{\phi'}{2}\right)}$$

$$L_{BC} = \frac{Be^{\frac{\pi}{2}\tan\phi'}}{2\cos\left(\dfrac{\pi}{4} + \dfrac{\phi'}{2}\right)} = L_{OC}$$

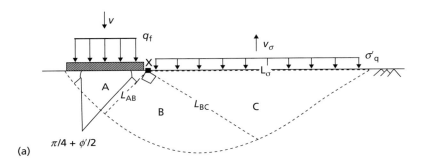

(a)

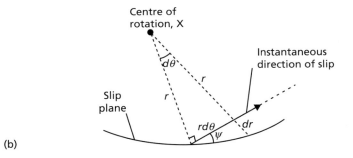

(b)

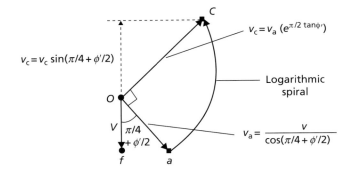

(c)

Figure 8.15 Upper bound mechanism in drained soil: (a) geometry of mechanism, (b) geometry of logarithmic spiral, (c) hodograph.

Applications in geotechnical engineering

The area per unit length over which the surcharge acts on the mechanism L_σ can then be found from trigonometry

$$L_\sigma = Be^{\frac{\pi}{2}\tan\phi'} \cdot \tan\left(\frac{\pi}{4} + \frac{\phi'}{2}\right) \tag{8.24}$$

The hodograph for the mechanism is shown in Figure 8.15(c). This is similar to the hodograph for the undrained case (UB-2) shown in Figure 8.6(d); in the drained case, however, the curved line joining points a and c is a logarithmic spiral rather than a circular arc (so the magnitude of v_c can be found from v_a using Equation 8.23, interchanging radii for velocity magnitudes).

As a result of the normality principle, there is no energy dissipated by shearing within the soil mass so $\Sigma E_i = 0$. As for the undrained case, the footing and surcharge pressures still do work and the computations for the drained case are shown in Table 8.4.

Applying Equation 8.3 then gives

$$\sum W_i = 0$$

$$q_f Bv - \sigma'_q Bv e^{\pi \tan\phi'} \tan^2\left(\frac{\pi}{4} + \frac{\phi'}{2}\right) = 0 \tag{8.25}$$

$$q_f = \left[e^{\pi \tan\phi'} \tan^2\left(\frac{\pi}{4} + \frac{\phi'}{2}\right)\right]\sigma'_q$$

Analysis using the lower bound theorem

The proposed stress state considered is the same as LB-2 for the undrained case i.e. considering a fan zone of frictional stress discontinuities. This is shown in Figure 8.16(a). The change in direction of the major principal stresses across a frictional discontinuity depends on the frictional strength along the discontinuity as before (τ_d, Figure 8.16(b)). However the envelope bounding the Mohr circles in zones 1 and 2 is now of the form $\tau_f = \sigma' \tan\phi'$ for the drained case, and $\tau_d = \sigma'_d \tan\phi'_{mob}$, where ϕ'_{mob} is the mobilised friction angle along the discontinuity. This is shown in Figure 8.16(c). The mean effective stress in each zone is represented by s'.

As with mechanism LB-2 the circles will touch at a point where $\tau = \tau_d$ as shown in the figure defining the relative position of the two circles. In crossing the discontinuity, the major principal stress will rotate by an amount $\delta\theta$. From Figure 8.16(b) it can be determined that

$$\delta\theta = \frac{\pi}{2} - \Delta \tag{8.26}$$

Considering the shear strength at the crossing point of the two Mohr circles in Figure 8.16(c)

Table 8.4 Work done by the external pressures, mechanism UB-1

Component	Pressure	Area, B_i	Relative velocity, v_i	Work done, W_i
Footing pressure	q_f	B	v	$q_f Bv$
Surcharge	σ'_q	$Be^{(\pi/2 \tan\phi)} \times \tan(\pi/4 + \phi/2)$	$v_2 = -ve^{(\pi/2 \tan\phi)} \times \tan(\pi/4 + \phi/2)$	$-\sigma'_q Bv\, e^{(\pi \tan\phi')} \times \tan^2(\pi/4 + \phi/2)$

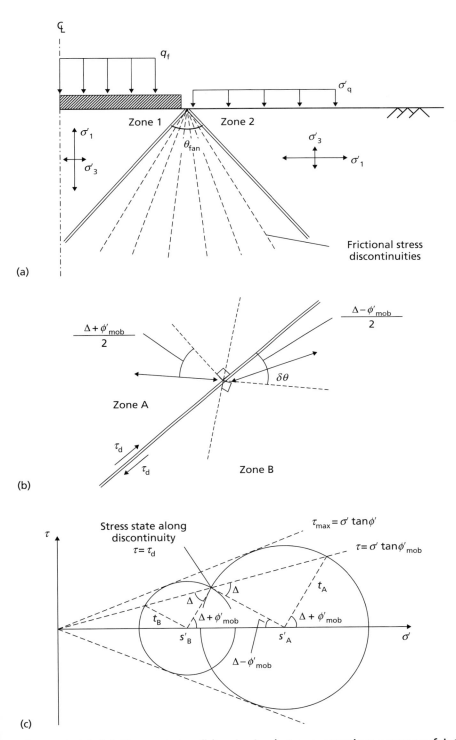

Figure 8.16 (a) Stress state, (b) principal stress rotation across a frictional stress discontinuity, (c) Mohr circles.

Applications in geotechnical engineering

$$\tau_d = t_B \sin(\Delta - \phi'_{mob}) = t_A \sin(\Delta + \phi'_{mob})$$

$$\frac{t_B}{t_A} = \frac{\sin(\Delta + \phi'_{mob})}{\sin(\Delta - \phi'_{mob})} \tag{8.27}$$

The radii of the Mohr circles (t_A and t_B) can also be described by $t = s'\sin\phi'$ for a cohesionless soil as shown in Figure 5.6. This condition means that $s'_B/s'_A = t_B/t_A$. Applying this and substituting Equation 8.26, Equation 8.27 becomes

$$\frac{s'_A}{s'_B} = \frac{s'_B + \delta s'}{s'_B} = \frac{\cos(\phi'_{mob} - \delta\theta)}{\cos(\phi'_{mob} + \delta\theta)} \tag{8.28}$$

Setting $s'_B = s'$, as the strength of the discontinuity approaches the strength of the soil ($\phi'_{mob} \to \phi'$) Equation 8.28 can be simplified to

$$1 + \frac{\delta s'}{s'} = 1 + 2\delta\theta \tan\phi'$$

$$\frac{\delta s'}{s'} = 2\delta\theta \tan\phi' \tag{8.29}$$

for small $\delta\theta$. For a fan zone of frictional stress discontinuities subtending an angle θ_{fan}, Equation 8.29 may be integrated from zone 1 to zone 2, i.e.

$$\int_{s'_2}^{s'_1} \frac{\delta s'}{s'} = \int_0^{\theta_{fan}} 2 \tan\phi' \delta\theta$$

$$\frac{s'_1}{s'_2} = e^{2\theta_{fan} \tan\phi'} \tag{8.30}$$

From Figure 8.16(c), $s'_1 = q_f - s'_1 \sin\phi'$ in zone 1 and $s'_2 = \sigma'_q + s'_2 \sin\phi'$ in zone 2. The principal stress rotation required in the fan is $\theta_{fan} = \pi/2$ (90°) giving from Equation 8.30

$$\frac{q_f}{(1 + \sin\phi')} \cdot \frac{(1 - \sin\phi')}{\sigma'_q} = e^{\pi \tan\phi'}$$

$$\therefore q_f = \left[\frac{(1 + \sin\phi')}{(1 - \sin\phi')} e^{\pi \tan\phi'}\right] \sigma'_q \tag{8.31}$$

Bearing capacity factors

The upper and lower bound solutions to the bearing capacity of a strip foundation on a weightless cohesionless soil (Equations 8.25 and 8.31 respectively) give the same answer for q_f as it can be shown mathematically that

$$\frac{(1 + \sin\phi')}{(1 - \sin\phi')} = \tan^2\left(\frac{\pi}{4} + \frac{\phi'}{2}\right)$$

As for the undrained case, the bearing capacity may be written in a generalised form as

$$q_f = s_q N_q \sigma'_q$$

Shallow foundations

where N_q is a bearing capacity factor relating to surcharge applied around a foundation under drained conditions and s_q is a shape factor. For the undrained case, it was shown that inclusion of the soil unit weight γ in the lower bound analyses did not influence the magnitude of q_f. This is because the amount of soil moving downwards with gravity beneath the foundation was equal to the amount of soil moving upwards against gravity, so that there is no net work done due to the soil weight. The same is not true for the mechanism shown in Figure 8.15, where the size of the upward moving block beneath the surcharge (doing negative work) is greater than the downward moving block beneath the footing (doing positive work). In this case, therefore, there will be an additional amount of resistance due to the additional net negative work input as a result of the self-weight. Any cohesion c' in the soil will also increase the bearing capacity. As a result, the bearing capacity in drained materials is usually expressed by

$$q_f = s_q N_q \sigma_q' + \frac{1}{2}\gamma B s_\gamma N_\gamma + s_c N_c c' \tag{8.32}$$

where N_γ is the bearing capacity factor relating to self-weight, N_c is the factor relating to cohesion, and s_γ and s_c are further shape factors. Values of N_q were found previously by limit analysis, and are given in closed-form by:

$$N_q = \frac{(1+\sin\phi')}{(1-\sin\phi')} e^{\pi \tan\phi'} \tag{8.33}$$

Parameter N_c can be similarly derived for soil with non-zero c' to give

$$N_c = \frac{N_q - 1}{\tan\phi'} \tag{8.34}$$

The final bearing capacity factor, N_γ, is difficult to determine analytically, and is influenced by the roughness of the footing-soil interface (Kumar and Kouzer, 2007). Furthermore, the dilation of soil (and hence the representative value of ϕ' and the degree of associativity) is controlled by the average effective stress beneath the footing ($0.5\gamma B$ in Equation 8.32) as described in Section 5.4, such that there is also a size effect on the value of N_γ (Zhu et al., 2001). Salgado (2008) recommended that for an associative soil with a rough footing-soil interface,

$$N_\gamma = (N_q - 1)\tan(1.32\phi') \tag{8.35}$$

In EC7 the following expression is proposed:

$$N_\gamma = 2(N_q - 1)\tan\phi' \tag{8.36}$$

Values of N_q, N_c and N_γ are plotted in terms of ϕ' in Figure 8.17. It should be noted that the from Equation 8.34, $N_c \to 2+\pi$ as $\phi' \to 0$, which matches the value found in Section 8.3 for undrained conditions.

Lyamin et al. (2007) present the shape factors for rectangular foundations derived from rigorous limit analyses. Their results for s_q are shown in Figure 8.18(a), where it will be seen that s_q varies with ϕ' and B/L. Also shown in this figure are values using the expression recommended in EC7:

$$s_q = 1 + \frac{B}{L}\sin\phi' \tag{8.37}$$

It can be seen from Figure 8.18(a) that for low values of ϕ' (typical of the drained strength of fine-grained soils) Equation 8.37 overpredicts s_q, which would result in an overestimation of bearing capacity;

Applications in geotechnical engineering

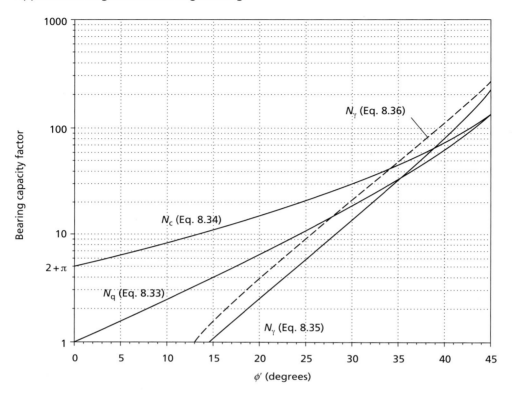

Figure 8.17 Bearing capacity factors for shallow foundations under drained conditions.

for higher values of $\phi' > 30°$ (typical of coarse-grained soils) Equation 8.37 underpredicts s_q, giving a conservative estimate of bearing capacity. Once s_q has been found, it can be shown analytically that

$$s_c = \frac{s_q N_q - 1}{N_q - 1} \tag{8.38}$$

Equation 8.38 is the approach recommended in EC7 for finding s_c.

Data for s_γ, also from Lyamin et al. (2007) are shown in Figure 8.18(b), which can similarly be seen to vary with ϕ' and B/L. Eurocode 7 recommends the use of the following expression:

$$s_\gamma = 1 - 0.3 \frac{B}{L} \tag{8.39}$$

which can be seen, from Figure 8.18(b), to form a lower bound to the data. Therefore, for $\phi' > 20°$, the use of Equation 8.39 will always give a conservative estimate of the bearing capacity of a rectangular foundation.

Depth factors d_c, d_q and d_γ can also be applied to the terms in Equation 8.32 for cases where the soil above the founding plane has non-negligible strength (depth effects for undrained soils were previously considered in Figure 8.10). These are a function of d/B, and recommended values may be found in Lyamin et al. (2007). However, they should only be used if it is certain that the shear strength of the soil above foundation level is, and will remain, equal (or almost equal) to that below foundation level. Indeed, EC7 does not recommend the use of depth factors (i.e. $d_c = d_q = d_\gamma = 1$).

Shallow foundations

For foundations under working load, the maximum shear strain within the supporting soil will normally be less than that required to develop peak shear strength in dense sands or stiff clays, as strains must be low enough to ensure that the settlement of the foundation is acceptable (Sections 8.6–8.8). The allowable bearing capacity or the design bearing resistance should be calculated, therefore, using the peak strength parameters corresponding to the appropriate stress levels. It should be recognised, however, that the results of bearing capacity calculations are very sensitive to the values assumed for the

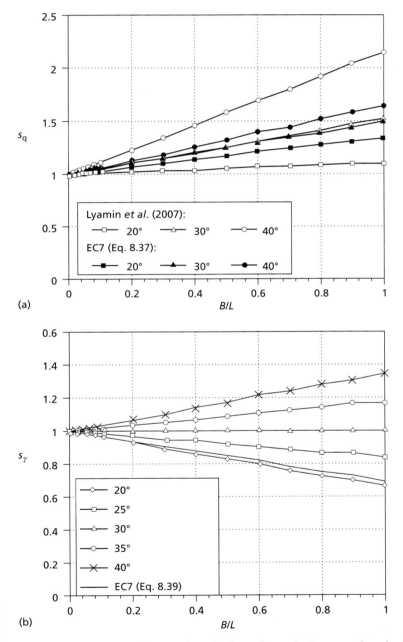

Figure 8.18 Shape factors for shallow foundations under drained conditions: (a) s_q, (b) s_γ.

Applications in geotechnical engineering

shear strength parameters, especially the higher values of ϕ'. Due consideration must therefore be given to the probable degree of accuracy of the parameters.

It is vital that the appropriate values of unit weight are used in the bearing capacity equation. In an effective stress analysis, three different situations should be considered:

i   If the water table is well below the founding plane, the bulk (total) unit weight (γ) is used in the first and second terms of Equation 8.32.
ii If the water table is at the founding plane, the **effective (buoyant) unit weight** ($\gamma' = \gamma - \gamma_w$) must be used in the second term (which represents the resistance due to the weight of the soil below foundation level), the bulk unit weight being used in the first term (resistance due to surcharge).
iii If the water table is at the ground surface or above (e.g. for soil beneath lakes/rivers or seabed soil), the effective unit weight must be used in both the first and second terms.

Example 8.2

A footing 2.25×2.25 m is located at a depth of 1.5 m in a sand for which $c' = 0$ and $\phi' = 38°$. Determine the bearing resistance (a) if the water table is well below foundation level and (b) if the water table is at the surface. The unit weight of the sand above the water table is $18 \, \text{kN/m}^3$; the saturated unit weight is $20 \, \text{kN/m}^3$.

Solution

For $\phi' = 38°$, the bearing capacity factors are $N_q = 49$ (Equation 8.33) and $N_\gamma = 75$ (Equation 8.36). The footing is square ($B/L = 1$), so the shape factors are $s_q = 1.62$ (Equation 8.37) and $s_\gamma = 0.70$ (Equation 8.39). The values of s_q and s_γ are both conservative (as $\phi' > 30°$). As $c' = 0$ in this case there is no need to compute N_c and s_c. For case (a), when the water table is well below the founding plane:

$$q_f = s_q N_q \gamma d + \frac{1}{2} \gamma B s_\gamma N_\gamma$$

$$= (1.62 \cdot 49 \cdot 18 \cdot 1.5) + (0.5 \cdot 18 \cdot 2.25 \cdot 0.70 \cdot 75)$$

$$= 3206 \text{ kPa}$$

When the water table is at the surface, the ultimate bearing capacity is given by

$$q_f = s_q N_q \gamma' d + \frac{1}{2} \gamma' B s_\gamma N_\gamma$$

$$= \left[1.62 \cdot 49 \cdot (20 - 9.81) \cdot 1.5\right] + \left[0.5 \cdot (20 - 9.81) \cdot 2.25 \cdot 0.70 \cdot 75\right]$$

$$= 1815 \text{ kPa}$$

Comparing these two results, it can be seen that changing the hydraulic conditions (pore pressures) within the ground has a significant effect on the bearing capacity.

8.5 Stresses beneath shallow foundations

Under typical working loads, the applied vertical bearing pressure applied by a shallow foundation to the underlying soil will be much less than the bearing capacity (so that the foundation is safe from collapse). Under these conditions, the soil will be in a state of elastic rather than plastic equilibrium. The constitutive behaviour of the soil is approximated as linear elastic (see Equation 5.4), though the stiffness (G) may vary with the confining stress (i.e. depth) or between soil layers. If the stresses beneath the foundation are known for an applied bearing pressure (q), then the induced strains (and hence movements of the foundation) can be determined from the elastic material properties (G, v). A range of solutions, suitable for determining the stresses below foundations, is given in this section.

Point load

The stresses within a semi-infinite, homogeneous, isotropic mass, with a linear stress–strain relationship, due to a point load on the surface, were determined by Boussinesq in 1885. The vertical, radial, circumferential and shear stresses at a depth z and a horizontal distance r from the point of application of the load were given. Referring to Figure 8.19(a), the stresses induced at X due to a point load Q on the surface are as follows:

$$\Delta\sigma_z = \frac{3Q}{2\pi z^2} \left[\frac{1}{1+\left(\frac{r}{z}\right)^2} \right]^{2.5} \tag{8.40}$$

$$\Delta\sigma_r = \frac{Q}{2\pi} \left[\frac{3r^2 z}{\left(r^2+z^2\right)^{2.5}} - \frac{1-2v}{r^2+z^2+z\sqrt{\left(r^2+z^2\right)}} \right] \tag{8.41}$$

$$\Delta\sigma_\theta = -\frac{Q}{2\pi}(1-2v)\left[\frac{z}{\left(r^2+z^2\right)^{1.5}} - \frac{1}{r^2+z^2+z\sqrt{\left(r^2+z^2\right)}} \right] \tag{8.42}$$

$$\Delta\tau_{rz} = \frac{3Q}{2\pi} \left[\frac{rz^2}{\left(r^2+z^2\right)^{2.5}} \right] \tag{8.43}$$

It should be noted that the stresses are dependent on position (r, z) and the magnitude of the load, but are independent of the soil stiffness (though the soil must be elastic and have constant stiffness with depth). When $v=0.5$, the second term in Equation 8.41 vanishes and Equation 8.42 gives $\Delta\sigma_\theta = 0$.

Of the four equations given above, Equation 8.40 is used most frequently in practice, and can be written in terms of an **influence factor** (relating to stresses) I_Q where

$$I_Q = \frac{3}{2\pi} \left[\frac{1}{1+\left(\frac{r}{z}\right)^2} \right]^{2.5} \tag{8.44}$$

Applications in geotechnical engineering

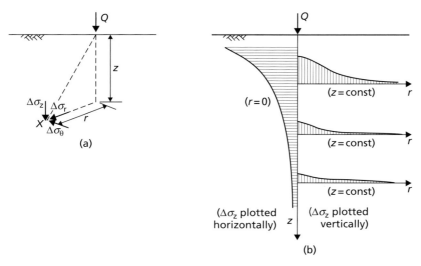

Figure 8.19 (a) Total stresses induced by point load, (b) variation of vertical total stress induced by point load.

Then,

$$\Delta \sigma_z = \left(\frac{Q}{z^2}\right) I_Q \tag{8.45}$$

Values of I_Q in terms of r/z are given in Table 8.5. The form of the variation of $\Delta \sigma_z$ with z and r is illustrated in Figure 8.19(b). The left-hand side of the figure shows the variation of $\Delta \sigma_z$ with z on the vertical through the point of application of the load Q (i.e. for $r=0$); the right-hand side of the figure shows the variation of $\Delta \sigma_z$ with r for three different values of z.

The stresses due to surface loads distributed over a particular area (e.g. a footing) can be obtained by integration from the point load solutions. The stresses at a point due to more than one surface load are obtained by superposition. In practice, loads are not usually applied directly on the surface, but the results for surface loading can be applied conservatively in problems concerning loads at a shallow depth.

Table 8.5 Influence factors (I_Q) for vertical stress due to point load

r/z	I_Q	r/z	I_Q	r/z	I_Q
0.00	0.478	0.80	0.139	1.60	0.020
0.10	0.466	0.90	0.108	1.70	0.016
0.20	0.433	1.00	0.084	1.80	0.013
0.30	0.385	1.10	0.066	1.90	0.011
0.40	0.329	1.20	0.051	2.00	0.009
0.50	0.273	1.30	0.040	2.20	0.006
0.60	0.221	1.40	0.032	2.40	0.004
0.70	0.176	1.50	0.025	2.60	0.003

Line load

Referring to Figure 8.20(a), the stresses induced at point X due to a line load of Q per unit length on the surface are as follows:

$$\Delta\sigma_z = \frac{2Q}{\pi}\left[\frac{z^3}{\left(x^2+z^2\right)^2}\right] \quad (8.46)$$

$$\Delta\sigma_x = \frac{2Q}{\pi}\left[\frac{x^2 z}{\left(x^2+z^2\right)^2}\right] \quad (8.47)$$

$$\Delta\tau_{xz} = \frac{2Q}{\pi}\left[\frac{xz^2}{\left(x^2+z^2\right)^2}\right] \quad (8.48)$$

Strip area carrying uniform pressure

Superposing a series of point loads acting over an area of width B, it can be shown that the stresses at point X due to a uniform pressure q on a strip area of width B and infinite length are given in terms of the angles α and β defined in Figure 8.20(b) as:

$$\Delta\sigma_z = \frac{q}{\pi}\left[\alpha + \sin\alpha \cdot \cos(\alpha + 2\beta)\right] \quad (8.49)$$

$$\Delta\sigma_x = \frac{q}{\pi}\left[\alpha - \sin\alpha \cdot \cos(\alpha + 2\beta)\right] \quad (8.50)$$

$$\Delta\tau_{xz} = \frac{q}{\pi}\left[\sin\alpha \cdot \sin(\alpha + 2\beta)\right] \quad (8.51)$$

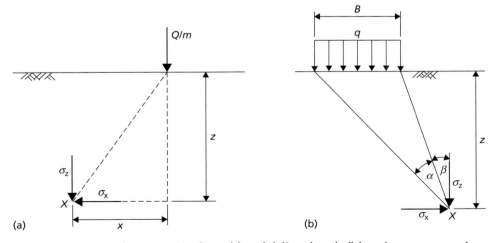

Figure 8.20 Total stresses induced by: (a) line load, (b) strip area carrying a uniform pressure.

Applications in geotechnical engineering

Rectangular area carrying uniform pressure

A solution has been obtained for the vertical stress induced at depth z under a corner of a rectangular area of dimensions mz and nz (Figure 8.21) carrying a uniform pressure q. The solution can be written in the form

$$\Delta \sigma_z = q I_{qr} \tag{8.52}$$

Values of the influence factor I_{qr} in terms of m and n are given in Figure 8.21, after Fadum (1948). The factors m and n are interchangeable. The chart can also be used for a strip area, considered as a rectangular area of infinite length ($n=\infty$), and for a square area ($n=m$). Superposition enables any area based on rectangles to be dealt with, and enables the vertical stress under any point within or outside the area to be obtained.

Contours of equal vertical stress in the vicinity of a strip area carrying a uniform pressure are plotted in Figure 8.22(a). The zone lying inside the vertical stress contour of value $0.2q$ is described as the **bulb of pressure**, and represents the zone of soil which is expected to contribute significantly to the settlement of the foundation under the applied bearing pressure. Contours of equal vertical stress in the vicinity of a square area carrying a uniform pressure are plotted in Figure 8.22(b).

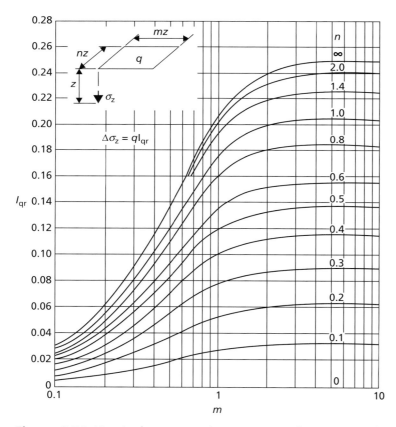

Figure 8.21 Vertical stress under a corner of a rectangular area carrying a uniform pressure (reproduced from R.E. Fadum (1948) *Proceedings of the 2nd International Conference of SMFE*, Rotterdam, Vol. 3 by permission of Professor Fadum).

Shallow foundations

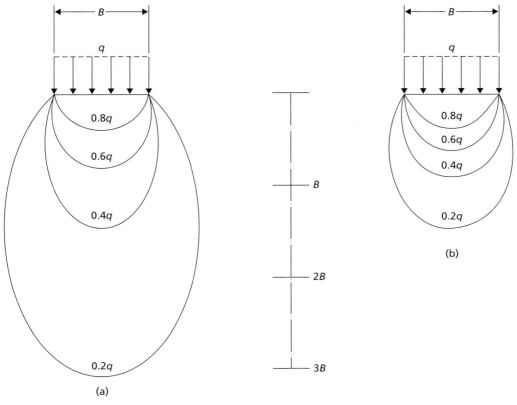

Figure 8.22 Contours of equal vertical stress: (a) under a strip area, (b) under a square area.

Example 8.3

A rectangular foundation 6 × 3 m carries a uniform pressure of 300 kPa near the surface of a soil mass. Determine the vertical stress at a depth of 3 m below a point (A) on the centre line 1.5 m outside a long edge of the foundation.

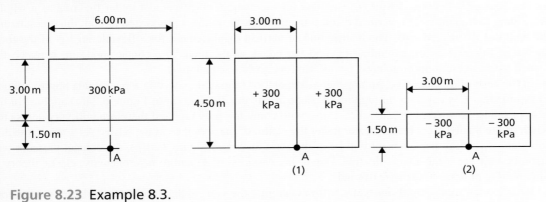

Figure 8.23 Example 8.3.

Applications in geotechnical engineering

Solution

Using the principle of superposition the problem is dealt with in the manner shown in Figure 8.23. For the two rectangles (1) carrying a positive pressure of 300 kPa, $m = 1.00$ and $n = 1.50$, and therefore

$$I_{qr} = 0.193$$

For the two rectangles (2) carrying a negative pressure of 300 kPa, $m = 1.00$ and $n = 0.50$, and therefore

$$I_{qr} = 0.120$$

Hence,

$$\Delta \sigma_z = \sum q I_{qr}$$
$$= 2 \cdot (300 \cdot 0.193) - 2 \cdot (300 \cdot 0.120)$$
$$= 44 \text{ kPa}$$

8.6 Settlements from elastic theory

The vertical displacement (s) under an area carrying a uniform pressure q on the surface of a semi-infinite, homogeneous, isotropic mass, with a linear stress–strain relationship, can be expressed as

$$s = \frac{qB}{E}(1 - v^2) I_s \tag{8.53}$$

where E is the Young's modulus of the soil and I_s is an influence factor depending on the shape of the loaded area. In the case of a rectangular area, B is the lesser dimension (the greater dimension being L); in the case of a circular area, B is the diameter. Values of influence factors are given in Table 8.6 for displacements under the centre and a corner (the edge in the case of a circle) of a **flexible** loaded area (i.e. having negligible bending stiffness), and also for the average displacement under the area as a whole. According to Equation 8.53, vertical displacement increases in direct proportion to both the pressure and the width of the loaded area. The distribution of vertical displacement for a flexible area is of the form shown in Figure 8.24(a), extending beyond the edges of the area. The contact pressure between the loaded area and the supporting mass is uniform (everywhere equal to q).

In the case of an extensive, homogeneous deposit of saturated clay, it is a reasonable approximation to assume that E is constant throughout the deposit and the distribution of Figure 8.24(a) applies. In the case of sands, however, the value of E varies with confining pressure, and therefore will vary across the width of the loaded area, being greater under the centre of the area than at the edges. As a result, the distribution of vertical displacement will be of the form shown in Figure 8.24(b); the contact pressure will again be uniform if the area is flexible. Due to the variation of E, and to heterogeneity, elastic theory is little used in practice in the case of sands.

If the loaded area is **rigid** (infinitely stiff in bending), the vertical displacement will be uniform across the width of the area and its magnitude will be only slightly less than the average displacement under a

Table 8.6 Influence factors (I_s) for vertical displacement under flexible and rigid areas carrying uniform pressure

Shape of area		I_s (flexible)			I_s (rigid)
		Centre	Corner	Average	Average
Square	(L/B = 1)	1.12	0.56	0.95	0.82
Rectangle	L/B = 2	1.52	0.76	1.30	1.20
Rectangle	L/B = 5	2.10	1.05	1.83	1.70
Rectangle	L/B = 10	2.54	1.27	2.25	2.10
Rectangle	L/B = 100	4.01	2.01	3.69	3.47
Circle		1.00	0.64	0.85	0.79

corresponding flexible area. Influence factors for rigid foundations are also given in Table 8.6. Under a rigid area, the contact pressure distribution is not uniform. For a circular area the forms of the distributions of contact pressure on clay and sand, respectively, are shown in Figures 8.25(a) and 8.25(b).

In most cases, in practice the soil deposit will be of limited thickness and will be underlain by a hard stratum (e.g. bedrock). Christian and Carrier (1978) proposed the use of results by Giroud (1972) and by Burland (1970) in such cases. For this case, the average vertical displacement under a flexible area carrying a uniform pressure q is given by

$$s = \mu_0 \mu_1 \frac{qB}{E} \tag{8.54}$$

where μ_0 depends on the depth of embedment and μ_1 depends on the layer thickness and the shape of the loaded area. Values of the coefficients μ_0 and μ_1 for Poisson's ratio equal to 0.5 are given in Figure 8.26.

Figure 8.24 Distributions of vertical displacement beneath a flexible area: (a) clay, and (b) sand.

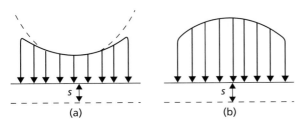

Figure 8.25 Contact pressure distribution beneath a rigid area: (a) clay, and (b) sand.

Applications in geotechnical engineering

The principle of superposition can be used in cases of a number of soil layers each having a different value of E (see Example 8.4).

It should be noted that, unlike the expressions for vertical stress ($\Delta\sigma_z$) given in Section 8.5, the expressions for vertical displacement are dependent on the values of elastic modulus (E) and Poisson's ratio (v) for the soil in question. Because of the uncertainties involved in obtaining these elastic parameters, values of vertical displacement calculated from elastic theory are less reliable than values of vertical stress. For most cases, in practice, simple settlement calculations are adequate provided that reliable values of soil parameters for the in-situ soil have been determined. It should be appreciated that the precision of settlement predictions is much more influenced by inaccuracies in the values of soil parameters than by shortcomings in the methods of analysis. Sampling disturbance can have a serious effect on the values of parameters determined in the laboratory. In settlement analysis the same degree of precision should not be expected as, for example, in structural calculations.

Determination of elastic parameters

In order to make use of Equations 8.53 or 8.54, the elastic properties E and v must be determined. These solutions are used mainly to estimate the **immediate settlement** $s = s_i$ (i.e. that occurring prior to consolidation) of foundations on fine-grained soils; such settlement occurs under undrained conditions, the appropriate value of Poisson's ratio being $v = v_u = 0.5$ (so that the values in Figure 8.26 are valid). The

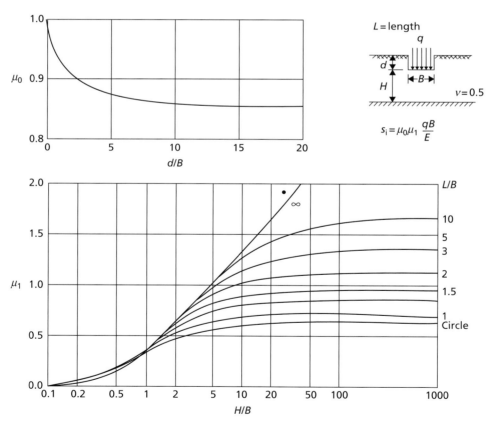

Figure 8.26 Coefficients μ_0 and μ_1 for vertical displacement (after Christian and Carrier, 1978).

value of the undrained modulus E_u is therefore required, and the main difficulty in predicting immediate settlement is in the determination of this parameter. Equation 5.6 can be used to determine E_u from a known value of shear modulus G. The shear modulus may be determined from (undrained) triaxial tests as described in Section 5.4, or from in-situ tests (seismic CPT or PMT, Sections 7.5 and 7.4 respectively). Care must be taken, however, regarding the strain at which the laboratory or in-situ value of G was determined (see Figure 5.4). For example, seismic CPT data will give G_0 (via V_s), but under working loads the operative stiffness beneath a foundation will usually be between 0.2 and $0.5G_0$ (Figure 5.4(c)), depending on how close the applied loads are to the bearing resistance. For most soils, the modulus increases with depth. Use of a constant value of E_u overestimates immediate settlement (i.e. gives a conservative value).

In principle, Equations 8.53 and 8.54 may be used to estimate the ultimate (drained) settlements of foundations on either coarse- or fine-grained soils using the drained Young's modulus E' and Poisson's ratio v'. However, v' (unlike v_u) is a material-dependent parameter, and therefore must be determined from triaxial testing with on-sample measurement (described in Section 5.4). Coupling the potential difficulties in obtaining reliable values of v' with the estimation of an appropriate strain level for determination of G (and hence E') gives drained elastic settlement values which are unreliable in practice. In place of this, for fine-grained soils consolidation settlements are computed and added to the immediate settlements calculated using elastic theory (see Section 8.7). For coarse-grained soils empirical methods are used based on either SPT or CPT data, which are described in Section 8.8.

Example 8.4

A foundation 4×2 m, carrying a uniform pressure of 150 kPa, is located at a depth of 1 m in a layer of clay 5 m thick for which the value of E_u is 40 MPa. This layer is underlain by a second clay layer 8 m thick for which the value of E_u is 75 MPa. A hard stratum lies below the second layer. Determine the average immediate settlement s_i under the foundation.

Solution

Now, $d/B = 0.5$, and therefore from Figure 8.26, $\mu_0 = 0.94$

1. Considering the upper clay layer, with $E_u = 40$ MPa:

$$H/B = 4/2 = 2, \quad L/B = 2$$
$$\therefore \mu_1 = 0.60$$

Hence, from Equation 8.54

$$s_{i1} = 0.94 \times 0.60 \times \frac{150 \times 2}{40} = 4.2 \,\text{mm}$$

2. Considering the two layers combined, with $E_u = 75$ MPa:

$$H/B = 12/2 = 6, \quad L/B = 2$$
$$\therefore \mu_1 = 0.85$$

$$s_{i2} = 0.94 \times 0.85 \times \frac{150 \times 2}{75} = 3.2 \,\text{mm}$$

Applications in geotechnical engineering

3   Considering the upper layer, with $E_u = 75\,\text{MPa}$:
$$H/B = 2, \quad L/B = 2$$
$$\therefore \mu_1 = 0.60$$
$$s_{i3} = 0.94 \times 0.60 \times \frac{150 \times 2}{75} = 2.3\,\text{mm}$$

Hence, using the principle of superposition, the settlement of the foundation is given by
$$s_i = s_{i1} + s_{i2} - s_{i3}$$
$$= 4.2 + 3.2 - 2.3$$
$$= 5\,\text{mm}$$

8.7 Settlements from consolidation theory

Predictions of consolidation settlement using the one-dimensional method can be made based on the results of oedometer tests using representative samples of the soil. Due to the confining ring in the oedometer the net lateral strain in the test specimen is zero, and for this condition the initial excess pore water pressure is equal theoretically to the increase in total vertical stress. In practice, the condition of zero lateral strain is satisfied approximately in the cases of thin clay layers and of layers under loaded areas which are large compared with the layer thickness (these situations were considered in Chapter 4). In many practical situations, however, significant lateral strain will occur, and the initial excess pore water pressure will depend on the in-situ stress conditions.

Consider an element of soil initially in equilibrium under total principal stresses σ_1, σ_2 and σ_3. If the major principal stress (σ_1) is increased by an amount $\Delta\sigma_1$ (as shown in Figure 8.27) due to a shallow foundation, for example, there will be an immediate increase Δu_1 in pore pressure. The increases in effective stress are

$$\Delta\sigma_1' = \Delta\sigma' - \Delta u_1$$
$$\Delta\sigma_3' = \Delta\sigma_2' = \Delta u_1$$

If the soil behaved as an elastic material during the initial application of load, then the reduction in volume of the soil skeleton would be

$$\frac{1}{3} C_s V \left(\Delta\sigma_1 - 3\Delta u_1 \right)$$

The reduction in volume of the pore space is

$$C_n n V \Delta u_1$$

Under undrained conditions, these two volume changes will be equal, i.e.

$$\frac{1}{3} C_s V \left(\Delta\sigma_1 - 3\Delta u_1 \right) = C_v n V \Delta u_1$$

Shallow foundations

Therefore,

$$\Delta u_1 = \frac{1}{3}\left(\frac{1}{1+n(C_v/C_s)}\right)\Delta \sigma_1$$

$$= \frac{1}{3}B\Delta\sigma_1$$

Soils, however, are not elastic, and the above equation is rewritten in the general form

$$\Delta u_1 = AB\Delta\sigma_1 \qquad (8.55)$$

where A is a pore pressure coefficient that can be determined experimentally. Coefficient B was previously described in Section 5.4 (Equation 5.35). In the case of fully saturated soil ($B=1$),

$$\Delta u_1 = A\Delta\sigma_1 \qquad (8.56)$$

Increasing σ_1 only represents a change in the deviatoric stress of $\Delta q = \Delta\sigma_1$ (as $\Delta\sigma_3 = \Delta\sigma_2 = 0$). Any general change in stress conditions within an element of soil may be represented by a deviatoric stress increment (above) and mean stress (i.e. an isotropic) increment. The case of an increment of mean stress was considered in Section 5.4 from which Equation 5.35 was obtained. Therefore, the general equation for the pore pressure response Δu to an isotropic stress increase $\Delta\sigma_3$ together with a deviatoric stress increase ($\Delta\sigma_1 - \Delta\sigma_3$) may be obtained by combining Equations 8.56 and 5.35:

$$\Delta u = \Delta u_3 + \Delta u_1$$
$$= B\left[\Delta\sigma_3 + A(\Delta\sigma_1 - \Delta\sigma_3)\right] \qquad (8.57)$$

In cases in which the lateral strain is not zero, there will be an immediate settlement, under undrained conditions, in addition to the consolidation settlement. Immediate settlement is zero if the lateral strain is zero, as assumed in the one-dimensional method of calculating settlement. In the Skempton–Bjerrum method (Skempton and Bjerrum, 1957), the total settlement (s) of a foundation on clay is given by

$$s = s_i + s_c$$

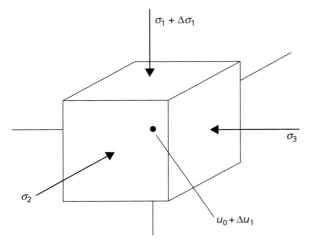

Figure 8.27 Soil element under major principal stress increment.

where s_i = immediate settlement, occurring under undrained conditions (see Section 8.6), and s_c = consolidation settlement, due to the volume reduction accompanying the gradual dissipation of excess pore water pressure.

If there is no change in static pore water pressure, the initial value of excess pore water pressure (denoted by u_i) at a point in the soil layer is given by Equation 8.57, with $B = 1$ for a fully saturated soil. Thus

$$u_i = \Delta\sigma_3 + A(\Delta\sigma_1 - \Delta\sigma_3)$$
$$= \Delta\sigma_1 \left[A + \frac{\Delta\sigma_3}{\Delta\sigma_1}(1-A) \right] \tag{8.58}$$

where $\Delta\sigma_1$ and $\Delta\sigma_3$ are the total principal stress increments due to surface loading. From Equation 8.58, it is seen that

$$u_i > \Delta\sigma_3$$

if A is positive. Note also that $u_i = \Delta\sigma_1$ if $A = 1$.

The in-situ effective stresses before loading, immediately after loading and after consolidation are represented in Figure 8.28(a), and the corresponding Mohr circles (A, B and C, respectively) in Figure 8.28(b). In Figure 8.28(b), abc is the effective stress path (ESP) for in-situ loading and consolidation, ab representing an immediate change of stress and bc a gradual change of stress as the excess pore water pressure dissipates. Immediately after loading there is a reduction in σ_3' due to u_i being greater than $\Delta\sigma_3$, and lateral expansion will occur. Subsequent consolidation will therefore involve lateral recompression. Circle D in Figure 8.28(b)

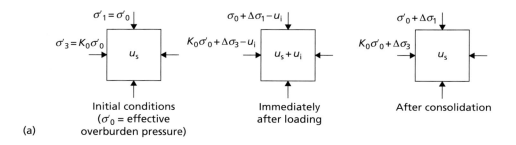

(a)

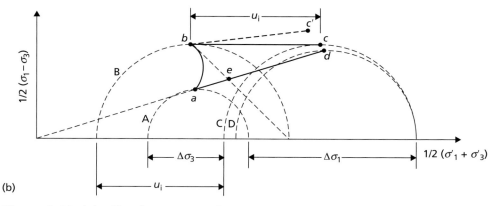

(b)

Figure 8.28 (a) Effective stresses for in-situ conditions and under a general total stress increment $\Delta\sigma_1$, $\Delta\sigma_3$, (b) stress paths.

Shallow foundations

represents the corresponding stresses in the oedometer test after consolidation and ad is the corresponding ESP for the oedometer test. As the excess pore water pressure dissipates, Poisson's ratio decreases from the undrained value (0.5) to the drained value at the end of consolidation. The decrease in Poisson's ratio does not significantly affect the vertical stress, but results in a small decrease in horizontal stress (point c would become c' in Figure 8.28(b)): this decrease is neglected in the Skempton–Bjerrum method.

Skempton and Bjerrum (1957) proposed that the effect of lateral strain be neglected in the calculation of consolidation settlement (s_c), thus enabling the oedometer test to be maintained as the basis of the method. It was admitted, however, that this simplification could involve errors of up to 20% in vertical settlements. However, the value of excess pore water pressure given by Equation 8.58 is used in the method.

By the one-dimensional method, consolidation settlement (equal to the total settlement) is given from Equation 4.8 as

$$s_{oed} = \int_0^H m_v \Delta \sigma_1 dz \quad (\text{i.e. } \Delta \sigma' = \Delta \sigma_1)$$

where H is the thickness of the soil layer. By the Skempton–Bjerrum method, consolidation settlement is expressed in the form

$$s_c = \int_0^H m_v u_i dz$$
$$= \int_0^H m_v \Delta \sigma_1 \left[A + \frac{\Delta \sigma_3}{\Delta \sigma_1}(1-A) \right] dz$$

The value of initial excess pore water pressure (u_i) should, in general, correspond to the in-situ stress conditions. A settlement coefficient μ_c is introduced such that

$$s_c = \mu_c s_{oed} \tag{8.59}$$

where

$$\mu_c = \frac{\int_0^H m_v \Delta \sigma_1 \left[A + \left(\frac{\Delta \sigma_3}{\Delta \sigma_1}\right)(1-A) \right] dz}{\int_0^H m_v \Delta \sigma_1 dz}$$

If it can be assumed that m_v and A are constant with depth (sub-layers can be used in analysis to account for modulus changing with depth, see Example 8.5), then μ_c can be expressed as

$$\mu_c = A + (1-A)\alpha \tag{8.60}$$

where

$$\alpha = \frac{\int_0^H \Delta \sigma_3 dz}{\int_0^H \Delta \sigma_1 dz}$$

The value of α depends only on the shape of the loaded area and the thickness of the soil layer in relation to the dimensions of the loaded area; thus α can be estimated from elastic theory.

Applications in geotechnical engineering

Determination of parameters A and μ_c

The value of A for a fully saturated soil can be determined from measurements of pore water pressure during the application of principal stress difference under undrained conditions in a triaxial test. The change in total major principal stress is equal to the value of the principal stress difference applied, and if the corresponding change in pore water pressure is measured the value of A can be calculated from Equation 8.56. The value of the coefficient at any stage of the test can be obtained, but the value at failure is of more interest.

For highly compressible soils such as normally consolidated clays, the value of A is usually within the range 0.5–1.0. In the case of clays of high sensitivity the increase in major principal stress may cause collapse of the soil structure, resulting in very high pore water pressures and values of A greater than 1. For soils of lower compressibility, such as lightly overconsolidated clays, the value of A typically lies within the range 0–0.5. If the clay is heavily overconsolidated there is a tendency for the soil to dilate as the major principal stress is increased, but under undrained conditions no water can be drawn into the element and a negative pore water pressure may result. The value of A for heavily overconsolidated soils may lie between −0.5 and 0.

The use of a value of A obtained from the results of a triaxial test on a cylindrical clay specimen is strictly applicable only for the condition of axial symmetry, i.e. for the case of settlement under the centre of a circular footing. However, a value of A so obtained will serve as a good approximation for

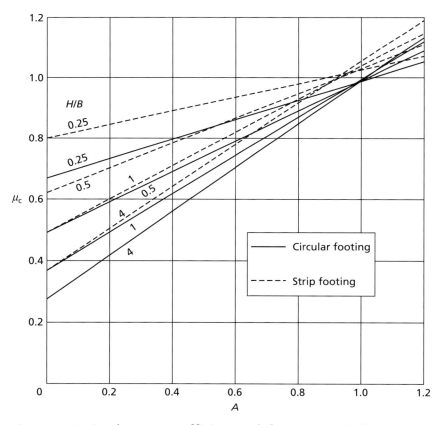

Figure 8.29 Settlement coefficient μ_c (after Scott, 1963).

the case of settlement under the centre of a square footing (using a circular footing of the same area). Under a strip footing, however, plane strain conditions apply, and the intermediate principal stress increment $\Delta\sigma_2$, in the direction of the longitudinal axis, is equal to $0.5(\Delta\sigma_1+\Delta\sigma_3)$. Scott (1963) has shown that the value of u_i appropriate in the case of a strip footing can be obtained by using a pore pressure coefficient A_s, where

$$A_s = 0.866A + 0.211 \tag{8.61}$$

The coefficient A_s replaces A (the coefficient for conditions of axial symmetry) in Equation 8.60 for the case of a strip footing, with the expression for α being unchanged.

Values of the settlement coefficient μ_c, for circular and strip footings, in terms of A and the ratio of layer thickness/breadth of footing (H/B) are given in Figure 8.29.

Example 8.5

A footing 6 m square, carrying a net pressure of 160 kPa, is located at a depth of 2 m in a deposit of stiff clay 17 m thick: a firm stratum lies immediately below the clay. From oedometer tests on specimens of the clay the value of m_v was found to be $0.13\,\text{m}^2/\text{MN}$, and from triaxial tests the value of A was found to be 0.35. The undrained Young's modulus for the clay is estimated to be 55 MPa. Determine the total settlement under the centre of the footing.

Solution

In this case there will be significant lateral strain in the clay beneath the footing (resulting in immediate settlement), and it is appropriate to use the Skempton–Bjerrum method. The section is shown in Figure 8.30.

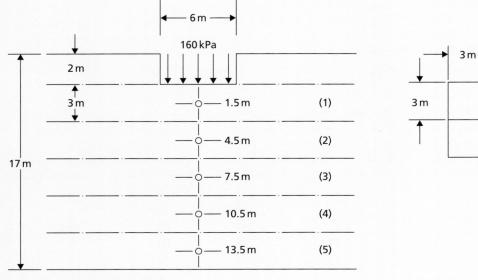

Figure 8.30 Example 8.5.

a *Immediate settlement.* The influence factors are obtained from Figure 8.26. Now

$$\frac{H}{B} = \frac{15}{6} = 2.5$$

$$\frac{d}{B} = \frac{2}{6} = 0.33$$

$$\frac{L}{B} = 1$$

$$\therefore \mu_0 = 0.95 \text{ and } \mu_1 = 0.55$$

Hence

$$s_i = \mu_0 \mu_1 \frac{qB}{E_u}$$

$$= 0.95 \times 0.55 \times \frac{160 \times 6}{55} = 9 \text{ mm}$$

b *Consolidation settlement.* In Table 8.7,

$$\Delta \sigma' = 4 \times 160 \times I_{qr} \quad (\text{kPa})$$

$$s_{oed} = 0.13 \times \Delta \sigma' \times 3 = 0.39 \Delta \sigma' \quad (\text{mm})$$

Now

$$\frac{H}{B} = \frac{15}{6.77} = 2.2$$

(equivalent diameter = 6.77 m) and $A = 0.35$. Hence, from Figure 8.29,

$$\mu_c = 0.55$$

Then

$$s_c = 0.55 \times 116.6 = 64 \text{ mm}$$

$$\text{Total settlement} = s_i + s_c$$

$$= 9 + 64$$

$$= 73 \text{ mm}$$

Table 8.7 Example 8.5

Layer	z (m)	m, n	I_{qr}	$\Delta \sigma'$ (kPa)	s_{oed} (mm)
1	1.5	2.00	0.233	149	58.1
2	4.5	0.67	0.121	78	30.4
3	7.5	0.40	0.060	38	14.8
4	10.5	0.285	0.033	21	8.2
5	13.5	0.222	0.021	13	5.1
					116.6

8.8 Settlement from in-situ test data

Due to the extreme difficulty of obtaining undisturbed sand samples for laboratory testing and to the inherent heterogeneity of sand deposits, foundation settlements on coarse-grained soils are normally estimated by means of correlations based on the results of in-situ tests. This section will describe methods recommended in EC7 based on SPT and CPT tests (these are described in Chapter 7).

Analysis using SPT data

Burland and Burbidge (1985) carried out a statistical analysis of over 200 settlement records of foundations on sands and gravels. A relationship was established between the compressibility of the soil (a_f), the width of the foundation (B) and the average value of SPT blowcount ($\overline{N}$) over the depth of influence (z_I) of the foundation. Evidence was presented which indicated that if N tends to increase with depth or is approximately constant with depth, then the ratio of the depth of influence to foundation width (z_I/B) decreases with increasing foundation width; values of z_I obtained from Figure 8.31 can be used as a guide in design. However, if N tends to decrease with depth, the value of z_I should be taken as $2B$, provided the stratum thickness exceeds this value. The compressibility is related to foundation width by a compressibility index (I_c), where

$$I_c = \frac{a_f}{B^{0.7}} \tag{8.62}$$

The compressibility index, in turn, is related to the average value of the corrected standard penetration resistance ($\overline{N} = \overline{N}_{60}$) by the expression

$$I_c = \frac{1.71}{\left(\overline{N}_{60}\right)^{1.4}} \tag{8.63}$$

Non-normalised N_{60} values are used as correcting for effective overburden pressure (Equation 7.2), has a major influence on both standard penetration resistance and compressibility; this influence should be eliminated from the correlation. The results of the analysis suggested that the influence of water table level is reflected in the measured N values. However, the position of the water table does influence settlement, so if the level were to fall subsequent to the determination of the N values then a greater settlement would be expected. In the case of fine sands and silty sands below the water table, the measured N value, if greater than 15, should be corrected for the increased resistance due to negative excess pore water pressure set up during driving and unable to dissipate immediately: the corrected value (N') is given by

$$N' = 15 + \frac{1}{2}(N - 15) \tag{8.64}$$

It was further proposed that in the case of gravels or sandy gravels the measured N values should be increased by 25%.

In a normally consolidated sand, the average settlement s (mm) at the end of construction for a foundation of width B (m) carrying a foundation pressure q (kPa) is given by

$$s = qB^{0.7}I_c \tag{8.65a}$$

Applications in geotechnical engineering

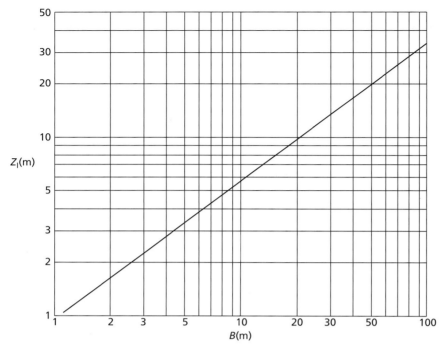

Figure 8.31 Relationship between depth of influence and foundation width (reproduced from J.B. Burland and M.C. Burbridge (1985) *Proceedings Institution of Civil Engineers*, Part 1, Vol. 78, by permission of Thomas Telford Ltd.).

If it can be established that the sand is overconsolidated and an estimate of preconsolidation pressure (σ'_{max}) can be made, the settlement is instead given by one or other of the following expressions:

$$s = \left(q - \frac{2}{3}\sigma'_{max}\right)B^{0.7}I_c \quad (\text{if } q > \sigma'_{max}) \tag{8.65b}$$

$$s = qB^{0.7}\frac{I_c}{3} \quad (\text{if } q < \sigma'_{max}) \tag{8.65c}$$

The analysis indicated that foundation depth had no significant influence on settlement for depth/breadth ratios less than 3. However, a significant correlation was found between settlement and the length/breadth ratio (L/B) of the foundation; accordingly, the settlement given by Equation 8.65 should be multiplied by a shape factor F_s where

$$F_s = \left(\frac{1.25 L/B}{L/B + 0.25}\right)^2 \tag{8.66}$$

It was tentatively proposed that if the thickness (H) of the sand stratum below foundation level is less than the depth of influence (z_I), the settlement should be multiplied by a factor F_l, where

$$F_l = \frac{H}{z_I}\left(2 - \frac{H}{z_I}\right) \tag{8.67}$$

Shallow foundations

Although it is normally assumed that settlement in sands is virtually complete by the end of construction and initial loading, the records indicated that continuing settlement (creep) can occur and it was proposed that the settlement should be multiplied by a factor F_t for time in excess of three years after the end of construction, where

$$F_t = \left[1 + R_3 + R_t \log\left(\frac{t}{3}\right)\right] \tag{8.68}$$

where R_3 is the time-dependent settlement, as a proportion of s, occurring during the first three years after construction, and R_t the settlement occurring during each log cycle of time in excess of three years. A conservative interpretation of the data indicates that after 30 years F_t can reach 1.5 for static loads and 2.5 for variable loads. An example of the use of this method in foundation design will be given in Section 8.9.

Analysis using CPT data

This method of settlement estimation was presented by Schmertmann (1970), and is based on a simplified distribution of vertical strain under the centre, or centre-line, of a shallow foundation, expressed in the form of a strain influence factor I_z. The vertical strain ε_z is written as

$$\varepsilon_z = \frac{q_n}{E} I_z$$

where q_n is the net pressure on the foundation and E the appropriate value of deformation modulus. The assumed distributions of strain influence factor with depth for square ($L/B=1$) and long ($L/B \geq 10$) or strip foundations are shown in Figure 8.32, depth being expressed in terms of the width of the foundations. The two cases correspond to conditions of axial symmetry and plane strain, respectively. These are simplified distributions, based on both theoretical and experimental results, in which it is assumed that strains become insignificant at a depth $z_{f0}=2B$ and $4B$, respectively, below the foundations. The peak value of strain influence factor I_{zp} in each case is given by the expression

$$I_{zp} = 0.5 + 0.1 \left(\frac{q_n}{\sigma'_p}\right)^{0.5} \tag{8.69}$$

where σ'_p is the effective overburden pressure at the depth of I_{zp}, and occurs at a depth $z_{fp}=0.25z_{f0}$. For rectangular foundations of dimensions B × L, z_{f0} and z_{fp} in Figure 8.32 are calculated after Lee et al. (2008) from

$$\frac{z_{f0}}{B} = 0.95 \cos\left\{\frac{\pi}{5}\left[\min\left(\frac{L}{B},6\right)-1\right]-\pi\right\}+3 \tag{8.70}$$

$$\frac{z_{fp}}{B} = 0.11\left[\min\left(\frac{L}{B},6\right)-1\right]+0.5 \leq 1 \tag{8.71}$$

It should be noted that the maximum vertical strains do not occur immediately below the foundations, as is the case with vertical stress. A correction can be applied to the strain distributions for the depth of the foundation below the surface. The correction factor for footing depth is given by

$$C_1 = 1 - 0.5\frac{\sigma'_q}{q_n} \tag{8.72}$$

where σ'_q = effective overburden pressure at foundation level and q_n = net foundation pressure.

Applications in geotechnical engineering

Although it is usually assumed that settlement in sands is virtually complete by the end of construction, some case records indicate continued settlement with time, thus suggesting a **creep effect**. This may be corrected for using Equation 8.73; however, this is usually ignored in routine design:

$$C_2 = 1 + 0.2 \log\left(\frac{t}{0.1}\right) \tag{8.73}$$

where t is the time in years at which the settlement is required.

The settlement of a footing carrying a net pressure q_n is written as

$$s = \int_0^{2B} \varepsilon_z \, dz$$

or, approximately,

$$s = C_1 C_2 q_n \sum_0^{2B} \frac{I_z}{E} \Delta z \tag{8.74}$$

Schmertmann obtained correlations, based on in-situ load tests, between deformation modulus and cone penetration resistance for normally consolidated sands as follows:

$E = 2.5 q_c$ for square foundations ($L/B = 1$)

$E = 3.5 q_c$ for long (strip) foundations ($L/B \geq 10$)

For overconsolidated sands, the above values should be doubled.

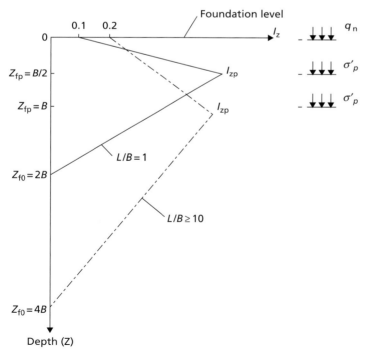

Figure 8.32 Distribution of strain influence factor.

Shallow foundations

To apply Equation 8.74, the q_c-depth profile, to a depth of either $2B$ or $4B$ (or an interpolated depth) below the foundation, is first divided into suitable layers of thicknesses Δz within each of which the value of q_c is assumed to be constant. The value of I_z at the centre of each layer is obtained from Figure 8.32. Equation 8.74 is then evaluated to give the settlement of the foundation.

Example 8.6

A footing 2.5×2.5 m supports a net foundation pressure of 150 kPa at a depth of 1.0 m in a deep deposit of normally consolidated fine sand of unit weight 17 kN/m³. The variation of cone penetration resistance with depth is given in Figure 8.33. Estimate the settlement of the footing.

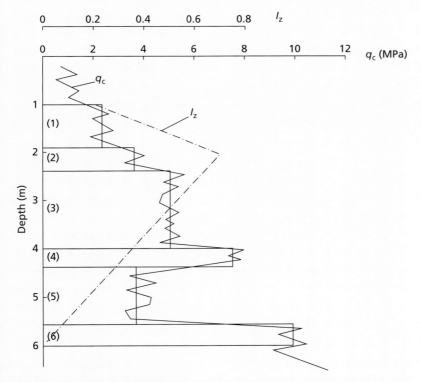

Figure 8.33 Example 8.6.

Solution

The q_c-z plot below foundation level is divided into a number of layers, of thicknesses Δz, for each of which the value of q_c can be assumed constant.

The peak value of strain influence factor occurs at a depth 2.25 m (i.e. $B/2$ below foundation level), and is given by

$$I_{zp} = 0.5 + 0.1 \left(\frac{150}{17 \times 2.25} \right)^{0.5}$$

$$= 0.70$$

Applications in geotechnical engineering

The distribution of strain influence factor with depth is superimposed on the q_c-z plot as shown in Figure 8.33, and the value of I_z is determined at the centre of each layer. The value of E for each layer is equal to $2.5 q_c$.

The correction factor for foundation depth (Equation 8.72) is

$$C_1 = 1 - \frac{0.5 \times 17}{150} = 0.94$$

The correction factor for creep (C_2) is taken as unity.

The calculations are set out in Table 8.8. The settlement is then given by Equation 8.74:

$$s = 0.94 \times 1.0 \times 150 \times 0.20$$
$$= 28 \, \text{mm}$$

Table 8.8 Example 8.6

	Δz (m)	q_c (MPa)	E (MPa)	I_z	$I_z \Delta z / E$ (m³/MN)
1	0.90	2.3	5.75	0.41	0.064
2	0.50	3.6	9.00	0.68	0.038
3	1.60	5.0	12.50	0.50	0.064
4	0.40	7.5	18.75	0.33	0.007
5	1.20	3.3	8.25	0.18	0.026
6	0.40	9.9	24.75	0.04	0.001
					0.200

8.9 Limit state design

Sections 8.3 and 8.4 described the basic analytical tools for determining the resistance of the soil at the onset of failure (bearing resistance). If the applied vertical load is equal to or greater than the bearing resistance, this will represent an ultimate limiting state causing collapse/failure. In practice, the principal aim is to ensure that the resistance of a shallow foundation is always greater than the applied loading by some margin so that the foundation will be safe. Sections 8.6–8.8 developed simple analytical tools for assessing the displacement response of a foundation under an applied load for loading conditions below the ultimate limiting state (i.e. at the serviceability limit state). In the material presented so far the emphasis has been on **analysis**, i.e. characterising the response of existing foundations (of known size/properties). From this it is possible to determine the amount of load that may be carried such that the foundation will neither collapse (ULS) nor exceed a specified settlement (SLS). This process is not the same as **design**, in which the applied loading is known (e.g. from a supported structure) and the aim is to size the foundation to provide a sufficient margin of safety against collapse and to avoid excessive movement. This section will describe how the aforementioned analytical tools may be used to design shallow foundations within a limit state design framework. This will be conducted within the framework of Eurocode 7, though the principles are directly transferable to other LSD frameworks with a change of partial factors.

Design at ULS

To satisfy the ultimate limit state, the sum of the applied **actions** (loads) on the foundation must be less than or equal to the available resistance. The (bearing) resistance, R, of a shallow foundation is the bearing capacity multiplied by the plan area of the foundation, A_f, (i.e. $R = q_f A_f$), and will be a function of various material properties (e.g. c_u for undrained conditions; ϕ', c' and γ for drained conditions). Defining the actions by Q and the material properties by X, the criterion that must be satisfied in design may be expressed as

$$\sum Q \leq R(X) \tag{8.75}$$

Equation 8.75 will just be satisfied when $\Sigma Q = R$; however this will leave no margin for error associated with the three terms in the equation, including assumptions implicit in the bearing capacity equation, potential variability in actual soil properties from those derived from laboratory, and the accurate determination of the magnitudes of the applied loads. As a result, partial factors (of safety) are used to modify the three terms in Equation 8.75 to give the design equation:

$$\sum \gamma_A Q \leq \frac{R\left(\dfrac{X}{\gamma_X}\right)}{\gamma_R} \tag{8.76}$$

where γ_A are partial factors applied to the actions Q, γ_X are partial factors applied to material properties X and γ_R are partial factors applied to the resistance R. Partial factors should not be confused with unit weights with which they share the same symbol. The parameters Q, R and X represent 'best estimates' of the actions, material properties and resistances – these are also known as **characteristic values**, and their determination will be discussed in Chapter 13. The partial factors all have a magnitude greater than or equal to 1.0, so γ_A will increase the magnitudes of the actions (i.e. the foundation must carry slightly more load than is expected), γ_X reduces the values of the material properties (i.e. the soil is weaker than measured) and γ_R reduces the resistance (i.e. the bearing capacity may be lower than predicted using the methods in Sections 8.3 and 8.4). A characteristic value which has been modified by a partial factor is known as a **design value** (i.e. $\gamma_A Q$ represents a design action, X/γ_X is a design material property and R/γ_R is the design resistance).

The three sets of partial factors are not necessarily all applied at the same time depending on the limit state design code which is being followed. In EC7, three possible design approaches are proposed:

- Design Approach 1 (DA1): (a) factoring actions only; (b) factoring materials only.
- Design Approach 2 (DA2): factoring actions and resistances (but not materials).
- Design Approach 3 (DA3): factoring structural actions only (geotechnical actions from the soil are unfactored) and materials.

Different design approaches may be adopted by different countries within the Eurocode zone, though it is prudent (and requires little extra effort) to check all of the design approaches and design the foundation to satisfy them all. This can easily be implemented using a spreadsheet to perform the calculations automatically for the different design approaches. In practice, values of partial factors are applied to every action, material property and resistance so that the same general equation can be used for all design approaches; properties which are unfactored are ascribed a partial factor of 1.0. It should be noted that the DA2 represents the approach used in LRFD (i.e. factoring the loads/actions and resistances).

Table 8.9 describes the sets of partial factors used in the three design approaches in EC7. The values of the partial factors for different sets which will be used throughout this book are given in Tables 8.10 and 8.11 for actions and material properties respectively. The factor $\gamma_R = \gamma_{Rv}$ for bearing resistance is 1.00

Applications in geotechnical engineering

Table 8.9 Selection of partial factors for use in ULS design to EC7

	Partial factors to be taken from set…		
	Actions (γ_A)	Resistances (γ_R)	Material properties (γ_X)
Design Approach 1a	A1	R1	M1
Design Approach 1b	A2	R1 (R4 for piles)	M2
Design Approach 2	A1	R2	M1
Design Approach 3	A2	R3	M2

Table 8.10 Partial factors on actions for use in ULS design to EC7

Action (Q)	Symbol	Set	
		A1	A2
Permanent unfavourable action	γ_A	1.35	1.00
Variable unfavourable action		1.50	1.30
Permanent favourable action		1.00	1.00
Variable favourable action		0	0
Accidental action		1.00	1.00

Table 8.11 Partial factors on material properties for use in ULS design to EC7

Material property (X)	Symbol	Set	
		M1	M2
$\tan \phi'$	$\gamma_{\tan\phi}$	1.00	1.25
Cohesion intercept, c'	γ_c	1.00	1.25
Undrained shear strength, c_u	γ_{cu}	1.00	1.40
Unit weight*, γ	γ_γ	1.00	1.00

Note: *EC7 uses the term weight density in place of unit weight.

for sets R1 and R3 (i.e. unfactored) and 1.40 for set R2. All of values given here and throughout the book are normative values recommended in the overarching Eurocode (EC7: Part 1) and will be used to demonstrate the principles of LSD; each country has its own National Annex for use within its borders which may suggest alternative values of the partial factors based on national experience. In design, the appropriate National Annex should always be followed. A quick-reference sheet with all of the (normative) partial factors and combination rules is provided on the Companion Website for use with the end-of-chapter problems in this and all subsequent chapters.

Regarding actions (Table 8.10), the mode of action must be determined. **Permanent actions** are those which always act on the foundation during its design life, e.g. from the dead load of the structure. **Variable actions** act only intermittently, e.g. wind or other environmental loading. An action is **unfavourable** if it will increase the total applied load – e.g. a downwards vertical load acting on a foundation. An

Shallow foundations

action is **favourable** if it reduces the total applied load. Regarding material properties (Table 8.11), the size of the partial factor depends on the accuracy with which a particular property can be determined and the likely variability in the characteristic value. Unit weight has a low partial factor, as the weight of a sample of soil can be determined very precisely using a balance. Undrained shear strength, on the other hand, is routinely determined either from laboratory tests, where sample disturbance may have significantly affected the value, or from in-situ tests (e.g. SPT, FVT, CPT) using empirical correlations which are fits to scattered data; the lower confidence in this value is represented by a larger partial factor.

Example 8.7

A foundation 2.0×2.0 m is located at a depth of 1.5 m in a layered clay of saturated unit weight 21 kN/m^3. The characteristic undrained shear strength is 160 kPa in the upper layer 2.5 m thick, and 80 kPa below. The foundation supports existing dead load of 1000 kN, and is subject to a variable load of 500 kN. Additional floors are to be added to the supported structure which will increase the dead load acting on the foundation. Determine the maximum allowable additional dead load which can be supported by the foundation under undrained conditions if it is to satisfy EC7 at ULS.

Solution

This is a problem of analysis (of an existing foundation). Defining the unknown increase in dead load as Q_1, the total applied action on the foundation is given by

$$\sum Q = \left[(1000 + Q_1) \cdot \gamma_{A1}\right] + \left[500 \cdot \gamma_{A2}\right]$$

The partial factors γ_{A1} and γ_{A2} are for permanent unfavourable and variable unfavourable conditions respectively. The bearing capacity is given by Equation 8.17, with N_c and s_c determined from Figure 8.11. $H/B = (2.5 - 1.5)/2.0 = 0.5$ and $c_{u1}/c_{u2} = 160/80 = 2.0$, giving $N_c = 3.52$ from Figure 8.11a (LB value, conservative) and $s_c = 1.41$ from Figure 8.11b. Hence

$$R = q_f A_f = \frac{\left[s_c N_c \left(\dfrac{c_{u1}}{\gamma_{cu}}\right) + \sigma_q\right] A_f}{\gamma_R}$$

$$= \frac{\left[1.41 \cdot 3.52 \cdot \left(\dfrac{160}{\gamma_{cu}}\right) + \left(\dfrac{21}{\gamma_\gamma} \cdot 1.5\right)\right] \cdot (2 \cdot 2)}{\gamma_R}$$

$$= \frac{\dfrac{3176.4}{\gamma_{cu}} + \dfrac{126}{\gamma_\gamma}}{\gamma_R}$$

Using DA1a as an example, $\gamma_{A1} = 1.35$, $\gamma_{A2} = 1.50$, $\gamma_{cu} = 1.00$, $\gamma_\gamma = 1.00$ and $\gamma_R = 1.00$ from Tables 8.9–8.11. Therefore, applying Equation 8.76 for ULS to be satisfied,

Applications in geotechnical engineering

$$(1000+Q_1)\gamma_{A1} + 500\gamma_{A2} \le \dfrac{\dfrac{3176.4}{\gamma_{cu}} + \dfrac{126}{\gamma_\gamma}}{\gamma_R}$$

$$2100 + 1.35Q_1 \le 3302.4$$

$$Q_1 \le 891\,\text{kN}$$

The results for the other design approaches are given in Table 8.12, from which it can be seen that DA2 is most critical and so the maximum additional (characteristic) deadload which can be applied is $Q_1 = 192\,\text{kN}$.

Table 8.12 Example 8.7

Design approach	Maximum Q_1 (kN)
DA1a	891
DA1b	745
DA2	192
DA3	745

Example 8.8

A concrete strip footing 0.7 m thick is to be designed to support a dead load of 500 kN/m and an imposed load of 300 kN/m at a depth of 0.7 m in a gravelly sand. Characteristic values of the shear strength parameters are $c' = 0$ and $\phi' = 40°$. Determine the required width of the footing to satisfy EC7 at the ULS, assuming that the water table may rise to foundation level. The unit weight of the sand above the water table is 17 kN/m³, and below the water table the saturated unit weight is 20 kN/m³. The bulk unit weight of the concrete is 24 kN/m³.

Solution

This is a problem of design (of a new, as yet unsized foundation). The weight of the foundation will apply an additional action (permanent, unfavourable) of $24dB = 16.8B\,\text{kN/m}$ for $d = 0.7\,\text{m}$. The total applied action on the foundation is therefore given by

$$\sum Q = \left[(500 + 16.8B)\cdot \gamma_{A1}\right] + \left[300\cdot \gamma_{A2}\right]$$

The partial factors γ_{A1} and γ_{A2} are for permanent unfavourable and variable unfavourable conditions, respectively, as in Example 8.7. The weight of the foundation is included as an additional dead load. The bearing capacity under drained conditions is given by Equation 8.32. The soil strength properties ϕ' and c' influence the bearing capacity through the bearing capacity and shape factors, so design values of these material properties must be used when determining the factors. The design value of ϕ' is given by

Shallow foundations

$$\phi'_{des} = \tan^{-1}\left(\frac{\tan\phi'}{\gamma_{\tan\phi}}\right) \qquad (8.77)$$

For DA1a (as an example) $\phi'_{des} = 40°$, giving $N_q = 64$ (Equation 8.33) and $N_\gamma = 106$ (Equation 8.36). The footing is a strip, so $s_q = s_\gamma = 1.00$. As $c' = 0$ in this case there is no need to compute N_c and s_c. For the worst case hydrological conditions when the water table is at the founding plane

$$R = q_f A_f = \frac{\left[s_q N_q \gamma d + \frac{1}{2}\gamma' B s_\gamma N_\gamma\right] A_f}{\gamma_R}$$

$$= \frac{\left[1.00 \cdot 64 \cdot \frac{17}{\gamma_\gamma} \cdot 0.7 + 0.5 \cdot \frac{(20-9.81)}{\gamma_\gamma} \cdot B \cdot 1.00 \cdot 106\right] B}{\gamma_R}$$

$$= \frac{\frac{761.6}{\gamma_\gamma} B + \frac{540.1}{\gamma_\gamma} B^2}{\gamma_R}$$

The area of a strip is B per unit length, and the partial factoring of the material properties are accounted for in the bearing capacity and shape factors. Using DA1a as an example, $\gamma_{A1} = 1.35$, $\gamma_{A2} = 1.50$, $\gamma_\gamma = 1.00$ and $\gamma_R = 1.00$ from Tables 8.9–8.11. Therefore applying Equation 8.76 for ULS to be satisfied

$$(500 + 16.8B)\gamma_{A1} + 300\gamma_{A2} \leq \frac{\frac{761.6}{\gamma_\gamma} B + \frac{540.1}{\gamma_\gamma} B^2}{\gamma_R}$$

$$1125 + 22.7B \leq 761.6B + 540.1B^2$$

$$0 \leq 540.1B^2 + 738.9B - 1125$$

This quadratic equation in B may be solved using standard methods (e.g. the quadratic formula) and taking the positive root. This gives $B \geq 0.91$ m for DA1a. The results for the other design approaches are given in Table 8.13, from which it can be seen that DA1b/3 is most critical with a required foundation width of $B \geq 1.46$ m.

Table 8.13 Example 8.8

Design approach	Minimum B (m)
DA1a	0.91
DA1b	1.46
DA2	1.16
DA3	1.46

Applications in geotechnical engineering

Design at SLS

For a shallow foundation to satisfy the serviceability limit state, the effect of the applied actions, E_A (also called an **action effect**), which will typically be a settlement calculated using the methods in Sections 8.6–8.8, must be less than or equal to a limiting value of the action effect, C_A (i.e. a limiting settlement). Mathematically this may be expressed as

$$E_A \leq C_A \tag{8.78}$$

In performing SLS calculations, characteristic values are used throughout as this limit state does not relate to the safety of the foundation, only its performance under working load. As a result, only a single calculation is typically required to demonstrate that this limiting state has been met. After a foundation has been sized based on satisfying the ULS, the settlement of the foundation ($s = E_A$) should be found. If this satisfies Equation 8.78 then the ULS governs the design, which is then complete. If Equation 8.78 is not satisfied, the calculation should be repeated making the width B variable as in ULS design (see Example 8.9), or by trial and error. This will result in a larger foundation which will govern the design (as both ULS and SLS must be satisfied). Enlarging the foundations may alter the actions applied, so a further ULS check should be undertaken.

For normal structures with isolated foundations, total (gross) settlements up to 50 mm are often acceptable, though in sands this may be reduced to 25 mm. Larger settlements may be acceptable, provided the total settlements do not cause problems with the services entering the structure, or cause tilting, etc. Zhang and Ng (2005) suggested that gross settlements of up to 125 mm in building foundations and 135 mm in bridge foundations may be tolerable, based on a probabilistic study of a large number of structures suffering various levels of serviceability damage. These guidelines concerning limiting settlements apply to normal, routine structures. They should not be applied to buildings or structures which are out of the ordinary or for which the loading intensity is markedly non-uniform.

Example 8.9

A square footing carrying an applied bearing pressure of 250 kPa is to be located at a depth of 1.5 m in a sand deposit, the water table being 3.5 m below the surface. Values of standard penetration resistance were determined as detailed in Table 8.14. Determine the minimum width of the foundation if the settlement is to be limited to 25 mm.

Table 8.14 Example 8.9

Depth (m)	N_{60}	σ'_v (kPa)	C_N	$(N_1)_{60}$
0.75	8	–	–	–
1.55	7	26	2.0	14
2.30	9	39	1.6	14
3.00	13	51	1.4	18
3.70	12	65	1.25	15
4.45	16	70	1.2	19
5.20	20	–	–	–

Solution

Due to the non-linearity involved in the empirical method for SPT data, a width B must first be estimated and the solution progressed through trial and error. Starting with $B=3$ m, the depth of influence (Figure 8.31) is 2.2 m, i.e. 3.7 m below the surface. The average of the measured N_{60} values between depths of 1.5 and 3.7 m is 10, hence the compressibility index (Equation 8.63) is given by

$$I_c = \frac{1.71}{10^{1.4}} = 0.068$$

Then, Equation 8.65a is used to evaluate the action effect ($E_A = s$), giving

$$E_A = qB^{0.7}I_c$$
$$= 250 \cdot B^{0.7} \cdot 0.068$$
$$= 17B^{0.7}$$

Applying Equation 8.78 for SLS to be satisfied gives

$$E_A \leq C_A$$
$$17B^{0.7} \leq 25$$
$$B \geq 1.73 \text{ m}$$

The calculations are then repeated using this new minimum value of B until there is no change in the value of B with subsequent iterations. The calculations are shown in Table 8.15, from which it can be seen that the solution converges after four iterations to give $B \geq 1.11$ m.

Table 8.15 Example 8.9 (contd.)

Iteration	B (m)	z_1 (m)	$\overline{N}_{60}$	I_c	B (m)
1	3.00	2.2	10	0.068	1.73
2	1.73	1.5	9	0.079	1.40
3	1.40	1.2	8	0.093	1.11
4	1.11	1.1	8	0.093	1.11

Summary

1 The application of load to a shallow foundation induces stresses within the underlying soil mass generating shear. As the foundation is loaded it will settle. If the shear stress reaches a condition of plastic equilibrium and a compatible failure mechanism can be formed, then the footing will suffer bearing capacity failure (the settlement will become infinite).

2 The condition of plastic failure within a soil mass may be analysed using limit analysis. Upper bound techniques involve postulating a compatible failure mechanism and performing an energy balance based on the movement within the mechanism. Lower bound techniques involve postulating a stress field which is in equilibrium with the applied external load. The true failure load will lie

Applications in geotechnical engineering

between the upper and lower bound solutions – if the upper and lower bounds are equal, then the true solution has been found. These methods have been applied to both undrained and drained soil conditions to determine bearing capacity. More advanced solutions given in the literature have extended these approaches to more complex ground conditions, to determine bearing capacity under more realistic conditions for use in foundation design.

3 The stresses induced beneath a loaded shallow foundation reduce in magnitude with depth and lateral distance from the foundation. Based on the induced total stresses, elasticity theory may be used to derive immediate (undrained) settlements in fine grained soil, while the consolidation theory outlined in Chapter 4 can be modified to predict the settlement occurring as the soil consolidates. The final settlement is the sum of these values.

4 Modern design codes use a limit state design philosophy, where limiting states are proposed representing an acceptable level of performance. These may relate to either the avoidance of catastrophic collapse (ULS) or the avoidance of damage in the supported structure (SLS). In order to ensure the foundation is safe, partial factors of safety are applied to the various design parameters in ULS calculations to account for uncertainty in the values of these parameters. SLS calculations are unfactored.

5 The bearing capacity equations developed from limit analysis (point 2) describe the failure condition for a shallow foundation, and therefore describe the behaviour at the ULS. If appropriate partial factors are applied to the parameters in these equations, the footing can be designed to ensure that the ULS is satisfied (the foundation will not collapse). The settlement equations developed from elasticity/consolidation theory (point 3) may similarly be used to ensure that the SLS is satisfied. For footings on coarse-grained soils, SPT or CPT data may alternatively/additionally be used to determine footing settlement and design to the SLS.

Problems

8.1 A strip footing 2 m wide is founded at a depth of 1 m in a stiff clay of saturated unit weight 21 kN/m³, the water table being at ground level. Determine the bearing capacity of the foundation (a) when $c_u = 105$ kPa and $\phi_u = 0$, and (b) when $c' = 10$ kPa and $\phi' = 28°$.

8.2 Determine the allowable design load on a footing 4.50×2.25 m at a depth of 3.50 m in a stiff clay to EC7 DA1a. The saturated unit weight of the clay is 20 kN/m³ and the characteristic shear strength parameters are $c_u = 135$ kPa and $\phi_u = 0$.

8.3 A footing 2.5×2.5 m carries a pressure of 400 kPa at a depth of 1 m in a sand. The saturated unit weight of the sand is 20 kN/m³ and the unit weight above the water table is 17 kN/m³. The design shear strength parameters are $c' = 0$ and $\phi' = 40°$. Determine the bearing capacity of the footing for the following cases:
   a the water table is 5 m below ground level;
   b the water table is 1 m below ground level;
   c the water table is at ground level and there is seepage vertically upwards under a hydraulic gradient of 0.2.

8.4 A strip footing is located at a depth of 0.75 m in a sand of unit weight 18 kN/m³, the water table being well below foundation level. The characteristic shear strength parameters are $c' = 0$ and $\phi' = 38°$. The footing supports a design load of 500 kN/m. Determine the required width of the foundation for the ultimate limit state to be satisfied to EC7 DA1b.

Shallow foundations

8.5 A square pad footing supports a permanent load of 4000 kN and a variable load of 1500 kN (both values characteristic) at a depth of 1.5 m in sand. The water table is at the surface, the saturated unit weight of the sand being 20 kN/m³. Characteristic values of the shear strength parameters are $c'=0$ and $\phi'=39°$. Determine the required size of the foundation for the ultimate limit state to be satisfied to EC7 DA1a.

8.6 A foundation 4×4 m is located at a depth of 1 m in a layer of saturated clay 13 m thick. Characteristic parameters for the clay are $c_u=100$ kPa, $\phi_u=0$, $c'=0$, $\phi'=32°$, $m_v=0.065$ m²/MN, $A=0.42$, $\gamma_{sat}=21$ kN/m³. Determine the allowable load for the foundation to ensure (a) the bearing resistance limit state is satisfied to EC7 DA1b, and (b) consolidation settlement does not exceed 30 mm.

8.7 A footing 3.0×3.0 m carries a net foundation pressure of 130 kPa at a depth of 1.2 m in a deep deposit of sand of unit weight 16 kN/m³, the water table being well below the surface. The variation of cone penetration resistance (q_c) with depth (z) is as follows:

TABLE R

z (m)	1.2	1.6	2.0	2.4	2.6	3.0	3.4	3.8	4.2	4.6	5.0	5.4	5.8	6.2	6.6	7.0	7.4	8.0
q_c (MPa)	3.2	2.1	2.8	2.3	6.1	5.0	3.6	4.5	3.5	4.0	8.1	6.4	7.6	6.9	13.2	11.7	12.9	14.8

Determine the settlement of the footing using Schmertmann's method.

8.8 A foundation 3.5×3.5 m is to be constructed at a depth of 1.2 m in a deep sand deposit, the water table being 3.0 m below the surface. The following values of standard penetration resistance were determined at the location:

TABLE S

Depth (m)	0.70	1.35	2.20	2.95	3.65	4.40	5.15	6.00
N_{60}	6	9	10	8	12	13	17	23

If the settlement is not to exceed 25 mm, determine the allowable load to satisfy the serviceability limit state.

8.9 A permanent load of 2500 kN and an imposed load of 1250 kN are to be supported on a square foundation at a depth of 1.0 m in a deposit of gravelly sand extending from the surface to a depth of 6.0 m. A layer of clay 2.0 m thick lies immediately below the sand. The water table may rise to foundation level. The unit weight of the sand above the water table is 17 kN/m³, and below the water table the saturated unit weight is 20 kN/m³. Characteristic values of the shear strength parameters for the sand are $c'=0$ and $\phi'=38°$. The coefficient of volume compressibility for the clay is 0.15 m²/MN. It is specified that the long-term settlement of the foundation due to consolidation of the clay should not exceed 20 mm. Determine the required size of the foundation for the ultimate and serviceability limit states to be satisfied (use EC7 DA1b at ULS).

References

Burland, J.B. (1970) Discussion, in *Proceedings of Conference on In-Situ Investigations in Soils and Rocks*, British Geotechnical Society, London, p. 61.

Burland, J.B. and Burbidge, M.C. (1985) Settlement of foundations on sand and gravel, *Proceedings ICE*, **1**(78), 1325–1381.

Christian, J.T. and Carrier III, W.D. (1978) Janbu, Bjerrum and Kjaernsli's chart reinterpreted, *Canadian Geotechnical Journal*, **15**(1), 123, 436.

Davis, E.H. and Booker, J.R. (1973) The effect of increasing strength with depth on the bearing capacity of clays, *Géotechnique*, **23**(4), 551–563.

EC7–1 (2004) *Eurocode 7: Geotechnical design – Part 1: General rules, BS EN 1997–1:2004*, British Standards Institution, London.

Fadum, R.E. (1948) Influence values for estimating stresses in elastic foundations, in *Proceedings of the 2nd International Conference of SMFE, Rotterdam*, Vol. 3, pp. 77–84.

Georgiadis, K. (2010) An upper-bound solution for the undrained bearing capacity of strip footings at the top of a slope, *Géotechnique*, **60**(10), 801–806.

Giroud, J.P. (1972) Settlement of rectangular foundation on soil layer, *Journal of the ASCE*, **98**(SM1), 149–54.

Kumar, J. and Kouzer, K.M. (2007) Effect of footing roughness on bearing capacity factor N_γ, *Journal of Geotechnical and Geoenvironmental Engineering*, **133**(5), 502–511.

Lee, J., Eun, J., Prezzi, M. and Salgado, R. (2008) Strain influence diagrams for settlement estimation of both isolated and multiple footings in sand. *Journal of Geotechnical and Geoenvironmental Engineering*, **134**(4), 417–427.

Lyamin, A.V., Salgado, R., Sloan, S.W. and Prezzi, M. (2007) Two- and three-dimensional bearing capacity of footings in sand, *Géotechnique*, **57**(8), 647–662.

Merifield, R.S. and Nguyen, V.Q. (2006) Two- and three-dimensional bearing-capacity solutions for footings on two-layered clays, *Geomechanics and Geoengineering: An International Journal*, **1**(2), 151–162.

Merifield, R.S., Sloan, S.W. and Yu, H.S. (1999) Rigorous plasticity solutions for the bearing capacity of two-layered clays, *Géotechnique*, **49**(4), 471–490.

Salgado, R. (2008). *The Engineering of Foundations*, McGraw-Hill, New York, NY.

Salgado, R., Lyamin, A.V., Sloan, S.W. and Yu, H.S. (2004) Two- and three-dimensional bearing capacity of foundations in clay, *Géotechnique*, **54**(5), 297–306.

Schmertmann, J.H. (1970) Static cone to compute static settlement over sand, *Proceedings ASCE*, **96**(SM3), 1011–1043.

Scott, R.E. (1963) *Principles of Soil Mechanics*, Addison-Wesley, Reading, MA.

Skempton, A.W. (1951) The bearing capacity of clays, *Proceedings of the Building Research Congress*, Vol. 1, pp. 180–189.

Skempton, A.W. and Bjerrum, L. (1957) A contribution to the settlement analysis of foundations on clay, *Géotechnique*, **7**(4) 168–178.

Zhang, L.M. and Ng, A.M.Y. (2005) Probabilistic limiting tolerable displacements for serviceability limit state design of foundations, *Géotechnique*, **55**(2), 151–161.

Zhu, F., Clark, J.I. and Philips, R. (2001) Scale effect of strip and circular footings resting on dense sand, *Journal of Geotechnical and Geoenvironmental Engineering*, **127**(7), 613–620.

Further reading

Frank, R., Bauduin, C., Driscoll, R., Kavvadas, M., Krebs Ovesen, N., Orr, T. and Schuppener, B. (2004) *Designers' Guide to EN 1997–1 Eurocode 7: Geotechnical Design – General rules*, Thomas Telford, London.

This book provides a guide to limit state design of a range of constructions (including shallow foundations) using Eurocode 7 from a designer's perspective and provides a useful companion to the Eurocodes when conducting design. It is easy to read and has plenty of worked examples.

For further student and instructor resources for this chapter, please visit the Companion Website at www.routledge.com/cw/craig

Chapter 9

Deep foundations

Learning outcomes

After working through the material in this chapter, you should be able to:

1 Understand the working principles behind deep foundations, how they are constructed/installed, and the advantages they offer over shallow foundations (Chapter 8);
2 Design a pile within a limit-state design framework (Eurocode 7), analytically (based on fundamental ground properties), directly from in-situ test data or from the results of a pile load test.

9.1 Introduction

When designing foundations, there are often situations where the use of shallow foundations is uneconomic or impractical. These include

- when the actions applied to the foundation are large (e.g. large concentrated loads);
- when near surface soils have low strength and or stiffness (i.e. low resistance);
- where large structures are situated on very heterogeneous deposits, or where the soil layers are inclined;
- for settlement-sensitive structures where displacements must be kept small;
- in marine environments where tidal, wave or flow actions may erode material from around a foundation near the ground surface (this process is known as **scour**).

The effect of all of the foregoing points will be to increase the plan area and/or embedment depth of a shallow foundation. These effects may also occur simultaneously such that the foundation would be too expensive or difficult to construct. In these circumstances, deep foundations may offer a more efficient and less costly design.

Where shallow foundations are wide compared to their depth, deep foundations are elements which are much smaller in plan but extend to greater depth within the ground. The most common type of deep foundation is the pile, which is a column of concrete, steel or timber installed within the ground (Figure 9.1). Piles may be circular or square in section, but will always have an (outside) diameter (D_0) or width (B_p) that is very much smaller than their length (L_p), i.e. $L_p \gg D_0$. A **pier** or **caisson** is another type of deep foundation which has a much larger diameter compared to its length, i.e. $L_p > D_0$, but which can be analysed in the same way as a pile. Caissons are often used as foundations for offshore structures.

Applications in geotechnical engineering

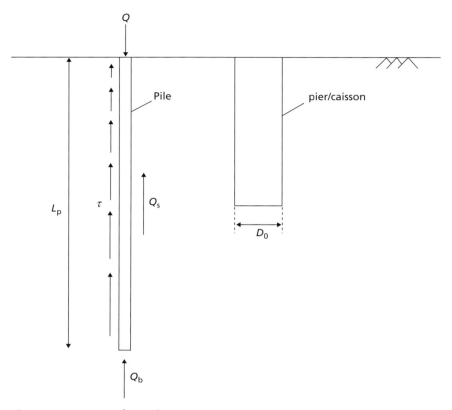

Figure 9.1 Deep foundations.

Piles generate some of their resistance at the base in bearing, and in this respect can be thought of as a very deeply embedded shallow foundation. This element of the resistance is also known as **base resistance**, Q_{bu}. Section 9.2 will describe how the bearing capacity theories developed in Chapter 8 may be applied to the determination of pile base capacity. As a result of their significant length (and therefore surface area), they can also generate significant additional resistance along the shaft due to interface friction between the pile material and the soil. This element of the resistance is also known as **shaft resistance** or **skin friction**. Piles are able to make use of this additional resistance, as their methods of installation (described in the following material) ensure that there is a good bond between the pile and the soil along the shaft. For shallow foundations this interface friction was neglected because the embedment depth $d \ll B$ such that the additional resistance was negligibly small compared to the bearing capacity. Figure 9.2 shows how the shaft capacity of a pile is found by first determining the interface shear strength τ_{int} along the pile length and then integrating this over the surface area of the pile to determine the shaft resistance, Q_{su}, giving

$$Q_{su} = \pi D_0 \int_0^{L_p} \tau_{int}(z) \, dz \quad \text{(circular cross-section)} \tag{9.1a}$$

$$Q_{su} = 4 B_p \int_0^{L_p} \tau_{int}(z) \, dz \quad \text{(square cross-section)} \tag{9.1b}$$

Deep foundations

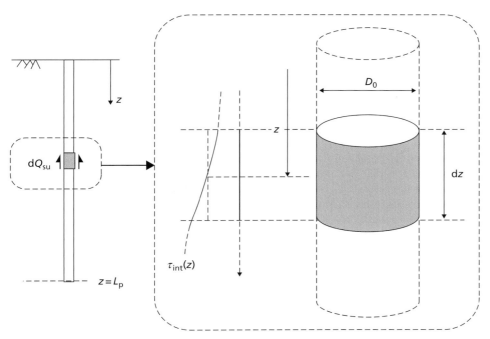

Figure 9.2 Determination of shaft resistance.

where z is the vertical distance along the pile measured from the surface. Section 9.2 describes how τ_{int} is determined for use in Equation 9.1. It should be noted that in Equation 9.1 the term before the integral sign (πD_0 or $4B_p$) is the perimeter around the cross-section of the pile. In some piles, the pile perimeter changes with depth due to a change in width or diameter; in these cases the constant D_0 or B_p must be replaced with $D_0(z)$ or $B_p(z)$ respectively and taken inside the integral. A pile where the diameter or width decreases with depth is known as a **tapered pile**.

Pile installation

The interface shear strength along the shaft of a pile is influenced not only by the exterior geometry (surface area), but also by the method of installation. Piles may be divided into two main categories according to their method of installation. The first category is **displacement piles**, as they involve displacement and disturbance of the soil around the pile. Examples of displacement piles include driven piles of steel or precast concrete and piles formed by driving tubes or shells which are fitted with a **driving shoe** (to ease penetration and protect the end of the pile from damage during driving), the tubes or shells being filled subsequently with concrete after driving (also known as a **shell pile**). In the case of steel H-piles and tubes without a driving shoe, however, soil displacement may be small, at least initially. With increased penetration of these piles, the horizontal stresses between the flanges of the H-pile or acting on the inside walls of a tubular pile are enhanced, increasing the interface friction on these surfaces. In some cases, these stresses can be so high that the action due to the base pressure acting upwards on the mass of soil within the pile is lower than the resistance due to the enhanced interface friction acting on the mass from the interior walls, such that soil can no longer continue to fill the void. When this occurs, the pile is said to be **plugged**. A plugged pile will have the resistance of a closed-ended pile of the same outside diameter. Also included in this category are piles formed by placing concrete as driven steel tubes are withdrawn (a **cast-in-situ pile**).

Applications in geotechnical engineering

The second category consists of piles which are installed without soil displacement (termed **non-displacement** or **bored piles**, or cast-in-situ piles). Soil is removed by boring or drilling to form a shaft, concrete then being cast in the shaft to form the pile. As such shafts can be very deep, depending on the type of soil, the shaft may be cased or bored under a drilling fluid such as bentonite, the fluid pressure from such a material acting to prevent the shaft collapsing until the concrete can be placed. The stability of excavations with and without fluid support is considered in Chapter 12. In clays, the shaft may be enlarged at its base by a process known as **under-reaming**; the resultant pile then has a larger base area in contact with the soil, increasing its base resistance. A **continuous flight auger (CFA) pile** is a type of bored pile in which a helical auger is drilled into the ground over the desired pile length in a single process. The plug of soil trapped between the flights is then withdrawn from the ground as concrete is pumped into the shaft through a tube running down the centre of the auger. These are now a more popular alternative to cast-in-situ piles. Figure 9.3 shows a continuous flight auger piling rig which demonstrates the length such tools must be if long piles are to be installed. The principal types of pile are summarised and illustrated in Figure 9.4.

Both displacement and non-displacement piling techniques have advantages and disadvantages. As driven piles are usually prefabricated, their structural integrity can be inspected before driving, and the soil displacement during installation can locally densify the soil around the pile in looser deposits, increasing capacity. However, in dense deposits dilation of the soil on shearing can cause heave of the ground around the pile, potentially damaging adjacent infrastructure; piling hammers (which drive piles

Figure 9.3 Pile installation: non-displacement piling (CFA) (image courtesy of Cementation Skanska Ltd.).

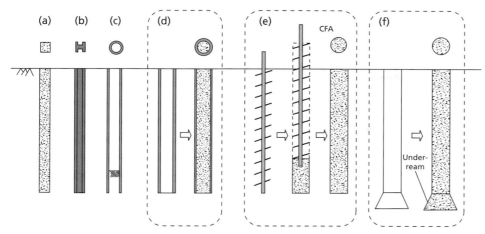

Figure 9.4 Principal types of pile: (a) precast RC pile, (b) steel H pile, (c) steel tubular pile (plugged), (d) shell pile, (e) CFA pile, (f) under-reamed bored pile (cast-in-situ).

via impact) are also very noisy, though use of modern suppressed piling hammers or push-in (**jacked**) piling techniques can ameliorate this. As such, displacement piles are not normally recommended for congested urban sites. They are, however, eminently suitable for use in offshore environments, where noise is not an issue and their simple method of installation is beneficial when working underwater.

In contrast, non-displacement piling techniques create minimal soil disturbance around the pile and are very quiet. They can thus be used very close to existing structures in congested urban areas. As they are cast in-situ, complex pile shapes can be formed, including adding an under-ream to enhance base capacity or adding flanges along the pile to increase shaft capacity. Being cast in-situ is also a disadvantage, however, as pile defects may occur due to poor compaction of concrete, voids, or wash-out of concrete if the pile passes through highly permeable layers through which seepage is occurring.

The installation method has a strong influence on the interface shear strength τ_{int}, and this will be described in greater detail in Section 9.2.

9.2 Pile resistance under compressive loads

Base resistance

As outlined in Section 9.1, the base resistance of piles and other deep foundations can be determined analytically by treating them as very deeply embedded shallow foundations. For soil under undrained conditions, Equation 8.17 describes the bearing pressure on the base of the pile at failure, so the base resistance is therefore

$$Q_{bu} = A_p \left(s_c N_c c_u + \sigma_q \right) \tag{9.2}$$

where A_p is the cross-sectional area of the base of the pile ($= \pi D_0^2/4$ or B_p^2 for circular or square piles respectively). In using Equation 9.2, all of the soil above the pile base level is treated as a surcharge, where $\sigma_q = \sigma_v$ at the level of the pile base. Therefore, even very complex soil layering around the shaft of the pile can be straightforwardly accounted for, provided that the total vertical stress from these layers can be computed (e.g. using Stress_CSM8.xls on the Companion Website, see Chapter 3). It is then

Applications in geotechnical engineering

possible to utilise the various solutions presented in Section 8.3 to account for the soil conditions around the base of the pile as shown in Figure 9.5. Circular piles are treated as square piles for the purposes of determining the shape factor s_c. With reference to Figure 9.5:

- for cases (a) and (b) Skempton's values of N_c are usually used from Figure 8.10, which for a square or circular pile limits $s_c N_c$ to 9.0;
- for case (c), N_c and s_c should be taken from Figure 8.11;
- for cases (d) and (e), the bearing capacity is based on Equation 8.21, i.e.

$$Q_{bu} = A_p s_c \left[(2+\pi) c_{u0} + \frac{CB}{4} \right] F_z + A_p \sigma_q \tag{9.3}$$

where F_z is found from Figure 8.13 and s_c is approximated using Equation 8.19, i.e. $s_c = 1.2$.

In drained soil, base resistance may be similarly determined by utilising Equation 8.32 in place of Equation 8.17. However, in the case of piles, the large embedment means that the values of the self-weight term and cohesion term are small compared to the surcharge term. Therefore, Equation 8.32 is normally simplified to

$$Q_{bu} = A_p \left(N_q \sigma'_q \right) \tag{9.4}$$

The values of N_q given by Equations 8.33 for shallow foundations are only approximate at very large L_p/D_0. Values given by Berezantsev et al. (1961) for circular piles should therefore be used in Equation 9.4, and these are given in Figure 9.6 (N.B. these values of N_q implicitly include the effect of shape).

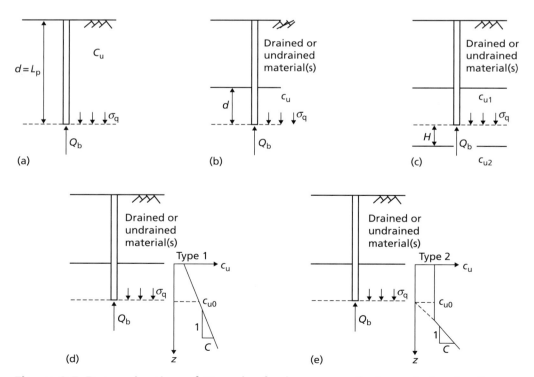

Figure 9.5 Determination of N_c and s_c for base capacity in undrained soil.

Deep foundations

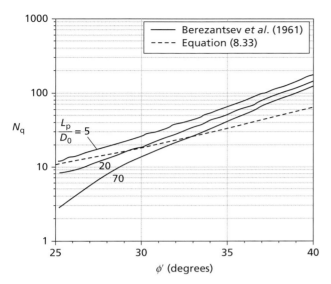

Figure 9.6 Bearing capacity factor N_q for pile base capacity.

Shaft resistance

To determine the shaft resistance of a pile using Equation 9.1, the interface shear strength must be determined. In undrained soil, this is represented by

$$\tau_{int} = \alpha c_u \tag{9.5}$$

where α is termed the **adhesion factor**, having a value between 0 and 1, $\alpha=1$ representing a perfectly rough interface which is as strong as the surrounding soil, and $\alpha=0$ representing a perfectly smooth interface (i.e. no stress can be transferred between pile and soil). The determination of shaft friction in undrained soil is often referred to as the **α-method** in pile design. The adhesion factor is a function both of the surface condition along the pile and the method of pile installation.

For displacement piles in fine-grained soils, it has been shown from field observations that a correlation with c_u/σ'_{v0} (also known as the **yield stress ratio**) exists, as shown in Figure 9.7(a), though there is much scatter. For the purposes of design, an averaging line can be defined after Randolph and Murphy (1985) and Semple and Rigden (1984) where

$$\begin{aligned} \alpha &= 0.5 F_p \left(\frac{c_u}{\sigma'_{v0}} \right)^{-0.5} \leq 1 \quad \text{for } \frac{c_u}{\sigma'_{v0}} \leq 1 \\ \alpha &= 0.5 F_p \left(\frac{c_u}{\sigma'_{v0}} \right)^{-0.25} \quad \text{for } \frac{c_u}{\sigma'_{v0}} \geq 1 \end{aligned} \tag{9.6}$$

where F_p is a factor related to the length of the pile (or length within a layer in the case of layered soil). For $L_p/D_0 < 50$ (e.g. caissons), $F_p = 1.0$; for $L_p/D_0 > 120$, $F_p = 0.7$; for values of L_p/D_0 between these limits, F_p is estimated by linear interpolation between 0.7 and 1.0. Kolk and van der Velde (1996) proposed an alternative approach in which

$$\alpha = 0.55 \left(\frac{40}{L_p/D_0} \right)^{0.2} \left(\frac{c_u}{\sigma'_{v0}} \right)^{-0.3} \tag{9.7}$$

333

Applications in geotechnical engineering

The parameter L_p/D_0 is also known as the **slenderness ratio** of the pile. Equations 9.6 and 9.7 are also plotted in Figure 9.7(a). While they approximate the field observations reasonably well, there is a considerable amount of scatter, such that care should be taken to select a conservative value of α in design.

For non-displacement piles in fine-grained soils, values of α are correlated with c_u only. A large database of pile test data is shown in Figure 9.7(b), where the data from Skempton (1959) were determined for piles in London Clay and the data from Weltman and Healy (1978) are for piles in glacial tills. The data show greater scatter than for displacement piles, though an averaging line can be developed which is described by

$$\begin{aligned}
&\alpha = 1 &&\text{for } c_u \leq 30 \\
&\alpha = 1.16 - \left(\frac{c_u}{185}\right) &&\text{for } 30 \leq c_u \leq 150 \\
&\alpha = 0.35 &&\text{for } c_u \geq 150
\end{aligned} \quad (9.8)$$

Equation 9.8 is plotted in Figure 9.7(b). As for displacement piles, conservative values of α should be selected for initial design.

In drained soil, the interface shear strength is represented by

$$\tau_{int} = K\sigma'_{v0} \tan \delta' \quad (9.9)$$

where δ' is an **interface friction angle** ($\delta' \leq \phi'$) and $K\sigma'_{v0}$ is the effective stress acting normal to the pile shaft. Equation 9.9 is therefore analogous to the Mohr–Coulomb failure criterion for a drained cohesionless material. Parameter δ' is a function of the pile roughness and soil properties; the case $\delta' = \phi'$ represents a perfectly rough interface, while $\delta' = 0$ represents a perfectly smooth interface. Parameter K is a horizontal earth pressure coefficient, and is a function of the soil properties and the installation method.

In coarse-grained materials (sands and gravels) the parameter K is usually expressed in terms of K_0, the coefficient of lateral earth pressure at rest. This parameter depends on the soil type and stress history (quantified by ϕ' and OCR), and is described in further detail in Section 11.3. For cast-in-situ piles $0.7 < K/K_0 < 1.0$, while for displacement piles K/K_0 may be as high as 2.0. The interface friction angle depends on the roughness of the pile material, and may be determined using an **interface shear test**. This is conducted in a direct shear apparatus (see Chapter 5) in which a plate/block of the pile material is placed in the bottom half of the shearbox, with soil being placed in the upper half. Tests are then conducted and interpreted in the same way as for the soil–soil tests in Chapter 5 by plotting the interface shear stress at failure against normal

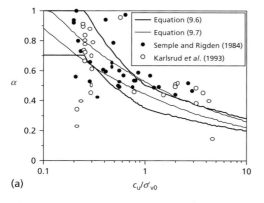
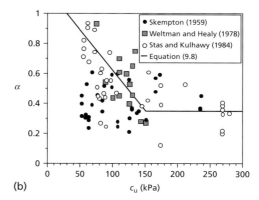

Figure 9.7 Determination of adhesion factor α in undrained soil: (a) displacement piles, (b) non-displacement piles.

effective stress; the gradient of the line of best fit is then $\tan\delta'$. Figure 9.8 shows data from interface shear tests collated from the literature (Uesugi and Kishida, 1986; Uesugi et al., 1990; Subba Rao et al., 1998; Frost et al., 2002) for a range of common pile materials in which δ' is expressed in terms of the angle of shearing resistance of the soil. These data are plotted against the parameter R_a/D_{50}, where R_a is the average height of the asperities on the surface of the pile material which gives the material its roughness and D_{50} is the mean particle size of the soil (see Chapter 1). It should be noted that mild steel oxidises (rusts) in the ground, so δ'/ϕ' is usually limited to a minimum of 0.5. The concrete values given in Figure 9.8 are for precast concrete (i.e. for driven piles); for concrete cast in-situ the roughness will be much larger, and $\delta' = \phi'$ is usually assumed.

In fine-grained materials under drained conditions, $K \approx K_0$ and $\delta' = \phi'$ is usually assumed. Alternatively, the parameters K and $\tan\delta'$ may be lumped together into a single factor $\beta = K\tan\delta'$, i.e.

$$\tau_{int} = \beta\sigma'_{v0} \tag{9.10}$$

The determination of shaft friction in drained soil is often referred to as the **β-method** in pile design. Burland (1993) showed that β correlates linearly with the yield stress ratio with a surprisingly small amount of scatter, as shown in Figure 9.9 (data points represent non-displacement piles). A best-fit line to this data gives

$$\beta = 0.52\left(\frac{c_u}{\sigma'_{v0}}\right) + 0.11 \tag{9.11}$$

Equation 9.11 has been found to provide reasonable estimates of shaft capacity for both displacement and non-displacement piles.

In the case of under-reamed piles, as a result of settlement there is a possibility that a small gap will develop between the top of the under-ream and the overlying soil. Accordingly, no skin friction should be taken into account below a level $2D_0$ above the top of the under-ream, and base resistance should be determined as if the base is not embedded (i.e. Figure 9.5 case (b) with $d=0$, $s_c N_c = 6.2$).

Pile resistance and limit state design

The resistance of a pile is the sum of its base and shaft capacities. In the limit state design framework, the combined (total) compressive resistance may be factored using a partial factor γ_{RC} to obtain the design resistance, i.e.

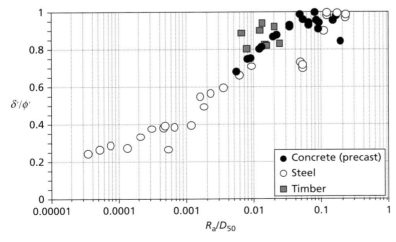

Figure 9.8 Interface friction angles δ' for various construction materials.

Applications in geotechnical engineering

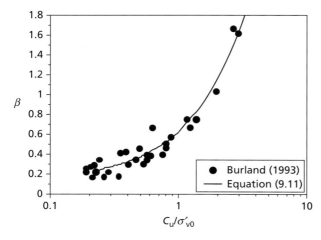

Figure 9.9 Determination of factor β in drained fine-grained soils (all pile types).

$$R = \frac{Q_{bu} + Q_{su}}{\gamma_{RC}} \qquad (9.12a)$$

Alternatively, the base and shaft resistances may be factored separately by γ_{Rb} and γ_{Rs} respectively to better account for the degree of accuracy to which Q_{bu} and Q_{su} are known, i.e.

$$R = \frac{Q_{bu}}{\gamma_{Rb}} + \frac{Q_{su}}{\gamma_{Rs}} \qquad (9.12b)$$

In Eurocode 7 DA1b, an alternative set of resistance factors is used (set R4). The normative values suggested for the partial resistance factors are shown in Table 9.1. Where there are two values given, the first is for displacement piles while the second is for non-displacement piles. It should be noted that the partial factors in Table 9.1 only relate to the degree of uncertainty in the calculation methods employed to determine the pile capacity; it is still necessary to be cautious when estimating empirically derived values such as α or β, as the variation in these is not accounted for in the partial resistance factors. As is noted in Chapter 8, the factors in Table 9.1 may be superseded by the values in a particular country's National Annex.

The resistance factors in Table 9.1 are included on the EC7 quick reference sheet on the Companion Website for use in the worked examples and end-of-chapter problems.

Table 9.1 Partial resistance factors for use in ULS pile design to EC7 (piles in compression only)

Resistance (R)	Symbol	Set			
		R1	R2	R3	R4
Total capacity (compression)	γ_{RC}	1.00/1.15	1.10	1.00	1.30/1.50
Base capacity Q_{bu}	γ_{Rb}	1.00/1.25	1.10	1.00	1.30/1.60
Shaft capacity Q_{su}	γ_{Rs}	1.00/1.00	1.10	1.00	1.30/1.30

Deep foundations

To satisfy the ULS, the resistance given by Equation 9.12 must be greater than the sum of the applied actions on the pile. This is the same condition as for shallow foundations, so Equation 8.76 represents the inequality that must be satisfied. The actions are factored using the values of γ_A from Table 8.10. As piles are often used in circumstances when high point loads are to be carried, the applied action at the top of the pile will normally be very much larger than the dead weight of the pile. The material properties used in the determination of Q_{bu} and Q_{su} are factored using the values in Table 8.11.

Example 9.1

A single steel tubular pile of outer diameter 0.3 m, wall thickness 10 mm and length 10 m is driven into dry loose sand. The soil has unit weight, $\gamma = 15$ kN/m³, $\phi' = 32°$ and $c' = 0$. Along the pile–soil interface, it may be assumed that $K = 1$ and $\delta' = 0.75\phi'$. Assuming that the pile is plugged and the weight of the soil inside the pile is negligible, determine the allowable (permanent) design load on the pile under EC7 DA1b.

Solution

As the soil is dry, there are no pore water pressures ($u = 0$) and the unit weight is the total value (γ). The effective vertical stress at the pile base is then $\sigma'_v(z = L_p) = \gamma L_p$. For DA1b, $\gamma_{\tan\phi} = 1.25$, the design value of the angle of shearing resistance is

$$\phi'_{des} = \tan^{-1}\left(\frac{\tan\phi'}{\gamma_{\tan\phi}}\right) = \tan^{-1}\left(\frac{\tan 32°}{1.25}\right) = 26.6°$$

from which $\delta'_{des} = 0.75\phi'_{des} = 20°$. The slenderness ratio $L_p/D_0 = 10/0.3 = 33$, so, from Figure 9.6, $N_q \approx 15$. The gross area of the pile $A_p = \pi \times 0.15^2 = 0.0707$ m² (pile is plugged), and for DA1b the partial resistance factor for a displacement pile $\gamma_{Rb} = 1.30$ from set R4. Therefore the design base capacity is

$$Q_{bu,\,des} = \frac{Q_{bu}}{\gamma_{Rb}} = \frac{A_p N_q \left(\dfrac{\gamma}{\gamma_\gamma}\right) L_p}{\gamma_{Rb}} = \frac{0.0707 \cdot 15 \cdot \left(\dfrac{15}{1.00}\right) \cdot 10}{1.30} = 122 \text{ kN}$$

From Equations 9.1a and 9.9, the design shaft capacity is

$$Q_{su,\,des} = \frac{Q_{su}}{\gamma_{Rs}} = \frac{\pi D_0 \int_0^{10} K\left(\dfrac{\gamma}{\gamma_\gamma}\right) z \tan\delta'_{des}\, dz}{\gamma_{Rs}}$$

$$= \frac{\pi \cdot 0.3 \cdot 1 \cdot \tan 20° \cdot \left(\dfrac{15}{1.00}\right) \cdot \left[\dfrac{z^2}{2}\right]_0^{10}}{1.30}$$

$$= 198 \text{ kN}$$

Therefore, the design capacity of the pile $R = 122 + 198 = 320$ kN. For ULS to be satisfied, the design value of the action applied to the pile $Q \leq R$. For a permanent unfavourable action $\gamma_A = 1.00$ for DA1b, therefore $Q = 320$ kN is the maximum allowable characteristic load that can be carried by the pile if ULS is to be satisfied to EC7.

Applications in geotechnical engineering

Example 9.2

A 0.75-m diameter bored concrete pile (weight density = 24 kN/m³) is to be formed in a two-layer deposit of clay with the water table at the ground surface. The upper layer of clay has saturated unit weight $\gamma = 18$ kN/m³, and constant undrained shear strength $c_u = 100$ kPa. Below this lies a thick lower layer of stronger clay starting at a depth of 15 m below the ground surface. This clay layer has $\gamma = 20$ kN/m³ and $c_u = 200$ kPa. All calculations are to be completed to EC7 DA1b.

a Determine the maximum allowable design load (permanent) which the pile can support under undrained conditions if it is to be made 15 m long.
b Determine the total length of pile required to support a (permanent) characteristic load of 3 MN under undrained conditions.

Solution

a If the pile is 15 m long, its shaft will be in the upper clay, while the base will rest at the surface of the lower clay (i.e. case (b) from Figure 9.5 with an embedment of $d = 0$ into the lower clay layer). The characteristic undrained shear strength for the base is then $c_u = 200$ kPa and $s_c N_c = 6.2$ (Figure 8.10, $d/B = 0$). The vertical total stress at the base of the pile $\sigma_q = \gamma L_p$. Under DA1b, $\gamma_{Rb} = 1.60$ for a bored pile (non-displacement) and $\gamma_{cu} = 1.40$, so

$$Q_{bu,\,des} = \frac{A_p \left[s_c N_c \left(\dfrac{c_u}{\gamma_{cu}} \right) + \gamma L_p \right]}{\gamma_{Rb}} = \frac{\dfrac{\pi \cdot 0.75^2}{4} \cdot \left[6.2 \cdot \left(\dfrac{200}{1.40} \right) + 18 \cdot 15 \right]}{1.60} = 319\,\text{kN}$$

Along the shaft, $c_u = 100$ kPa everywhere, so, from Figure 9.7(b) or Equation 9.8, $\alpha = 0.62$ everywhere. As τ_{int} is constant along the shaft, the integration in Equation 9.1a is trivial. Under DA1b, $\gamma_{Rs} = 1.30$ for a bored pile, so

$$Q_{su,\,des} = \frac{\pi D_0 L_p \alpha \left(\dfrac{c_u}{\gamma_{cu}} \right)}{\gamma_{Rs}} = \frac{\pi \cdot 0.75 \cdot 15 \cdot 0.62 \cdot \left(\dfrac{100}{1.40} \right)}{1.30} = 1204\,\text{kN}$$

Therefore, the design capacity of the pile $R = 319 + 1204 = 1523$ kN. The actions on the pile are the applied load (Q) and the self-weight of the pile $= 24 \times A_p \times L_p = 159$ kN. If the actions are permanent, then $\gamma_A = 1.00$ for DA1b. Then for ULS to be satisfied to EC7, $Q + 159 < 1523$, so the maximum design (and characteristic) value of the action that can be applied to the 15-m long pile $Q \leq 1.36$ MN.

b If the pile is to carry a permanent characteristic load of 3 MN it will have to be longer than 15 m, based on the answer to part (a). Under DA1b $\gamma_A = 1.00$, so the design load $Q = 3$ MN. Considering Figure 9.10(b), the pile tip will now extend deeper within the lower layer, so $s_c N_c$ will increase compared to the value in part (a). As an initial assumption $s_c N_c = 9.0$ is used (i.e. assuming d/B large in Figure 8.10; this will be checked later), so the base capacity is now written in terms of the unknown L_p:

Deep foundations

$$Q_{bu, des} = \frac{A_p \left[s_c N_c \left(\dfrac{c_u}{\gamma_{cu}} \right) + \gamma L \right]}{\gamma_{Rb}} = \frac{\dfrac{\pi \cdot 0.75^2}{4} \cdot \left[9.0 \cdot \left(\dfrac{200}{1.40} \right) + 20 L_p \right]}{1.60} = 354 + 5.5 L_p \text{ kN}$$

The design shaft capacity over the top 15 m of the pile was calculated in part (a). The additional shaft capacity from the section of pile in the lower clay layer is found similarly, but with $\alpha = 0.35$ for $c_u = 200$ kPa:

$$Q_{su, des}\big|_{layer 2} = \frac{\pi D_0 (L_p - 15) \alpha \left(\dfrac{c_u}{\gamma_{cu}} \right)}{\gamma_{Rs}} = \frac{\pi \cdot 0.75 \cdot (L_p - 15) \cdot 0.35 \cdot \left(\dfrac{200}{1.40} \right)}{1.30}$$

$$= 90.6 L_p - 1359 \text{ kN}$$

The total action applied to the pile (applied load + self-weight) in terms of L_p is now

$$1.00 \cdot \left[Q + \left(\frac{\pi \cdot 0.75^2}{4} \cdot 24 \cdot L_p \right) \right] = 3000 + 10.6 L_p$$

Then, to satisfy ULS,

$$Q + 10.6 L_p \leq R$$
$$3000 + 10.6 L_p \leq 354 + 5.5 L_p + 1204 + 90.6 L_p - 1359$$
$$L_p \geq \frac{2801}{85.5}$$
$$= 32.8 \text{ m}$$

At this pile length, the pile will be embedded by 18 m into the lower clay layer, so $d/B = 18/0.75 = 24$ and the use of $s_c N_c = 9.0$ was correct.

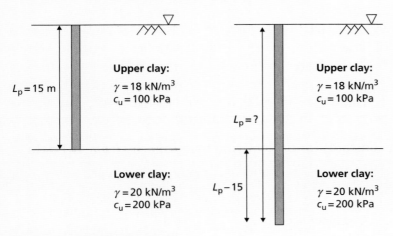

Figure 9.10 Example 9.2.

Applications in geotechnical engineering

9.3 Pile resistance from in-situ test data

Due to the difficulties in obtaining accurate values of the required parameters, empirical correlations, based on the results of pile loading tests and in-situ tests, are commonly used to provide alternative methods of obtaining pile resistance. From SPT data, the following correlation is generally used:

$$Q_{bu} = A_p C_b N_{60} \text{ (kN)} \tag{9.13}$$

where N_{60} is the value of standard penetration resistance in the vicinity of the pile base and C_b is a soil-dependent constant, the values of which are given in Table 9.2 after Poulos (1989). The correlation for shaft capacity is

$$Q_{su} = (\pi D_0 L_p) C_s \overline{N}_{60} \text{ (kN)} \tag{9.14}$$

where $\overline{N}_{60}$ is the average value of N_{60} along the length of the pile and C_s is a constant which in unknown soil conditions can be taken as $C_s \approx 2.0$ (Clayton, 1995).

The results of cone penetration testing (CPT) can also be directly used in pile design, particularly for displacement piles due to the similarity between the CPT and the method of pile installation. The limiting end bearing pressure ($q_b = Q_{bu}/A_p$) is related to the average cone resistance $\overline{q}_c$ in the vicinity of the pile base after

$$q_b = C_{cpt} \overline{q}_c \tag{9.15}$$

where C_{cpt} depends on pile and soil type. Suggested values of C_{cpt} are given in Table 9.3 based on the findings of Jardine et al. (2005) and Lee and Salgado (1999). Different procedures have been suggested for determining this average. The values of C_{cpt} in Table 9.3 apply if the q_c is averaged over $1.5D_0$ above

Table 9.2 Soil dependent constants for determining base capacity from SPT data

Pile type	Soil	C_b
Displacement (driven)	Sand	400–450
	Silt	350
	Glacial till	250
	Clay	75–100
Driven cast-in-situ	Cohesionless	150
Bored	Sand	100
	Clay	75–100

Table 9.3 Soil dependent constants for determining base capacity from CPT data

Pile type	Soil	C_{cpt}
Driven (closed)	Sand	0.4
	Clay (undrained)	0.8
	Clay (drained)	1.3
Bored pile	Sand	0.2

Deep foundations

and below the base of the pile. Correlations of CPT data with shaft friction parameters are notoriously unreliable, and it is suggested that basic soil properties (e.g. c_u, ϕ') are determined using the CPT data from which Q_{su} is then determined using the methods described in Section 9.2.

The values of Q_{bu} and Q_{su} determined from in-situ tests are characteristic resistances in terms of limit state design frameworks. If n tests have been conducted, the characteristic resistance ($R_k = Q_{bu} + Q_{su}$) is determined using

$$R_k = \min\left[\frac{R_{avg}}{\xi_3}, \frac{R_{min}}{\xi_4}\right] \qquad (9.16)$$

where ξ_3 and ξ_4 are **correlation factors** which depend on the number of tests undertaken. Normative values of the correlation factors are given in Table 9.4; as before, these values may be superseded by those published in National Annex documents. It should be noted that the values of ξ reduce with increasing numbers of tests to reflect the greater confidence in the derived resistance with more supporting tests. This characteristic value may be reduced using a **model factor** to account for the uncertainty in the correlations used to derive Q_{bu} and Q_{su}. The resulting value is then factored as a resistance as in Section 9.2.

9.4 Settlement of piles

As for shallow foundations, verification of the SLS involves ensuring that the settlement of the pile under the applied action will not adversely affect the supported structure. The settlement of a pile is much more difficult to compute than that for a shallow foundation for a number of reasons, including:

1. constitutive behaviour – the mechanisms of stress transfer at the base and along the shaft of the pile are very different;
2. layering – piles often pass through layers of soil with dissimilar stiffness;
3. pile slenderness (L_p/D_0) – because piles are long compared to the cross-sectional area, the compression (shortening) of the pile itself can be significant in magnitude;
4. elasto-plastic condition – loads carried by the pile are highest at the top, where the strength of the soil is weakest; therefore, soil may be at failure towards the top of the pile and elastic at depth while still being below the ULS (as this requires all of the soil to be at failure).

As a result of these difficulties, piles will always be load tested to the working load to verify that the SLS has been met. In cases where many piles of the same design will be used, only a proportion of the piles will need to be tested. Load testing will be described in greater detail in Section 9.6. However, in order to minimise any possible re-design, a good estimate can be made using analytical or numerical techniques, or based on previous experience of design in similar ground conditions, to produce a pile which can confidently be expected meet the SLS during the load test.

Table 9.4 Correlation factors for determination of characteristic resistance from in-situ tests to EC7

Factor	n = 1	2	3	4	5	10
ξ_3	1.40	1.35	1.33	1.31	1.29	1.25
ξ_4	1.40	1.27	1.23	1.20	1.15	1.08

Applications in geotechnical engineering

Three analytical techniques will be considered in this section. The first, known as the **Randolph and Wroth method** after its creators, considers the soil to behave elastically (see Section 8.6 for shallow foundations) and the pile to be axially rigid in comparison. Simple closed form solutions for the vertical stiffness of a pile can be derived, from which the displacement under a given load can be determined.

In the second, known as the **T–z method**, the pile is split up into a number of discrete sections, to each of which a spring is attached which represents the soil–pile interaction. The properties of these springs may be determined analytically using the elastic models of the Randolph and Wroth model, defining the reaction on the pile T for a given relative pile–soil displacement z. Having defined the T–z springs, the resulting set of equations can be solved iteratively using a Finite Difference Scheme; the method is thus amenable to computer analysis using a spreadsheet, a tool for which is provided on the Companion Website. The use of a finite difference scheme allows this method to account for pile compressibility and variation in soil or pile properties along the length of the pile.

The third method is based on the observation that the overall pile load–settlement curve can in almost all cases be reasonably approximated by a hyperbolic curve. This method is therefore known as the **hyperbolic method**, and requires knowledge of Q_{bu} and Q_{su} (Section 9.2) in addition to some additional fitting parameters determined empirically from a large database of pile load tests (to which the method is fairly insensitive). This is also amenable to solution using a spreadsheet, and has the added advantage that data from a subsequent load test (Section 9.6) can be used to update the soil parameters used in the analysis for subsequent application to other piles in the same unit of soil which might be carrying different working loads.

Randolph and Wroth method

The soil is considered to respond in a linear elastic way until failure (at Q_{bu} or Q_{su} as appropriate). The pile base is treated as an embedded shallow foundation as in Section 9.2, resting on elastic soil (below the pile base). Equation 8.53 may then be used to define the base stiffness $K_{bi} = T_{bi}/s_b$, where T_{bi} is the load carried at the pile base and s_b is the settlement of the pile base. In the case of piles, the stiffness is usually defined in terms of shear modulus G instead of Young's modulus E ($E = 2G(1+v)$ from Equation 5.6). For a circular pile assumed to be rigid at its base, as L_p/D_0 is large, the base pressure $q = 4T_{bi}/\pi D_0^2$ and $I_s = 0.79$ from Table 8.6. Recognising that $(1-v^2) = (1-v)(1+v)$,

$$s_b = \frac{4T_{bi}D_0(1-v^2)I_s}{\pi D_0^2 2G(1+v)}$$

$$\therefore K_{bi} = \frac{T_{bi}}{s_b} = \frac{\pi}{2I_s}\left(\frac{D_0 G}{1-v}\right) \quad (9.17a)$$

$$= 2.00\left(\frac{D_0 G}{1-v}\right)$$

For a square pile, $q = T_{bi}/B_p^2$ and $I_s = 0.82$ from Table 8.6, giving

$$\therefore K_{bi} = \frac{T_{bi}}{s_b} = \frac{2}{I_s}\left(\frac{B_p G}{1-v}\right)$$

$$= 2.44\left(\frac{B_p G}{1-v}\right) \quad (9.17b)$$

Along the pile shaft, if the interface friction has not been exceeded at the pile–soil interface, the pile will impose a shear mode of deformation on the surrounding annuli of soil as shown in Figure 9.11. For the soil annulus shown in Figure 9.11 to be in vertical equilibrium,

Deep foundations

$$(2\pi r L_{si}) \cdot \bar{\tau} = \left(2\pi \frac{D_0}{2} L_{si}\right) \cdot \bar{\tau}_0$$

$$\therefore \bar{\tau} = \left(\frac{D_0}{2r}\right) \bar{\tau}_0$$

(9.18)

where $\bar{\tau}_0$ is the average shear stress acting over the element length L_{si}. The shear strain in the soil annulus is approximated by $\gamma = ds/dr$ (assuming that any radial strain in the soil is negligible), so that from Equation 5.4

$$\gamma = \frac{\bar{\tau}}{\bar{G}_{si}}$$

$$\frac{ds}{dr} = \left(\frac{D_0 \bar{\tau}_0}{2\bar{G}_{si}}\right) \frac{1}{r}$$

(9.19)

where $\bar{G}_{si}$ is the average shear modulus of the soil element of length L_{si}. From Equation 9.19, the settlement of the soil will reduce with increasing r, such that $s=0$ at $r=\infty$. It is usually sufficient in practice to define a maximum radius (r_m) beyond which settlement is negligible, i.e. $s \approx 0$ at $r=r_m$. At the pile shaft, $s=s_s$ (settlement at pile shaft) at $r=D_0/2$. Integrating Equation 9.19 between these limits then gives

$$\int_{s_s}^{0} ds = \left(\frac{D_0 \bar{\tau}_0}{2\bar{G}_{si}}\right) \int_{D_0/2}^{r_m} \frac{1}{r} dr$$

$$s_s = \left(\frac{D_0 \bar{\tau}_0}{2\bar{G}_{si}}\right) \ln\left(\frac{2r_m}{D_0}\right)$$

(9.20)

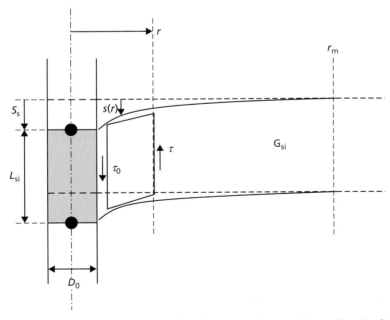

Figure 9.11 Equilibrium of soil around a settling pile shaft.

Applications in geotechnical engineering

The force along the pile shaft that results in s_s is simply

$$T_{si} = \pi D_0 L_{si} \bar{\tau}_0 \tag{9.21}$$

By dividing Equation 9.21 by Equation 9.20, the shaft stiffness can be determined:

$$K_{si} = \frac{T_{si}}{s_s} = \frac{2\pi L_{si} \bar{G}_{si}}{\ln\left(\dfrac{2r_m}{D_0}\right)} \tag{9.22}$$

The value of r_m in Equation 9.22 is usually approximated by L_p.

If the pile is rigid, Equations 9.17 and 9.22 can be used to determine the pile head settlement under a given load analytically, as shown by Randolph and Wroth (1978). For a rigid pile the settlements are the same all the way along the pile, i.e. $s_b = s_s = s_r$, where s_r is the overall settlement at the top of the rigid pile. The applied load at the top of the pile $Q = T_{bi} + T_{si}$, therefore

$$\begin{aligned} Q &= T_{bi} + T_{si} \\ &= K_{bi} s_r + K_{si} s_r \\ \therefore s_r &= \frac{Q}{K_{bi} + K_{si}} \end{aligned} \tag{9.23}$$

If the pile is compressible $s_b \neq s_s \neq s_r$, such that Equation 9.23 no longer applies. Randolph and Wroth (1978) also present a more sophisticated solution for the case of a compressible pile (accounting for the effect of an under-ream, if present). However, this case (and, indeed, more complex cases) are generally quicker to solve numerically using the T–z method.

T–z method

In the T–z method, the pile is split up into a series of sections as shown in Figure 9.12, to each of which is attached a linear elastic spring, representing the soil–pile interaction.

The i'th element is loaded by the force transmitted from the sections of pile above (F_i). Under this load the pile element, which is considered to be linear elastic with Young's modulus E_{pi}, cross-sectional area A_{pi} and length L_{si}, may compress by an amount $\Delta z_i - \Delta z_{i+1}$ (the method therefore accounts for the pile slenderness effect). The average vertical movement of the pile element relative to the surrounding soil $(\Delta z_i + \Delta z_{i+1})/2$ generates a resistive force in the T–z spring (T_{si}). This force, plus the force from the pile elements below (F_{i+1}), act to resist F_i. If the pile is below the ULS then it must be in a condition of equilibrium, so

$$F_i = F_{i+1} + K_{si}\left(\frac{\Delta z_i + \Delta z_{i+1}}{2}\right) \tag{9.24}$$

Determination of the T–z spring stiffness (K_{si}) for an elastic soil is accomplished using Equation 9.22. The displacements in Equation 9.24 are related to each other through the elastic compression of the pile element under the average force in the element, i.e.

$$\Delta z_i - \Delta z_{i+1} = \frac{L_{si}}{E_{pi} A_{pi}}\left(\frac{F_i + F_{i+1}}{2}\right) \tag{9.25}$$

Deep foundations

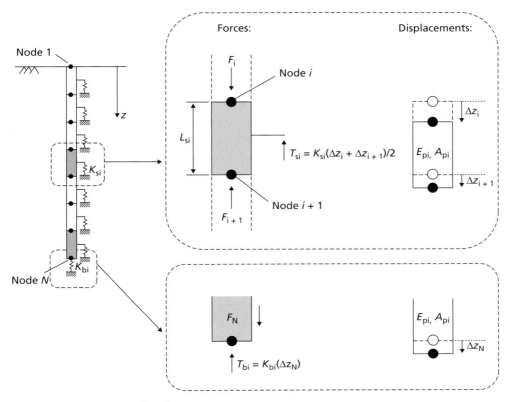

Figure 9.12 *T–z* method.

The aim of the analysis is to find the unknown distributions of forces and displacements (i.e. the values of F_i and Δz_i along the pile. The method of solution is to make an initial guess of the pile head settlement (at node 1). The displacements of the nodes below (for the rest of the pile) can then be found using Equation 9.25. The displacement at the node at the base of the pile (Δz_N) then corresponds to the compression of the elastic spring representing the base behaviour and can therefore be used to find the force at this node using

$$F_N = K_{bi} \Delta z_N \qquad (9.26)$$

where K_{bi} is the stiffness at the pile base (determined using Equation 9.17). As different soil springs can be defined at the base (K_{bi}) and along the shaft (K_{si}), the method therefore accounts for the constitutive behaviour effect described earlier in this section.

From the known force at the pile base (node N), the forces for each node above are found using Equation 9.24. Unless the correct pile head settlement was guessed, the resulting pile load recovered for node 1 at the top of the pile will not equal the applied working load. An iterative technique is therefore required to find the pile head settlement giving the applied working load at the pile head. The values of F_i and Δz_i at each node depend on each other (Equations 9.24 and 9.25) just as in the case of the nodes in the two-dimensional seepage problems of Section 2.7. The solution procedure can therefore be automated using a spreadsheet. The Companion Website includes an implementation of this – PileTz_CSM8.xls – which is provided with a comprehensive user manual describing how general pile settlement problems may be solved.

Applications in geotechnical engineering

It should be noted that different pile properties (E_{pi}, A_{pi} and L_{si}) can be defined for each individual element without making the calculations any more difficult, such that the method can also account for soil layering and piles with changing cross-section (e.g. tapered piles).

A rigid pile can be modelled using PileTz_CSM8.xls by setting E_{pi} to a very large value. In this case, the solution will then be identical to that from the Randolph and Wroth method.

Example 9.3

A 21.5-m long under-reamed bored pile is to be constructed in a unit of clay for which $G = 5 + 0.5z$ MPa, where z is the depth in metres below ground level and $v' = 0.2$. The pile shaft has a length of 20 m above the under-ream, and a diameter of 1.5 m. The under-ream has a diameter of 2.5 m at the base, and the pile may be assumed to be rigid. Determine the long-term settlement of the pile and the load carried by the shaft and the base of the pile under a working load of 5 MN.

a using the Randolph and Wroth method;
b using the T–z method.

Determine also the settlement and load distribution values if the pile is compressible with $E_{pi} = 30$ GPa.

Solution

a As the pile has an under-ream, shaft friction is only accounted for to a depth of $2D_0$ (= 3 m) above the top of the under-ream. The shear modulus is 5 MPa at the surface and 13.5 MPa at 17-m depth, so that the average shear modulus over the top 17 m of the pile $\overline{G} = 9.25$ MPa. Then, from Equation 9.22:

$$K_{si} = \frac{2\pi L_{si} \overline{G}_{si}}{\ln\left(\frac{2r_m}{D_0}\right)} = \frac{2 \cdot \pi \cdot 17 \cdot 9.25}{\ln\left(\frac{2 \cdot 21.5}{1.5}\right)} = 294.4 \text{ MN/m}$$

At the pile base, $D_0 = 2.5$ m and $G = 15.75$ MPa, so from Equation 9.17a

$$K_{bi} = 2.00\left(\frac{D_0 G}{1-v}\right) = 2 \cdot \left(\frac{2.5 \cdot 15.75}{1 - 0.2}\right) = 98.4 \text{ MN/m}$$

The pile is rigid and $Q = 5$ MN, so from Equation 9.23

$$s_r = \frac{Q}{K_{bi} + K_{si}} = \frac{5}{98.4 + 294.4} = 0.0127 \text{ m} = 12.7 \text{ mm}$$

The load carried at the base is then $T_b = K_{bi} s_r = 1.25$ MN, and that along the shaft $T_s = K_{si} s_r = 3.74$ MN (as a check, these sum to 4.99 MN).

b The pile is divided into 1-m long elements along the shaft (20) plus an additional 1.5-m long element for the under-ream. The shear modulus is then evaluated at the mid-depth of each element to a depth of 17 m (e.g. at $z = 0.5$ m, 1.5 m ... to 16.5 m) from which values of K_{si} are found for each spring as in part (a), the calculations being automated in a spreadsheet.

A complete worksheet for this example using PileTz_CSM8.xls is available on the Companion Website. To approximately model a rigid pile, $E_{pi} = 5 \times 10^{27}$ was used.

The resulting pile head settlement is 12.8 mm and is constant along the length of the pile. This compares favourably with the answer from part (a). Figure 9.13 shows the load distribution along the length of the pile which could not be obtained from part (a). The forces at the base and along the shaft are then $T_b = 1.25$ and $T_s = 3.75$ MN, respectively.

For the case of the compressible pile, $E_{pi} = 30$ GPa is set in PileTz_CSM8 and the optimisation re-run. The new pile head settlement is 13.6 mm and the load distribution is practically unchanged, as shown in Figure 9.13.

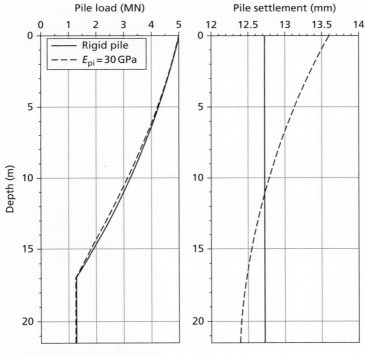

Figure 9.13 Example 9.3.

Hyperbolic method

The hyperbolic model is described by Fleming (1992). The settlement at the top of a rigid pile under an applied load Q is given by the solution of the following quadratic equation:

$$as_r^2 + bs_r + c = 0 \tag{9.27}$$

where

$$a = \eta(Q - \alpha) - \beta$$
$$b = Q(\delta + \lambda\eta) - \alpha\delta - \beta\lambda$$
$$c = \lambda\delta Q$$

Applications in geotechnical engineering

and

$$\alpha = Q_{su}$$
$$\beta = D_b Q_{bu} E_b$$
$$\delta = 0.6 Q_{bu}$$
$$\lambda = M_s D_s$$
$$\eta = D_b E_b$$

The parameters D_b and D_s represent the diameter of the pile at the base and shaft respectively (so that an under-ream can be modeled if necessary), E_b is an operative elastic modulus of the soil beneath the pile base (an initial estimate of which can be made using the approximate values in Figure 9.14), and M_s is a shaft–soil flexibility factor. Based on a large database of pile load tests, M_s values between 0.001 and 0.0015 have been empirically determined; the analysis is relatively insensitive to this value such that $M_s = 0.001$ can usually be assumed. In order to use the hyperbolic model, the following procedure is therefore followed:

1. Determine Q_{bu} and Q_{su} (Section 9.2);
2. Estimate E_b;
3. Determine $\alpha, \beta, \delta, \lambda$ and η;
4. Determine a, b and c;
5. Solve Equation 9.27 for s_r (this is a standard quadratic equation and may be solved using the quadratic formula; only the positive root is taken).

These calculations may be input into a spreadsheet to automate the analysis. The procedure given above is ideal for the analysis of a foundation of known dimensions. In principle, if Q_{bu} and Q_{su} are provided as a function of L_p, then Equation 9.27 can be used to determine the minimum pile length required to satisfy a certain limiting settlement in SLS design. In practice, this is difficult to do by hand; however, by using

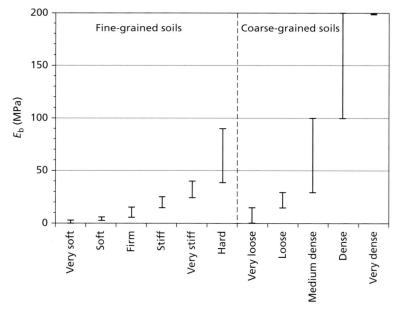

Figure 9.14 Approximate values of E_b for preliminary design purposes.

a spreadsheet, an optimisation routine (e.g. Solver in Microsoft Excel) can be used to do precisely this. Furthermore, if a pile load test has been conducted, the known dimensions of the test pile and measured settlement at the applied load can be entered and the optimisation performed to back-calculate the value of E_b (i.e. using the pile load test as an in-situ soil test).

Equation 9.27 applies to a rigid pile only. If the pile is compressible, there will be an additional amount of elastic shortening s_e, so that the overall pile settlement at the pile head is

$$s = s_r + s_e \qquad (9.28)$$

Fleming (1992) presented additional equations to determine s_e, which can account for a region (of length L_0) at the top of the pile where a negligible amount of shaft load transfer occurs. This may be used to model de-bonding towards the top of the pile as a result of the construction procedure (e.g. a smooth steel casing used to support a bored shaft which will ultimately have a rough surface).

$$s_e = \frac{4Q\left[L_0 + K_e\left(L_p - L_0\right)\right]}{\pi D_s^2 E_p} \qquad \text{for } Q \leq Q_{su}$$

$$s_e = \frac{4\left[QL_p - Q_{su}\left(1 - K_e\right)\left(L_p - L_0\right)\right]}{\pi D_s^2 E_p} \qquad \text{for } Q \geq Q_{su} \qquad (9.29)$$

Fleming (1992) provides discussion on the value of K_e which should be selected; for most situations, $K_e = 0.4$ is sufficiently accurate.

The hyperbolic model has been programmed into a spreadsheet, PileHyp_CSM8.xls, which can be found on the Companion Website, accompanied by a comprehensive user manual. This implementation of the hyperbolic method additionally calculates the settlement at a range of loads from $Q = 0$ to $Q = Q_{bu} + Q_{su}$ such that the full load deflection curve for the pile can be estimated.

9.5 Piles under tensile loads

While piles are most commonly used to carry compressive loading, there are a number of situations where piles may be used to carry tensile loads. These include:

- when used as part of a pile group supporting a structure to which horizontal or moment loading is applied;
- when used as reaction piles to provide reaction for pile load tests (Section 9.6);
- to provide anchorage against uplift forces (Section 10.2).

When loaded in tension it is assumed that the pile resistance is due to shaft friction alone, the base being lifted away from the soil beneath. In fine-grained soils, there may be additional suction pressures acting on the pile base under undrained (rapid) loading providing some additional tensile capacity, though it is conservative to neglect this as far as the stability of a pile under tension is concerned.

In undrained soil, the shaft friction is calculated using the methods described in Section 9.2 without modification. Figure 9.15(a) shows values of α determined from a database of tension tests on non-displacement piles in fine-grained soils collated by Kulhawy and Phoon (1993). Equation 9.8 (which was developed based on pile compression tests) is also shown in this figure, from which it can be seen that tensile values are approximately 70% of the compression values. In coarse-grained soil, the stress reversals that occur between installation (under compressive load) followed by loading (tensile) have a more significant effect on the soil at the interface of the pile, generally leading to a contraction of the soil in

Applications in geotechnical engineering

this zone and a shaft friction which is lower in tension than in compression. De Nicola and Randolph (1993) proposed the following expression:

$$\frac{Q_{su}|_{\text{tension}}}{Q_{su}|_{\text{compression}}} \approx \left[1 - 0.2 \log_{10}\left(\frac{100}{L_p/D_0}\right)\right]\left[1 - 8\left(v\frac{L_p\overline{G}}{D_0 E_p}\right) + 25\left(v\frac{L_p\overline{G}}{D_0 E_p}\right)^2\right] \quad (9.30)$$

where $\overline{G}$ is the average shear modulus of the soil over the pile length L_p, and v its Poisson's ratio. The tensile shaft resistance given by Equation 9.30 is usually 70–85% of the compressive shaft resistance for typical values of the soil and pile parameters. Figure 9.15(b) shows measured shaft resistance as a function of SPT resistance N_{60} from a database of tension tests on non-displacement piles in coarse-grained soils collated by Rollins *et al.* (2005). Equation 9.8 developed for compressive loading is also shown, and a 70% reduction from these values is seen to provide a good lower bound to the data (for conservative estimates of shaft resistance).

When factoring the tensile resistance of a pile for checking the ULS, alternative partial factors are used in Eurocode 7. The partial fator γ_{RT} is used, the '*T*' denoting tension (as opposed to γ_{RC} for compression from Section 9.2). The normative values, applicable to both displacement and non-displacement piles, are γ_{RT} = 1.25 (set R1), 1.15 (Set R2), 1.10 (Set R3) and 1.60 (Set R4).

9.6 Load testing

It is common practice in the design of piling to verify the capacity of a pile design using a load test. By doing this:

1 the uncertainty associated with using empirical properties in calculations (e.g. α and β) is reduced;
2 it can be verified that the proposed construction technique is acceptable and allows the integrity of cast-in-situ piles formed using the proposed method to be checked;
3 it can be verified that the ULS and SLS will be met by the proposed design.

Load testing may be conducted on **trial piles** (also known as **test piles**) – these are piles which are constructed solely for the purposes of load testing, usually before the main piling works commence. If sufficient load can be applied, these piles can be tested to the ULS (i.e. failed) to verify the pile capacity, as

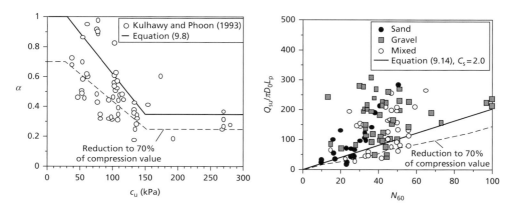

Figure 9.15 Shaft friction in tension: (a) α for non-displacement piles in fine-grained soil, (b) shaft resistance for non-displacement piles in coarse-grained soil.

Deep foundations

they will not subsequently be used to support the proposed structure. Load tests may also be carried out on **working piles** (also known as **contract piles**). These are piles that will be part of the final foundation, and as such will not be loaded to failure. A typical maximum load in such a test would be 150% of the working load that the pile will ultimately carry, allowing for the SLS to be verified with an allowance for possible load redistribution from other piles within the foundation.

Static load testing

Static load testing is the most common form of pile testing, and the method that is most similar to the loading regime in the completed foundation. Figure 9.16 shows the set-up of a static load test. A hydraulic jack is used to push the pile under test into the ground (for a conventional compression test), using either the dead weight of **kentledge** (typically blocks of precast concrete or iron, Figure 9.16(a)) or a series of tension piles/anchors (Figure 9.16(b)) to provide the reaction. If kentledge is used, the weight must be at least equal to the maximum test load, though this is normally increased by 20% to account for variability in the predicted capacity. In the case of tension piles the tensile resistance is less easy to predict with certainty, so the pile or anchor system should be proof-tested before use, usually to 130% of the required test load. An in-line load cell is used to measure the force applied at the pile head, while the displacement of the pile head may be measured either using local displacement transducers or by remote measurement using precision levelling equipment. The former method is generally more accurate, though will be affected by any ground settlements around the test pile.

Static load tests are usually conducted in one of two modes. **Constant rate of penetration tests** (CRP) are used for trial piles in which a penetration rate of 0.5–2 mm/min in compression is used to displace the pile until either a steady ultimate load is reached or the settlement exceeds 10% of the pile diameter (or width for a square pile). This test is essentially a very large CPT test, using the pile instead of the CPT probe. CRP tests may also be conducted on tension piles (the reaction piles then being loaded in compression, Figure 9.16(b)), in which case the penetration rate is reduced to 0.1–0.3 mm/min as the pile will mobilise its tension capacity at much smaller displacements than in compression.

Maintained load tests (MLT) are used for working piles. This involves applying load to the pile through the jack which is then maintained for a period of time. A series of loading stages are normally applied as detailed below:

1. Load to 100% of the design (working) load, also called the **design verification load (DVL)** in 25% increments;
2. Unload fully in 25% increments;

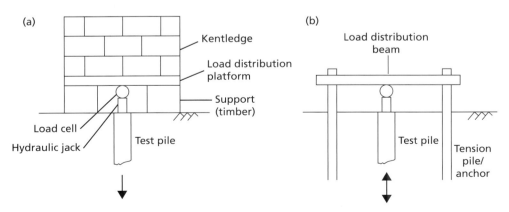

Figure 9.16 Static load testing of piles: (a) using kentledge, (b) using reaction piles.

Applications in geotechnical engineering

3 Reload directly to 100% DVL, then load to 150% of the working load (also called the **proof load**) in 25% increments;
4 Unload fully in 25% increments.

The settlement recorded under the proof load is used to verify that the SLS has been met for the pile. This settlement and the proof load may also be input into the PileHyp_CSM8.xls spreadsheet and used to determine the operative value of E_b to refine the serviceability calculations or for use in the design of similar piles on the site.

For piles which are long, of large diameter, in strong soil, having an under-ream or with a combination of any of these features, pile load tests may not be continued to failure, due to the cost involved and/or the relatively large settlement required. However, a number of methods of extrapolating test data to ultimate failure have been proposed; these methods have been summarised by Fellenius (1980). A popular method is that originally proposed by Chin (1970) based on the assumption of a hyperbolic shape to the load settlement curve. By plotting s/Q against s, the test data will tend towards a straight line where the gradient is $1/R$, as shown in Figure 9.17. Further information on CRP and MLT testing and data interpretation may also be found in Fleming *et al.* (2009).

The pile capacity determined from a CRP load test is an in-situ measurement of the characteristic resistance (R_k) of the tested pile at ULS. This resistance can be used to provide an alternative value of the design load by appropriate factoring. In Eurocode 7 the load test result is not used directly, R_k being determined by dividing the average measured resistance by a correlation factor (ξ) depending on the number of tests. If a series of tests are undertaken, the average and minimum recorded resistances are factored separately and the minimum value taken as R_k, i.e.

$$R_k = \min\left[\frac{R_{avg}}{\xi_1}, \frac{R_{min}}{\xi_2}\right] \tag{9.31}$$

Normative values of the correlation factors are given in Table 9.5; as before, these values may be superseded by those published in National Annex documents.

The values of ξ in Table 9.5 reduce with increasing numbers of tests to reflect the greater confidence in the derived resistance with more supporting tests. Bearing resistance should be based either on calculations validated by load tests, or on load tests alone.

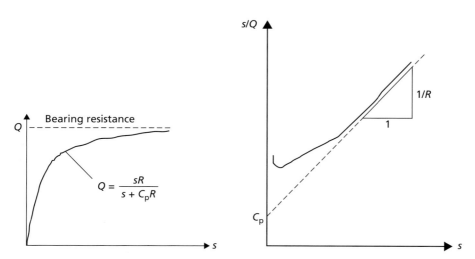

Figure 9.17 Interpretation of pile capacity using Chin's method.

Table 9.5 Correlation factors for determination of characteristic resistance from static load tests to EC7

Factor	n=1	2	3	4	5+
ξ_1	1.40	1.30	1.20	1.10	1.00
ξ_2	1.40	1.20	1.05	1.00	1.00

Other pile testing methods

To overcome the difficulties in providing sufficient reaction for static load tests of piles which are long, of large diameter, in strong soils or with under-reams, alternative methods of testing have been derived. In **dynamic tests**, an impulse loading (usually from an instrumented hammer) is applied to the top of the pile. From analysing the wave propagation through the pile, particularly the wave reflected from the base of the pile, the pile capacity can be determined. In a **Statnamic test**, a combustion chamber filled with solid propellant fuel is placed between the top of the pile and a reaction mass. The test consists of igniting the fuel, which leads to explosive combustion, the high pressure gases from which drive the mass upwards against gravity, and the pile downwards. The acceleration imparted to the mass is usually between 10–$20\,g$ – as this is 10–20 times the gravity used in a conventional static test, only 5–10% of the maximum pile load is required as a reaction mass.

In both of the aforementioned procedures, the stresses applied to the pile on test are very rapid, and dynamic effects need to be taken into account when interpreting the test data (e.g. damping in the pile and soil). Further information may be found in Fleming *et al.* (2009).

9.7 Pile groups

A pile foundation may consist of a group of piles installed fairly close together (at a spacing S which is typically $2D_0$–$4D_0$) and joined by a slab, known as the **pile cap**, cast on top of the piles. The cap is usually in contact with the soil, in which case part of the structural load is carried directly on the soil immediately below the surface. If the cap is clear of the ground surface, the piles in the group are referred to as **freestanding**.

Design at ULS

In general, the ultimate load which can be supported by a group of n piles is not equal to n times the ultimate load of a single isolated pile of the same dimensions in the same soil. The ratio of the average load per pile in a group at failure to the resistance of a single pile is defined as the **efficiency** (η_g) of the group. In most fine-grained soils under undrained conditions, the efficiency will be close to 1.0. In some sensitive clays, the efficiency may be slightly lower than this. Displacement piles in all soils will generally have efficiencies greater than 1.0 due to the compactive effort imparted to the soil during installation, which will tend to increase the shaft capacity around the piles (though the base capacity is likely to remain unchanged). For example, the driving of a group of piles into loose or medium-dense sand causes compaction of the sand between the piles, provided that the spacing is less than about $8D_0$; consequently, the efficiency of the group is greater than unity. A value of 1.2 is often used in design. It is generally assumed that the distribution of load between the piles in an axially loaded group is uniform. However, experimental evidence indicates that under working load conditions for a group in sand, the piles at the centre of the group carry greater loads than those on the perimeter; in clay, on the other hand, the piles on the perimeter of the group carry greater loads than those at the centre (see Figure 8.25 for shallow foundations).

Applications in geotechnical engineering

As a result, verification of the ULS for a pile group can be achieved by ensuring that each individual pile satisfies ULS under the load which is distributed to it from the pile cap, using the methods outlined in Section 9.2. This is shown as Mode 1 in Figure 9.18. It can generally be assumed that at the ULS the piles all carry the same amount of load ($=Q_{pg}/n$, where Q_{pg} is the load carried by the group), as this distribution satisfies the lower bound theorem (provided that the pile cap is ductile enough to allow load redistribution to take place from the working condition). However, there is an alternative failure mechanism which must also be checked, known as **block failure** (Mode 2 in Figure 9.18). This is when the whole block of soil beneath the pile cap and enclosed by the piles fails as one large pier.

The capacity of the block is determined by treating it as a single pier with length L_p and cross-sectional area $B_r \times L_r$ as shown in Figure 9.18, using the methods outline in Section 9.2. When determining the shaft capacity, $\alpha = 1$ should be used in undrained conditions while $\delta' = \phi'$ should be used in drained conditions, in each case as the interface shear along the walls of the block is almost entirely soil–soil. As, generally, $Q_{bu} > Q_{su}$, block failure is much more likely for closely-spaced groups of long, slender piles (high L_p/D_0) than for widely-spaced groups of stocky piles (low L_p/D_0). Block failure is also more likely for pile groups in fine-grained soils, rather than in coarse-grained soil, as Q_{bu}/Q_{su} is generally smaller in the former case. Tests of model pile groups in clays by Whitaker (1957) and De Mello (1969) suggest that the failure mode does not transition to block failure until $S < 2$–$3D_0$.

Design at SLS

When piles are installed in closely spaced groups, the settlement of any the piles in the group will cause settlement in the surrounding soil. Provided that the pile behaviour is still within the elastic range (i.e. below ULS), the settlement of a given pile i in the group is therefore equal to the settlement under its own load (Section 9.4) plus a small amount of additional settlement induced by each of the other piles in the group. The amount of additional settlement induced by a nearby pile j will reduce with distance from the pile of interest (pile i). For a group of n identical piles, the settlement of pile i is then given by

$$s_i = \left(\frac{1}{K_{pile}}\right)\left[Q_i + \sum_{j=1}^{n} \alpha_j Q_j\right] \tag{9.32}$$

where K_{pile} is the overall pile-head stiffness of pile i (as determined in Section 9.4), Q_i is the sum of the actions on pile i, Q_j is the sum of the actions on pile j, and α_j is an interaction factor describing the influence of pile j on the settlement of pile i. This approach was originally presented by Poulos and Davis (1980),

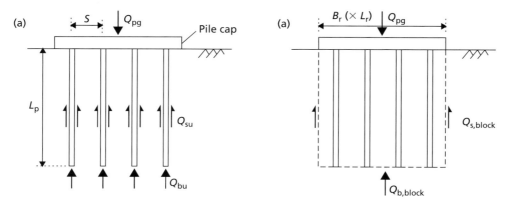

Figure 9.18 Failure modes for pile groups at ULS: (a) mode 1, individual pile failure, (b) mode 2, block failure.

Deep foundations

who also presented simplified design charts for α_j which depended on E_{pi}, G, L_p, D_0 and S. Mylonakis and Gazetas (1998) presented revised interaction factors which also incorporated the reinforcing effects of adjacent piles, and which are therefore more accurate for use in design. These factors were defined by:

$$\alpha_j = \frac{\ln\left(\dfrac{r_m}{S}\right)}{\ln\left(\dfrac{2r_m}{D_0}\right)} F_\alpha \tag{9.33}$$

As in Section 9.4, r_m can be approximated by L_p. The value of F_α (also known as a diffraction factor) is determined using Equation 9.34 or graphically from Figure 9.19:

$$F_\alpha = \frac{2\mu L_p + \sinh(2\mu L_p) + \Omega^2\left[\sinh(2\mu L_p) - 2\mu L_p\right] + 2\Omega\left[\cosh(2\mu L_p) - 1\right]}{\left[2 + 2\Omega^2\right]\sinh(2\mu L_p) + 4\Omega\cosh(2\mu L_p)} \tag{9.34}$$

where

$$\mu = \sqrt{\frac{2\pi \overline{G}}{\ln(2r_m/D_0) E_p A_p}}$$

and

$$\Omega = \frac{K_{bi}}{E_p A_p \mu}$$

Equation 9.34 is plotted for a wide range of possible combinations of pile properties using the parameters μ and Ω defined above in Figure 9.19.

The procedure for determining the settlements and load distribution within a pile group is as follows:

1. For each pile, determine the interaction factors α_j for all of the surrounding piles. This is best done in tabular format, making use of symmetry in the pile layout (see Example 9.4).
2. Use Equation 9.32 to form an equation for the settlement at each pile in terms of the unknown forces in the piles.
3. If the pile cap is perfectly flexible, the total load carried by the group will be equally distributed amongst the piles in the group. These known pile loads are then used in the equations from step (2) to determine the maximum possible settlements of each pile (as the pile cap will in reality have some stiffness).
4. If the pile cap is perfectly rigid, the settlements of each of the piles will be the same. This condition can be used with the equations from step (2) to determine the load distribution amongst the piles in the group.

Steps (3) and (4) represent the two extreme cases of pile cap flexibility. In reality, the finite stiffness of the pile cap will mean that the settlements of and loads carried by each pile will be somewhere between those calculated in steps (3) and (4). The foregoing approach to analysis is therefore very useful in providing bounds on the pile responses which can be used in design.

To determine the actual pile settlements and load distribution, a more advanced version of the analysis presented above must be undertaken using commercially available computer software which includes the pile cap, modelled as an elastic beam. Furthermore, the interaction factors described by Equation 9.34 are only applicable for piles in a uniform soil with G constant with depth, and Equation 9.32 is only valid if the piles are identical. Computer-based approaches can account for more complex variations of stiffness with depth, as well as large groups (with many tens or even hundreds of piles), irregular layouts and different length piles within the group.

Applications in geotechnical engineering

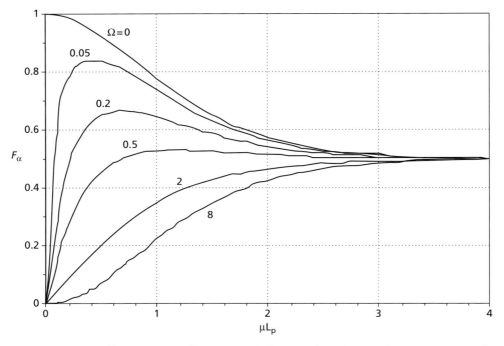

Figure 9.19 Diffraction coefficient F_α (after Mylonakis and Gazetas, 1998).

Example 9.4

A 4×2 pile group is formed from eight of the piles considered in Example 9.3. The pile group carries a total load of 40 MN, and the piles are spaced at $5D_0$ (centre-to-centre). Determine the settlement of each of the piles in the group.

Solution

From symmetry considerations, the four corner piles in the groups will settle identically, as will the four central piles. Therefore, calculations only need to be conducted for these two piles defined as pile type A (corner) and pile type B (centre). To apply Equation 9.32, the interaction factors must be found. From Equation 9.34,

$$\mu = \sqrt{\frac{2 \cdot \pi \cdot 9.25 \times 10^6}{\ln\left(\frac{2 \cdot 21.5}{1.5}\right) \cdot 30 \times 10^9 \cdot \left(\frac{\pi \cdot 1.5^2}{4}\right)}} = 0.018$$

and

$$\Omega = \frac{98.4 \times 10^6}{30 \times 10^9 \cdot \left(\frac{\pi \cdot 1.5^2}{4}\right) \cdot 0.018} = 0.103$$

Deep foundations

so $\mu L_p = 0.018 \times 21.5 = 0.39$ and $F_a = 0.75$. Then, to use Equation 9.33, the pile-to-pile spacing between each pair of piles must be found for piles A and B as shown in Figure 9.20. The interaction factors can then be found and used in Equation 9.32. The calculations for piles A and B are shown in Tables 9.6 and 9.7, respectively, from which

$$s_A = \left(\frac{1}{K_{pile}}\right)(1.86Q_A + 1.38Q_B)$$

$$s_B = \left(\frac{1}{K_{pile}}\right)(1.38Q_A + 2.24Q_B)$$

If the pile cap is perfectly flexible, then $Q_A = Q_B = 40/8 = 5$ MN. From Example 9.3, an individual pile settles 13.6 mm under this load, so $K_{pile} = 5/0.0136 = 385$ MN/m. Inserting these values into the foregoing equations gives $s_A = 42$ mm and $s_B = 47$ mm.

If the pile cap is perfectly rigid then $s_A = s_B$, giving $0.48Q_A = 0.86Q_B$, or $Q_A = 1.79Q_B$. From equilibrium, $4Q_A + 4Q_B = 40$ MN. Substituting Q_A into this expression gives $Q_B = 3.58$ MN and $Q_A = 6.42$ MN, and $s_A = s_B = 44$ mm.

It should be noted that, irrespective of the flexibility of the pile cap, the settlement of a group of eight piles is very much larger than the settlement of a single pile under the same nominal load (from Example 9.3). When designing pile groups, the individual piles must therefore be designed to be very stiff, particularly when n is large.

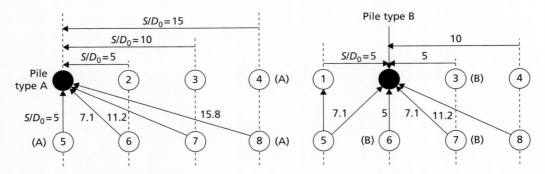

Figure 9.20 Example 9.4.

Table 9.6 Example 9.4 – calculations for pile type A

Pile type: A	Effect of pile...							
	j = 1	2	3	4	5	6	7	8
S/D_0	0	5	10	15	5	7.1	11.2	15.8
α_j	N/A	0.44	0.30	0.22	0.44	0.36	0.28	0.20
Q_j	Q_A	Q_B	Q_B	Q_B	Q_A	Q_A	Q_B	Q_B	Q_A
Notes:	Pile i							

Applications in geotechnical engineering

Table 9.7 Example 9.4 – calculations for pile type B

Pile type: B	Effect of pile...							
	$j=1$	2	3	4	5	6	7	8
SD_0	5	0	5	10	7.1	5	7.1	11.2
α_j	0.44	N/A	0.44	0.30	0.36	0.44	0.36	0.28
Q_j	Q_A	Q_B	Q_B	Q_A	Q_A	Q_B	Q_B	Q_A
Notes:		Pile i						

9.8 Negative skin friction

Negative skin friction can occur on the perimeter of a pile driven through a layer of clay undergoing consolidation (e.g. due to a fill recently placed over the clay) into a firm bearing stratum (Figure 9.21). The consolidating layer exerts a downward drag on the pile, and therefore the direction of skin friction in

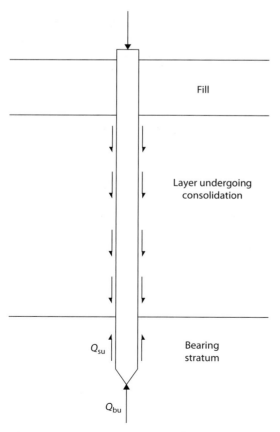

Figure 9.21 Negative skin friction.

this layer is reversed. The force due to this downward or negative skin friction is thus an additional action (and factored accordingly) instead of helping to support the external load applied at the head of the pile. It is conservative to assume that the whole consolidating layer applies negative skin friction and only the soil below contributes to the resistance of the pile, though in reality the cross-over point between negative and positive skin friction (also called the **neutral plane**) will occur within the consolidating soil, where the settlements of the soil and the pile are equal. Negative skin friction increases gradually as consolidation of the clay layer proceeds, the effective overburden pressure σ'_{v0} gradually increasing as the excess pore water pressure dissipates. Equation 9.10 can also be used to represent negative skin friction. In normally consolidated clays, present evidence indicates that a value of β of 0.25 represents a reasonable upper limit to negative skin friction for preliminary design purposes. It should be noted that there will be a reduction in effective overburden pressure adjacent to the pile in the bearing stratum due to the transfer of part of the overlying soil weight to the pile; if the bearing stratum is sand, this will result in a reduction in bearing capacity.

Summary

1 Deep foundations derive their resistance from a combination of end-bearing at the base of the piles and shaft friction along their length. They are advantageous when large concentrated loads are applied by the structure, or when the soil near the ground surface is unsuitable for shallow foundations. Pre-formed piles may be installed by driving or jacking, displacing the surrounding soil. Alternatively, piles may be cast from reinforced concrete in-situ into a pre-formed hole (non-displacement piles). The method of installation and construction procedure has a large influence of the soil–shaft interface and, consequently, on the shaft capacity.

2 At ULS, the base capacity of a deep foundation is determined using modified versions of the bearing capacity equations from Chapter 8. Shaft capacity depends on the pile–soil interface friction, and may be determined using the α-method (undrained conditions) or β-method (drained conditions). Pile capacity may also be determined directly from SPT or CPT data, or from pile load tests taken to failure on trial piles. At SLS, pile settlement may be determined computationally using the T–z method or the hyperbolic method. In the former case, elasticity theory may be used to define pile–soil stiffness (though more sophisticated non-linear relationships are available in the literature), and complex problems may be solved using a finite difference scheme. Pile load tests on working piles may be used to verify SLS calculations.

Problems

9.1 A single square precast concrete pile of 0.5×0.5 m in cross-section, and of length 10 m, is driven into dry loose sand. The soil has unit weight $\gamma = 15$ kN/m³, $\phi' = 32°$ and $c' = 0$. Along the pile–soil interface, it may be assumed that $K = 1$ and $\delta' = 0.75\phi'$. Determine the characteristic resistance of the pile and the allowable design load (permanent) on the pile under EC7 DA1b. Hence, determine the equivalent overall factor of safety that has been designed into the pile by using the EC7 design framework in this case.

Applications in geotechnical engineering

9.2 A bored concrete pile (unit weight $23.5\,kN/m^3$) with an enlarged base is to be installed in a stiff clay, the characteristic undrained strength increasing with depth from $80\,kPa$ at the ground surface to $220\,kPa$ at 22 m and then remaining constant below this depth. The saturated unit weight of the clay is $21\,kN/m^3$. The diameters of the pile shaft and base are 1.05 and 3.00 m, respectively. The pile extends from a depth of 4 m to a depth of 22 m, the top of the under-ream being at a depth of 20 m. Determine the design load of the pile under short-term conditions if the pile is to satisfy ULS to EC7 DA1a.

9.3 The turbines for a new offshore windfarm are to be installed using 1.5 m (outside) diameter steel tubular monopiles which are driven into the seabed. The soil at the site is a soft normally consolidated (NC) clay with $c_u = 1.5z$ (kPa), where z is the depth below the seabed, and saturated unit weight of $16\,kN/m^3$. The turbine superstructure has a total mass of 236 tonnes above the seabed, and the plugged monopile can be assumed have the same mass as the volume of soil it displaces. Determine the minimum length of the monopile required to support the turbine safely under short-term conditions according to Eurocode 7, using Design Approach DA1b.

9.4 For the situation described in Problem 9.3, determine the minimum length of the monopile required to support the turbine safely under long-term conditions according to Eurocode 7, using the same design approach. The clay has $\phi' = 24°$, $\delta' = \phi' - 5°$.

9.5 A series of four static pile load tests has been conducted in a stiff heavily overconsolidated clay as part of the foundation design for a major new airport structure. The bored concrete piles are 39 m long and 1.05 m in diameter. The load penetration data are shown in Figure 9.22. Early on in the design process, only test 1 had been completed. Determine the characteristic resistance of the pile design using only data from test 1. Subsequently, three additional tests were conducted, which failed at slightly lower loads. Using all of the data in Figure 9.22, determine the revised characteristic resistance of the pile design.

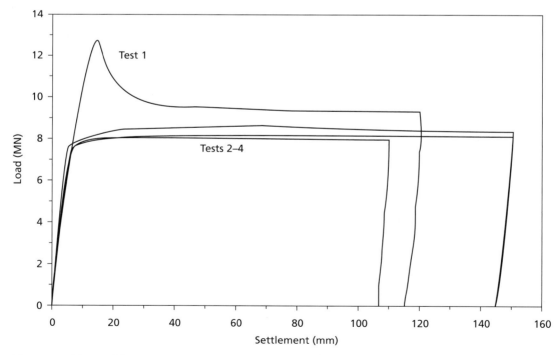

Figure 9.22 Problem 9.5.

9.6 A single closed-ended steel tubular pile of outer diameter 0.4 m and wall thickness 5 mm is driven 8 m into soil consisting of 5 m of loose sand over a thick layer of medium dense sand. The water table (WT) is 1.5 m below the ground surface. The upper layer of sand has unit weight $\gamma = 16$ kN/m^3 above the WT and $\gamma = 17.9$ kN/m^3 below the WT, $\phi' = 31°$, and $c' = 0$. The lower deposit of sand has $\gamma = 20.5$ kN/m^3, $\phi' = 40°$ and $c' = 0$. Along the pile–soil interface, it may be assumed that $K = 1.2 K (1 - \sin \varphi)_0$ and $\delta' = 0.75 \phi'$, and the pile may be assumed to be weightless. Determine the characteristic shaft and base resistances of the pile. Hence, if the base stiffness $E_b = 2 \times 10^5$ kPa and $M_s = 0.001$, determine the settlement at the pile head under an applied vertical load of 600 kN using the hyperbolic method.

9.7 At a particular site, the soil profile consists of a layer of soft clay underlain by a depth of sand. The values of standard penetration resistance at depths of 0.75, 1.50, 2.25, 3.00 and 3.75 m in the sand are 18, 24, 26, 34 and 32, respectively. Nine precast concrete piles, in a square group, are driven through the clay and 2 m into the sand. The piles are 0.25 × 0.25 m in section, and are spaced at 0.75-m centres. The pile group supports a permanent load of 2 MN and an imposed load of 1 MN. Neglecting skin friction in the clay, determine whether the ULS is satisfied according to Eurocode 7 (DA1a).

9.8 A 3 × 3 pile group consists of nine solid circular concrete piles with $E_p = 30$ GPa, 300 mm in diameter and 7.6 m long installed in firm clay, for which $G = 9.6$ MPa, $v = 0.25$, both of which are constant with depth. The piles are spaced at 1.5 m centre-to-centre, and the group is to carry a vertical working load of 4.5 MN. A load test conducted on one of the piles gave 15-mm settlement at 500-kN applied load. Treating the pile cap as flexible, determine the settlement of each pile in the group.

References

Berezantzev, V.G., Khristoforov, V.S. and Golubkov, V.N. (1961) Load bearing capacity and deformation of piled foundations, in *Proceedings of the 5th International Conference on Soil Mechanics and Foundation Engineering, Paris, France*, pp. 11–15.

Burland, J.P. (1993) Closing address, in *Proceedings of Recent Large-Scale Fully Instrumented Pile Tests in Clay, Institute of Civil Engineers, London*, pp. 590–595.

Chin, F.K. (1970) Estimation of the ultimate load of piles not carried to failure, in *Proceedings of the 2nd South East Asia Conference on Soil Engineering*, pp. 81–90.

Clayton, C.R.I. (1995) The Standard Penetration Test (SPT): methods and use, *CIRIA Report 143*, CIRIA, London.

De Mello, V.F.B. (1969) Foundations of buildings on clay. State of the Art report, in *Proceedings of the 7th International Conference on Soil Mechanics and Foundation Engineering*, Vol. 1, pp. 49–136.

De Nicola, A. and Randolph, M.F. (1993). Tensile and compressive shaft capacity of piles in sand. *Journal of the Geotechnical Engineering Division, ASCE*, **119**(12), 1952–1973.

EC7-1 (2004) *Eurocode 7: Geotechnical Design – Part 1: General Rules, BS EN 1997–1:2004*, British Standards Institution, London.

Fellenius, B.H. (1980) The analysis of results from routine pile load tests, *Ground Engineering*, **13**(6), 19–31.

Fleming, W.G.K. (1992). A new method for single pile settlement prediction and analysis, *Géotechnique*, **42**(3), 411–425.

Fleming, W.G.K., Weltman, A., Randolph, M.R. and Elson, K. (2009). *Piling Engineering* (3rd edn), Taylor & Francis, Abingdon, Oxon.

Frost, J.D., DeJong, J.T. and Recalde, M. (2002). Shear failure behavior of granular-continuum interfaces, *Engineering Fracture Mechanics*, **69**(17), 2029–2048.

Jardine, R., Chow, F.C., Overy, R. and Standing, J. (2005). *ICP Design Methods for Driven Piles in Sands and Clays*. Thomas Telford, London.

Karlsrud, K., Hansen, S.B., Dyvik, R. and Kalsnes, B. (1993) NGI's pile tests at Tilbrook and Pentre – review of testing procedures and results, in *Large-scale Pile Tests in Clay*, Thomas Telford, London, pp. 549–583.

Kolk, H.J. and van der Velde, E. (1996) A reliable method to determine friction capacity of piles driven into clays, in *Proceedings of the Offshore Technology Conference, Houston*, Paper OTC 7993.

Kulhawy, F.H. and Phoon, K.-K. (1993) Drilled shaft side resistance in clay soil to rock, in *Design and Performance of Deep Foundations: Piles and Piers in Soil and Rock, ASCE GSP No. 38*, pp. 172–183.

Lee, J.H. and Salgado, R. (1999) Determination of pile base resistance in sands, *Journal of Geotechnical and Geoenvironmental Engineering*, **125**(8), 673–683.

Mylonakis, G. and Gazetas, G. (1998) Settlement and additional internal forces of grouped piles in layered soil, *Géotechnique*, **48**(1), 55–72.

Poulos, H.G. (1989) Pile behaviour – theory and application, *Géotechnique*, **39**(3), 363–416.

Poulos, H.G. and Davis, E.H. (1980) *Pile Foundation Analysis and Design*, John Wiley & Sons, New York, NY.

Randolph, M.F. and Murphy, B.S. (1985) Shaft capacity of driven piles in clay, in *Proceedings of the 17th Annual Offshore Technology Conference, Houston*, Vol. 1, pp. 371–378.

Randolph, M.F. and Wroth, C.P. (1978) Analysis of deformation of vertically loaded piles, *Journal of the Geotechnical Engineering Division, ASCE*, **104**(GT12), 1465–1488.

Rollins, K.M., Clayton, R.J., Mikesell, R.C. and Blaise, B.C. (2005) Drilled shaft side friction in gravelly soils, *Journal of Geotechnical and Geoenvironmental Engineering, ASCE*, **131**(8), 987–1003.

Semple, R.M. and Rigden, W.J. (1984) Shaft capacity of driven piles in clay, in *Proceedings of the Symposium on Analysis and Design of Pile Foundations, San Francisco*, pp. 59–79.

Skempton, A.W. (1959) Cast in-situ bored piles in London Clay, *Géotechnique*, **9**(4) 153–173.

Stas, C.V. and Kulhawy, F.H. (1984) *Critical Evaluation of Design Methods for Foundations under Axial Uplift and Compression Loading*, EPRI Report EL-3771, Palo Alto, CA, 198 pp.

Subba Rao, K.S., Allam, M.M. and Robinson, R.G. (1998) Interfacial friction between sands and solid surfaces, *Proceedings ICE – Geotechnical Engineering*, **131**(2) 75–82.

Uesugi, M. and Kishida, H. (1986) Frictional resistance at yield between dry sand and mild steel, *Soils and Foundations*, **26**(4), 139–149.

Uesugi, M., Kishida, H. and Uchikawa, Y. (1990) Friction between dry sand and concrete under monotonic and repeated loading, *Soils and Foundations*, **30**(1), 115–128.

Weltman, A.J. and Healy, P.R. (1978) Piling in 'Boulder Clay' and other glacial tills, *CIRIA Report PG5*, CIRIA, London.

Whitaker, T. (1957) Experiments with models piles in groups, *Géotechnique*, **7**(4), 147–167.

Further reading

Fleming, W.G.K., Weltman, A., Randolph, M.R. and Elson, K. (2009) *Piling Engineering* (3rd edn), Taylor & Francis, Abingdon, Oxon.

Along with the book by Tomlinson and Woodward (see below), this book is the definitive reference volume for all aspects of piling engineering, including ground investigation requirements, analysis, design, construction and testing.

Frank, R., Bauduin, C., Driscoll, R., Kavvadas, M., Krebs Ovesen, N., Orr, T. and Schuppener, B. (2004) *Designers' Guide to EN 1997–1 Eurocode 7: Geotechnical Design – General Rules*, Thomas Telford, London.

This book provides a guide to limit state design of a range of constructions (including deep foundations) using Eurocode 7 from a designer's perspective and provides a useful companion to the Eurocodes when conducting design. It is easy to read and has plenty of worked examples.

Poulos, H.G. and Davis, E.H. (1980) *Pile Foundation Analysis and Design*. John Wiley & Sons, New York, NY.

A slightly older book, but contains a plethora of analytical solutions for the analysis of piles and other deep foundations under a range of loading conditions, many of which are still referenced by modern design codes.

Randolph, M.F. (2003) Science and empiricism in pile foundation design, *Géotechnique*, **53**(10), 847–875.

A state-of-the-art review of the behaviour of piled foundations and design methodologies, which also highlights key unknowns and limitations that still remain in piling engineering.

Tomlinson, M.J. and Woodward, J. (2008). *Pile Design and Construction Practice* (5th edn), Taylor & Francis, Abingdon, Oxon.

Along with the book by Fleming et al. *(see above), this book is the definitive reference volume for all aspects of piling engineering, including ground investigation requirements, analysis, design, construction and testing.*

For further student and instructor resources for this chapter, please visit the Companion Website at www.routledge.com/cw/craig

Chapter 10

Advanced foundation topics

> **Learning outcomes**
>
> After working through the material in this chapter, you should be able to:
>
> 1 Understand how piles and shallow foundations may be used as elements of larger foundation systems, including pile groups, rafts, piled rafts and deep basements, and be able to design such systems;
> 2 Design shallow and deep foundation elements which are subjected to combined loading (vertical, horizontal, moment), using limit analysis techniques (ULS) and elastic solutions (SLS), within a limit-state design framework.

10.1 Introduction

The foundation elements that have been explored in Chapter 8 (shallow foundations) and Chapter 9 (piles and pile groups) are not always used in isolation. Indeed, Figure 8.1 shows a simple structure in which two adjacent strip footings are used to support the columns on either side of a structure. Design of the individual foundation elements will still require the use of the techniques described in Chapters 8 and 9; however, the load-sharing between the different elements of the system must also be considered, either to derive the loading conditions for design of the individual elements, or for determining any modifications to the element response due to its incorporation in the system. Section 10.2 will consider the performance of a range of different foundation systems under vertical load.

A second key factor is that while most foundations exist mainly to carry vertical load, there are certain applications in which significant horizontal or moment loading may be applied to the foundation, in addition to the vertical load. Indeed, Figure 8.1 showed only vertical actions acting on the foundation system; there may, however, be horizontal actions due to wind-loading acting on the side of the structure. This will clearly add a horizontal action to the foundation; if the columns of the structure bend as a result of the loading, then bending moments may additionally be applied at the column-footing connection and rotation of the structure may transfer additional vertical loading onto the foundation elements. It is therefore necessary to understand how combinations of actions influence the stability (ULS) and movements (SLS) of a foundation.

In most building structures the horizontal and moment actions are small compared to the vertical loads, and do not need to be considered explicitly in design. Examples where there may be significant horizontal and moment actions include:

Applications in geotechnical engineering

- foundations for wind turbines and electricity distribution towers;
- coastal and offshore foundations;
- foundation elements used for anchoring (e.g. to carry the tension of a suspension bridge cable which is applied at an angle to the foundation);
- seismic actions on any foundation.

Section 10.3 will develop the limit analysis techniques from Chapter 8 to consider the stability of shallow foundations under combined loading (ULS), and present new elastic solutions for determining foundation displacements under such loading (SLS). Section 10.4 will then consider deep foundations at both ULS and SLS.

10.2 Foundation systems

Consider a building structure with a large plan area (**footprint**). There are a number of possible foundation designs which might be considered (shown schematically in Figure 10.1), including:

- individual pad footings underneath each column;
- adjacent strip footings, each supporting a row of columns;
- a single shallow foundation extending over the whole footprint (raft);
- individual piles beneath each column (the columns may be cast directly into the top of the piles – this is known as **plunge column** construction);
- a **piled raft**, where the foundation consists of piles connected by a raft acting as the pile cap.

There are advantages and disadvantages to each of the aforementioned foundation systems, though it will generally be possible to produce a foundation which satisfies both ULS and SLS requirements using any of them, selection of the final design being based on cost and practical issues relating to construction. This section will explore some of the issues related to these different foundation systems. In designing foundation systems for large-area or long-span structures (e.g. bridges), the serviceability criteria tend to relate less to the gross settlement of the individual foundation elements and more towards the **differential settlements**, Δ, between different parts of the structure. This is all the more important if the ground properties are very variable over the footprint of the structure. Damage due to differential settlements will therefore also be discussed in this section, including tolerable limits for various classes of structure.

Differential settlement and structural damage

Damage due to differential settlement may be classified as architectural, functional or structural. In the case of framed building structures, settlement damage is usually confined to the cladding and finishes (i.e. architectural damage); such damage is due only to the settlement occurring subsequent to the application of the cladding and finishes. In some cases, structures can be designed and constructed in such a way that a certain degree of movement can be accommodated without damage; in other cases, a certain degree of minor damage may be inevitable if the structure is to be economic. It may be that damage to services, and not to the structure, will be the limiting criterion. Based on observations of damage in buildings, Skempton and MacDonald (1956) proposed limits for maximum differential settlement at which damage could be expected, and related maximum settlement to **angular distortion**, β_d. The angular distortion (also known as relative rotation) between two points under a structure is equal to the differential settlement between the points divided by the distance between them (and is therefore dimensionless), as shown in Figure 10.2. No damage was observed where the angular distortion was less than

Advanced foundation topics

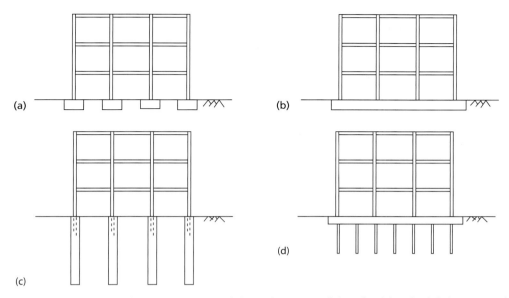

Figure 10.1 Foundation systems: (a) pads/strips, (b) raft, (c) piled (plunge column), (d) piled raft.

1/300. Angular distortion limits were subsequently proposed by Bjerrum (1963) as a general guide for a number of structural situations; these are given in Table 10.1. It is recommended that the safe limit to avoid cracking in the panel walls of framed structures should be 1/500. Structures may also **tilt** as well as angularly distorting (Figure 10.2), and this must also be minimised in design. Tilt limits have been collated by Charles and Skinner (2004) as a general guide for a number of structural situations; these are given in Table 10.2. Angular distortion and tilt can be determined as shown in Figure 10.2 once the settlement at each point of the foundation system is known.

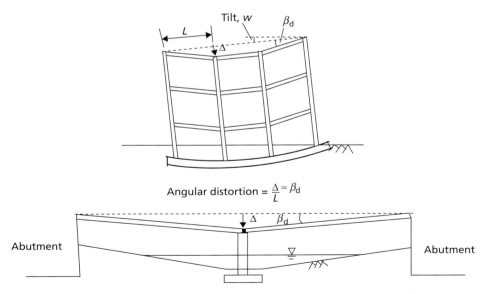

Figure 10.2 Differential settlement, angular distortion and tilt.

Applications in geotechnical engineering

Table 10.1 Angular distortion limits for building structures

1/150	Structural damage of general buildings expected
1/300	Cracking in panel walls expected
	Difficulties with overhead cranes
1/500	Limit for buildings in which cracking is not permissible
1/600	Overstressing of structural frames with diagonals
1/750	Difficulties with machinery sensitive to settlement

Table 10.2 Tilt limits for building structures

1/50	Building is likely to be structurally unsound, requiring urgent re-levelling or demolition
1/100	Floor drainage may not work, stacking of goods dangerous
1/250	Tilting of high-rise buildings (e.g. chimneys and towers) may be visible
1/333	Difficulties with overhead cranes
1/400	Design limit value for low-rise housing
1/500	Maximum limit for monolithic concrete tanks
1/2000	Difficulties with high racking in warehouses
1/5000	Maximum limit for machine foundations (e.g. power station turbines)

The bell tower of the Cathedral of St Mary of the Assumption in Pisa, Italy (colloquially known as the 'Leaning Tower of Pisa') is arguably the most famous example of a structure suffering from excessive tilt (Figure 10.3). Prior to stabilisation works in 1990 the tilt was 5.5° (approx. 1/10). These values are significantly larger than the limiting value of 1/250 (the tilt is certainly noticeable!) and the 1/50 intervention limit from Table 10.2. This occurred due to differential settlement in the highly compressible soil beneath the tower. The foundation system can definitely be classified as having failed to meet serviceability criteria, as construction of the upper levels had to be modified to prevent further tilt and the structure now has a distinctive curve. Stabilisation was achieved by geotechnical engineers carefully removing soil from beneath the higher side of the foundation, causing the tower to rotate back towards vertical such that the tower is currently at 4° to the vertical (a tilt of approximately 1/15).

Based on a probabilistic study of over 300 steel and reinforced concrete framed buildings with a range of foundation types having minor damage to varying degrees, Zhang and Ng (2005) suggested limiting tolerable angular distortions of 1/360. In the same study they also considered over 400 steel and reinforced concrete highway bridge structures of simply supported or continuous construction, finding limiting tolerable angular distortions over an individual span of 1/160 for steel bridges and 1/130 for concrete bridges.

The approach to settlement limits given above is empirical, and is intended to be only a general guide for simple structures. A more fundamental damage criterion is the limiting tensile strain at which visible cracking occurs in a given material. This is particularly important for masonry structures (either those with load-bearing masonry walls or framed structures with masonry infill walls). Burland and Wroth (1975) presented a semi-empirical analysis for such structures, considering the masonry wall as an elastic deep beam in bending. They considered two limiting modes of failure within the masonry, namely bending (where tensile cracking starts at the extreme edge of the wall) and shear (where diagonal tension cracking occurs), as shown in Figure 10.4. Cases where the settlement was highest around the centre of the building (sagging mode) and around the edge of the building (hogging mode) were considered. The tensile strain induced in bending is given by

Advanced foundation topics

Figure 10.3 The 'Leaning Tower of Pisa': an example of excessive tilt (image courtesy of Guy Vanderelst/Photographer's Choice/Getty Images).

$$\varepsilon_{t,\,bend} = \frac{\dfrac{\Delta}{L_w}}{\dfrac{L_w}{6H_w} + \dfrac{H_w}{4L_w}\left(\dfrac{E}{G}\right)} \tag{10.1a}$$

in sagging, and

$$\varepsilon_{t,\,bend} = \frac{\dfrac{\Delta}{L_w}}{\dfrac{L_w}{12H_w} + \dfrac{H_w}{2L_w}\left(\dfrac{E}{G}\right)} \tag{10.1b}$$

in hogging. In Equation 10.1, L_w and H_w are the dimensions of the wall (Figure 10.4), E is the Young's modulus of the masonry and G its shear modulus. From Equation 5.6, $E/G = 2(1+v) \approx 2.6$ for masonry ($v \approx 0.3$). In the shear mode of failure, the limiting tensile strain is given by

$$\varepsilon_{t,\,shear} = \frac{\dfrac{\Delta}{L_w}}{1 + \dfrac{2}{3}\left(\dfrac{H_w}{L_w}\right)^2 \left(\dfrac{E}{G}\right)} \tag{10.2a}$$

Applications in geotechnical engineering

in sagging, and

$$\varepsilon_{t,\text{shear}} = \frac{\dfrac{\Delta}{L_w}}{1 + \dfrac{1}{6}\left(\dfrac{H_w}{L_w}\right)^2\left(\dfrac{E}{G}\right)} \qquad (10.2\text{b})$$

in hogging. Figure 10.4 shows damage observed for a database of structures with load bearing walls in which the differential settlements were in both sagging and hogging modes. Equations 10.1 and 10.2 are also plotted for the case of $\varepsilon_t = 0.075\%$. This value of limiting strain approximately separates the cases of no damage from slight and severe damage. It can be seen that at low values of L_w/H_w shear is the critical mechanism (lowest Δ/L), while for high values of L_w/H_w bending becomes the controlling failure mechanism. The aspect ratios at which there is a change in failure mode are $L_w/H_w = 0.6$ and 1.3 for sagging and hogging, respectively. Also plotted in this figure are the limiting values derived empirically in a separate study by Polshin and Tokar (1957).

Figure 10.5 shows similar comparisons for framed structures which almost always behave in a sagging mode. As before, $\varepsilon_t = 0.075\%$ divides 'no damage' from 'slight damage', while $\varepsilon_t = 0.25\%$ divides 'slight' and 'severe' damage. This latter result is particularly noteworthy, as this limiting strain is approximately equal to the limiting tensile strain at failure for masonry/concrete.

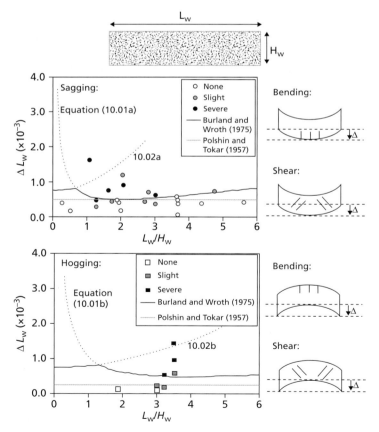

Figure 10.4 Damage to load-bearing masonry walls (after Burland and Wroth, 1975).

Advanced foundation topics

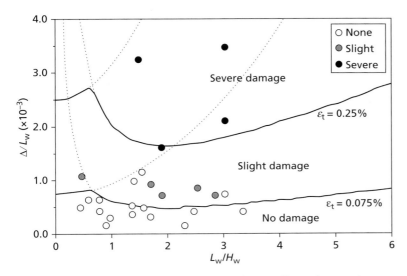

Figure 10.5 Damage to masonry infill walls in framed structures.

Pad/strip foundation systems

When pads or strips are situated close together, they may interact with each other. At ULS this interaction is always beneficial, the capacity of adjacent footings being larger than that of an isolated footing. In coarse-grained soil there will be negligible interaction for $S>3B$ ($\phi'=30°$) to $S>5B$ ($\phi'=40°$), where S is the centre-to-centre spacing, as demonstrated by Stuart (1962) and Kumar and Ghosh (2007). In undrained soil, Gourvenec and Steinepreis (2007) showed that the beneficial interaction reduces more rapidly with spacing, with the footings having the same capacity as isolated footings for $S>B$. Therefore, for ULS design of pad/strip foundation systems, the critical condition that must be checked is the stability of the isolated footings within the system.

At the SLS, the effects of interaction are detrimental, with adjacent footings increasing the overall settlement of the foundation, much as for piles acting in a group (Section 9.7). There is little information on this in the literature, and no interaction factors comparable to those described by Equation 9.33 for piles. However, Lee et al. (2008) recently presented a method based on Schmertmann's strain influence diagram and CPT data (see Section 8.8). In the case of two adjacent identical footings with edge-to-edge spacing S_e,

$$\frac{z_{f0}}{B} = 0.22 e^{1.1\left(0.6-\frac{S_e}{B}\right)} + 2 \tag{10.3}$$

replaces Equation 8.70, with z_{fp} and I_{zp} remaining unaltered (determined using Equations 8.71 and 8.69 respectively). For the case of three identical adjacent footings with equal edge-to-edge spacing between each foundation, z_{f0} for the central foundation is found using

$$\frac{z_{f0}}{B} = 0.50 e^{1.2\left(0.6-\frac{S_e}{B}\right)} + 2 \tag{10.4}$$

with the edge foundations being analysed using Equation 10.3. These values are then used to construct a new strain influence diagram from which the settlement can be determined from CPT data, as demonstrated in Example 8.6.

Applications in geotechnical engineering

Raft foundation systems

As shown in Figure 10.1(b), a raft is a shallow foundation with a large breadth to thickness ratio (B_r/t_r) which extends under the whole structure. Unlike for pad/strip foundation systems considered previously, where differential settlement occurs across the structure due to interaction between the adjacent footings, in a raft the large B_r/t_r makes the raft flexible such that the settlement will increase towards the centre of the raft. This effect can be seen in Table 8.6, where the influence factor is much higher at the centre of a perfectly flexible area than at the corner. This, however, represents an extreme case, as the bending stiffness of the raft will help to reduce the differential settlement. Horikoshi and Randolph (1997) detemined the maximum differential settlement between the centre and edge of rafts of various stiffnesses for use in SLS design, which are shown in Figure 10.6 as a function of the normalised raft–soil stiffness K_{rs} given by

$$K_{rs} = 5.57 \left(\frac{E_r}{E_s}\right)\left(\frac{1-v_s^2}{1-v_r^2}\right)\left(\frac{B_r}{L_r}\right)^{0.5}\left(\frac{t_r}{L_r}\right)^3 \tag{10.5}$$

where parameters with a subscript 'r' relate to the raft and those with a subscript 's' relate to the soil. It should be noted that with $K_{rs}=0$ the normalised differential settlements are the same as those computed using the methods from Section 8.6 (flexible case). It can be seen from Figure 10.6 that as the raft is made stiffer, the differential settlement is greatly reduced. To make the most of this effect **cellular rafts** may be used, which contain voids in the raft to allow the raft to be thicker for the same foundation weight (and therefore cost). By increasing t_r, K_{rs} is increased from Equation 10.5, reducing differential settlements.

Horikoshi and Randolph (1997) also presented results for the maximum bending moment, M_{max}, induced in the raft (at its centre) to allow the foundation to be designed structurally; these were found to be dependent on the aspect ratio when plotted as a function of K_{rs}, and values for $L_r/B_r=1.0$ (square) and $L_r/B_r=10$ ($\approx$ strip) are given in Figure 10.7 as a function of qL_r^2, where q is the average pressure applied to the raft, as in Section 8.6.

For design at the ULS, the raft is treated as a shallow foundation using the approaches detailed in Sections 8.3 and 8.4.

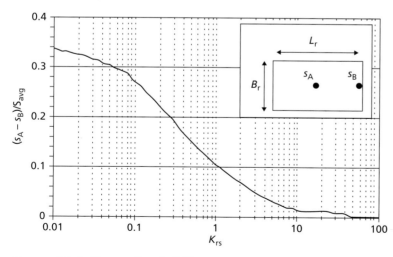

Figure 10.6 Normalised differential settlement in rafts (after Horikoshi and Randolph, 1997).

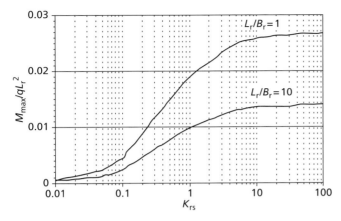

Figure 10.7 Normalised maximum bending moment at the centre of a raft (after Horikoshi and Randolph, 1997).

Example 10.1

A raft foundation is to be used to support a building with a footprint of 20×20 m applying a bearing pressure of 100 kPa on sandy soil with $E' = 30$ MPa and $v' = 0.3$. The raft is to be 1.5 m thick and made of concrete ($E = 30$ GPa, $v = 0.15$). Determine the differential settlement and angular distortion between the centre and edge of the raft:

a treating the raft as flexible;
b accounting for the actual flexural stiffness of the raft.

If sensitive internal load-bearing masonry walls run through the centre of the foundation across its width, determine whether they will be damaged.

Solution

a If the raft is flexible, it may be analysed following the principles outlined in Section 8.6. The settlement at the centre can be found using Equation 8.53 with $I_s = 1.12$ (Table 8.6):

$$s = \frac{100 \cdot 20}{30\,000} \cdot (1 - 0.3^2) \cdot 1.12$$

$$= 0.0679\,\text{m}$$

$$= 68\,\text{mm}$$

To find the settlement at the edge, the footprint is divided into two rectangular areas as shown in Figure 10.8 (each having $L/B = 2$, $B = 10$ m and $I_s = 0.76$), and the principle of superposition employed:

$$s = 2 \cdot \left[\frac{100 \cdot 10}{30\,000} \cdot (1 - 0.3^2) \cdot 0.76 \right]$$

$$= 0.0461\,\text{m}$$

$$= 46\,\text{mm}$$

Therefore, the differential settlement $\Delta = 68 - 46 = 22$ mm, giving an angular distortion of $\beta_d = 0.022/10 = 2.2 \times 10^{-3}$ (approximately 1/500).

b From Equation 10.5,

$$K_{rs} = 5.57 \cdot \left(\frac{30\,000}{30}\right)\left(\frac{1-0.3^2}{1-0.15^2}\right)\left(\frac{20}{20}\right)^{0.5}\left(\frac{1.5}{20}\right)^3$$

$$= 2.19$$

Then, from Figure 10.6, $\Delta/s_{avg} \approx 0.06$. A conservative estimate of the average settlement can be found using Equation 8.53 with $I_s = 0.95$ (flexible, average):

$$s_{avg} = \frac{100 \cdot 20}{30\,000} \cdot (1 - 0.3^2) \cdot 0.95$$

$$= 0.0576 \text{ m}$$

$$= 58 \text{ mm}$$

so that $\Delta = 0.06 \times 58 = 3.5\,\text{mm}$, giving an angular distortion of $\beta_d = 0.0035/10 = 0.35 \times 10^{-3}$ (approximately 1/1400).

The load bearing walls are shown in Figure 10.8. As the central settlement is larger than that at the edge, the walls behave in a sagging mode. From Figure 10.4, the walls would be severely damaged using the prediction in (a) but not in (b), showing the importance of accounting for actual raft stiffness in SLS design.

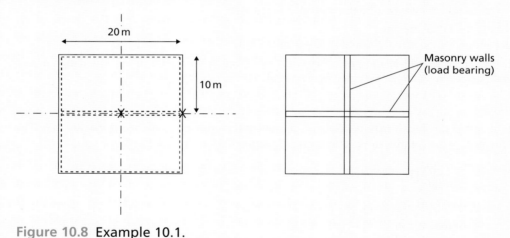

Figure 10.8 Example 10.1.

Piled foundation systems

As described in chapter 9, piles may be used individually (e.g. a single monopile beneath a wind turbine, or to support a plunge column in a building structure) or in groups. Small groups of piles may be used to support point loads from the structure (e.g. from columns) as an alternative to the plunge column approach, and these groups would typically be interconnected by ground beams to provide some additional stiffness against differential settlement. If the loading is uniformly distributed over the footprint of the structure, a raft is often a good option; however, as described above, the differential settlements in a raft can be large if the raft is relatively flexible (low K_{rs}). In some circumstances it may be more effective or economical to reduce the differential settlements of the raft by installing piles underneath, rather than trying to increase K_{rs}. The resulting foundation is known as a piled raft, and differs slightly from a pile group as the plan area of the pile cap (the raft) is very large compared to the length of the piles ($B_r \gg L_p$), such that the raft can play a more important role in the overall behaviour of the foundation.

The overall stiffness of a piled raft (K_f) arises due to a combination of the vertical stiffness of the raft (K_r) and vertical stiffness of the piles acting as a group (K_{pg}). Randolph (1983) proposed that

$$K_f = \frac{K_{pg} + K_r\left(1 - 2\alpha_{rp}\right)}{1 - \alpha_{rp}^2 \left(\dfrac{K_r}{K_{pg}}\right)} \tag{10.6}$$

where

$$\alpha_{rp} \approx 1 - \frac{\ln\left(0.5\dfrac{S}{D_0}\right)}{\ln\left(2\dfrac{L_p}{D_0}\right)} \tag{10.7}$$

for piles uniformly distributed below the raft at centre-to-centre spacing of S. As a result of the pile-to-pile interaction described in Section 9.7, the stiffness of a group of n piles will be lower than the combined stiffness of n individual piles. Rather than have to undertake the laborious calculations detailed in Section 9.7 for large numbers of piles in piled rafts, the overall group stiffness can be approximated using

$$K_{pg} = \eta_g n K_{pile}$$

where η_g is the pile group efficiency. It has been shown by Butterfield and Douglas (1981) that

$$\eta_g \approx n^{-e_p}$$

where e_p is a value typically in the range 0.5–0.6, such that

$$K_{pg} \approx n^{(1-e_p)} K_{pile} \tag{10.8}$$

The stiffness of the individual piles (K_{pile}) is determined using the methods outlined in Section 9.4. For large piled rafts, the raft can be assumed to be very flexible such that K_r can be estimated using the methods outlined in Section 8.6 (using I_s values for average conditions).

The total load carried by the foundation (Q_f) will be distributed between the raft (Q_r) and the group of piles (Q_{pg}) according to

$$\frac{Q_r}{Q_f} = \frac{K_r\left(1 - \alpha_{rp}\right)}{K_{pg} + K_r\left(1 - 2\alpha_{rp}\right)} \tag{10.9}$$

and

$$\frac{Q_{pg}}{Q_f} = 1 - \frac{Q_r}{Q_f} \tag{10.10}$$

Applications in geotechnical engineering

As an example, a square piled raft of dimensions $B_r \times B_r$ is considered resting on uniform soil (G constant with depth). If the piles are considered to be rigid and circular, then, from Equations 9.17a, 9.22 and 9.23,

$$K_{\text{pile}} = \left[2\left(\frac{D_0}{L_p}\right) + \frac{2\pi(1-v)}{\ln\left(2\frac{L_p}{D_0}\right)} \right] \frac{GL_p}{(1-v)}$$

From Equation 8.53, using $I_s = 0.95$ (average settlement, square),

$$K_r = \left[2.1\left(\frac{B_r}{L_p}\right) \right] \frac{GL_p}{(1-v)}$$

These expressions for K_{pile} and K_r are then used with Equation 10.7 and 10.8 in Equation 10.6, from which it can be found that the foundation stiffness is a function of L_p/D_0 (pile slenderness), S/D_0 (normalised pile spacing), n, B_r/L_p (which describes the overall geometry of the piled raft) and $GL_p/(1-v)$. If the foundation stiffness K_f is then expressed in terms of the stiffness of the non-piled raft (i.e. K_f/K_r), then the latter parameter cancels. Figure 10.9(a) plots the ratio K_f/K_r as a function of B_r/L_p for $L_p/D_0 = 25$, $S/D_0 = 5$, $e_p = 0.6$ and $v = 0.5$ (i.e. undrained case). It can be seen that for 'small' pile rafts ($B_r/L_p < 1$) the stiffness of the foundation is dominated by the stiffness of the grouped piles – these cases are essentially

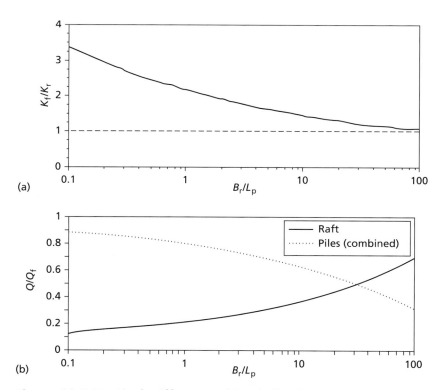

Figure 10.9 Vertical stiffness and load distribution in a square piled raft ($L_p/D_0 = 25$, $S/D_0 = 5$, $v = 0.5$).

Advanced foundation topics

the pile groups that were considered in Section 9.7, where the pile cap has very little influence over the stiffness of the foundation. For 'large' piled rafts ($B_r/L_p > 1$), the stiffness of the foundation approaches the stiffness of the raft, such that the piles are being very inefficient (this would represent a very costly foundation design). Figure 10.9(b) shows the load carried by the raft and the piles calculated using Equations 10.9 and 10.10, where it can be seen that for the small rafts the piles carry the majority of the load, while the reverse is true for the larger piled rafts.

Although it is clear form the forgoing discussion that piled rafts are stiffer than non-piled rafts, and will therefore have lower average (gross) settlements, their main advantage is that the piles reduce the differential settlement within the foundation. As this settlement is highest at the centre of the foundation (Figure 10.6), it is often only necessary to include piles beneath the central area of the raft to compensate (provided that the gross settlement is acceptable without piling the whole raft). This is shown in Figure 10.10. An approximate design procedure then becomes:

1. Determine the differential settlement of the raft (without any piles).
2. Add a small pile group to the centre of the raft, calculate its vertical stiffness (K_{pg}) and determine the load distribution between the raft and the group of piles (Equations 10.9 and 10.10).
3. Determine the settlement of the group of piles under the load Q_{pg} (Section 9.7).
4. Re-evaluate the differential settlement between the exterior of the raft (governed by raft behaviour, settlement from step 1) and the centre of the raft (governed by the pile behaviour, settlement from step 4).
5. If differential settlement is still too high, increase the number of piles at the centre and repeat steps 2–5.

Usually only a small fraction of the raft area around the centre of the foundation needs to be supported by piles for differential settlements to be acceptable. In this way, an efficient/optimised design is achieved where piles are only added in areas where they are most effective.

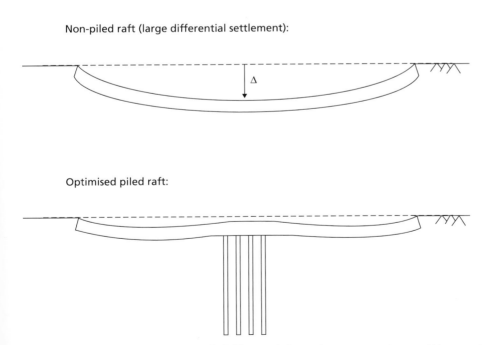

Figure 10.10 Minimisation of differential settlements using settlement-reducing piles.

Applications in geotechnical engineering

Deep basements

Deep basements are used when a structure is to have additional usable space below ground level. They are often used in the foundations of tall buildings where the space may be used to create underground car parks or retail space, or as stations for underground mass-transit systems. They can therefore be very efficient, acting simultaneously as a foundation and as usable space. In design, a deep basement is treated as a deeply embedded cellular raft; if required, piles may also be added below the basement to provide additional load-carrying capacity at ULS or to reduce gross/differential settlement at SLS. The walls of a basement must also be designed to resist the horizontal earth pressures from the surrounding soil at both ULS and SLS; this is addressed in Chapter 11. As deep basements normally include a significant amount of empty space, the total weight of the basement may be less than the weight of the soil that was removed to create it. In saturated ground with a high water table, the basement can then become buoyant – a failure mode known as **uplift**.

Verification of the ULS relating to uplift involves ensuring that the total vertical force (mainly due to self weight) acting downwards on the basement is greater than the uplift force due to the pore water pressure (U), i.e.

$$\sum Q \geq U \qquad (10.11)$$

In all of the previous ULS design calculations in this and the previous two chapters, geotechnical (GEO) or structural (STR) modes of failure have been considered, for each of which it has been possible to define a resistance (R_k). In contrast, uplift (UPL) is an example of a failure mode relating to a loss of overall equilibrium between two sets of actions (Q and U) requiring a different approach to partial factoring. In EC7, loads that are favourable act to stabilise the construction, while unfavourable loads act to destabilise the construction. Table 10.3 gives the normative values recommended by EC7 for the ULS of uplift.

If Equation 10.11 cannot be satisfied under the weight of the structure and basement, tension piles (Section 9.5) or ground anchors (Section 11.8) may be incorporated to provide an additional permanent stabilising action. It should be noted that interface friction between the walls of the basement and the soil is usually ignored (which is conservative).

Table 10.3 Partial factors on actions for verification of ULS against uplift according to EC7

Action (Q, U)	Symbol	Value
Permanent unfavourable action	$\gamma_{A,dst}$	1.00
Variable unfavourable action	$\gamma_{A,dst}$	1.50
Permanent favourable action	$\gamma_{A,stb}$	0.90

Example 10.2

An underground bookstore is to be constructed as part of a library extension. The store is a box-type structure with outer dimensions of 20×20 m in plan and 10 m deep. The wall thickness is 0.5 m throughout, and the concrete has a unit weight of 24 kN/m³. The upper surface of the store roof is 2 m below ground level, and the surrounding soil has $\gamma = 19$ kN/m³. Determine the minimum depth of the water table below the ground surface at which the ULS (uplift) is satisfied:

a immediately after construction when the store is empty;
b during service, when half of the store's internal volume is filled with books having a unit weight of $3.5\,\text{kN/m}^3$.

If the water table can rise to the ground surface, determine the additional restraining force that tension piles would need to provide to satisfy the ULS.

Solution

a The volume of the concrete in the bookstore is $(20 \times 20 \times 10) - (19 \times 19 \times 9) = 751\,\text{m}^3$. The vertical load due to the concrete is therefore $751 \times 24 = 18\,024\,\text{kN} = 18.0\,\text{MN}$. The weight of the soil above the roof of the store is $(20 \times 20 \times 2) \times 19 = 15\,200\,\text{kN} = 15.2\,\text{MN}$. If the water table is at a depth z below the ground surface, then the pore water pressure acting on the base of the store is $u = \gamma_w(10+2-z) = 117.7 - 9.8z\,\text{kPa}$, giving an uplift force of $U = 47.1 - 3.9z\,\text{MN}$.

The vertical actions from the concrete and soil weights are permanent favourable actions (hence factored by $\gamma_{A,stb} = 0.90$ from Table 10.3), while the uplift force U is a permanent unfavourable action ($\gamma_{A,dst} = 1.00$). Therefore, applying Equation 10.11 gives:

$$\sum Q \geq U$$
$$0.90 \cdot (18.0 + 15.2) \geq 1.00 \cdot (47.1 - 3.9z)$$
$$29.9 \geq 47.1 - 3.9z$$
$$z \geq 4.4\,\text{m}$$

b The internal volume of the store is $(19 \times 19 \times 9) = 3249\,\text{m}^3$, so that the additional vertical force due to the books is $0.5 \times 3249 \times 3.5 = 5686\,\text{kN} = 5.7\,\text{MN}$. Equation 10.11 then becomes:

$$\sum Q \geq U$$
$$0.90 \cdot (18.0 + 15.2 + 5.7) \geq 1.00 \cdot (47.1 - 3.9z)$$
$$35.0 \geq 47.1 - 3.9z$$
$$z \geq 3.1\,\text{m}$$

From the answers to (a) and (b), the immediate condition after construction when the applied loads are lower is more critical (the reverse of bearing failure). Defining the additional restraining force by T, if the water table rises to the surface, $z = 0$ and Equation 10.11 becomes:

$$\sum Q \geq U$$
$$0.90 \cdot (18.0 + 15.2 + T) \geq 1.00 \cdot (47.1)$$
$$29.9 + 0.9T \geq 47.1$$
$$T \geq 19.1\,\text{MN}$$

Applications in geotechnical engineering

10.3 Shallow foundations under combined loading

Most foundations are subjected to a horizontal component of loading (H) in addition to the vertical actions (V) considered in Chapter 8. If this is relatively small in relation to the vertical component, it need not be considered in design – e.g. the typical wind loading on a building structure can normally be carried safely by a foundation designed satisfactorily to carry the vertical actions. However, if H (and/or any applied moment, M) is relatively large (e.g. a tall building under hurricane wind loading), the overall stability of the foundation under the combination of actions must be verified.

For a foundation loaded by actions V, H and M (usually abbreviated to V–H–M), the following limit states must be met:

> **ULS-1** The resultant vertical action on the foundation V must not exceed the bearing resistance of the supporting soil (i.e. Equation 8.75 must be satisfied);
> **ULS-2** Sliding must not occur between the base of the wall and the underlying soil due to the resultant lateral action, H;
> **ULS-3** Overturning of the wall must not occur due to the resultant moment action, M;
> **SLS** The resulting foundation movements due to any settlement, horizontal displacement and rotation must not cause undue distress or loss-of-function in the supported structure.

Limit analysis techniques may be applied to determine stability at the ULS and elastic solutions may be used to determine foundation movements at the SLS, similar to those undertaken in Chapter 8 for pure vertical loading.

Foundation stability from limit analysis (ULS)

Before the general case of V–H–M loading is considered, the stability of foundations under simpler combinations of V–H loading will be considered initially to introduce the key concepts. This builds on the lower bound limit analysis techniques from Chapter 8.

The addition of horizontal load H to a conventional vertical loading problem ($q_f = V/A_f$) will induce an additional shear stress $\tau_f = H/A_f$ at the soil footing interface as shown in Figure 10.11(a). It is assumed in this analysis that the footing is perfectly rough. This will serve to rotate the major principal stress direction in zone 1. For an undrained material, this rotation will be $\theta = \Delta/2$ from the vertical (Figure 10.11(b)). The stress conditions in zone 2 are unchanged from those shown in Figure 8.9(a). Therefore, the overall rotation of principal stresses across the fan zone is now $\theta_{fan} = \pi/2 - \Delta/2$, such that from Equation 8.15

$$s_1 - s_2 = c_u \left(\pi - \Delta \right) \tag{10.12}$$

From Figure 10.11(b),

$$\sin \Delta = \frac{\tau_f}{c_u} = \frac{H}{A_f c_u} \tag{10.13}$$

In zone 2, $s_2 = \sigma_q + \gamma z + c_u$ as in Section 8.3 (unchanged), while in zone 1, $s_1 = q_f + \gamma z - c_u \cos \Delta$ from Figure 10.11(b). Substituting these relationships into Equation 10.12 and rearranging gives

$$q_f = \frac{V}{A_f} = c_u \left(1 + \pi - \Delta + \cos \Delta \right) + \sigma_q$$
$$= c_u N_c + \sigma_q \tag{10.14}$$

Advanced foundation topics

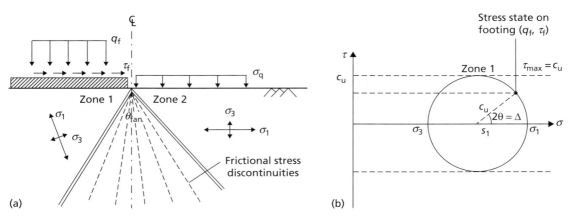

Figure 10.11 (a) Stress state for V–H loading, undrained soil, (b) Mohr circle in zone 1.

For all possible values of H ($0 \leq H/A_f c_u \leq 1$), Δ can be found from Equation 10.13 and $V/A_f c_u$ (= bearing capacity factor N_c) found from Equation 10.14. These are plotted in Figure 10.12 for the case of no surcharge ($\sigma_q = 0$). When $H/A_f c_u = 0$ (i.e. purely vertical load), $\Delta = 0$ and $V/A_f c_u = 2 + \pi$ (see Equation 8.16); when $H/A_f c_u = 1$, the shear stress $\tau_f = c_u$ and the footing will slide horizontally, irrespective of the value of V. The resulting curve represents the **yield surface** for the foundation under V–H loading. Combinations of V and H which lie within the yield surface will be stable, while those lying outside the yield surface will be unstable (i.e. result in plastic collapse). If $V \gg H$, collapse will be predominantly in bearing (vertical, ULS-1); if $V \ll H$, collapse will be predominantly by sliding (horizontal translation, ULS-2). For intermediate states, a combined mechanism resulting in significant vertical and horizontal components will occur.

In Eurocode 7 and many other design specifications worldwide, a different approach is adopted. This consists of applying an additional **inclination factor** i_c to the standard bearing capacity equation (Equation 8.17), where

$$i_c = \frac{1}{2}\left(1 + \sqrt{1 - \frac{H}{A_f c_u}}\right) \tag{10.15}$$

For the case of no surcharge and a strip footing ($s_c = 1$), from Equation 8.17

$$q_f = \frac{V}{A_f} = i_c (2 + \pi) c_u$$

$$\frac{V}{A_f c_u} = \frac{(2 + \pi)}{2}\left(1 + \sqrt{1 - \frac{H}{A_f c_u}}\right) \tag{10.16}$$

Equation 10.16 is also plotted in Figure 10.12 where it is practically indistinguishable from the rigorous plasticity solution represented by Equation 10.14.

Gourvenec (2007) presented a yield surface for the general case of V–H–M loading on undrained soil, where

$$\left[\frac{1.29\frac{H}{R}}{0.25 - \left(\frac{V}{R} - 0.5\right)^2}\right]^2 + \left[\frac{2.01\frac{M}{BR}}{\frac{V}{R} - \left(\frac{V}{R}\right)^2}\right]^2 = 1 \tag{10.17}$$

Applications in geotechnical engineering

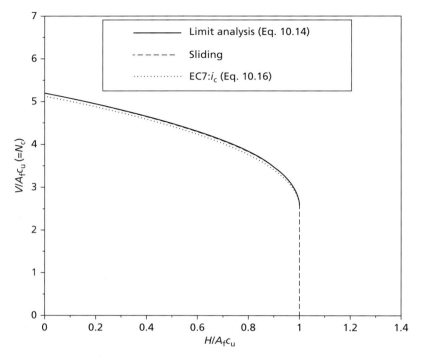

Figure 10.12 Yield surface for a strip foundation on undrained soil under V–H loading.

In Equation 10.17, R is the vertical resistance of the foundation under pure vertical loading, ($H=M=0$), as found in Chapter 8, and B is the breadth of the foundation. As $R=(2+\pi)A_f c_u$, Equations 10.14 and 10.16 can be rewritten in terms of V/R and H/R by substituting for $A_f c_u$. Figure 10.13(a) compares the lower bound solution, Eurocode 7 approach and the full yield surface (Equation 10.17) for the case of V–H loading ($M=0$). Using this alternative normalisation, it can be seen that the maximum horizontal action which can be sustained is approximately 19% of the vertical resistance.

When $M \neq 0$, the yield surface becomes a three-dimensional surface (a function of V, H and M). Figure 10.13(b) shows contours of V/R for combinations of H and M under general loading for use in ULS design. The presence of moments allows for overturning (rotation) of the foundation when $M \gg V, H$. The yield surface represented in Figure 10.13(b) assumes that tension cannot be sustained along the soil–footing interface, i.e. the foundation will **uplift** if the overturning effect is strong. Provided that the combination of V, H and M lies within the yield surface, the foundation will not fail in bearing, sliding or overturning such that ULS-1–ULS-3 will all be satisfied (and can be checked simultaneously), showing the power of the yield surface concept.

A lower bound analysis may also be conducted for a foundation on a weightless drained material (Figure 10.14(a)). Here, $H/V = \tau_f/q_f = \tan\beta$. From Figure 10.14(b), the rotation of the major principal stress direction in zone 1 is $\theta = (\Delta + \beta)/2$ from the vertical. The stress conditions in zone 2 are unchanged from those shown in Figure 8.16(a). Therefore, the overall rotation of principal stresses across the fan zone is now $\theta_{fan} = \pi/2 - (\Delta + \beta)/2$, such that from Equation 8.30

$$\frac{s'_1}{s'_2} = e^{(\pi - \Delta - \beta)\tan\phi'} \tag{10.18}$$

Advanced foundation topics

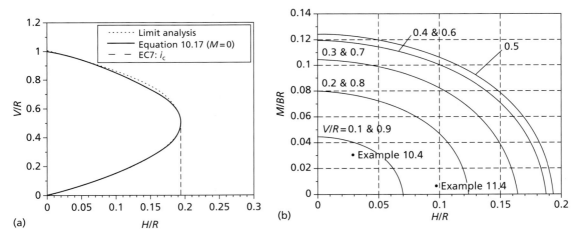

Figure 10.13 Yield surfaces for a strip foundation on undrained soil under (a) V–H loading; (b) V–H–M loading.

From Figure 10.14(b),

$$\sin \Delta = \frac{\sin \beta}{\sin \phi'} \qquad (10.19)$$

In zone 2, $s'_2 = \sigma_q + s'_2 \sin \phi'$ as in Section 8.4 (unchanged), while in zone 1, $s'_1 = q_f - s'_1 \sin \phi' \cos(\Delta + \beta)$ from Figure 10.14(b). Substituting these relationships into Equation 10.18 and rearranging gives

$$q_f = \frac{V}{A_f} = \left[\frac{1 + \sin \phi' \cos(\Delta + \beta)}{1 - \sin \phi'}\right] e^{(\pi - \Delta - \beta) \tan \phi'} \cdot \sigma_q$$

$$= N_q \sigma_q \qquad (10.20)$$

Values of β may be found for any combination of V and H, from which Δ may be found from Equation 10.19 and values of N_q from Equation 10.20. These are plotted in Figure 10.15. If the footing is perfectly rough ($\delta' = \phi'$) then sliding will occur if $H/V \geq \tan \phi'$.

As for the undrained case, Eurocode 7 and many other design specifications worldwide adopt an alternative approach. This consists of applying an additional inclination factor i_q to the standard bearing capacity equation (Equation 8.31), where

$$i_q = \left(1 - \frac{H}{V}\right)^2 \qquad (10.21)$$

For the case of a strip footing on cohesionless soil ($c' = 0$), from Equation 8.31

$$N_q = \frac{q_f}{\sigma_q} = i_q \left[\frac{(1 + \sin \phi')}{(1 - \sin \phi')} e^{\pi \tan \phi'}\right] \qquad (10.22)$$

Equation 10.22 is also plotted in Figure 10.15 where it is practically indistinguishable from the rigorous plasticity solution represented by Equation 10.20, though it should be noted that Equation 10.21 is only valid when $H/V \leq \tan \phi'$ to account for sliding.

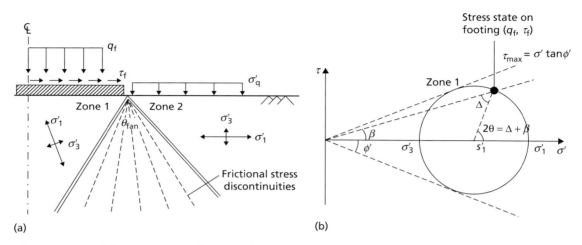

Figure 10.14 (a) Stress state for V–H loading, drained soil, (b) Mohr circle in zone 1.

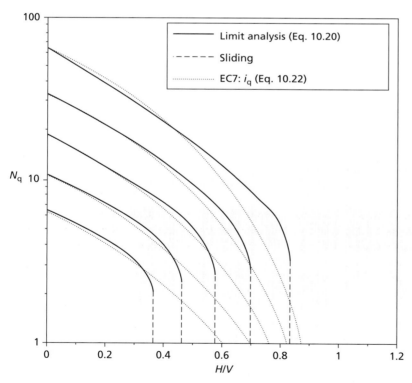

Figure 10.15 N_q for a strip foundation on drained soil under V–H loading.

Advanced foundation topics

Butterfield and Gottardi (1994) presented a yield surface for the general case of V–H–M loading on drained soil, where

$$\frac{3.70\left(\frac{H}{R}\right)^2 - 2.42\left(\frac{H}{R}\right)\left(\frac{M}{BR}\right) + 8.16\left(\frac{M}{BR}\right)^2}{\left[\frac{V}{R}\left(1-\frac{V}{R}\right)\right]^2} = 1 \qquad (10.23)$$

Recognising that V/R at any value of H/V is the value of N_q from Equation 10.22 divided by the value of N_q at $H/V=0$ (Equation 8.31), and that $H/R=(H/V) \times (V/R)$, Equations 10.20 and 10.22 can be expressed in terms of V/R and H/R for the case of $M=0$. Figure 10.16(a) compares the lower bound solution, Eurocode 7 approach and the full yield surface for the case of V–H loading ($M=0$). Using this alternative normalisation, it can be seen that the maximum horizontal thrust which can be sustained is approximately 13% of the vertical resistance, lower than for the undrained case.

When $M \ne 0$ the yield surface becomes a three-dimensional surface, as before. Figure 10.16(b) shows contours of V/R for combinations of H and M under general loading for use in ULS design from Equation 10.23. For the drained case, Eurocode 7 is also able to account for moment effects through the use of a reduced footing width $B'=B-2e_m$ in Equation 8.31, where e_m is the eccentricity of the vertical load from the centre of the footing which creates a moment of magnitude M, i.e. $e_m=M/V$. For a strip footing, the footing–soil contact area is therefore B' per metre length under V–H–M loading and

$$q_f = \frac{V}{B'} = i_q N_q \sigma_q \qquad (10.24)$$

Under pure vertical loading V (where $V=R$ at bearing capacity failure),

$$q_f = \frac{R}{B} = N_q \sigma_q \qquad (10.25)$$

Dividing Equation 10.24 by 10.25 and substituting for i_q (Equation 10.21), B' and e_m gives

$$\frac{V}{R} = i_q\left(\frac{B'}{B}\right) = \left(1-\frac{H}{V}\right)^2\left(1-\frac{2M}{BV}\right) \qquad (10.26)$$

Equation 10.26 may be plotted out as a yield surface for comparison with Equation 10.23, from which it can be seen that for low values of $V/R \le 0.3$ Equation 10.26 will provide an unconservative estimation of foundation stability. For all values of V/R, Equation 10.26 also overestimates the capacity at low H/R (i.e. where overturning is the predominant failure mechanism), making it less suitable for checking ULS-3.

Foundation displacement from elastic solutions (SLS)

If the combination of actions V–H–M applied to a shallow foundation lies within the yield surface, it is still necessary to check that the foundation displacements under the applied actions are tolerable (SLS). Whilst in Chapter 8 settlement s (vertical displacement) was the action effect associated with the action V, under multi-axial loading the footing may additionally displace horizontally by h (under the action of H) and rotate by θ (under the action of M). The relationship between the action and the action effect in each case may be related by elastic stiffness $K_v = V/s$ (vertical stiffness), $K_h = H/h$ (horizontal stiffness) and $K_\theta = M/\theta$ (rotational stiffness).

The elastic solution for vertical settlement given as Equation 8.53 may be re-expressed as a vertical stiffness K_v, by recognising that the bearing pressure $q=V/BL$. It is common to express foundation

Applications in geotechnical engineering

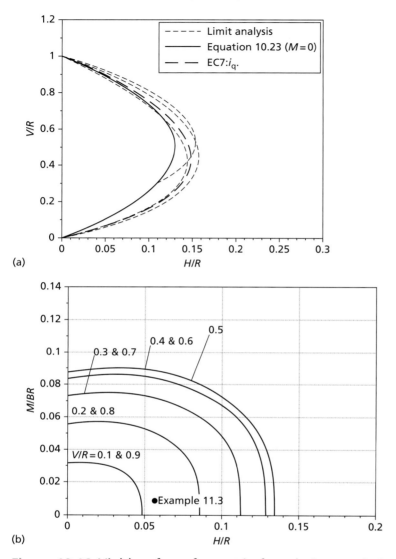

Figure 10.16 Yield surfaces for a strip foundation on drained soil under (a) V–H loading, (b) V–H–M loading.

stiffness in terms of shear modulus G, rather than Young's Modulus E, so that consideration of undrained or drained conditions only involves changing the value of v. By using Equation 5.6, then, Equation 8.53 may be rearranged as:

$$K_v = \frac{V}{s} = \left(\frac{2L}{I_s}\right)\frac{G}{(1-v)} \tag{10.27}$$

The horizontal stiffness of a shallow foundation was derived by Barkan (1962) as

$$K_h = \frac{H}{h} = 2G(1+v)F_h\sqrt{BL} \tag{10.28}$$

Advanced foundation topics

where F_h is a function of L/B as shown in Figure 10.17. The rotational stiffness of a shallow foundation was derived by Gorbunov-Possadov and Serebrajanyi (1961) as

$$K_\theta = \frac{M}{\theta} = \frac{G}{1-v} F_\theta BL^2 \qquad (10.29)$$

where F_θ is also a function of L/B as shown in Figure 10.17. The three foregoing equations 10.27–10.29 assume that there is no coupling between the different terms.

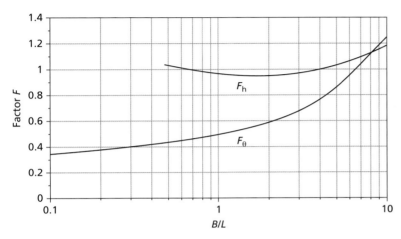

Figure 10.17 Non-dimensional factors F_h and F_θ for foundation stiffness determination.

Example 10.3

An offshore wind turbine is to be installed on a square gravity-base foundation as shown in Figure 10.18 (a gravity base is a large shallow foundation). The subsoil is clay with $c_u = 20\,\text{kPa}$ (constant with depth). The weight of the turbine structure is 2.6 MN, and the gravity base is neutrally buoyant (i.e. it is hollow, with its weight balancing the resultant uplift force form the water pressure). Determine the required width of the foundation to satisfy ULS to DA1a if the horizontal (environmental) loading is 20% of the vertical load and acts at the level of the seabed.

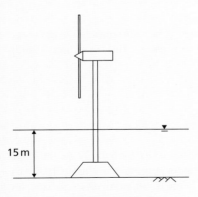
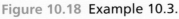
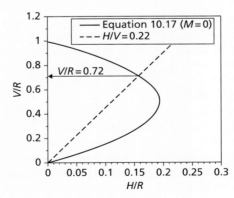

Figure 10.18 Example 10.3.

387

Applications in geotechnical engineering

Solution

If the horizontal loading is a variable unfavourable action ($\gamma_A=1.50$) which acts rapidly (undrained) and the vertical load a permanent unfavourable action ($\gamma_A=1.35$), then the ratio of the design loads is

$$\frac{H}{V} = 0.2 \cdot \frac{1.50}{1.35} = 0.22$$

This plots as a straight line with gradient $1/0.22=4.5$ on Figure 10.13(a), as shown in Figure 10.18. This intersects the yield surface at $V/R=0.72$. The design load $V=1.35 \times 2.6=3.51$ MN (neutral buoyancy means that the gravity base does not apply any net load). The vertical resistance R is given by Equation 8.17 as

$$\frac{R}{B^2} = \frac{s_c N_c \left(\dfrac{c_u}{\gamma_{cu}}\right)}{\gamma_{Rv}}$$

where $s_c=1.2$ (Equation 8.19), $N_c=5.14$, $c_u=20$ kPa and $\gamma_{cu}=\gamma_{Rv}=1.00$ for DA1a. Substituting these values gives $R=123.4B^2$ (kN). Then, for stability,

$$\frac{V}{R} \leq 0.72 \quad \therefore 3510 \leq 0.72 \cdot 123.4B^2$$

$$B \geq 6.29 \text{ m}$$

Example 10.4

The gravity base for the wind turbine in Example 10.3 was subsequently constructed 15×15 m square. If the horizontal action now acts at sea level, determine whether the foundation satisfies the ULS to EC7 DA1a. If the seabed soil has $E_u/c_u=500$, determine the displacements of the foundation under the applied actions.

Solution

At ULS, the vertical design load $V=3.51$ MN as before. The horizontal design load $H=1.50 \times 0.2 \times 2.6=0.78$ MN. As this acts at sea level (15 m above the founding plane), $M=15 \times H=11.7$ MNm. The design resistance is now

$$R = \frac{s_c N_c \left(\dfrac{c_u}{\gamma_{cu}}\right)}{\gamma_{Rv}} B^2 = \frac{1.2 \cdot 5.14 \cdot \left(\dfrac{20}{1.00}\right)}{1.00} \cdot 15^2 = 27.8 \text{ MN}$$

The normalised parameters for use with Figure 10.13(b) are then: $H/R=0.78/27.8=0.03$, $M/BR=11.7/(15 \times 27.8) =0.03$. The point defined by these parameters is plotted in Figure 10.13(b), which shows that $0.07 < V/R < 0.93$. As $V/R=3.51/27.8=0.13$, the foundation does satisfy the ULS.

The undrained Young's Modulus $E_u = 500 \times 20 = 10\,\text{MPa}$, so that, from Equation 5.6,

$$G = \frac{E_u}{2(1+v_u)} = \frac{10}{2(1+0.5)} = 3.3\,\text{MPa}$$

Treating the gravity base as rigid, $I_s = 0.82$ from Table 8.6 (square) so that, from Equation 10.27,

$$\frac{V}{s} = \left(\frac{2L}{I_s}\right)\frac{G}{(1-v)} \quad \therefore s = \frac{2.60}{\left(\frac{2\cdot 15}{0.82}\right)\frac{3.3}{(1-0.5)}} = 0.0107\,\text{m}$$

noting that V is now the characteristic load for SLS calculations. From Figure 10.17, $F_h = 0.95$ and $F_\theta = 0.5$, so that from Equations 10.28 and 10.29,

$$\frac{H}{h} = 2G(1+v)F_h\sqrt{BL} \quad \therefore h = \frac{(0.2 \cdot 2.6)}{2 \cdot 3.3 \cdot 1.5 \cdot 0.95 \cdot \sqrt{15 \cdot 15}} = 0.0037\,\text{m}$$

$$\frac{M}{\theta} = \frac{G}{1-v}F_\theta BL^2 \quad \therefore \theta = \frac{(0.2 \cdot 2.6 \cdot 15)}{\left(\frac{3.3}{1-0.5}\right)\cdot 0.5 \cdot 15^3} = 7.0\times 10^{-4}\,\text{radians}$$

The gravity base will therefore displace vertically by 10.7 mm and horizontally by 3.7 mm, and rotate by 0.04°, under the applied actions.

10.4 Deep foundations under combined loading

Piles resist lateral actions due to the resistance of the adjacent soil, lateral stresses in the soil increasing in front of the pile and decreasing behind as load is applied. As with shallow foundations (Section 10.3), if the horizontal action is relatively small in relation to the vertical action, it need not be considered explicitly in design. If H (and/or M) is relatively large, the lateral resistance of the pile should be determined. Very large lateral actions may require the installation of inclined (or **raked**) piles, which can make use of some of the available axial resistance to carry a proportion of the horizontal action. Lateral loading on piles can also be induced by soil movement, e.g. lateral movement of the soil below an embankment behind a piled bridge abutment.

As under lateral loading the actions and soil reactions applied to the pile are horizontal, the pile acts like a beam in bending. The ULS which must be satisfied therefore consists of ensuring that the pile does not fail structurally by forming a plastic hinge. This is not normally accomplished by changing the length of the pile (this being selected based on vertical considerations as described in Chapter 9) but by selecting an appropriate section size to give the required moment capacity (in concrete piles, this can include the detailing of the reinforcing steel). At the SLS, movement of the pile head (as a result of settlement, horizontal displacement and rotation) must be within tolerable limits, as for shallow foundations.

Pile capacity under horizontal and moment loading

The maximum lateral loading which can be applied to a pile will occur when the soil fails (yields) around the pile. Close to the surface (at a depth not exceeding the diameter/width of the pile), failure in the soil is assumed to be analogous to the formation of a passive wedge in front of a retaining wall (see Chapter 11), the soil surface being pushed upwards. In an undrained soil of uniform undrained shear

Applications in geotechnical engineering

strength c_u, a generalised relationship between the ultimate or limiting value of lateral pressure (p_1) acting on the pile and depth has been proposed by Fleming *et al.* (2009) in which p_1 increases linearly from a value of $2c_u$ at the surface to $9c_u$ at a depth of $3D_0$, where D_0 is the (outside) diameter or width of the pile, and remains at a constant value of $9c_u$ at depths below $3D_0$. In a drained cohesionless soil, p_1 increases linearly with depth and can be approximated to

$$p_1 = K_p^2 \sigma_v'$$

ignoring three-dimensional effects, where K_p is the passive earth pressure coefficient, the determination of which is described in Chapter 11, and σ_v' is the effective vertical stress at the depth in question.

The mode of failure of a pile under lateral load depends on its length and whether or not it is restrained at its head by a pile cap. A relatively short, rigid, unrestrained pile will rotate about a point B near the bottom, as shown in Figure 10.19(a). In the case of a relatively long flexible pile, a plastic hinge will develop at some point D along the length of the pile, as shown in Figure 10.19(b), and only above this point will there be significant displacement of the pile and soil.

The horizontal force which would result in failure within the soil is denoted by H_{max}. The moment acting on the pile is expressed as a ratio of the horizontal load, i.e. by the parameter M/H. For a 'short' unrestrained pile (Figure 10.19(a)), the depth of the point of rotation is written as z_B. The limit forces on the front of the pile above the point of rotation and on the back of the pile below the point of rotation are denoted by P_{AB} and P_{BC}, respectively. The limit forces are calculated by integrating the limit pressure p_1 over the lengths AB and BC of the pile below the ground surface, respectively. These forces act at depths of h_{AB} and h_{BC}, respectively. Then, for horizontal equilibrium

$$H_{max} = P_{AB} - P_{BC} \tag{10.30a}$$

and for moment equilibrium

$$H_{max}\left(z_B + \frac{M}{H}\right) = P_{AB}\left(z_B - h_{AB}\right) + P_{BC}\left(h_{BC} - z_B\right) \tag{10.30b}$$

The unknown quantities are H_{max} and z_B, M/H being the known ratio of the applied horizontal and moment actions.

For a 'long' unrestrained pile (Figure 10.19(b)) a plastic hinge develops at depth z_D, and at this point the bending moment will be a maximum, of value M_p (the moment capacity of the pile), and the shear force will be zero. Only the forces above the hinge, i.e. over the length AD, need be considered. Then, for horizontal equilibrium

$$H_{max} = P_{AD} \tag{10.31a}$$

and for moment equilibrium

$$H_{max}\left[\frac{M}{H} + z_D - (z_D - h_{AD})\right] = M_p$$

$$H_{max}\left[\frac{M}{H} + h_{AD}\right] = M_p \tag{10.31b}$$

Solving Equations 10.30 or 10.31 can be quite awkward as the value of p_1 used to find the limit forces in the equations depends on z_B, which is unknown and therefore requires an iterative approach. As an

Advanced foundation topics

alternative method of solution, Fleming et al. (2009) present design charts for various values of L_p/D_0, M_p and M/H, for both undrained and drained soil conditions, which are shown in Figures 10.20 and 10.21, respectively. Whether the 'short' or 'long' pile failure mechanism occurs depends on which mechanism gives the lower lateral capacity H_{max} (as this will occur first). This has been analysed for a range of pile properties and dimensions to produce the charts shown in Figure 10.22(a) for undrained conditions and 10.22(b) for drained conditions. This figure can be used to determine which the critical failure mode is, and therefore which of the plots in Figures 10.20 and 10.21 should be used to find the lateral capacity.

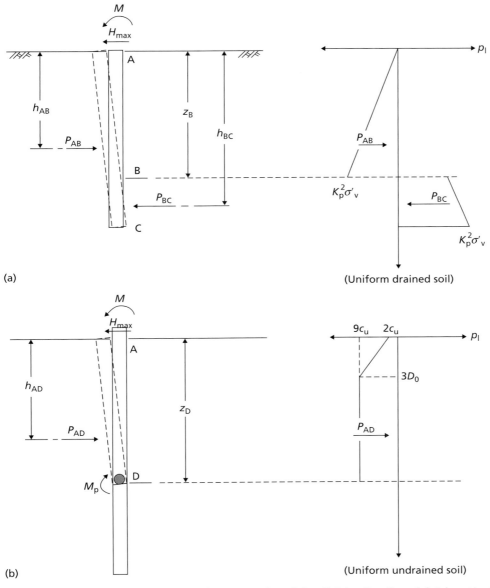

Figure 10.19 Lateral loading of unrestrained (individual) piles: (a) 'short' pile, (b) 'long' pile.

Applications in geotechnical engineering

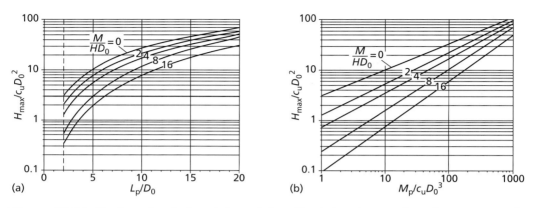

Figure 10.20 Design charts for determining the lateral capacity of an unrestrained pile under undrained conditions: (a) 'short' pile, (b) 'long' pile.

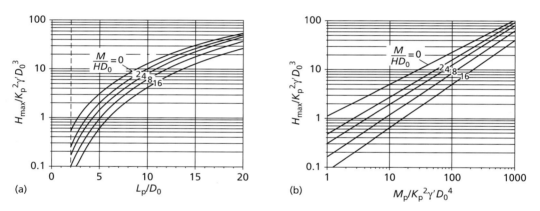

Figure 10.21 Design charts for determining the lateral capacity of an unrestrained pile under drained conditions: (a) 'short' pile, (b) 'long' pile.

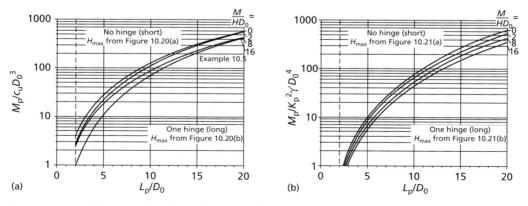

Figure 10.22 Determination of critical failure mode, unrestrained piles: (a) undrained conditions, (b) drained conditions.

Advanced foundation topics

For piles surmounted by a pile cap which restrains the pile heads from rotating, there are three possible modes of failure. A short rigid pile will undergo translational displacement as shown in Figure 10.23(a). H_{max} in this case is found by integrating the limit pressure p_l over the full length of the pile.

A restrained pile of intermediate length will develop a plastic hinge at cap level and rotate about a point near the bottom of the pile, as illustrated in Figure 10.23(b). The pile–soil interaction in this case is the same as for the short unrestrained pile (Figure 10.19(a)), though there is an additional moment of M_p acting at the top of the pile, so that Equation 10.30b becomes

$$H_{max}\left(z_B + \frac{M}{H}\right) = P_{AB}(z_B - h_{AB}) + P_{BC}(h_{BC} - z_B) + M_p \tag{10.32}$$

The equation of horizontal equilibrium is unaltered.

A long restrained pile will develop plastic hinges at cap level and at a point along the length of the pile as indicated in Figure 10.23(c). The pile–soil interaction in this case is the same as for the long unrestrained pile (Figure 10.19(b)), though as before, there is an additional moment of M_p acting at the top of the pile, so that Equation 10.31b becomes

$$H_{max}\left[\frac{M}{H} + h_{AD}\right] = 2M_p \tag{10.33}$$

The equation of horizontal equilibrium is unaltered.

As for unrestrained piles, Fleming et al. (2009) present design charts for restrained piles for various values of L_p/D_0 and M_p, for both undrained and drained soil conditions, which are shown in Figures 10.24 and 10.25, respectively. Whether the 'short', 'intermediate' or 'long' pile failure mechanism occurs depends on which mechanism gives the lower lateral capacity H_{max} (as this will occur first). This has been analysed for a range of pile properties and dimensions to produce the charts shown in Figure 10.26(a) for undrained conditions and 10.26(b) for drained conditions. This figure can be used to determine which the critical failure mode is, and therefore which of the lines in Figures 10.24 and 10.25 should be used to find the lateral capacity.

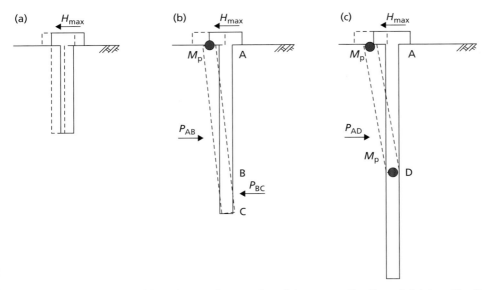

Figure 10.23 Lateral loading of restrained (grouped) piles: (a) 'short' pile, (b) 'intermediate' pile, (c) 'long' pile.

Applications in geotechnical engineering

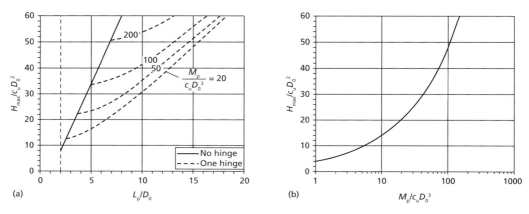

Figure 10.24 Design charts for determining the lateral capacity of a restrained pile under undrained conditions: (a) 'short' and 'intermediate' piles, (b) 'long' pile.

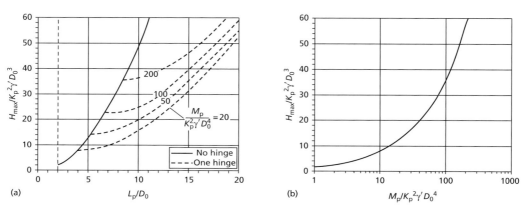

Figure 10.25 Design charts for determining the lateral capacity of a restrained pile under drained conditions: (a) 'short' and 'intermediate' piles, (b) 'long' pile.

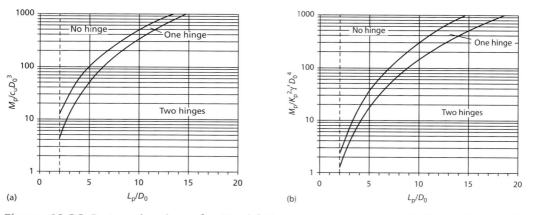

Figure 10.26 Determination of critical failure mode, restrained piles: (a) undrained conditions, (b) drained conditions.

Advanced foundation topics

Pile capacity under combined loading (ULS)

In many cases, piles subject to horizontal actions will also be carrying vertical loads. Levy *et al.* (2008) demonstrated that the yield surface under combined *V–H* loading can be approximated as shown in Figure 10.27, where *R* is the resistance of the pile under pure vertical loading, calculated using the methods in Chapter 9, and H_{max} is determined from Figures 10.20–10.26. In contrast to shallow foundations, the lateral resistance of a pile is normally greater than its axial resistance (i.e. $R/H_{max}<1$). With reference to Figure 10.27, this means that under most circumstances the vertical and horizontal limit states are independent of each other and can be checked separately.

Foundation displacement from elastic solutions (SLS)

Horizontal and moment actions which lie within the yield surface (i.e. satisfying ULS) will generate horizontal displacement and/or rotation at the head of a pile. Most of the deflections of the pile will occur close to the ground surface, reducing with depth until they become zero at a depth known as the **critical length** (L_c). For the general case of soil shear modulus increasing linearly with depth, $G(z)=G_{gl}+G_r z$, the critical depth can be found after Randolph (1981):

$$L_c = D_0 \left(\frac{E_p}{G_c} \right)^{\frac{2}{7}} \tag{10.34}$$

where

$$G_c = \bar{G}_{Lc} \cdot [1+0.75v] \tag{10.35}$$

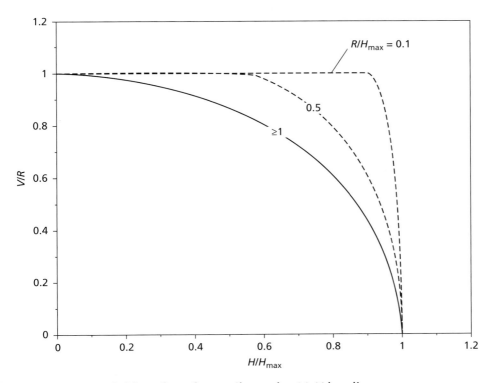

Figure 10.27 Yield surface for a pile under *V–H* loading.

Applications in geotechnical engineering

and E_p is the equivalent Young's Modulus of a pile of solid circular cross-section, to account for pile shape, i.e.

$$E_p I_{circle} = EI_{pile}$$
$$E_p \left(\frac{\pi D_0^2}{64} \right) = EI_{pile} \qquad (10.36)$$
$$E_p = \frac{64 EI_{pile}}{\pi D_0^2}$$

In Equation 10.35, $\overline{G}_{Lc}$ is the median value of shear modulus over the critical length, i.e. the value of G at a depth of $L_c/2$. The definition of L_c and $\overline{G}_{Lc}$ are shown schematically in Figure 10.28.

The horizontal displacement at the pile head is then given by:

$$h = \frac{(E_p/G_c)^{1/7}}{\rho_c G_c} \left[\frac{H}{1.85 L_c} + \frac{M}{0.83 L_c^2} \right] \qquad (10.37)$$

where ρ_c is a homogeneity factor describing the variation of G with depth, calculated using:

$$\rho_c = \frac{G(z = L_c/4)}{G(z = L_c/2)} = \frac{G(z = L_c/4)}{\overline{G}_{Lc}} \qquad (10.38)$$

The rotation at the pile head is given by

$$\theta = \frac{(E_p/G_c)^{1/7}}{\rho_c G_c} \left[\frac{H}{0.83 L_c^2} + \frac{M}{0.16 \rho_c^{-0.5} L_c^3} \right] \qquad (10.39)$$

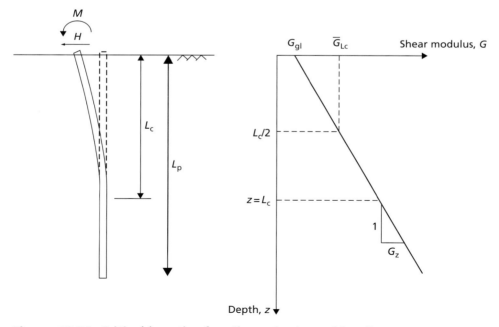

Figure 10.28 Critical length of a pile under lateral loading.

Advanced foundation topics

From Equations 10.37 and 10.39 it can be seen that there is significant coupling between the effects of H and M; the settlement, however, is largely independent of these effects, and can be calculated using the techniques described in Chapter 9. In the case of piles with restrained heads under pure lateral loading, the pile cap will prevent rotation. By setting $\theta = 0$ in Equation 10.39, M can be found as a function of H; this can then be substituted into Equation 10.37 to give the lateral displacement of a restrained pile under a lateral action H:

$$h = \frac{(E_p/G_c)^{1/7}}{\rho_c G_c} \left(0.54 - \frac{0.22}{\rho_c^{0.5}} \right) \frac{H}{L_c} \tag{10.40}$$

Example 10.5

An alternative foundation for the wind turbine in Example 10.4 has been proposed, consisting of a single steel tubular monopile, 40 m long, having an outer diameter of 2 m and a wall thickness of 20 mm (EI = 12 GNm²; M_p = 28 MNm). If the actions applied to the foundation are the same as those in Example 10.4, determine whether the foundation satisfies the ULS to EC7 DA1a and determine the displacements of the foundation under the applied actions (it may be assumed that the monopile already satisfies the vertical ULS).

Solution

The pile has no restraint at the head. DA1a only factors actions, so the parameters required to use Figures 10.20 and 10.22 are $L_p/D_0 = 40/2 = 20$; $M_p/c_u D_0^3 = 28/(0.02 \times 2^3) = 175$; $M/HD_0 = 11.7/(0.78 \times 2) = 7.5$. From Figure 10.22(a), it can be seen that the pile is 'long', so that Figure 10.20(b) should be used to determine the horizontal capacity. This is shown in Figure 10.29, from which

$$\frac{H_{max}}{c_u D_0^2} \approx 18 \quad \therefore H_{max} = 18 \cdot 0.02 \cdot 2^2 = 1.44 \text{ MN}$$

This is much greater than the applied design horizontal load (0.78 MN), so the pile satisfies ULS (i.e. it will not fail structurally).

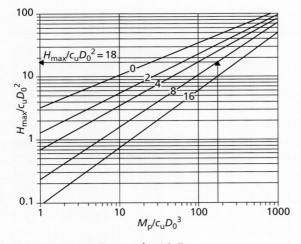

Figure 10.29 Example 10.5.

Applications in geotechnical engineering

At SLS, the equivalent Young's modulus is found from Equation 10.36:

$$E_p = \frac{64 \cdot 12}{\pi \cdot 2^2} = 61\,\text{GPa}$$

As the soil shear modulus is constant with depth $\rho_c = 1$ (Equation 10.38) and $\overline{G}_{Lc} = 3.3\,\text{MPa}$ so that from Equation 10.35:

$$G_c = 3.3 \cdot \left[1 + (0.75 \cdot 0.5)\right]$$
$$= 4.54\,\text{MPa}$$

Then, from Equation 10.34:

$$L_c = 2 \cdot \left(\frac{61000}{4.54}\right)^{\frac{2}{7}} = 30.2\,\text{m}$$

The horizontal movement and rotation are then found from Equations 10.37 and 10.39 respectively:

$$h = \frac{(61000/4.54)^{1/7}}{1 \cdot 4.54}\left[\frac{0.52}{1.85 \cdot 30.2} + \frac{7.8}{0.83 \cdot 30.2^2}\right] = 0.0168\,\text{m}$$

$$\theta = \frac{(61000/4.54)^{1/7}}{1 \cdot 4.54}\left[\frac{0.52}{0.83 \cdot 30.2^2} + \frac{7.8}{0.16 \cdot 1 \cdot 30.2^3}\right] = 2.1 \times 10^{-3}\,\text{radians}$$

The monopile will therefore displace horizontally by 16.8 mm and rotate by 0.12° at the pile head under the applied actions. The foundation movements are larger than for the gravity base (Example 10.4), which in this case appears to be a better choice of foundation for the wind turbine (though cost has not been considered here).

Summary

1 Piles may be grouped together and connected by a stiff pile cap to form a pile group. Shallow foundation elements may be used to support structures as a series of pads/strips. In both cases the critical ULS will be failure of the individual foundation element, while at SLS much larger settlements will occur due to interaction between adjacent foundation elements. For structures covering a large area, differential settlements often govern the SLS. The limiting differential settlement for use in design depends on the type of structure supported, values for which have been presented. Rafts may be used to support such structures, and these may be piled to reduce both gross and differential settlement. Deep basements may be used to provide usable underground space as well as supporting the structure – these must be designed against bearing failure, lateral earth pressures and uplift.

2 The vertical bearing resistance (and therefore the carrying capacity) of shallow foundations is reduced if significant horizontal and moment actions are present.

Advanced foundation topics

Limit analysis can be used to derive a yield surface for use in ULS design, which can efficiently check the overall stability of the foundation (i.e. against bearing capacity failure, sliding and overturning). Elastic solutions, similar to those in Chapter 8 for vertical loading, have been presented for determining the additional horizontal movement and rotation at the SLS. In deep foundations lateral resistance is usually much larger than axial resistance, and the effects of horizontal and moment actions can usually be considered independently of the vertical actions. At the ULS, lateral resistance is controlled by the structural strength of the foundation and the relative soil–foundation strength. At SLS, a deep foundation will only actively displace over a limited depth (its critical length) and an elastic solution can be used to determine the movements at the top of the foundation, depending chiefly on the relative soil–foundation stiffness. For both ULS and SLS conditions, solutions for both single piles and piles in groups have been presented.

Problems

10.1 A raft is to be used to support a warehouse structure applying a bearing pressure of 150 kPa over a plan area of 100 × 50 m. The concrete raft is to be 1.5 m thick with Young's Modulus of 30 GPa and $v = 0.25$. The soil has $E' = 25$ MPa and $v' = 0.3$. The cranes and tall shelving in the warehouse are sensitive to differential settlement, and will become unusable if the angular distortion exceeds 1/300. Determine the maximum differential settlement between the edge of the foundation and the centre and determine whether the proposed raft will meet the serviceability limit state.

10.2 The raft described in Problem 10.1 is to be piled to reduce the differential settlements, by installing piles around the centre of the raft. If a 20 × 20 group of piles is to be used, with $D_0 = 300$ mm, $L_p = 10$ m, $S = 1.5$ m and $e_p = 0.6$, determine the revised differential settlement between the edge of the raft and its centre.

10.3 A shallow tunnel, formed from pre-fabricated concrete box sections ($\gamma = 23.5$ kN/m³), is to be installed in water-bearing ground. The tunnel has exterior dimensions of 15 m deep by 30 m wide, and has walls, roof and ceiling 1.5 m thick. It will be installed such that the top surface of the box is level with the ground surface. During construction the water table is drawn-down to below the underside of the box, but this will be relaxed once construction is complete.

  a Determine the depth of the water table at which the structure will start to float, to EC7 (assuming that the walls of the box are smooth).

  b If tension piles are installed beneath the box, determine the design resistance that the piling must be able to provide.

10.4 Determine the maximum lateral load which can be applied to the pile described in Problem 9.1 if pure lateral loading is applied and $M_p = 200$ kNm, performing all calculations to EC7 DA1b. ($K_p = 3$ may be assumed).

10.5 Determine the maximum horizontal load that can be applied to a 2 × 2 group of the 33 m long piles considered in Example 9.2(b) ($EI = 230$ MNm²) if the horizontal foundation movement is not to exceed 20 mm under short-term conditions. It may be assumed that $E_u/c_u = 250$.

10.6 Determine the maximum horizontal load that can be applied to the wind turbine of Problem 9.3 if the rotation of the foundation is not to exceed 0.1° for the following cases:

  a The horizontal load is applied at the base of the wind turbine (head of the pile).

  b The horizontal load is applied to the wind turbine 20 m above the seabed.

  It may be assumed that $E_u/c_u = 100$ for the clay, $L_p = 34$ m and $EI = 5$ GNm².

References

Barkan, D.D. (1962) *Dynamic Bases and Foundations*, McGraw-Hill Book Company, New York, NY.

Bjerrum, L. (1963) Discussion, in *Proceedings of the European Conference on SMFE, Wiesbaden*, Vol. 3, pp. 135–137.

Burland, J.B. and Wroth, C.P. (1975) Settlement of buildings and associated damage, in *Proceedings of Conference on Settlement of Structures (British Geotechnical Society)*, Pentech Press, London, pp. 611–653.

Butterfield, R. and Douglas, R.A. (1981) *Flexibility Coefficients for the Design of Piles and Pile Groups*, CIRIA Technical Note 108, Construction Industry Research and Information Association.

Butterfield, R. and Gottardi, G. (1994) A complete three-dimensional failure envelope for shallow footings on sand, *Géotechnique*, **44**(1), 181–184.

Charles, J.A. and Skinner, H.D. (2004) Settlement and tilt of low-rise buildings, *Proceedings ICE – Geotechnical Engineering*, **157**(GE2), 65–75.

Fleming, W.G.K., Weltman, A., Randolph, M.R. and Elson, K. (2009) *Piling Engineering* (3rd edn), Taylor & Francis, Abingdon, Oxon.

Gorbunov-Possadov, M.I. and Serebrajanyi, R.V. (1961) Design of structures upon elastic foundations, in *Proceedings of the 5th International Conference on Soil Mechanics and Foundation Engineering*, Vol. 1, pp. 643–648.

Gourvenec, S. (2007) Shape effects on the capacity of rectangular footings under general loading, *Géotechnique*, **57**(8), 637–646.

Gourvenec, S. and Steinepreis, M. (2007) Undrained limit states of shallow foundations acting in consort, *International Journal of Geomechanics*, **7**(3), 194–205.

Horikoshi, K. and Randolph, M.F. (1997) On the definition of raft–soil stiffness ratio, *Géotechnique* **47**(5), 1055–1061.

Kumar, J and Ghosh, P.(2007) Ultimate bearing capacity of two interfering rough strip footings, *International Journal of Geomechanics*, **7**(1), 53–62.

Lee, J., Eun, J., Prezzi, M. and Salgado, R. (2008) Strain influence diagrams for settlement estimation of both isolated and multiple footings in sand, *Journal of Geotechnical and Geoenvironmental Engineering*, **134**(4), 417–427.

Levy, N.H., Einav, I. and Randolph, M.F. (2008) Numerical modeling of pile lateral and axial load interaction, in *Proceedings of the 2nd BGA International Conference on Foundations, ICOF 2008*, IHS BRE Press, Bracknell, Berkshire.

Polshin, D.E. and Tokar, R.A. (1957) Maximum allowable non-uniform settlement of structures, in *Proceedings of the 4th International Conference SMFE, London*, Vol. 1, pp. 402–405.

Randolph, M.F. (1981) The response of flexible piles to lateral loading, *Géotechnique*, **31**(2), 247–259.

Randolph, M.F. (1983) Design of piled raft foundations, in *Proceedings of the International Symposium on Recent Developments in Laboratory and Field Tests and Analysis of Geotechnical Problems, Bangkok*, pp. 525–537.

Skempton, A.W. and MacDonald, D.H. (1956) Allowable settlement of buildings, *Proceedings ICE*, **5**(3), 727–768.

Stuart, J.G. (1962) Interference between foundations, with special reference to surface footings in sand, *Géotechnique* **12**(1), 15–22.

Zhang, L.M. and Ng, A.M.Y. (2005) Probabilistic limiting tolerable displacements for serviceability limit state design of foundations, *Géotechnique*, **55**(2), 151–161.

Further reading

Tomlinson, M.J. (2001) *Foundation Design and Construction* (7th edn), Prentice Hall, Pearson Education Ltd, Harlow.

A comprehensive reference volume, covering various applications of foundation systems and their design, which is much broader in scope than the material presented herein.

For further student and instructor resources for this chapter, please visit the Companion Website at www.routledge.com/cw/craig

Chapter 11

Retaining structures

> **Learning outcomes**
>
> After working through the material in this chapter, you should be able to:
>
> 1 Use limit analysis and limit equilibrium techniques to determine the limiting lateral earth pressures acting on retaining structures;
> 2 Determine in-situ lateral stresses based on fundamental soil properties and understand how limiting earth pressures are mobilised from these values by relative soil–structure movement;
> 3 Determine the lateral stresses induced on a retaining structure due to external loads and construction procedures;
> 4 Design a gravity retaining structure, an embedded wall, a braced excavation or a reinforced soil retaining structure within a limit-state design framework (Eurocode 7).

11.1 Introduction

It is often necessary in geotechnical engineering to retain masses of soil (Figure 11.1). Such applications may be permanent, e.g.

- retaining unstable soil next to a road or railway,
- raising a section of ground with minimal land-take,
- creation of underground space;

or temporary, e.g.

- creating an excavation to install service pipes/cables or to repair existing services.

In permanent applications, a structural element is usually used to support the retained mass of soil. This will typically be either a **gravity retaining wall**, which keeps the retained soil stable due to its mass, or a **flexible retaining wall** which resists soil movement by bending. In both cases it is essential to determine the magnitude and distribution of lateral pressure between the soil mass and the adjoining retaining structure to check the stability of a gravity wall against sliding and overturning or to undertake

Applications in geotechnical engineering

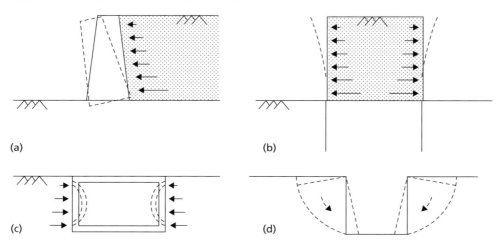

Figure 11.1 Some applications of retained soil: (a) restraint of unstable soil mass, (b) creation of elevated ground, (c) creation of underground space, (d) temporary excavations.

the structural design of a flexible retaining wall. As with foundations, such permanent structures must be designed to satisfy both ultimate and serviceability limit states. These will be discussed in greater detail in Section 11.4 and 11.7. Soil may also be retained by reinforcing the soil mass itself; this will be described in Section 11.11.

Excavations may be self-supporting if undrained strength can be mobilised, and in these cases lateral earth-pressure theory may be used to determine the maximum depth to which such excavations can safely be made (an ultimate limiting state). Supported excavations will be discussed in greater detail in Section 11.9, while unsupported excavations will be discussed in Chapter 12.

Section 11.2 introduces the basic theories of lateral earth pressure using limit analysis techniques, as in Chapter 8. Rigorous lower bound solutions will be derived for both undrained and drained conditions. As previously, it is assumed that the stress–strain behaviour of the soil can be represented by the rigid–perfectly plastic idealisation, shown in Figure 8.3. Conditions of plane strain are also assumed (as for strip footings), i.e. strains in the longitudinal direction of the structure are assumed to be zero due to the length of most retaining structures.

11.2 Limiting earth pressures from limit analysis

Limiting lateral earth pressures

Figure 11.2(a) shows the stress conditions in soil on either side of an embedded retaining wall, where major principal stresses are defined by σ_1, σ'_1 (total and effective, respectively) and minor principal stresses are defined by σ_3, σ'_3. If the wall were to fail by moving horizontally (translating) in the direction shown, the horizontal stresses within the **retained soil** behind the wall will reduce. If the movement is large enough, the value of horizontal stress decreases to a minimum value such that a state of plastic equilibrium develops for which the major principal total and effective stresses are vertical. This is known as the **active condition**. In the soil on the other side of the wall there will be lateral compression of the soil as the wall displaces, resulting in an increase in the horizontal stresses until a state of plastic equilibrium is reached, such that the major principal total and effective stresses are horizontal. This is known as the **passive condition**. Mohr circles at the point of failure are shown in Figure 11.2(b) for undrained soil and Figure 11.2(c) for drained material.

Retaining structures

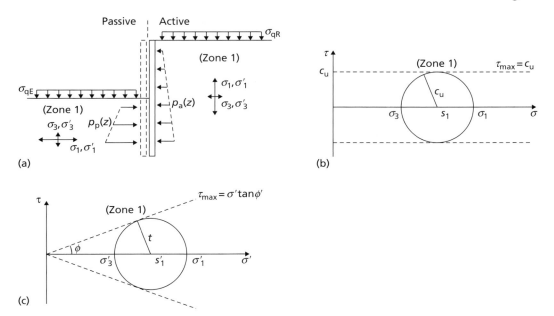

Figure 11.2 Lower bound stress field: (a) stress conditions under active and passive conditions, (b) Mohr circle, undrained case, (c) Mohr circle, drained case.

In the active case for an undrained cohesive material at failure, $\sigma_1(z) = \sigma_v(z) = \gamma z + \sigma_{qR}$ (total vertical stress) and σ_3 is the horizontal stress in the soil, which by equilibrium must also act on the wall. Therefore, $\sigma_3 = \sigma_h = p_a$, where p_a is the **active earth pressure**. From Figure 11.2(b),

$$\sigma_3 = \sigma_1 - 2c_u$$
$$p_a(z) = \sigma_v(z) - 2c_u \tag{11.1}$$

In the passive case, $\sigma_3(z) = \sigma_v(z) = \gamma z + \sigma_{qE}$ and $\sigma_1 = p_p$, where p_p is the **passive earth pressure**. From Figure 11.2(b),

$$\sigma_1 = \sigma_3 + 2c_u$$
$$p_p(z) = \sigma_v(z) + 2c_u \tag{11.2}$$

In the active case for a drained cohesionless material, $\sigma'_1(z) = \sigma'_v(z)$ (effective vertical stress) and $\sigma'_3(z) = \sigma'_h(z)$ is the horizontal effective stress in the soil. From Figure 11.2(c),

$$\sin \phi' = \frac{t}{s'_1} = \frac{\sigma'_1 - \sigma'_3}{\sigma'_1 + \sigma'_3} \tag{11.3}$$

Substituting for σ'_1 and σ'_3 in Equation 11.3 and rearranging gives

$$\frac{\sigma'_h}{\sigma'_v} = \frac{1 - \sin \phi'}{1 + \sin \phi'} \tag{11.4}$$

Applications in geotechnical engineering

The ratio σ'_h/σ'_v is termed the earth pressure coefficient, K (see Chapter 7). As Equation 11.4 was derived for active conditions,

$$K_a = \frac{1-\sin\phi'}{1+\sin\phi'} \tag{11.5}$$

where K_a is the **active earth pressure coefficient**. The active earth pressure acting on the wall ($p_a = \sigma_h$, a total stress) is then found using Terzaghi's Principle ($\sigma_h = \sigma'_h + u$)

$$p_a(z) = \sigma_h(z) = K_a\sigma'_v(z) + u(z) \tag{11.6}$$

In the passive case for a drained material, $\sigma'_3(z) = \sigma'_v(z)$ (effective vertical stress) and $\sigma'_1(z) = \sigma'_h(z)$ is the horizontal effective stress in the soil. Substituting for σ'_1 and σ'_3 in Equation 11.3 and rearranging gives

$$\frac{\sigma'_h}{\sigma'_v} = \frac{1+\sin\phi'}{1-\sin\phi'} = K_p \tag{11.7}$$

where K_p is the **passive earth pressure coefficient**. The passive earth pressure acting on the wall is then

$$p_p(z) = \sigma_h(z) = K_p\sigma'_v(z) + u(z) \tag{11.8}$$

Rankine's theory of earth pressure (general ϕ', c' material)

Rankine developed a lower bound solution based on the **Method of Characteristics** for the case of a soil with general strength parameters c' and ϕ'. The Mohr circle representing the state of stress at failure in a two-dimensional element is shown in Figure 11.3, the relevant shear strength parameters being denoted by c' and ϕ'. Shear failure occurs along a plane at an angle of $45° + \phi'/2$ to the major principal plane. If the soil mass as a whole is stressed such that the principal stresses at every point are in the same directions then, theoretically, there will be a network of failure planes (known as a slip line field) equally inclined to the principal planes, as shown in Figure 11.3. The two sets of planes are termed α- and β-characteristics, from where the method gets its name. It should be appreciated that the state of plastic equilibrium can be developed only if sufficient deformation of the soil mass can take place (see Section 11.3).

A semi-infinite mass of soil with a horizontal surface is considered as before, having a vertical boundary formed by a smooth wall surface extending to semi-infinite depth, as represented in Figure 11.4(a). Referring to Figure 11.3,

$$\sin\phi' = \frac{(\sigma'_1 - \sigma'_3)}{(\sigma'_1 + \sigma'_3 + 2c'\cot\phi')}$$

$$\therefore \sigma'_3(1+\sin\phi') = \sigma'_1(1-\sin\phi') - 2c'\cos\phi'$$

$$\therefore \sigma'_3 = \sigma'_1\left(\frac{1-\sin\phi'}{1+\sin\phi'}\right) - 2c'\left(\frac{\sqrt{1-\sin^2\phi'}}{1+\sin\phi'}\right) \tag{11.9}$$

$$\therefore \sigma'_3 = \sigma'_1\left(\frac{1-\sin\phi'}{1+\sin\phi'}\right) - 2c'\left(\sqrt{\frac{1-\sin\phi'}{1+\sin\phi'}}\right)$$

Alternatively, $\tan^2(45° - \phi'/2)$ can be substituted for $(1-\sin\phi')/(1+\sin\phi')$.

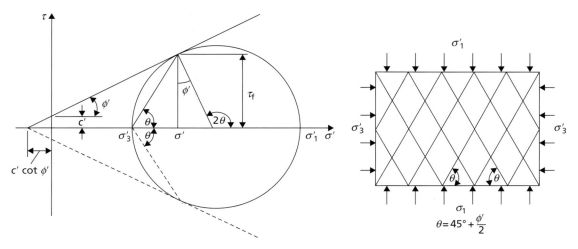

Figure 11.3 State of plastic equilibrium.

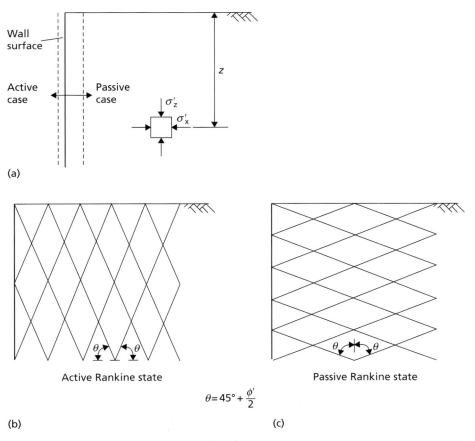

Figure 11.4 Active and passive Rankine states.

Applications in geotechnical engineering

As before, in the active case $\sigma_1' = \sigma_v'$ and $\sigma_3' = \sigma_h'$. When the horizontal stress becomes equal to the active pressure the soil is said to be in the active Rankine state, there being two sets of failure planes each inclined at $\theta = 45° + \phi'/2$ to the horizontal (the direction of the major principal plane) as shown in Figure 11.4(b). Using Equation 11.5 and Terzaghi's Principle, Equation 11.9 may be written as

$$p_a(z) = \sigma_h(z) = K_a \sigma_v'(z) - 2c'\sqrt{K_a} + u(z) \qquad (11.10)$$

If $c' = 0$, Equation 11.10 reduces to Equation 11.6; if $\phi' = 0$ and $c' = c_u$, $K_a = 1$ and Equation 11.10 reduces to Equation 11.1.

In the passive case $\sigma_1' = \sigma_h'$ and $\sigma_3' = \sigma_v'$. When the horizontal stress becomes equal to the passive pressure the soil is said to be in the passive Rankine state, there being two sets of failure planes each inclined at $\theta = 45° + \phi'/2$ to the vertical (the direction of the major principal plane) as shown in Figure 11.4(c). Rearranging Equation 11.9 gives

$$\sigma_1' = \sigma_3'\left(\frac{1+\sin\phi'}{1-\sin\phi'}\right) + 2c'\left(\sqrt{\frac{1+\sin\phi'}{1-\sin\phi'}}\right) \qquad (11.11)$$

Using Equation 11.5 and Terzaghi's Principle, Equation 11.11 may be written as

$$p_p(z) = \sigma_h(z) = K_p \sigma_v'(z) + 2c'\sqrt{K_p} + u(z) \qquad (11.12)$$

If $c' = 0$, Equation 11.12 reduces to Equation 11.8; if $\phi' = 0$ and $c' = c_u$, $K_a = 1$ and Equation 11.12 reduces to Equation 11.2.

Example 11.1

The soil conditions adjacent to a sheet pile wall are given in Figure 11.5, a surcharge pressure of 50 kPa being carried on the surface behind the wall. For soil 1, a sand above the water table, $c' = 0$, $\phi' = 38°$ and $\gamma = 18$ kN/m³. For soil 2, a saturated clay, $c' = 10$ kPa, $\phi' = 28°$ and $\gamma_{sat} = 20$ kN/m³. Plot the distributions of active pressure behind the wall and passive pressure in front of the wall.

Solution

For soil 1,

$$K_a = \frac{1-\sin 38°}{1+\sin 38°} = 0.24, \quad K_p = \frac{1}{0.24} = 4.17$$

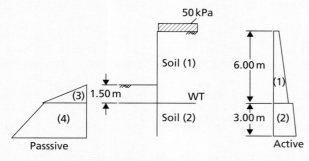

Figure 11.5 Example 11.1.

For soil 2,

$$K_a = \frac{1-\sin 28°}{1+\sin 28°} = 0.36, \quad K_p = \frac{1}{0.36} = 2.78$$

The pressures in soil 1 are calculated using $K_a = 0.24$, $K_p = 4.17$ and $\gamma = 18\,\text{kN/m}^3$. Soil 1 is then considered as a surcharge of $(18 \times 6)\,\text{kPa}$ on soil 2, in addition to the surface surcharge. The pressures in soil 2 are calculated using $K_a = 0.36$, $K_p = 2.78$ and $\gamma' = (20-9.8) = 10\,\text{kN/m}^3$ (see Table 11.1). The active and passive pressure distributions are shown in Figure 11.5. In addition, there is equal hydrostatic pressure on each side of the wall below the water table.

Table 11.1 Example 11.1

Soil	Depth (m)	Pressure (kPa)	
Active pressure:			
(1)	0	0.24×50	$= 12.0$
(1)	6	$(0.24 \times 50) + (0.24 \times 18 \times 6) = 12.0 + 25.9$	$= 37.9$
(2)	6	$0.36[50 + (18 \times 6)] - (2 \times 10 \times \sqrt{0.36}) = 56.9 - 12.0$	$= 44.9$
(2)	9	$0.36[50 + (18 \times 6)] - (2 \times 10 \times \sqrt{0.36}) + (0.36 \times 10.2 \times 3) = 56.9 - 12.0 + 11.0$	$= 55.9$
Passive pressure:			
(3)	0	0	
(3)	1.5	$4.17 \times 18 \times 1.5$	$= 112.6$
(4)	1.5	$(2.78 \times 18 \times 1.5) + (2 \times 10 \times \sqrt{2.78}) = 75.1 + 33.3$	$= 108.4$
(4)	4.5	$(2.78 \times 18 \times 1.5) + (2 \times 10 \times \sqrt{2.78}) + (2.78 \times 10.2 \times 3) = 75.1 + 33.3 + 85.1$	$= 193.5$

Effect of wall properties (roughness, batter angle)

In most practical cases the wall will not be smooth such that shear stresses may be generated along the soil–wall interface, which may also not be vertical but slope at an angle w to the vertical. This additional shear will cause a rotation of the principal stresses close to the wall, while in the soil further away the major principal stresses will still be vertical (active case) or horizontal (passive case) as before. In order to ensure equilibrium throughout the soil mass, frictional stress discontinuities (see Sections 8.3 and 8.4) must be used to rotate the principal stresses between zone 1 and zone 2 as shown in Figure 11.6.

The amount by which the principal stresses rotate depends on the magnitude of the shear stress that can be developed along the soil–wall interface, τ_w (the interface shear strength, of Chapter 9). In undrained materials $\tau_w = \alpha c_u$ is assumed, while in drained materials $\tau_w = \sigma' \tan \delta'$ is used. Both of these approaches are described in Section 9.2.

In an undrained material, the stress conditions in zone 1 are still represented by the Mohr circle shown in Figure 11.2(b). The Mohr circle for zone 2 is shown in Figure 11.7(a) for the active case when the soil in zone 2 is at plastic failure. The major principal stress σ_1 acts on a plane which is rotated by $2\theta = \pi - \Delta_2$ from the stress state on the wall, which is itself at an angle w to the vertical. As σ_1 in zone 1 is vertical, the rotation in principal stress direction from zone 1 to 2 is $\theta_{\text{fan}} = \Delta_2/2 - w$ from Figure 11.7(a). The magnitude of s_1 reduces moving from zone 1 to zone 2, so, from Equation 8.15,

Applications in geotechnical engineering

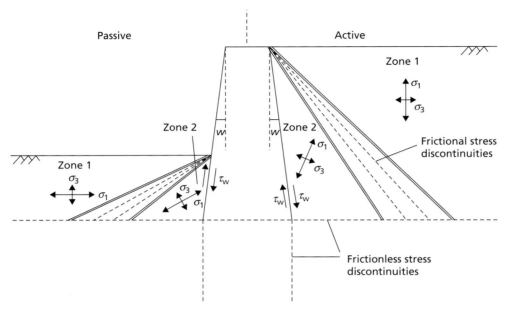

Figure 11.6 Rotation of principal stresses due to wall roughness and batter angle (only total stresses shown).

$$s_1 - s_2 = c_u (\Delta_2 - 2w) \tag{11.13}$$

In zone 1 (from Figure 11.2(b)),

$$s_1 = \sigma_v - c_u \tag{11.14}$$

In zone 2, the total stress acting normal to the wall (the active earth pressure) from Figure 11.7(a) is given by

$$p_a = s_2 - c_u \cos \Delta_2 \tag{11.15}$$

The Mohr circle for zone 2 is shown in Figure 11.7(b) for the passive case when the soil in zone 2 is at undrained plastic failure. The stress state on the wall represents the stresses in the vertical direction. The major principal stress σ_1 acts on a plane which is rotated by $2\theta = \Delta_2$ from the stress state on the wall, which is itself at an angle of w to the vertical. As σ_1 in zone 1 was horizontal, the rotation in principal stress direction is $\theta_{fan} = \Delta_2/2 - w$. The magnitude of s_1 increases moving from zone 1 to zone 2, so, from Equation 8.15,

$$s_2 - s_1 = c_u (\Delta_2 - 2w) \tag{11.16}$$

In zone 1 (from Figure 11.2(b)),

$$s_1 = \sigma_v + c_u \tag{11.17}$$

In zone 2, the total stress acting normal to the wall (the active earth pressure) from Figure 11.7(b) is given by

$$p_p = s_2 + c_u \cos \Delta_2 \tag{11.18}$$

Retaining structures

From Figure 11.7, it is clear that

$$\sin \Delta_2 = \frac{\tau_w}{c_u} = \alpha \tag{11.19}$$

To determine the active or passive earth pressures acting on a given wall, the following procedure should be followed:

1. Find the mean stress in zone 1 from Equations 11.14 or 11.17 for active or passive conditions respectively;
2. Determine Δ_2 from Equation 11.19;
3. Find the mean stress in zone 2 from Equation 11.13 or 11.16;
4. Evaluate earth pressures using Equation 11.15 or 11.18.

It should be noted that for the special case of a smooth vertical wall, $\alpha = w = 0$, so $\Delta_2 = 0$ (Equation 10.19) and Equations 11.15 and 11.19 reduce to Equations 11.1 and 11.2 respectively.

The Mohr circle for zone 2 in drained soil is shown in Figure 11.8(a) for the active case when the soil in zone 2 is at plastic failure. The major principal stress σ'_1 acts on a plane which is rotated by $2\theta = \pi - (\Delta_2 - \delta')$ from the stress state on the wall, which is itself at an angle w to the vertical. The stress conditions in zone 1 are still represented by the Mohr circle shown in Figure 11.2(b). As σ'_1 in zone 1 is vertical, the rotation in principal stress direction is $\theta_{fan} = (\Delta_2 - \delta')/2 - w$ from Figure 11.8(a). The magnitude of s'_1 reduces moving from zone 1 to zone 2, so, from Equation 8.30,

$$\frac{s'_1}{s'_2} = e^{(\Delta_2 - \delta' - 2w)\tan\phi'} \tag{11.20}$$

In zone 1 (from Figure 11.2(b)),

$$s'_1 = \frac{\sigma'_v}{1 + \sin\phi'} \tag{11.21}$$

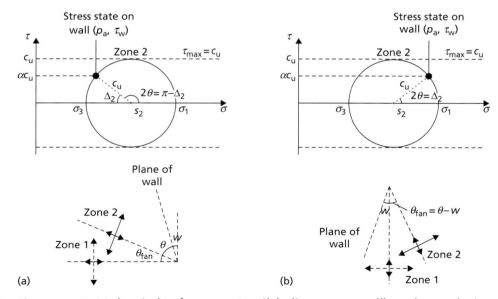

Figure 11.7 Mohr circles for zone 2 soil (adjacent to wall) under undrained conditions: (a) active case, (b) passive case.

Applications in geotechnical engineering

In zone 2, the effective stress acting normal to the wall (σ'_n) from Figure 11.8(a) is given by

$$\sigma'_n = s'_2 - s'_2 \sin\phi' \cos(\Delta_2 - \delta')$$
$$= s'_2 \left[1 - \sin\phi' \cos(\Delta_2 - \delta')\right] \quad (11.22)$$

Then, defining the active earth pressure coefficient in terms of the normal rather than horizontal effective stress (as the wall is battered by w)

$$K_a = \frac{\sigma'_n}{\sigma'_v} = \frac{s'_2}{s'_1} \cdot \frac{1 - \sin\phi' \cos(\Delta_2 - \delta')}{1 + \sin\phi'}$$
$$= \frac{1 - \sin\phi' \cos(\Delta_2 - \delta')}{1 + \sin\phi'} e^{-(\Delta_2 - \delta' - 2w)\tan\phi'} \quad (11.23)$$

The Mohr circle for zone 2 is shown in Figure 11.8(b) for the passive case when the soil in zone 2 is at plastic failure. The major principal stress σ'_1 acts on a plane which is rotated by $2\theta = \Delta_2 + \delta'$ from the stress state on the wall, which is itself at an angle of w to the vertical. As σ'_1 in zone 1 is horizontal, the rotation in principal stress direction is $\theta_{fan} = (\Delta_2 + \delta')/2 - w$. The magnitude of s'_1 increases moving from zone 1 to zone 2, so, from Equation 8.30,

$$\frac{s'_2}{s'_1} = e^{(\Delta_2 + \delta' - 2w)\tan\phi'} \quad (11.24)$$

In zone 1 (from Figure 11.2(b)),

$$s'_1 = \frac{\sigma'_v}{1 - \sin\phi'} \quad (11.25)$$

In zone 2, the effective stress acting normal to the wall from Figure 11.8(b) is given by

$$\sigma'_n = s'_2 + s'_2 \sin\phi' \cos(\Delta_2 + \delta')$$
$$= s'_2 \left[1 + \sin\phi' \cos(\Delta_2 + \delta')\right] \quad (11.26)$$

Then, defining the passive earth pressure coefficient in terms of the normal rather than horizontal effective stress (as the wall is battered by w)

$$K_p = \frac{\sigma'_n}{\sigma'_v} = \frac{s'_2}{s'_1} \cdot \frac{1 + \sin\phi' \cos(\Delta_2 + \delta')}{1 - \sin\phi'}$$
$$= \frac{1 + \sin\phi' \cos(\Delta_2 + \delta')}{1 - \sin\phi'} e^{(\Delta_2 + \delta' - 2w)\tan\phi'} \quad (11.27)$$

In both Equations 11.23 and 11.27

$$\sin\Delta_2 = \frac{\sin\delta'}{\sin\phi'} \quad (11.28)$$

To determine the active or passive earth pressures for a given wall, the following procedure should be followed:

1 Find the vertical effective stresses σ'_v in zone 1;
2 Determine Δ_2 from Equation 11.28;

Retaining structures

3 Find the earth pressure coefficient from Equation 11.23 or 11.27;
4 Evaluate earth pressures using Equation 11.6 or 11.8.

It should be noted that for the special case of a smooth vertical wall, $\delta' = w = 0$, so $\Delta_2 = 0$ (Equation 11.28) and Equations 11.23 and 11.27 reduce to Equations 11.5 and 11.7 respectively.

Sloping retained soil

In many cases the retained soil behind the wall may not be level but may slope upwards at an angle β to the horizontal, as shown in Figure 11.9(a). In this case, the major principal stress in the retained soil in zone 1 will no longer be vertical as the slope will induce a permanent static shear stress within the soil mass, rotating the principal stress direction. The stress conditions (normal and shear stresses) on a plane parallel to the surface at any depth may be found by considering equilibrium as shown in Figure 11.9(b). The component of the block's self-weight W, acting normal to the inclined plane per metre length of the soil mass, is then

$$W = \gamma z A \cos \beta \tag{11.29}$$

Considering equilibrium parallel to the plane,

$$\tau_{mob} = \frac{W}{A} \sin \beta = \gamma z \cos \beta \sin \beta \tag{11.30}$$

Considering equilibrium perpendicular to the plane,

$$\sigma' = \sigma - u = \frac{W}{A} \cos \beta - u = \gamma z \cos^2 \beta - u \tag{11.31}$$

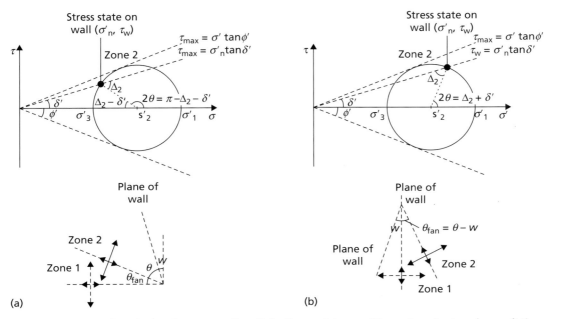

Figure 11.8 Mohr circles for zone 2 soil (adjacent to wall) under drained conditions: (a) active case, (b) passive case.

Applications in geotechnical engineering

where u is the pore pressure on the plane. The stress conditions in the retained soil may be represented by the stress ratio τ_{mob}/σ' given by

$$\frac{\tau_{mob}}{\sigma'} = \frac{\gamma z \cos\beta \sin\beta}{\gamma z \cos^2\beta - u} = \tan\phi'_{mob} \tag{11.32a}$$

Equation 11.32 applies for drained conditions. For undrained conditions where only total stresses are required,

$$\frac{\tau_{mob}}{\sigma} = \frac{\gamma z \cos\beta \sin\beta}{\gamma z \cos^2\beta} = \tan\beta \tag{11.32b}$$

The Mohr circle for undrained conditions in zone 1 in the inclined case is shown in Figure 11.10(a). It can be seen that the major principal stress σ_1 is rotated $2\theta = \Delta_1$ from the stress state on the inclined plane in the retained soil, which is itself at an angle of β to the horizontal. The resultant rotation of the principal stresses in zone 1 compared to the level ground case is $\theta = \Delta_1/2 - \beta$. As a result, the value of θ_{fan} in Equation 11.13 should be modified to

$$\theta_{fan} = \left(\frac{\Delta_2}{2} - w\right) - \left(\frac{\Delta_1}{2} - \beta\right) \tag{11.33}$$

In zone 1 (from Figure 11.10(a)),

$$s_1 = \sigma - c_u \cos\Delta_1 \tag{11.34}$$

which replaces Equation 11.14. The difference in mean stress between zones 1 and 2 (active case) for the general case of sloping backfill retained by a battered rough wall is then given by

$$s_1 - s_2 = c_u \left(\Delta_2 - 2w - \Delta_1 + 2\beta\right) \tag{11.35}$$

where

$$\sin\Delta_1 = \frac{\tau_{mob}}{c_u} \tag{11.36}$$

Equation 11.35 replaces Equation 11.13 for the case of sloping backfill and undrained conditions; the solution procedure remains unchanged.

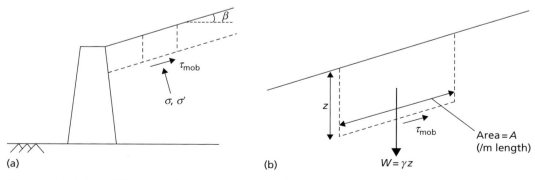

Figure 11.9 Equilibrium of sloping retained soil.

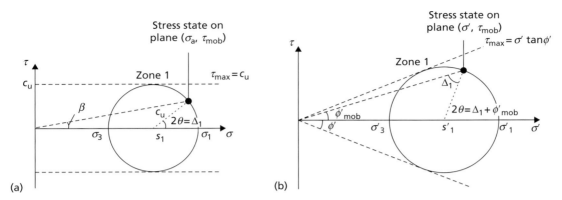

Figure 11.10 Mohr circles for zone 1 soil under active conditions: (a) undrained case, (b) drained case.

The Mohr circle for drained conditions in zone 1 is shown in Figure 11.10(b). It can be seen that the major principal effective stress σ'_1 is rotated $2\theta = \Delta_1 + \phi'_{mob}$ from the stress state on the inclined plane in the retained soil, which is itself at an angle of β to the horizontal. The resultant rotation of the principal stresses in zone 1 compared to the level ground case is then $\theta = (\Delta_1 + \phi'_{mob})/2 - \beta$. As a result, the value of θ_{fan} in Equation 11.20 should be modified to

$$\theta_{fan} = \left(\frac{\Delta_2 - \delta}{2} - w\right) - \left(\frac{\Delta_1 + \phi'_{mob}}{2} - \beta\right) \tag{11.37}$$

From Figure 11.10(b) it can also be seen that

$$\begin{aligned}\sigma' &= s'_1 + s'_1 \sin\phi' \cos(\Delta_1 + \beta) \\ &= s'_1\left[1 + \sin\phi' \cos(\Delta_1 + \beta)\right]\end{aligned} \tag{11.38}$$

so that the active earth pressure coefficient for the general case of sloping backfill retained by a battered rough wall is given by

$$K_a = \frac{\sigma'_n}{\sigma'} = \frac{1 - \sin\phi' \cos(\Delta_2 - \delta')}{1 + \sin\phi' \cos(\Delta_1 + \beta)} e^{-(\Delta_2 - \delta' - 2w - \Delta_1 - \phi'_{mob} + 2\beta)\tan\phi'} \tag{11.39}$$

where

$$\sin\Delta_1 = \frac{\sin\phi'_{mob}}{\sin\phi'} \tag{11.40}$$

Equation 11.39 replaces Equation 11.23 for the case of sloping backfill and drained conditions; the solution procedure and all other equations remain unchanged.

11.3 Earth pressure at rest

It has been shown that active pressure is associated with lateral expansion of the soil at failure and is a minimum value; passive pressure is associated with lateral compression of the soil at failure and is a maximum value. The active and passive values are therefore referred to as **limit pressures**. If the lateral strain in the soil is zero, the corresponding lateral pressure, p'_0, acting on a retaining structure is called the **earth pressure at-rest**, and is usually expressed in terms of effective stress by the equation

$$p'_0 = K_0 \sigma'_v \tag{11.41}$$

where K_0 is the coefficient of earth pressure at-rest in terms of effective stress. In the absence of a retaining structure p'_0 is the horizontal effective stress in the ground, σ'_h, such that Equation 11.41 becomes a restatement of Equation 7.42.

Since the at-rest condition does not involve failure of the soil, the Mohr circle representing the vertical and horizontal stresses does not touch the failure envelope and the horizontal stress cannot be determined analytically through limit analysis. The value of K_0, however, can be determined experimentally by means of a triaxial test in which the axial stress and the all-round pressure are increased simultaneously such that the lateral strain in the specimen is maintained at zero; this will generally require the use of a stress-path cell.

The value of K_0 may also be estimated from in-situ test data, notably using the pressuremeter (PMT) or CPT. Of these methods, the PMT is most reliable as the required parameters are directly measured, while the CPT relies on empirical correlation. Using the PMT, the total in-situ horizontal stress σ_{h0} in any type of soil is determined from the lift-off pressure as shown in Figures 7.14 and 7.16. The in-situ vertical total stress σ_{v0} is then determined from bulk unit weight (from disturbed samples taken from the borehole), and pore pressures (u_0) are determined from the observed depth of the water table in the borehole or other piezometric measurements (see Chapter 6). Then

$$K_0 = \frac{\sigma'_{h0}}{\sigma'_{v0}} = \frac{\sigma_{h0} - u_0}{\sigma_{v0} - u_0} \tag{11.42}$$

From CPT data, K_0 is determined empirically from the normalised cone tip resistance using Equation 7.43 (note: this only applies to fine-grained soils).

For normally consolidated soils, the value of K_0 can also be related approximately to the strength parameter ϕ' by the following formula proposed by Jaky (1944):

$$K_{0,NC} = 1 - \sin \phi' \tag{11.43a}$$

For overconsolidated soils the value of K_0 depends on the stress history and can be greater than unity, a proportion of the at-rest pressure developed during initial consolidation being retained in the soil when the effective vertical stress is subsequently reduced. Mayne and Kulhawy (1982) proposed the following correlation for overconsolidated soils during expansion (but not recompression):

$$K_0 = (1 - \sin \phi') \cdot \text{OCR}^{\sin \phi'} \tag{11.43b}$$

In Eurocode 7 it is proposed that

$$K_0 = (1 - \sin \phi') \cdot \sqrt{\text{OCR}} \tag{11.43c}$$

Values of K_0 from Equation 11.43 are shown in Figure 11.11, where they are compared with values for a range of soils derived from laboratory tests collated by Pipatpongsa et al. (2007), Mayne (2007) and Mayne and Kulhawy (1982). While the analytical expressions do fit the data, it should be noted that the scatter suggests that there is likely to be a significant amount of uncertainty associated with this parameter.

Generally, for any condition intermediate to the active and passive states, the value of the lateral stress is unknown. Figure 11.12 shows the form of the relationship between strain and the lateral pressure coefficient. The exact relationship depends on the initial value of K_0 and on whether excavation

Retaining structures

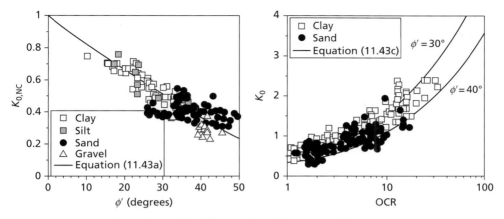

Figure 11.11 Estimation of K_0 from ϕ' and OCR, and comparison to in-situ test data.

or backfilling is involved in construction of the retained soil mass. The strain required to mobilise the passive pressure is considerably greater than that required to mobilise the active pressure.

In the lower bound limit analyses presented in Section 11.2, the entire soil mass was subjected to lateral expansion (active case) or compression (passive case). However, the movement of a retaining wall of finite dimensions cannot develop the active or passive state in the soil mass as a whole. The active state, for example, would be developed only within a wedge of soil between the wall and a failure plane passing through the lower end of the wall and at an angle of $45° + \phi'/2$ to the horizontal, as shown in Figure 11.13(a); the remainder of the soil mass would not reach a state of plastic equilibrium. A specific (minimum) value of lateral strain would be necessary for the development of the active state within this wedge. A uniform strain within the wedge would be produced by a rotational movement (A′B) of the wall, away from the soil, about its lower end, and a deformation of this type, of sufficient magnitude, constitutes the minimum deformation requirement for the development of the active state. Any deformation configuration enveloping A′B, for example, a uniform translational movement A′B′, would also

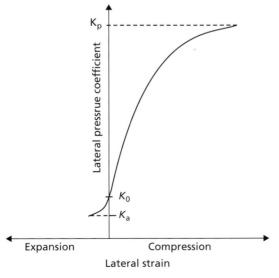

Figure 11.12 Relationship between lateral strain and lateral pressure coefficient.

Applications in geotechnical engineering

result in the development of the active state. If the deformation of the wall were not to satisfy the minimum deformation requirement, the soil adjacent to the wall would not reach a state of plastic equilibrium and the lateral pressure would be between the active and at-rest values.

In the passive case, the minimum deformation requirement is a rotational movement of the wall, about its lower end, into the soil. If this movement were of sufficient magnitude, the passive state would be developed within a wedge of soil between the wall and a failure plane at an angle of $45° + \phi'/2$ to the vertical, as shown in Figure 11.13(b). In practice, however, only part of the potential passive resistance would normally be mobilised. The relatively large deformation necessary for the full development of passive resistance would be unacceptable, with the result that the pressure under working conditions would be between the at-rest and passive values, as indicated in Figure 11.12 (and consequently providing a factor of safety against passive failure).

Experimental evidence indicates that the mobilisation of full passive resistance requires a wall movement of the order of 2–4% of embedded depth in the case of dense sands and of the order of 10–15% in the case of loose sands. The corresponding percentages for the mobilisation of active pressure are of the order of 0.25 and 1%, respectively.

11.4 Gravity retaining structures

The stability of gravity (or freestanding) walls is due to the self-weight of the wall, perhaps aided by passive resistance developed in front of the toe of the wall. The traditional gravity wall (Figure 11.14(a)), constructed of masonry or mass concrete, is uneconomic because the material is used only for its dead weight. Reinforced concrete cantilever walls (Figure 11.14(b)) can be more economic because the backfill itself, acting on the base, is employed to provide most of the required dead weight. Other types of gravity structure include gabion and crib walls (Figures 11.14(c) and (d)). Gabions are cages of steel mesh, rectangular in plan and elevation, filled with particles generally of cobble size, the units being used as the building blocks of a gravity structure. Cribs are open structures assembled from precast concrete or timber members and enclosing coarse-grained fill, the structure and fill acting as a composite unit to form a gravity wall.

Limit state design

At the ultimate limit state (ULS) failure will occur with the retained soil under active conditions as the wall moves towards the excavation, as the generation of passive pressures (higher than K_0, Figure 11.12) would require additional forcing towards the retained soil to induce slip in this direction. A gravity

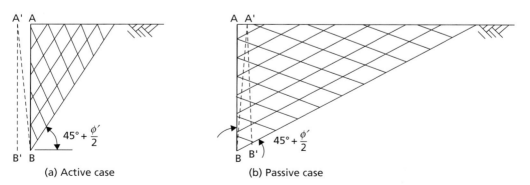

Figure 11.13 Minimum deformation conditions to mobilise: (a) active state, (b) passive state.

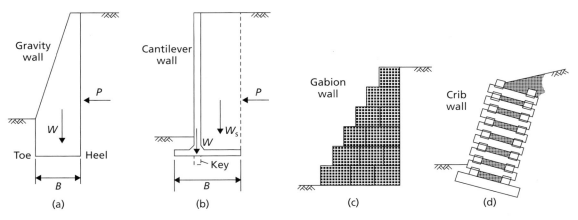

Figure 11.14 Gravity retaining structures.

retaining wall is more complex than the uniaxially loaded foundations of Chapters 8 and 9, with vertical actions (wall dead load), horizontal actions (active earth pressures), potential seepage effects and uneven ground levels. Ultimate limit states which must be considered in wall design are shown schematically in Figure 11.15 and described as follows:

> ULS-1 Base pressure applied by the wall must not exceed the ultimate bearing capacity of the supporting soil;
> ULS-2 Sliding between the base of the wall and the underlying soil due to the lateral earth pressures;
> ULS-3 Overturning of the wall due to horizontal earth pressure forces when the retained soil mass becomes unstable (active failure);
> ULS-4 The development of a deep slip surface which envelops the structure as a whole (analysed using methods which will be described in Chapter 12);
> ULS-5 Adverse seepage effects around the wall, internal erosion or leakage through the wall: consideration should be given to the consequences of the failure of drainage systems to operate as intended (Chapter 2);
> ULS-6 Structural failure of any element of the wall or combined soil/structure failure.

Serviceability criteria must also be met at the serviceability limit state (SLS) as follows:

> SLS-1 Soil and wall deformations must not cause adverse effects on the wall itself or on adjacent structures and services;
> SLS-2 Excessive deformations of the wall structure under the applied earth pressures must be avoided (typically this is only significant in the case of slender cantilever walls, Figure 11.14b, which will be considered separately later in this chapter).

Applications in geotechnical engineering

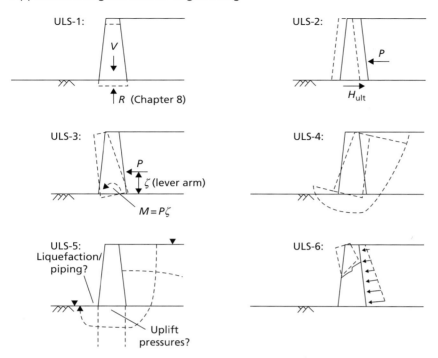

Figure 11.15 Failure modes for gravity retaining structures at ULS.

The first step in design is to determine all the actions/earth pressures acting on the wall from which the resultant thrusts can be determined as described later in this section. Soil and water levels should represent the most unfavourable conditions conceivable in practice. Allowance must be made for the possibility of future (planned or unplanned) excavation in front of the wall known as **overdig**, a minimum depth of 0.5 m being recommended; accordingly, passive resistance in front of the wall is normally neglected.

In terms of ULS design, this chapter will chiefly focus on limit states ULS-1 to ULS-3. Once the earth pressure thrusts are known, the resultant vertical, horizontal and moment components acting on the base of the wall (V, H and M, respectively) are obtained. ULS-1 to ULS-3 can then be checked using the yield surface concepts from Section 10.3, treating the gravity wall as a shallow foundation under combined loading. ULS-2 requires an additional check if the wall–soil interface is not perfectly rough (as assumed in Section 10.3). In this case, the resultant active earth pressure force from the retained soil represents the action that is inducing failure. With passive resistance in front of the wall neglected, the resistance to sliding per metre length of wall (H_{ult}) comes from interface friction along the base of the wall. For a rough wall sliding on undrained soil the design resistance is

$$H_{ult} = \frac{\alpha c_u B}{\gamma_{Rh}} \tag{11.44a}$$

while for sliding on drained soil is

$$H_{ult} = \frac{V \tan \delta'}{\gamma_{Rh}} \tag{11.44b}$$

where α and δ' represent the adhesion and interface friction angle respectively, as before. H_{ult} is a resistance, so parameter γ_{Rh} is a partial resistance factor for sliding, analogous to γ_{Rv} for the bearing resistance of shallow foundations (Chapter 8). In Eurocode 7, the normative value of $\gamma_{Rh}=1.00$ for sets R1 and R3 and 1.10 for set R2. Then, to satisfy ULS-2,

$$H \leq H_{ult} \tag{11.45}$$

Of the remaining ULS conditions, ULS-4 will be considered in Chapter 12 (stability of unsupported soil masses), as in this case the failure bypasses the wall completely (Figure 11.15). The effects of seepage on retaining wall stability have been partially considered in earlier chapters (ULS-5), and ULS-6 is not considered herein as it relates solely to the structural strength of the wall itself under the lateral earth pressures, which must be determined using structural/continuum mechanics principles. It should be noted that if seepage is occurring, then the effects of seepage should also be accounted for in the other ULS failure modes (e.g. uplift pressures will reduce the normal contact stress in ULS-2, reducing sliding resistance).

Resultant thrust

In order to check the overall stability of a retaining structure, the active and passive earth pressure distributions are integrated over the height of the wall to determine the **resultant thrust** (a force per unit length of wall), which is used to define the horizontal action. If the retaining structure is smooth, the resultant thrust will act normal to the wall (this may not be horizontal if the wall has a battered back). For the case of retained undrained soil having uniform bulk density γ (so that $\sigma = \gamma z \cos^2 \beta$), combining Equations 11.35, 11.15 and 11.34 gives

$$p_a = \gamma z \cos^2 \beta - c_u \left(1 + \Delta_2 + \cos \Delta_2 - 2w - \Delta_1 - 2\beta\right) \tag{11.46a}$$

while combining Equations 11.16, 11.17 and 11.18 gives

$$p_p = \gamma z + c_u \left(1 + \Delta_2 + \cos \Delta_2 - 2w\right) \tag{11.46b}$$

Equation 11.46 is linear in z and these total pressure distributions are shown in Figure 11.16.

In the active case, the value of p_a is zero at a particular depth z_0. From Equation 11.46a, with $p_a=0$,

$$z_0 = \frac{c_u \left(1 + \Delta_2 + \cos \Delta_2 - 2w - \Delta_1 - 2\beta\right)}{\gamma} \tag{11.47}$$

This means that in the active case the soil is in a state of tension between the surface and depth z_0. In practice, however, this tension cannot be relied upon to act on the wall, since cracks are likely to develop within the tension zone and the part of the pressure distribution diagram above depth z_0 should be neglected. The total active thrust (P_a) acting normal to a wall of height h with a batter angle of w is then

$$P_a = \int_{z_0}^{h} p_a \frac{dz}{\cos w} \tag{11.48}$$

The force P_a acts at a distance of $\frac{1}{3}(h-z_0)$ above the bottom of the wall. If the wall is rough there is additionally a traction force (T_a) due to the interface shearing, acting downwards along the wall surface at the same point of

$$T_a = \int_{z_0}^{h} \tau_w \frac{dz}{\cos w} \tag{11.49}$$

where $\tau_w = \alpha c_u$. Equation 11.49 neglects any interface shear in the tension zone above z_0 (i.e. it is assumed that a crack opens between the soil and the wall in this zone), so the force acts at a distance of $\frac{1}{2}(h-z_0)$ above the bottom of the wall.

Applications in geotechnical engineering

Active:

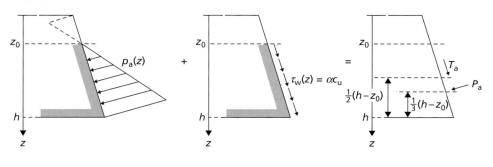

Passive:

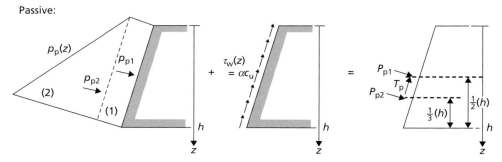

Figure 11.16 Pressure distributions and resultant thrusts: undrained soil.

In the passive case p_p is always positive so that the total passive thrust (P_p) acting normal to a wall of height h with a batter angle of w is

$$P_p = \int_0^h p_p \frac{dz}{\cos w} \tag{11.50}$$

The two components of P_p act at distances of ⅓h and ½h, respectively, above the bottom of the wall surface, as shown in Figure 11.16. If the wall is rough there is additionally a traction force (T_p) acting upwards along the wall surface of

$$T_p = \int_0^h \tau_w \frac{dz}{\cos w} \tag{11.51}$$

In the case of a drained cohesionless material ($c'=0$), the lateral pressures acting on the retaining structure should be separated into the component due to the effective stress in the soil ($p'=K\sigma'$) and the component due to any pore water pressure (u). The effective thrust (P') acting normal to a wall of height h with a batter angle of w is

$$P'_a = \int_0^h K_a \sigma' \frac{dz}{\cos w} \quad \text{(active)} \tag{11.52a}$$

$$P'_p = \int_0^h K_p \sigma' \frac{dz}{\cos w} \quad \text{(passive)} \tag{11.52b}$$

and resultant pore water pressure thrust (U)

$$U = \int_0^h u \frac{dz}{\cos w} \tag{11.53}$$

Retaining structures

If the wall is rough, there will be an additional interface shear force acting downwards along the wall surface on the active side and upwards along the wall surface on the passive side. This is found using Equation 11.51 on both active and passive sides (as there is no tension zone); however, the interface shear strength is $\tau_w = \sigma' \tan \delta'$. The normal and shear components of the thrust P' and T are often combined into a single resultant force which acts at an angle of δ' to the wall normal, as shown in Figure 11.17.

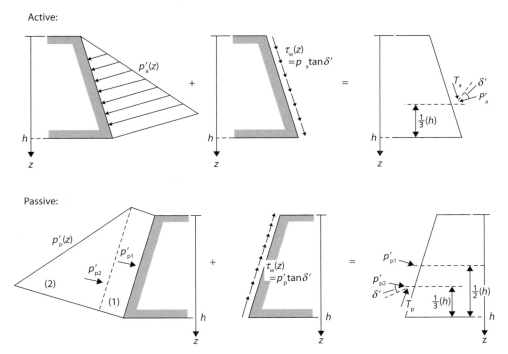

Figure 11.17 Pressure distributions and resultant thrusts: drained soil.

Example 11.2

a Calculate the total active thrust on a smooth vertical wall 5 m high retaining a sand of unit weight 17 kN/m³ for which $\phi' = 35°$ and $c' = 0$; the surface of the sand is horizontal and the water table is below the bottom of the wall.

b Determine the thrust on the wall if the water table rises to a level 2 m below the surface of the sand. The saturated unit weight of the sand is 20 kN/m³.

Solution

a $w = \delta' = \beta = 0$, so using either Equation 11.39, 11.23 or 11.5 gives

$$K_a = \frac{1 - \sin \phi'}{1 + \sin \phi'} = 0.27$$

423

Applications in geotechnical engineering

With the water table below the bottom of the wall, $\sigma' = \sigma'_v = \gamma_{dry} z$, so, from Equation 11.52a,

$$P'_a = \int_0^h K_a \gamma_{dry} z \, dz$$

$$= \frac{1}{2} K_a \gamma_{dry} h^2$$

$$= \frac{1}{2} \cdot 0.27 \cdot 17 \cdot 5^2$$

$$= 57.5 \, \text{kN/m}$$

b The pressure distribution on the wall is now as shown in Figure 11.18, including hydrostatic pressure on the lower 3 m of the wall. The thrust components are:

(1) $P'_a = \int_0^2 K_a \gamma_{dry} z \, dz = \frac{1}{2} \cdot 0.27 \cdot 17 \cdot 2^2 = 9.2 \, \text{kN/m}$

(2) $P'_a = 0.27 \cdot 17 \cdot 2 \cdot 3 = 27.6 \, \text{kN/m}$

(3) $P'_a = \int_{2-2}^{5-2} K_a (\gamma_{sat} z - \gamma_w z) \, dz = \frac{1}{2} \cdot 0.27 \cdot (20 - 9.81) \cdot 3^2 = 12.4 \, \text{kN/m}$

(4) $U_a = \int_{2-2}^{5-2} \gamma_w z \, dz = \frac{1}{2} \cdot 9.81 \cdot 3^2 = 44.1 \, \text{kN/m}$

Summing the four components of thrust gives a total thrust = 93.3 kN/m.

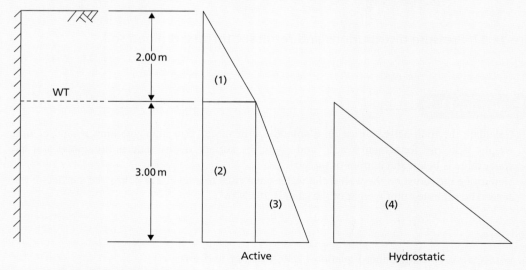

Figure 11.18 Example 11.2.

Retaining structures

> **Example 11.3**

Details of a cantilever retaining wall are shown in Figure 11.19, the water table being below the base of the wall. The unit weight of the backfill is $17\,\text{kN/m}^3$ and a surcharge pressure of $10\,\text{kPa}$ acts on the surface. Characteristic values of the shear strength parameters for the backfill are $c'=0$ and $\phi'=36°$. The angle of friction between the base and the foundation soil is $27°$ (i.e. $\delta'=0.75\phi'$). Is the design of the wall satisfactory at the ultimate limit state to EC7, DA1b?

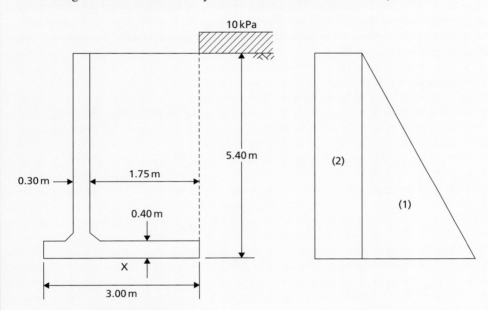

Figure 11.19 Example 11.3.

Solution

The unit weight of concrete will be taken as $23.5\,\text{kN/m}^3$. The active thrust from the retained soil acts on the vertical plane through the soil shown by the dotted line in Figure 11.19 (also known as the **virtual back**), so $\delta'=\phi'$ and $w=0$ on this interface. The value of K_a can then be found from Equation 11.39 with $w=\beta=\Delta_1=\phi'_{\text{mob}}=0$, or from Equation 11.23 (as the retained soil is not sloping). For DA1b (Tables 8.9 and 8.11), the design value of $\phi'=\tan^{-1}(\tan 36°/1.25)=30°$. Therefore $\delta'=30°$ ($\pi/6$) and $\Delta_2=\pi/2$, giving $K_a=0.27$. The soil is dry, so $\sigma'=\sigma'_v=\gamma z+\sigma_q$, giving $p_a=K_a\sigma'_v$ (Equation 11.6). Then, from Equation 11.53 $U_a=0$ and from Equation 11.52

$$P_a = P'_a = \int_0^h K_a(\gamma z + \sigma_q)\,dz = \frac{1}{2}K_a\gamma h^2 + K_a\sigma_q h$$

The first term in P_a relates to (1), which has a triangular distribution with depth, and will therefore act at $h/3$ above the base of the wall. The second term relates to the surcharge (2) which is uniform with depth, so that this component acts at $h/2$ above the base of the wall. Action (1) is a permanent unfavourable action (partial factor=1.00, Table 8.10), while the surcharge effect (2) is

Applications in geotechnical engineering

a variable unfavourable action (partial factor = 1.30, Table 8.10). For using the yield surface in Figure 10.16b, the vertical forces (downwards positive) due to the weight of the wall (stem + base) and soil above the base are considered as permanent favourable actions for $V/R < 0.5$ (where sliding and overturning dominate) and permanent unfavourable actions for $V/R \geq 0.5$ (when bearing failure dominates). For DA1b this does not affect the value of the partial factors, which in both cases are 1.00. The vertical (downwards) traction force $T_a = P'_a \tan \delta'$ from the interface shear on the virtual back is factored similarly. The moments induced by the vertical and horizontal actions are determined about the centre of the footing base (point X in Figure 11.19), with anti-clockwise moments positive (these act to overturn the wall). The calculations of the actions acting on the wall are set out in Table 11.2.

Table 11.2 Example 11.3

	Force (kN/m)		Lever arm (m)	Moment (kNm/m)
(1)	$\frac{1}{2} \cdot 0.27 \left(\frac{17}{1.00}\right) \cdot 5.40^2 \cdot 1.00$	= 66.9	1.80	= 120.4
(2)	$0.27 \cdot 10 \cdot 5.40 \cdot 1.30$	= 19.0	2.70	= 51.3
		H = 85.9		
(Stem)	$5.00 \cdot 0.30 \cdot \left(\frac{23.5}{1.00}\right) \cdot 1.00$	= 35.3	0.40	= 14.1
(Base)	$0.40 \cdot 3.00 \cdot \left(\frac{23.5}{1.00}\right) \cdot 1.00$	= 28.2	0.00	= 0.0
(Soil)	$5.00 \cdot 1.75 \cdot \left(\frac{17}{1.00}\right) \cdot 1.00$	= 148.8	−0.63	= −93.7
(Virtual back)	$(66.9 + 19.0) \tan 30$	= 49.6	−1.50	= −74.4
		V = 261.9		M = 17.7

The bearing resistance of the wall is calculated assuming that the failure occurs in front of the wall (worst case), so, from Equation 8.32,

$$R = \frac{q_f A}{\gamma_R} = \frac{\frac{1}{2} \gamma B^2 N_\gamma}{\gamma_R} = \frac{\frac{1}{2} \cdot \left(\frac{17}{1.00}\right) \cdot 3^2 \cdot 20.1}{1.00} = 1.54 \, \text{MN/m}$$

where N_γ is from Equation 8.36. Therefore, $H/R = 0.056$ and $M/BR = 0.004$. This point is plotted in Figure 10.16b, from where it can be seen that, to lie within the yield surface, $0.12 < V/R < 0.88$ (interpolating between contours). In this case, $V/R = 0.11$, so ULS is not satisfied. Checking ULS-2 explicitly (Equations 11.44b and 11.45) $H_{ult} = V \tan \delta' = 261.9 \times \tan 30 = 151.2 \, \text{kN/m}$, which is greater than H so the wall does not slide. Also, $V \ll R$, so ULS-1 is also satisfied. It is the overturning that is therefore causing the problem.

If the wall is required to be 5.40 m tall, the width of the base would need to be extended (into the retained soil). This would increase the value of V and reduce the value of M (by increasing the restoring moment from the soil above the base). It should be noted that passive resistance in front of the wall has been neglected to allow for unplanned excavation.

Retaining structures

Example 11.4

Details of a mass concrete gravity-retaining wall are shown in Figure 11.20, the unit weight of the wall material being $23.5 \, \text{kN/m}^3$. The unit weight of the dry retained soil is $18 \, \text{kN/m}^3$, and the characteristic shear strength parameters are $c'=0$ and $\phi'=33°$. The value of δ' between wall and retained soil and between wall and foundation soil is $26°$. The wall is founded on uniform clay with $c_u = 120 \, \text{kPa}$ and $\alpha = 0.8$ between the base of the wall and the clay. Is the design of the wall satisfactory at the ultimate limit state to EC7 DA1a?

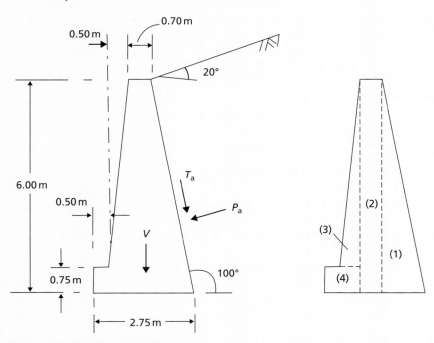

Figure 11.20 Example 11.4.

Solution

As the back of the wall and the soil surface are both inclined, the value of K_a will be calculated from Equation 11.39. The design values of the angles of shearing resistance in this equation are $\phi' = 33°$, $\delta' = 26°$ for DA1a, $w = 100 - 90 = 10°$, $\Delta_2 = 53.6°$ (Equation 11.28), $\beta = 20°$, $\phi'_{mob} = \beta = 20°$ (Equation 11.32), and $\Delta_1 = 38.9°$ (Equation 11.40), giving $K_a = 0.46$. From Equation 11.53 $U_a = 0$, and from Equations 11.52 and 11.31

$$P_a = P'_a = \int_0^h K_a \sigma' \frac{dz}{\cos w} = \int_0^h \frac{K_a \gamma z \cos^2 \beta}{\cos w} dz = \frac{K_a \gamma \cos^2 \beta}{2 \cos w} h^2$$

This force acts normal to the back of the wall at a vertical distance of $h/3$ above the base of the wall and is a permanent unfavourable action (partial factor = 1.35, Table 8.10). The (downwards) interface friction force along the back of the wall is given by

$$T_a = \int_0^h (K_a \sigma') \tan \delta' \frac{dz}{\cos w} = P'_a \tan \delta' = \frac{K_a \gamma z \cos^2 \beta}{2 \cos w} h^2 \tan \delta'$$

Applications in geotechnical engineering

This is similarly a permanent unfavourable action. Evaluating these two expressions and applying partial factors gives design values of the actions: $P_a = 180.4$ kN/m, $T_a = 88.0$ kN/m. These actions are shown in Figure 11.20. The vertical action due to the dead weight of the retaining wall is determined by splitting the wall into smaller, simpler sections as shown in Figure 11.20. This is also a permanent unfavourable action. Moments are considered about the centre of the base of the wall (anticlockwise positive), the calculations being set out in Table 11.3.

Table 11.3 Example 11.4

	Force (kN/m)		Lever arm (m)	Moment (kNm/m)
$P_a \cos 10°$	180.4 cos 10	= 239.8	2.00	= 479.6
$T_a \sin 10°$	−88.0 sin 10	= −20.6	2.00	= −41.2
		H = 219.2		
$P_a \sin 10°$	180.4 sin 10	= 42.3	−2.40	= −101.5
$T_a \cos 10°$	88.0 cos 10	= 117.0	−2.40	= −280.8
Wall (1)	$\frac{1}{2} \cdot 1.05 \cdot 6.00 \cdot \left(\frac{23.5}{1.00}\right) \cdot 1.35$	= 99.9	−0.68	= −67.9
(2)	$0.70 \cdot 6.00 \cdot \left(\frac{23.5}{1.00}\right) \cdot 1.35$	= 133.2	0.03	= 4.0
(3)	$\frac{1}{2} \cdot 0.50 \cdot 5.25 \cdot \left(\frac{23.5}{1.00}\right) \cdot 1.35$	= 41.6	0.54	= 22.5
(4)	$1.00 \cdot 0.75 \cdot \left(\frac{23.5}{1.00}\right) \cdot 1.35$	= 23.8	0.88	= 20.9
		V = 457.8		M = 35.6

The bearing resistance of the wall is calculated assuming that the failure occurs in front of the wall (worst case), so, from Equation 8.17,

$$R = \frac{q_f A}{\gamma_R} = \frac{N_c c_u B}{\gamma_R} = \frac{5.14 \cdot \left(\frac{120}{1.00}\right) \cdot 2.75}{1.00} = 1.70 \text{ MN/m}$$

where $N_c = 5.14$. Therefore, $H/R = 0.129$ and $M/BR = 0.008$. This point is plotted in Figure 10.13b, from where it can be seen that to lie within the yield surface $0.21 < V/R < 0.79$. In this case, $V/R = 0.25$, so ULS is satisfied. A further check should be made for sliding (ULS-2) in this case, as the yield surface approach assumes perfect bonding between the wall and the clay (i.e. $\alpha = 1$). From Equations 11.44a and 11.45, $H_{ult} = (\alpha c_u B)/\gamma_{Rh} = (0.8 \times 120 \times 2.75)/1.00 = 264$ kN/m, which is greater than H so the wall will not slide.

Retaining structures

11.5 Coulomb's theory of earth pressure

While the rigorous limit analysis solutions presented in Section 11.2 are applicable to a wide range of retaining structure problems, they are by no means the only way of determining lateral earth pressures. An alternative method of analysis, known as **limit equilibrium**, involves consideration of the stability, as a whole, of the wedge of soil between a retaining wall and a trial failure plane. The force between the wedge and the wall surface is determined by considering the equilibrium of forces acting on the wedge when it is on the point of sliding either up or down the failure plane, i.e. when the wedge is in a condition of limiting equilibrium. Friction between the wall and the adjacent soil can be taken into account as for the limit analyses in Section 11.2, represented by $\tau_w = \alpha c_u$ in undrained soil and $\tau_w = \sigma' \tan \delta'$ in drained soil. The method was first developed for retaining structures by Coulomb (1776), and is popular for use in design.

The limit equilibrium theory is now interpreted as an upper bound plasticity solution (although analysis is based on force equilibrium and not on the work–energy balance defined in Chapter 8), collapse of the soil mass above the chosen failure plane occurring as the wall moves away from or into the soil. Thus, in general, the theory underestimates the total active thrust and overestimates the total passive resistance (i.e. provides upper bounds to the true collapse load).

Active case

Figure 11.21(a) shows the forces acting on the soil wedge between a wall surface AB, inclined at angle x to the horizontal, and a trial failure plane BC, at angle θ_p to the horizontal. The soil surface AC is inclined at angle β to the horizontal. The shear strength parameter c' is initially taken as zero, though this limitation can subsequently be relaxed for a general ϕ', c' material. For the failure condition, the soil wedge is in equilibrium under its own weight (W), the reaction to the force (P) between the soil and the wall, and the resultant reaction (R_s) on the failure plane. Because the soil wedge tends to move down the plane BC at failure, the reaction P acts at angle δ' below the normal to the wall. (If the wall were to settle more than the backfill, the reaction P would act at angle δ' above the normal.) At failure, when the shear strength of the soil has been

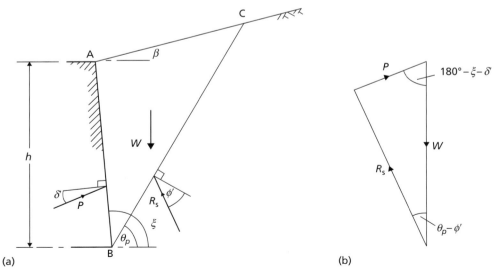

Figure 11.21 Coulomb theory: active case with $c' = 0$: (a) wedge geometry, (b) force polygon.

Applications in geotechnical engineering

fully mobilised, the direction of R_s is at an angle ϕ' below the normal to the failure plane. The directions of all three forces, and the magnitude of W, are known, and therefore a triangle of forces (Figure 11.21(b)) can be drawn and the magnitude of P determined for the trial in question.

A number of trial failure planes would have to be selected to obtain the maximum value of P, which would be the total active thrust on the wall. However, using the sine rule, P can be expressed in terms of W and the angles in the triangle of forces. Then the maximum value of P, corresponding to a particular value of θ_p, is given by $\partial P/\partial \theta = 0$. This leads to the following solution for P'_a in dry soil:

$$P'_a = \frac{1}{2} K_a \gamma H^2 \tag{11.54}$$

where

$$K_a = \left[\frac{\dfrac{\sin(\xi - \phi')}{\sin \xi}}{\sqrt{\sin(\xi + \delta')} + \sqrt{\dfrac{\sin(\phi' + \delta')\sin(\phi' - \beta)}{\sin(\xi - \beta)}}} \right]^2 \tag{11.55}$$

The point of application of the total active thrust is not given by the Coulomb theory but is assumed to act at a distance of $\tfrac{1}{3}h$ above the base of the wall as considered previously. Had Equation 11.55 been used to determine the values of K_a in Examples 11.3 and 11.4, values of 0.30 and 0.48 would have been obtained respectively, which compare favourably with the values from lower bound limit analysis of 0.27 and 0.46.

The analysis can be extended to dry soil cases in which the shear strength parameter c' is greater than zero. It is assumed that tension cracks may extend to a depth z_0, the trial failure plane (at angle θ_p to the horizontal) extending from the heel of the wall to the bottom of the tension zone, as shown in Figure 11.22. The forces acting on the soil wedge at failure are then as follows:

1. the weight of the wedge (W);
2. the reaction (P) between the wall and the soil, acting at angle δ' below the normal;
3. the force due to the constant component of shearing resistance on the wall ($T_a = \tau_w \times EB$);
4. the reaction (R_s) on the failure plane, acting at angle ϕ' below the normal;
5. the force on the failure plane due to the constant component of shear strength ($C = c' \times BC$).

The directions of all five forces are known together with the magnitudes of W, T_a and C, and therefore the value of P can be determined from the force diagram for the trial failure plane. Again, a number of trial failure planes would be selected to obtain the maximum value of P. The resultant thrust is then expressed as

$$P'_a = \frac{1}{2} K_a \gamma h^2 - 2 K_{ac} c' h \tag{11.56}$$

where

$$K_{ac} = 2 \sqrt{K_a \left(1 + \frac{\tau_w}{c'}\right)} \tag{11.57}$$

If the soil is saturated, there will be additional forces acting due to the pore water. Under these conditions it is best to determine the resultant active thrust using a force diagram that includes the resultant pore water thrust U_a acting on the slip plane. This general technique will then allow for the determination of the resultant thrust on a retaining structure under either static or seepage conditions (see Example 11.5).

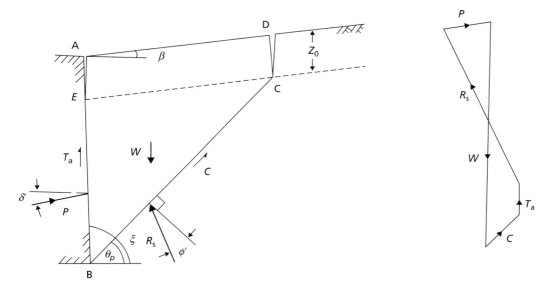

Figure 11.22 Coulomb theory: active case with $c' > 0$.

Passive case

In the passive case, the reaction P acts at angle δ' above the normal to the wall surface (or δ' below the normal if the wall were to settle more than the adjacent soil) and the reaction R_s at angle ϕ' above the normal to the failure plane. In the triangle of forces, the angle between W and P is $180° - \xi + \delta'$ and the angle between W and R_s is $\theta_p + \phi'$. The total passive resistance, equal to the minimum value of P, is given by

$$P'_p = \frac{1}{2} K_a \gamma H^2 \tag{11.58}$$

where

$$K_p = \left[\frac{\dfrac{\sin(\xi + \phi')}{\sin \xi}}{\sqrt{\sin(\xi - \delta')} - \sqrt{\dfrac{\sin(\phi' + \delta')\sin(\phi' + \beta)}{\sin(\xi - \beta)}}} \right]^2 \tag{11.59}$$

Care must be exercised when using Equation 11.59 as it is known to overestimate passive resistance, seriously so for the higher values of ϕ', representing an error on the unsafe side.

As before, the analysis can be extended to dry soil cases in which the shear strength parameter c' is greater than zero, giving

$$P_p = \frac{1}{2} K_p \gamma h^2 - 2 K_{pc} c' h \tag{11.60}$$

where

$$K_{pc} = 2\sqrt{K_p \left(1 + \frac{\tau_w}{c'}\right)} \tag{11.61}$$

Applications in geotechnical engineering

Example 11.5

Details of a retaining structure, with a vertical drain adjacent to the back surface, are shown in Figure 11.23(a), the saturated unit weight of the retained soil being 20 kN/m³. The design strength parameters for the soil are $c'=0$, $\phi'=38°$ and $\delta'=15°$. Assuming an active failure plane at $\theta_p = 45° + \phi'/2$ to the horizontal (see Figure 11.13), determine the total horizontal thrust on the wall under the following conditions:

a when the backfill becomes fully saturated due to continuous rainfall, with steady-state seepage towards the drain;
b if the vertical drain were replaced by an inclined drain below the failure plane at 45° to the horizontal;
c if there were no drainage system behind the wall (i.e. the drains in (a) or (b) become blocked).

Solution

This problem demonstrates how the effects of seepage may be accounted for in earth pressure problems. In each case resultant forces due to the pore water must be determined. This may be achieved by drawing a flow net (Chapter 2); however, only the pore water pressures are required in order to determine the resultant pore water thrust, so the Spreadsheet tool Seepage_CSM8.xls can be used more rapidly to determine the values of head for the different cases. The results of this approach are shown below; the modelling of this problem using the spreadsheet tool is detailed in an appendix to the accompanying User Manual which may be found on the Companion Website.

a The equipotentials for seepage towards the vertical drain from the spreadsheet tool are shown in Figure 11.23(a). Since the permeability of the drain must be considerably greater than that of the backfill, the drain remains unsaturated and the pore water pressure at every point within the drain is zero (atmospheric). Thus, at every point on the boundary between the drain and the backfill, total head is equal to elevation head (this can easily be included as a boundary condition within the FDM spreadsheet). The equipotentials, therefore, must intersect this boundary at points spaced at equal vertical intervals Δh as shown. The boundary itself is neither a flow line nor an equipotential.

The values of total head and elevation head may be determined at the points of intersection of the equipotentials with the failure plane. The pore water pressures at these points are then determined using Equation 2.1 and integrated numerically along the slip plane to give the pore water thrust on the slip plane, $U = 36.8$ kN/m. The pore water forces on the other two boundaries of the soil wedge are zero.

The total weight (W) of the soil wedge is now calculated, i.e.

$$W = \frac{1}{2} \cdot 6.00 \cdot \left[\frac{6.00}{\tan(45+19)°}\right] \cdot 20 = 176 \text{ kN/m}$$

The forces acting on the wedge are shown in Figure 11.23(b). Since the directions of the four forces are known, together with the magnitudes of W and U, the force polygon can be drawn to scale as shown, from which $P_a = 104$ kN/m can be measured from the diagram. The horizontal thrust on the wall is then given by

$$P_a \cos \delta' = 101 \text{ kN/m}$$

Other failure surfaces would have to be chosen in order that the maximum value of total active thrust can be determined (most critical case).

Retaining structures

b For the inclined drain shown in Figure 11.23(c), the equipotentials above the drain are horizontal, the total head being equal to the elevation head in this zone. Thus at every point on the failure plane the pore water pressure is zero, and hence $U=0$ also. This form of drain is preferable to the vertical drain. In this case, from Figure 11.23(d), $P_a = 80$ kN/m such that the horizontal thrust on the wall is then given by

$$P_a \cos \delta' = 77 \text{ kN/m}$$

c For the case of no drainage system behind the wall, the pore water is static, i.e. the pore water pressure at each point on the slip plane is $\gamma_w z$ (Figure 11.23(e)). This distribution can again be integrated numerically to give $U = 196.5$ kN/m. From Figure 11.23(f), $P_a = 203$ kN/m such that the horizontal thrust on the wall is then given by

$$P_a \cos \delta' = 196 \text{ kN/m}$$

This example demonstrates how important it is to keep the drainage behind retaining structures well maintained, as the thrust on the wall is greatly increased (and the stability at ULS therefore greatly reduced) when the pore water cannot drain.

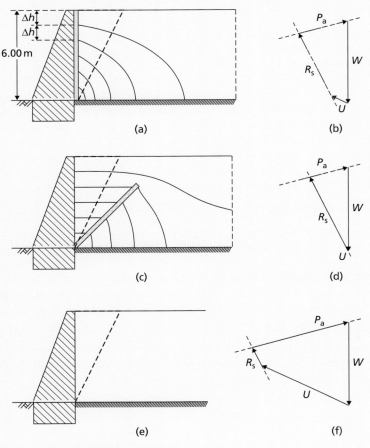

Figure 11.23 Example 11.5.

Applications in geotechnical engineering

11.6 Backfilling and compaction-induced earth pressures

If retaining walls are used to create elevated ground (Figure 11.1(b)), it is common to construct the retaining structure to the desired level first and subsequently **backfill** behind the structure with fill. This soil is known as backfill, and is commonly compacted in-situ to achieve optimum density (and, hence, engineering performance; see Section 1.7). The resultant lateral pressure against the retaining structure is influenced by the compaction process, an effect that was not considered in the earth pressure theories described previously. During backfilling, the weight of the compaction plant produces additional lateral pressure on the wall. Pressures significantly in excess of the active value can result near the top of the wall, especially if it is restrained by propping during compaction. As each layer is compacted, the soil adjacent to the wall is pushed downwards against the frictional resistance on the wall surface (τ_w). When the compaction plant is removed the potential rebound of the soil is restricted by wall friction, thus inhibiting reduction of the additional lateral pressure. Also, the lateral strains induced by compaction have a significant plastic component which is irrecoverable. Thus, there is a residual lateral pressure on the wall. A simple analytical method of estimating this residual lateral pressure has been proposed by Ingold (1979).

Compaction of backfill behind a retaining wall is normally effected by rolling. The compaction plant can be represented approximately by a line load equal to the weight of the roller. If a vibratory roller is employed, the centrifugal force due to the vibrating mechanism should be added to the static weight. Referring to Figure 11.24, the stresses at point X due to a line load of Q per unit length on the surface are as follows:

$$\sigma_z = \frac{2Q}{\pi} \frac{z^3}{\left(x^2 + z^2\right)^2} \qquad (11.62)$$

$$\sigma_x = \frac{2Q}{\pi} \frac{x^2 z}{\left(x^2 + z^2\right)^2} \qquad (11.63)$$

$$\tau_{xz} = \frac{2Q}{\pi} \frac{xz^2}{\left(x^2 + z^2\right)^2} \qquad (11.64)$$

From Equation 11.62, the vertical stress immediately below the line load is

$$\sigma_z = \frac{2Q}{\pi z}$$

Then the lateral pressure on the wall at depth z is given by

$$p_c = K_a \left(\gamma z + \sigma_z\right)$$

When the stress σ_z is removed, the lateral stress may not revert to the original value ($K_a \gamma z$). At shallow depth the residual lateral pressure could be high enough, relative to the vertical stress γz, to cause passive failure in the soil. Therefore, assuming there is no reduction in lateral stress on removal of the compaction plant, the maximum (or critical) depth (z_c) to which failure could occur is given by

$$p_c = K_p \gamma z_c$$

Thus

$$K_a \left(\gamma z_c + \sigma_z\right) = \frac{1}{K_a} \gamma z_c$$

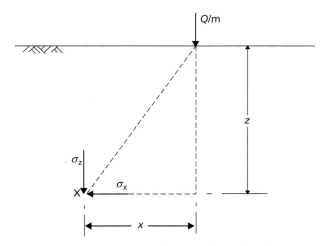

Figure 11.24 Stresses due to a line load.

If it is assumed that γz_c is negligible compared to σ_z, then

$$z_c = \frac{K_a^2 \sigma_z}{\gamma}$$

$$= \frac{K_a^2 2Q}{\gamma \pi z_c}$$

Therefore

$$z_c = K_a \sqrt{\frac{2Q}{\pi \gamma}}$$

The maximum value of lateral pressure (p_{max}) occurs at the critical depth, therefore (again neglecting γz_c)

$$p_{max} = \frac{2QK_a}{\pi z_c}$$

$$= \sqrt{\frac{2Q\gamma}{\pi}} \qquad (11.65)$$

Backfill is normally placed and compacted in layers. Assuming that the pressure p_{max} is reached, and remains, in each successive layer, a vertical line can be drawn as a pressure envelope below the critical depth. Thus, the distribution shown in Figure 11.25 represents a conservative basis for design. However, at a depth z_a the active pressure will exceed the value p_{max}. The depth z_a, being the limiting depth of the vertical envelope, is obtained from the equation

$$K_a \gamma z_a = \sqrt{\frac{2Q\gamma}{\pi}}$$

Thus

$$z_a = \frac{1}{K_a} \sqrt{\frac{2Q}{\pi \gamma}} \qquad (11.66)$$

Applications in geotechnical engineering

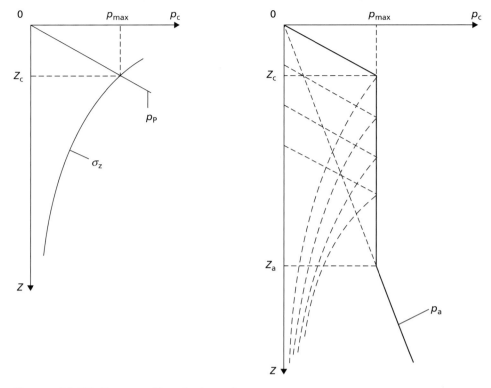

Figure 11.25 Compaction-induced pressure.

11.7 Embedded walls

Cantilever walls

Walls of this type are mainly of steel sheet piling, and are used only when the retained height of soil is relatively low. In sands and gravels these walls may be used as permanent structures, but in general they are used only for temporary support. The stability of the wall is due entirely to passive resistance mobilised in front of the wall. The principal limit states considered previously for gravity retaining structures are then replaced by:

> ULS-1 Horizontal translation of the wall;
> ULS-2 Rotation of the wall;
> ULS-3 Bearing failure of the wall (acting as a pile) under its own self weight and interface shear forces from the retained soil.

Limit states ULS-4 to ULS-6 listed for gravity walls (Section 11.4) should also be considered, as should serviceability limit states SLS-1 and SLS-2. The mode of failure of an embedded wall is by rotation about a point O near the lower end of the wall, as shown in Figure 11.26(a). Consequently, passive resistance acts in front of the wall above O and behind the wall below O, as shown in Figure 11.26(b),

Retaining structures

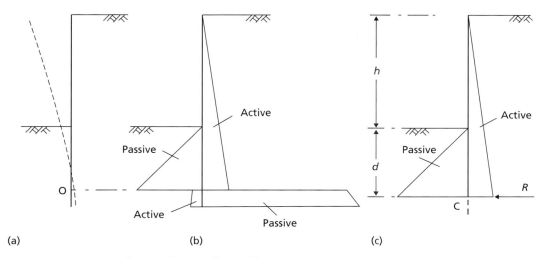

Figure 11.26 Cantilever sheet pile wall.

thus providing a fixing moment. However, this pressure distribution is an idealisation, as there is unlikely to be a complete change in passive resistance from the front to the back of the wall at point O. To allow for over-excavation, it is recommended that the soil level in front of the wall should be reduced by 10% of the retained height, subject to a maximum of 0.5 m (overdig).

Design is generally based on the simplification shown in Figure 11.26(c), it being assumed that the net passive resistance below point O is represented by a concentrated force R acting at a point C, slightly below O, at depth d below the lower soil surface. ULS-1 is verified by ensuring that the resultant passive thrust is greater than the resultant active thrust. ULS-2 is verified by ensuring that the (clockwise) restoring moment about O due to the resultant passive thrust of the soil in front of the wall is greater than the (anticlockwise) driving moment due to the resultant active thrust of the soil behind the wall. The methods outlined in Chapters 9 and 10 should be used to verify ULS-3. Given that a cantilever wall is flexible, the check of the internal structural stability and performance of the wall (ULS-6 and SLS-2) is of greater importance than for gravity structures, so it is typical in design to determine the shear force and bending moment distributions in the wall for subsequent structural checks (which are beyond the scope of this book).

Anchored and propped walls

Generally, structures of this type are either of steel sheet piling or reinforced concrete diaphragm walls, the construction of which is described in Section 11.10; however, secant piling may also be used to form a wall which would be anchored or propped. Additional support to embedded walls is provided by a row of **tie-backs** (anchors) or props near the top of the wall, as illustrated in Figure 11.27(a). Tie-backs are normally high-tensile steel cables or rods, anchored in the soil some distance behind the wall. Walls of this type are used extensively in the support of deep excavations and in waterfront construction. In the case of sheet pile walls there are two basic modes of construction. Excavated walls are constructed by driving a row of sheet piling, followed by excavation or dredging to the required depth in front of the wall. Backfilled walls are constructed by partial driving, followed by backfilling to the required height behind the piling (see Section 11.6). In the case of diaphragm walls, excavation takes place in front of the wall after it has been cast in-situ. Stability is due to the passive resistance developed in front of the wall together with the supporting forces in the ties or props.

The limit states to be considered include those listed above for cantilever walls. In addition, the following two limit states must be considered:

Applications in geotechnical engineering

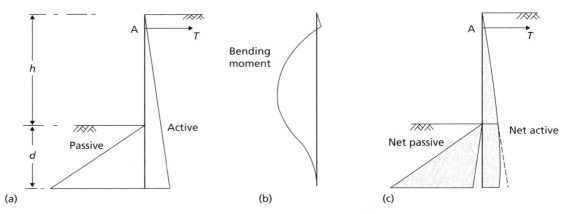

Figure 11.27 Anchored sheet pile wall: free earth support method.

> ULS-7 Failure of anchors/ties by pull-out from the soil (anchored walls with ties in tension only). This is essentially a check of the soil–tie interface strength.
> ULS-8 Structural failure of ties/props. For props which are loaded in compression, failure would be by buckling; in anchors tie failure would be by fracture. This is essentially a check of the structural strength of the tie/prop itself.

An additional serviceability limit state is

> SLS-3 Yield of ties/props should be minimal.

These additional limiting states are described in greater detail in Section 11.8.

The consequences of not meeting these limiting states can be severe. On 20 April 2004, a section of a 33-m deep propped excavation collapsed in Singapore (Figure 11.28). The failure of the wall led to collapse of a zone of soil 150×100 m in plan and 30 m deep, making the adjacent Nicoll Highway impassable for 8 months after the collapse. Of the construction workers who were in the excavation at the time of the collapse, four were killed and a further three injured. The causes of the collapse were ultimately identified as underestimation of the thrusts acting on the wall (ULS-1/2) coupled with inadequate structural design of the prop–wall connection (ULS-8). Excessive deformation of the wall was observed prior to collapse, but was not acted upon.

Free earth support analysis for tied/propped walls

It is assumed that the depth of embedment below excavation level is insufficient to produce fixity at the lower end of the wall. Thus, the wall is free to rotate at its lower end, the bending moment diagram being of the form shown in Figure 11.27(b). To satisfy the rotation limit state ULS-2, the sum of the restoring moments (ΣM_R factored as resistances) about the anchor or prop must be greater than or equal to the sum of the overturning moments (ΣM_A factored as actions), i.e.

$$\sum M_A \leq \sum M_R \tag{11.67}$$

Retaining structures

Figure 11.28 Nicoll Highway collapse, Singapore.

ULS-1 is then verified by considering equilibrium of the horizontal forces. For this limiting state the active thrusts behind the wall are still the actions (causing the wall to move outwards), while the prop/tie force and passive thrusts are resistances. This gives the minimum resistance that the prop/tie must be able to provide to satisfy ULS-1. These prop forces are then the characteristic actions in the tie/prop for verification of ULS-7 and ULS-8. The net earth pressure distribution on the wall using the final propping/tie system is then determined under working load conditions (all partial factors = 1.00) to provide the characteristic loadings for verification of the wall's structural stability (ULS-6). Finally, if appropriate, the vertical forces on the wall are calculated and checked using the methods in Chapter 9, it being a requirement that the downward force (e.g. the component of the force in an inclined tie-back) should not exceed the (upward) frictional resistance available between the wall and the soil on the passive side minus the (downward) frictional force on the active side (ULS-3).

When applying free earth support analysis in ULS design, the active (effective) earth pressures which result in overturning moments are considered as actions and factored accordingly (as for gravity retaining structures). The passive earth pressures acting in front of the wall which result in restoring moments are treated as resistances. The partial factors for earth pressure resistance $\gamma_{Re} = 1.00$ for sets R1 and R3, and 1.40 for set R2. These factors are included on the EC7 quick reference sheet on the Companion Website. The net pore pressure distribution on an embedded wall will always be an overturning moment, and is therefore treated as an action.

It should be realised that full passive resistance is only developed under conditions of limiting equilibrium, i.e. when the structure is at the pint of failure. Under working conditions, analytical and experimental work has indicated that the distribution of lateral pressure is likely to be of the form shown in Figure 11.29, with passive resistance being fully mobilised close to the lower surface. The extra depth of embedment required to provide adequate safety results in a partial fixing moment at the lower end of the wall and, consequently, a lower maximum bending moment than the value under limiting equilibrium or collapse conditions. In view of the uncertainty regarding the pressure

Applications in geotechnical engineering

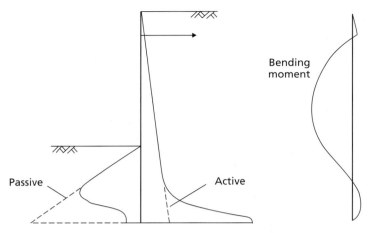

Figure 11.29 Anchored sheet pile wall: pressure distribution under working conditions.

distributions under working conditions, it is recommended that bending moments and tie or prop force under limiting equilibrium conditions should be used in the structural design of the wall. The tie or prop force thus calculated should be increased by 25% to allow for possible redistribution of pressure due to arching (see below). Bending moments should be calculated on the same basis in the case of cantilever walls.

The behaviour of an anchored wall is also influenced by its degree of flexibility or stiffness. In the case of flexible sheet pile walls, experimental and analytical results indicate that redistributions of lateral pressure take place. The pressures on the yielding parts of the wall (between the tie and excavation level) are reduced and those on the unyielding parts (in the vicinity of the tie and below excavation level) are increased with respect to the theoretical values, as illustrated in Figure 11.30. These redistributions of lateral pressure are the result of the phenomenon known as **arching**. No such redistributions take place in the case of stiff walls, such as concrete diaphragm walls (Section 11.10).

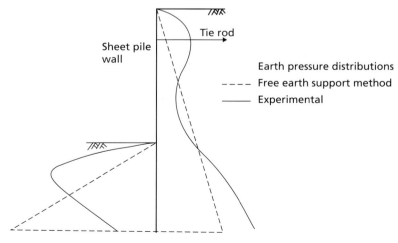

Figure 11.30 Arching effects.

Retaining structures

Arching was defined by Terzaghi (1943) in the following way:

> If one part of the support of a soil mass yields while the remainder stays in place, the soil adjoining the yielding part moves out of its original position between adjacent stationary soil masses. The relative movement within the soil is opposed by shearing resistance within the zone of contact between the yielding and stationary masses. Since the shearing resistance tends to keep the yielding mass in its original position, the pressure on the yielding part of the support is reduced and the pressure on the stationary parts is increased. This transfer of pressure from a yielding part to adjacent non-yielding parts of a soil mass is called the arching effect. Arching also occurs when one part of a support yields more than the adjacent parts.

The conditions for arching are present in anchored sheet pile walls when they deflect. If yield of an anchor takes place (ULS-7), arching effects are reduced to an extent depending on the amount of yielding. On the passive side of the wall, the pressure is increased just below excavation level as a result of larger deflections into the soil. In the case of backfilled walls, arching is only partly effective until the fill is above tie level. Arching effects are much greater in sands than in silts or clays, and are greater in dense sands than in loose sands.

Generally, redistributions of earth pressure result in lower bending moments than those obtained from the free earth support method of analysis; the greater the flexibility of the wall, the greater the moment reduction. However, for stiff walls, such as diaphragm walls, formed by excavation in soils having a high K_0 value (in the range 1–2), such as overconsolidated clays, Potts and Fourie (1984, 1985) showed that both maximum bending moment and prop force could be significantly higher than those obtained using the free earth support method.

Pore water pressure distribution

Embedded walls are normally analysed in terms of effective stresses. Care is therefore required in deciding on the appropriate distribution of pore water pressure. Several different situations are illustrated in Figure 11.31.

If the water table levels are the same on both sides of the wall, the pore water pressure distributions will be hydrostatic and will balance (Figure 11.31(a)); they can thus be eliminated from the calculations, as the resultant thrusts and moments due to the pore water on either side of the wall will balance.

If the water table levels are different and if steady seepage conditions have developed and are maintained, the distributions on the two sides of the wall will be unbalanced. The pressure distributions on each side of the wall may be combined into a single net pressure distribution because no earth pressure coefficient is involved. The net distribution on the back of the wall could be determined from a flow net, as illustrated in Example 2.2, or using the FDM spreadsheet accompanying Chapter 2. However, in many situations an approximate distribution, ABC in Figure 11.31(b), can be obtained by assuming that the

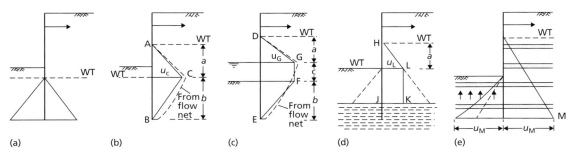

Figure 11.31 Various pore water pressure distributions.

Applications in geotechnical engineering

total head is dissipated uniformly along the back and front wall surfaces between the two water table levels. The maximum net pressure occurs opposite the lower water table level and, referring to Figure 11.31(b), is given by

$$u_C = \frac{2ba}{2b+a}\gamma_w \tag{11.68}$$

In general, this approximate method will underestimate net water pressure, especially if the bottom of the wall is relatively close to the lower boundary of the flow region (i.e. if there are large differences in the sizes of curvilinear squares in the flow net approaching the bottom of the wall). The approximation should not be used in the case of a narrow excavation between two lines of sheet piling where curvilinear squares are relatively small (and seepage pressure relatively high) approaching the base of the excavation. In these cases, the FDM spreadsheet or a flow net should be used.

In Figure 11.31(c), a depth of water is shown in front of the wall, the water level being below that of the water table behind the wall. In this case the approximate distribution DEFG should be used in appropriate cases, the net pressure at G being given by

$$u_G = \frac{(2b+c)a}{2b+c+a}\gamma_w \tag{11.69}$$

If $c=0$, then Equation 11.69 reduces to Equation 11.68.

A wall constructed mainly in a soil of relatively high permeability but penetrating a layer of low permeability layer (typically clay) is shown in Figure 11.31(d). This may be done to reduce the amount of seepage underneath the wall by using the low permeability soil as a natural barrier to seepage. If undrained conditions apply within the clay, the pore water pressure in the overlying soil would be hydrostatic and the net pressure distribution would be HJKL as shown, where

$$u_L = a\gamma_w \tag{11.70}$$

A wall constructed in a low permeability soil (e.g. clay) which contains thin layers or partings of high permeability material (e.g. sand or silt) is shown in Figure 11.31(e). In this case, it should be assumed that the sand or silt allows water at hydrostatic pressure to reach the back surface of the wall. This implies pressure in excess of hydrostatic, and consequent upward seepage in front of the wall (see Section 3.7).

For short-term situations for walls in clay (e.g. during and immediately after excavation), there exists the possibility of tension cracks developing or fissures opening. If such cracks or fissures fill with water, hydrostatic pressure should be assumed over the depth in question. The water in the cracks or fissures would also result in softening of the clay. Softening would also occur near the soil surface in front of the wall as a result of stress relief on excavation. An effective stress analysis would ensure a safe design in the event of rapid softening of the clay taking place, or if work were delayed during the temporary stage of construction.

Under conditions of steady seepage, use of the approximation that total head is dissipated uniformly along the wall has the consequent advantage that the seepage pressure is constant. For the conditions shown in Figure 11.31(b), for example, the seepage pressure at any depth is

$$j = \frac{a}{2b+a}\gamma_w \tag{11.71}$$

The effective unit weight of the soil below the water table, therefore, would be increased to $\gamma'+j$ behind the wall, where seepage is downwards, and reduced to $\gamma'-j$ in front of the wall, where seepage is upwards. These values should be used in the calculation of effective active and passive pressures, respectively, if groundwater conditions are such that steady seepage is maintained. Thus, active pressures are increased and passive pressures are decreased relative to the corresponding static values.

Example 11.6

The sides of an excavation 2.25 m deep in sand are to be supported by a cantilever sheet pile wall, the water table being 1.25 m below the bottom of the excavation. The unit weight of the sand above the water table is 17 kN/m³, and below the water table the saturated unit weight is 20 kN/m³. Characteristic strength parameters are $c'=0$, $\phi'=35°$ and $\delta'=0$. Allowing for a surcharge pressure of 10 kPa on the surface, determine the required depth of embedment of the piling to satisfy the rotational (ULS-2) and translational (ULS-1) limit states to Eurocode 7, DA1b.

Solution

Below the water table, the effective unit weight of the soil is $(20-9.81)=10.2$ kN/m³. To allow for possible over-excavation the soil level should be reduced by 10% of the retained height of 2.25 m, i.e. by 0.225 m. The depth of the excavation therefore becomes 2.475 m, say 2.50 m, and the water table will be 1.00 m below this level.

The design dimensions and the earth pressure diagrams are shown in Figure 11.32. The distributions of hydrostatic pressure on the two sides of the wall balance and can be eliminated from the calculations (there is no seepage occurring). For applying partial factors:

- the active earth pressures on the retained side (2–4 in Figure 11.32) are unfavorouable permanent actions (causing the wall to rotate or slide);
- the horizontal earth pressure due to the surcharge (1 in Figure 11.32) is an unfavourable variable action;
- the passive earth pressures in front of the wall (5–7 in Figure 11.32) are treated as resistances (as they prevent the wall rotating or sliding) and factored by γ_{Re}.

The procedure is to check the rotational stability first by applying Equation 11.67 about point C. The design value of ϕ' for DA1b is $\tan^{-1}(\tan 35°/1.25)=29°$. The corresponding values of K_a and K_p are 0.35 and 2.88, respectively, using Equations 11.5 and 11.7, or 11.23 and 11.27. The design (factored) values of forces, lever arms and moments are set out in Table 11.4.

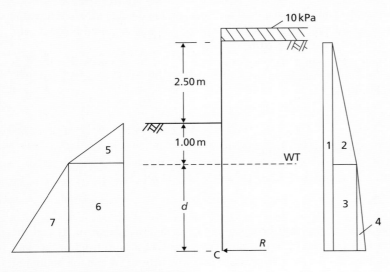

Figure 11.32 Example 11.6.

Table 11.4 Example 11.6

	Force (kN/m)	Lever arm (m)	Moment (kNm/m)
	Actions (H_A, M_A):		
(1)	$0.33 \cdot 10 \cdot (d+3.5) \cdot 1.30 = 4.29d + 15.02$	$\dfrac{d+3.5}{2}$	$2.15d^2 + 15.02d + 26.29$
(2)	$\frac{1}{2} \cdot 0.33 \cdot 17 \cdot 3.5^2 \cdot 1.00 = 34.36$	$d + \dfrac{3.5}{3}$	$34.36d + 40.09$
(3)	$0.33 \cdot 17 \cdot 3.5 \cdot d \cdot 1.00 = 19.64d$	$\dfrac{d}{2}$	$9.82d^2$
(4)	$\frac{1}{2} \cdot 0.33 \cdot 10.2 \cdot d^2 \cdot 1.00 = 1.68d^2$	$\dfrac{d}{3}$	$0.56d^3$
	Resistances (H_R, M_R):		
(5)	$\dfrac{\frac{1}{2} \cdot 2.88 \cdot 17 \cdot 1.0^2}{1.00} = 24.48$	$d + \dfrac{1.0}{3}$	$24.48d + 8.16$
(6)	$\dfrac{2.88 \cdot 17 \cdot 1 \cdot d}{1.00} = 48.96d$	$\dfrac{d}{2}$	$24.48d^2$
(7)	$\dfrac{\frac{1}{2} \cdot 2.88 \cdot 10.2 \cdot d^2}{1.00} = 14.69d^2$	$\dfrac{d}{3}$	$4.90d^3$

Applying Equation 11.67 to check ULS-2, the minimum depth of embedment will just satisfy ULS, so:

$$\sum M_A = \sum M_R$$
$$0.56d^3 + 11.97d^2 + 49.38d + 66.38 = 4.90d^3 + 24.48d^2 + 24.48d + 8.16$$
$$0 = 4.34d^3 + 12.51d^2 - 24.9d - 58.22$$

The resulting cubic equation can be solved by standard methods, giving $d = 2.27$ m. Therefore, the required depth of embedment accounting for the additional depth of wall required below C and o-verdig $= 1.2(2.27 + 1.00) + 0.25 = 4.18$ m, say 4.20 m.

To check ULS-1, horizontal equilibrium is considered with $d = 2.27$ m to check that the R is sufficient for fixity, compared with the net passive resistance available over the additional 20% embedment depth. From Figure 11.32,

$$R + \sum H_A = \sum H_R$$
$$R = 24.48 + (48.96 \cdot 2.27) + (14.69 \cdot 2.27^2) -$$
$$(4.29 \cdot 2.27 + 15.02) - 34.36 - (19.64 \cdot 2.27) - (1.68 \cdot 2.27^2)$$
$$= 99.0 \text{ kN/m}$$

Passive pressure acts on the back of the wall between depths of 5.77 m (depth of R) and the bottom of the wall at 6.70 m (see Figure 11.26b), while active pressure acts in front of the wall over the same distance. Therefore, the net passive pressure half way between R and the bottom of the wall (6.24 m) is

$$p'_p - p'_a = \left[2.88 \cdot 10 \cdot 6.24 + 2.88 \cdot 17 \cdot 3.5 + 2.88 \cdot 10.2 \cdot (6.24 - 3.5)\right] -$$
$$\left[0.35 \cdot 17 \cdot 1 + 0.35 \cdot 10.2 \cdot (6.24 - 3.5)\right]$$
$$= 415.8 \, \text{kPa}$$

The net passive resistance available over the additional embedded depth ($P_{p,net}$) must then be greater than or equal to R to satisfy ULS-1:

$$P_{p,\,net} = 415.8 \cdot (6.70 - 5.77)$$
$$= 386.7 \, \text{kN/m}$$

$P_{p,net} > R$, so ULS-1 is satisfied.

The procedure detailed above could then be repeated for any other design approaches as required by altering the partial factors applied in Table 11.4.

Example 11.7

A propped cantilever wall supporting the sides of an excavation in stiff clay is shown in Figure 11.33. The saturated unit weight of the clay (above and below the water table) is $20 \, \text{kN/m}^3$. The design values of the active and passive coefficients of lateral earth pressure are 0.30 and 4.2, respectively. Assuming conditions of steady-state seepage, determine the required depth of embedment for the wall to be stable (use EC7 DA1b). Determine also the force in each prop.

Solution

The distributions of earth pressure and net pore water pressure (assuming uniform decrease of total head around the wall as shown in Figure 11.31b) are shown in Figure 11.33.

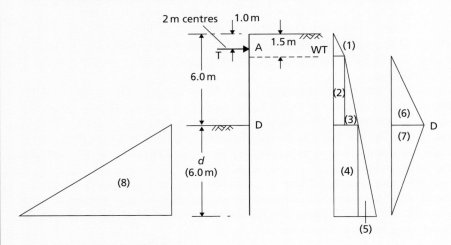

Figure 11.33 Example 11.7.

The maximum net water pressure at level D from Equation 11.68 is:

$$u_D = \frac{2 \cdot d \cdot (6.0-1.5)}{2 \cdot d + (6.0-1.5)} \cdot 9.81 = \frac{88.3d}{2d+4.5} \text{ kPa}$$

and the average seepage pressure is

$$j = \frac{(6.0-1.5)}{2 \cdot d + (6.0-1.5)} \cdot 9.81 = \frac{44.1}{2d+4.5} \text{ kPa}$$

Thus, below the water table, active forces are calculated using an effective unit weight of

$$(\gamma' + j) = 10.2 + \frac{44.1}{2d+4.5} \text{ kN/m}^3$$

and passive forces are calculated using an effective unit weight of

$$(\gamma' + j) = 10.2 - \frac{44.1}{2d+4.5} \text{ kN/m}^3$$

If the forces, lever arms and moments are expressed in terms of the unknown embedment depth d, complex algebraic expressions would result; thus it is preferable to assume a series of trial values of d and check ULS-2 for each. If the ULS is satisfied at the initial trial value of d, further trials should be made, reducing d until the ULS is just not satisfied. Equally, if the ULS is not satisfied on the first trial, further trials should increase d until ULS is just satisfied. In either case, the final value at which ULS is just satisfied represents the minimum value of d. To avoid the unknown prop force, T, ULS-2 is verified by taking moments about point A through which T acts.

Following this procedure, a trial value of $d=6.0$ m is first selected. Then, $u_D=32.1$ kPa, $(\gamma'+j)=12.9$ kN/m³ and $(\gamma'-j)=7.5$ kN/m³. The active thrusts generate disturbing moments and are factored as unfavourable permanent actions as before. The net pressure due to seepage also acts in this direction and is factored in the same way. The passive thrusts are factored as resistances. For DA1b, all of these factors are 1.00; they are, however, included in the calculations below for completeness. The calculations for $d=6.0$ m are then shown in Table 11.5, from which $\Sigma M_A = 3068.3$ kNm/m $> \Sigma M_R = 5103.0$, satisfying ULS-2 and suggesting that d can be reduced to produce a more efficient design.

The calculations may be input into a spreadsheet in a table similar to the above, but as a function of d. An optimisation tool (e.g. Solver in MS Excel) can then be used to find the value of d which makes ULS-2 just satisfied. An example of this approach is provided on the Companion Website, from which $d=4.57$ m (for DA1b). The use of a spreadsheet makes it straightforward to consider other design approaches, as it is only necessary to change the partial factors appropriately and re-run the optimisation.

The load carried in the propping should then be calculated from limiting (horizontal) equilibrium. The spreadsheet which was used to find the optimum value of d can also be used to check this, from which $T=122.2$ kN/m. Multiplying the calculated value by 1.25 to allow for arching, the force in each prop when spaced at 2-m centres is

$$1.25 \cdot 2 \cdot 122.2 = 306 \text{ kN}$$

Retaining structures

Table 11.5 Example 11.7 (case $d = 6.0$ m)

	Force (kN/m)		Lever arm (m)	Moment (kNm/m)
	Actions (H_A, M_A):			
(1)	$\frac{1}{2} \cdot 0.30 \cdot \left(\frac{20}{1.00}\right) \cdot 1.5^2 \times 1.00$	= 6.8	0.00	= 0.0
(2)	$0.30 \cdot \left(\frac{20}{1.00}\right) \cdot 1.5 \cdot 4.5 \cdot 1.00$	= 40.5	2.75	= 111.4
(3)	$\frac{1}{2} \cdot 0.30 \cdot \left(\frac{12.9}{1.00}\right) \cdot 4.5^2 \times 1.00$	= 39.2	3.50	= 137.2
(4)	$0.30 \cdot \left[\left(\frac{20}{1.00} \cdot 1.5\right) + \left(\frac{12.9}{1.00} \cdot 4.5\right)\right] \cdot 6.0 \cdot 1.00$	= 158.2	8.00	= 1265.6
(5)	$\frac{1}{2} \cdot 0.30 \cdot \left(\frac{12.9}{1.00}\right) \cdot 6.0^2 \times 1.00$	= 69.7	9.00	= 627.3
(6)	$\frac{1}{2} \cdot 32.1 \cdot 4.5 \cdot 1.00$	= 72.2	3.50	= 252.7
(7)	$\frac{1}{2} \cdot 32.1 \cdot 6.0 \cdot 1.00$	= 96.3	7.00	= 674.1
	Resistances (H_R, M_R):			
(8)	$\dfrac{\frac{1}{2} \cdot 4.2 \cdot \left(\frac{7.5}{1.00}\right) \cdot 6.0^2}{1.00}$	= 567.0	9.00	= 5103.0

11.8 Ground anchorages

Tie rods are normally anchored in beams, plates or concrete blocks (known as dead-man anchors) some distance behind the wall (Figure 11.34(a)). The tie rod force from a free earth support analysis (T) is resisted by the passive resistance developed in front of the anchor, reduced by the active pressure acting on the back. To avoid the possibility of progressive failure of a line of ties, it should be assumed that any single tie could fail either by fracture or by becoming detached and that its load could be redistributed safely to the two adjacent ties. Accordingly, it is recommended that a load factor of at least 2.0 should be applied to the tie rod force, in addition to any partial factors involved in the ULS checks. The following sections describe the calculation models for determining the pull-out resistance of anchorages (T_f) for verification of ULS-7. The normative values of the partial factor which should be applied to this resistance according to Eurocode 7 are $\gamma_{Ra} = 1.10$ for sets R1 and R2, and 1.00 for set R3. These factors are included on the EC7 quick reference sheet on the Companion Website. For ULS-7 to be verified:

$$T \leq T_f \tag{11.72}$$

Plate anchors

If the width (b) of the anchor is not less than half the depth (d_a) from the surface to the bottom of the anchor, it can be assumed that passive resistance is developed over the depth d_a. The anchor must be

Applications in geotechnical engineering

located beyond the plane YZ (Figure 11.34(a)) to ensure that the passive wedge of the anchor does not encroach on the active wedge behind the wall.

The equation of equilibrium for a ground anchor at failure (for ULS-7) is

$$T_f = \frac{1}{2}(K_p - K_a)\gamma d_a^2 l - K_a \sigma_q d_a l \tag{11.73}$$

where l = length of anchor per tie and σ_q = surface surcharge pressure.

Ground anchors

Tensioned cables, attached to the wall and anchored in a mass of cement grout or grouted soil (Figure 11.34(b)), are another means of support. These are known as **ground anchors**. A ground anchor normally consists of a high-tensile steel cable or bar, called the tendon, one end of which is held securely in the soil by a mass of cement grout or grouted soil; the other end of the tendon is anchored against a bearing plate on the structural unit to be supported. While the main application of ground anchors is in the construction of tie-backs for diaphragm or sheet pile walls, other applications are in the anchoring of any structure subjected to overturning, sliding or buoyancy, or in the provision of reaction for in-situ load tests (e.g. pile load testing in Chapter 9). Ground anchors can be constructed in sands (including gravelly sands and silty sands) and stiff clays, and they can be used in situations where either temporary or permanent support is required.

The grouted length of tendon, through which force is transmitted to the surrounding soil, is called the **fixed anchor length**. The length of tendon between the fixed anchor and the bearing plate is called the **free anchor length**; no force is transmitted to the soil over this length. For temporary anchors, the tendon is normally greased and covered with plastic tape over the free anchor length. This allows for free movement of the tendon and gives protection against corrosion. For permanent anchors, the tendon is normally greased and sheathed with polythene under factory conditions; on site, the tendon is stripped and degreased over what will be the fixed anchor length.

The ultimate load which can be carried by an anchor depends on the soil resistance (principally skin friction) mobilised adjacent to the fixed anchor length. This, of course, assumes that there will be no prior failure at the grout–tendon interface, or of the tendon itself (i.e. ULS-7 will be achieved before ULS-8). Anchors are usually prestressed in order to reduce the lateral displacement required to mobilise soil resistance, and to minimise ground movements in general. Each anchor is subjected to a test loading after installation, temporary anchors usually being tested to 1.2 times the working load and permanent anchors to 1.5 times the working load. Finally, prestressing of the anchor takes place. Creep

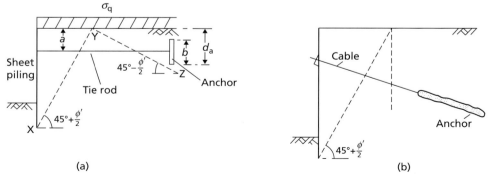

Figure 11.34 Anchorage types: (a) plate anchor, (b) ground anchor.

Retaining structures

displacements under constant load will occur in ground anchors. A creep coefficient, defined as the displacement per unit log time, can be determined by means of a load test.

A comprehensive ground investigation is essential in any location where ground anchors are to be employed. The soil profile must be determined accurately, any variations in the level and thickness of strata being particularly important. In the case of sands, the particle size distribution should be determined in order that permeability and grout acceptability can be estimated.

Design of ground anchors in coarse-grained soils

In general, the sequence of construction is as follows. A cased borehole (diameter usually within the range 75–125 mm) is advanced through the soil to the required depth. The tendon is then positioned in the hole, and cement grout is injected under pressure over the fixed anchor length as the casing is withdrawn. The grout penetrates the soil around the borehole, to an extent depending on the permeability of the soil and on the injection pressure, forming a zone of grouted soil, the diameter of which can be up to four times that of the borehole (Figure 11.35(a)). Care must be taken to ensure that the injection pressure does not exceed the overburden pressure of the soil above the anchor, otherwise heaving or fissuring may result. When the grout has achieved adequate strength, the other end of the tendon is anchored against the bearing plate. The space between the sheathed tendon and the sides of the borehole, over the free anchor length, is normally filled with grout (under low pressure); this grout gives additional corrosion protection to the tendon.

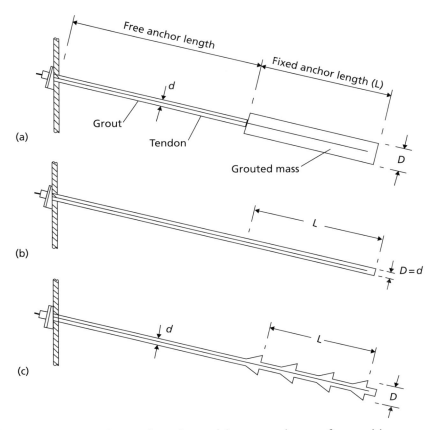

Figure 11.35 Ground anchors: (a) grouted mass formed by pressure injection, (b) grout cylinder, and (c) multiple under-reamed anchor.

449

Applications in geotechnical engineering

The ultimate resistance of an anchor to pull-out (ULS-7) is equal to the sum of the side resistance and the end resistance of the grouted mass. Considering the anchor to act as a pile, the following theoretical expression has been proposed:

$$T_f = A\sigma'_v \pi DL \tan\phi' + B\gamma'h \frac{\pi}{4}(D^2 - d^2) \tag{11.74}$$

where T_f=pull-out capacity of anchor, A=ratio of normal pressure at interface to effective overburden pressure (essentially an earth pressure coefficient), σ'_v=effective overburden pressure adjacent to the fixed anchor and B=bearing capacity factor.

It was suggested that the value of A is normally within the range 1–2. The factor B is analogous to the bearing capacity factor N_q in the case of piles, and it was suggested that the ratio N_q/B is within the range 1.3–1.4, using the N_q values of Berezantzev et al. (1961) which can be found in Figure 9.6. However, Equation 11.74 is unlikely to represent all the relevant factors in a complex problem. The ultimate resistance also depends on details of the installation technique, and a number of semi-empirical formulae have been proposed by specialist contractors, suitable for use with their particular technique. An example of such a formula is

$$T_f = Ln \tan\phi' \tag{11.75}$$

The value of the empirical factor n is normally within the range 400–600 kN/m for coarse sands and gravels, and within the range 130–165 kN/m for fine to medium sands.

Design of ground anchors in fine-grained soils

The simplest construction technique for anchors in stiff clays is to auger a hole to the required depth, position the tendon and grout the fixed anchor length using a tremie pipe (Figure 11.35(b)). However, such a technique would produce an anchor of relatively low capacity because the skin friction at the grout–clay interface would be unlikely to exceed $0.3c_u$ (i.e. $\alpha \leq 0.3$).

Anchor capacity can be increased by the technique of gravel injection. The augered hole is filled with pea gravel over the fixed anchor length, then a casing, fitted with a pointed shoe, is driven into the gravel, forcing it into the surrounding clay. The tendon is then positioned and grout is injected into the gravel as the casing is withdrawn (leaving the shoe behind). This technique results in an increase in the effective diameter of the fixed anchor (of the order of 50%) and an increase in side resistance: a value of $\alpha \approx 0.6$ can be expected. In addition, there will be some end resistance. The borehole is again filled with grout over the free anchor length.

Another technique employs an expanding cutter to form a series of enlargements (or under-reams) of the augered hole at close intervals over the fixed anchor length (Figure 11.35(c)); the cuttings are generally removed by flushing with water. The cable is then positioned and grouting takes place. A value of $\alpha \approx 1$ can normally be assumed along the cylindrical surface through the extremities of the enlargements.

The following design formula (analogous to Equation 11.74) can be used for anchors in undrained soil conditions (at ULS-7):

$$T_f = \pi DL\alpha c_u + \frac{\pi}{4}(D^2 - d^2)c_u N_c \tag{11.76}$$

where T_f=pull-out capacity of anchor, L=fixed anchor length, D=diameter of fixed anchor, d=diameter of borehole, α=skin friction coefficient and N_c=bearing capacity factor (generally assumed to be 9). Resistance at the grout–clay interface along the free anchor length may also be taken into account.

Retaining structures

Example 11.8

Details of an anchored sheet pile wall are given in Figure 11.36, the design ground and water levels being as shown. The ties are spaced at 2.0-m centres. Above the water table the unit weight of the soil is $17\,\text{kN/m}^3$, and below the water table the saturated unit weight is $20\,\text{kN/m}^3$. Characteristic soil parameters are $c'=0$, $\phi'=36°$, and δ' is taken to be $1/2\phi'$. Determine the required depth of embedment and the minimum capacity of each tie to satisfy ULS to Eurocode 7 (DA1b). Design a continuous anchor to support the ties.

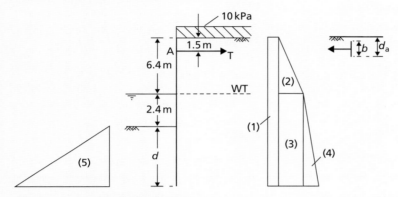

Figure 11.36 Example 11.8.

Solution

For DA1b, the design value of $\phi'=\tan^{-1}(\tan 36°/1.25) = 30°$. Hence (for $\delta'=1/2\phi'$) $K_a=0.29$ and $K_p=4.6$. The lateral pressure diagrams are shown in Figure 11.36. The water levels on the two sides of the wall are equal, therefore the hydrostatic pressure distributions are in balance and can be eliminated from the calculations. The forces and their lever arms are set out in Table 11.6. Force (1) in Table 11.6 is multiplied by the partial factor 1.30, surcharge being a variable unfavourable action. The partial factor for all other forces, being permanent unfavourable actions, is 1.00.

Applying Equation 11.67 to check ULS-2, the minimum depth of embedment will just satisfy ULS, so:

$$\sum M_A = \sum M_R$$
$$0.99d^3 + 32.02d^2 + 312.6d + 893.12 = 15.63d^3 + 171.11d^2$$
$$0 = 14.64d^3 + 139.09d^2 - 312.6d - 893.12$$

The resulting cubic equation can be solved by standard methods, giving $d=3.19\,\text{m}$. For ULS-1 to be satisfied,

$$\sum H_A = \sum H_R$$
$$1.48d^2 + 42.41d + 218.38 = 23.44d^2 + T$$
$$T = -21.96 \cdot (3.19)^2 + 42.41 \cdot (3.19) + 218.38$$
$$= 130.2\,\text{kN/m}$$

Applications in geotechnical engineering

Table 11.6 Example 11.8

Force (kN/m)			Lever arm (m)
	Actions (H_A, M_A):		
(1)	$1.30 \cdot 0.29 \cdot 10 \cdot (d+8.8)$	$= 3.77d + 33.18$	$\frac{1}{2}d + 2.9$
(2)	$1.00 \cdot \frac{1}{2} \cdot 0.29 \cdot \left(\frac{17}{1.00}\right) \cdot 6.4^2$	$= 100.97$	2.77
(3)	$1.00 \cdot 0.29 \cdot \left(\frac{17}{1.00}\right) \cdot 6.4 \, (d + 2.4)$	$= 31.55d + 75.72$	$\frac{1}{2}d + 6.1$
(4)	$1.00 \cdot \frac{1}{2} \cdot 0.29 \cdot \left(\frac{20-9.81}{1.00}\right) \cdot 6.4 \, (d+2.4)^2$	$= 1.48d^2 + 7.09d + 8.51$	$\frac{2}{3}d + 6.5$
	Resistances (H_R, M_R):		
(5)	$1.00 \cdot \frac{1}{2} \cdot 4.6 \cdot \left(\frac{20-9.81}{1.00}\right) \cdot d^2$	$= 23.44d^2$	$\frac{2}{3}d + 7.3$
Tie		$= T$	0

Hence the force in each tie $= 2 \times 130.2 = 260$ kN. The design load to be resisted by the anchor is 130.2 kN/m. Therefore, the minimum value of d_a is given from Equation 11.73 as

$$\frac{T_f}{l} = \frac{1}{2}(K_p - K_a)\gamma d_a^2 - K_a \sigma_q d_a$$

$$130.2 = \frac{1}{2} \cdot (4.6 - 0.29) \cdot \left(\frac{17}{1.00}\right) \cdot d_a^2 - 0.29 \cdot 10 \cdot d_a$$

$$0 = 36.64 d_a^2 - 2.90 d_a - 130.2$$

$$\therefore d_a = 1.93 \, \text{m}$$

Then the vertical dimension (*b*) of the anchor $= 2(1.93 - 1.5) = 0.86$ m.

11.9 Braced excavations

Sheet piling or timbering is normally used to support the sides of deep, narrow excavations, stability being maintained by means of struts acting across the excavation, as shown in Figure 11.37(a). The piling is usually driven first, the struts being installed in stages as excavation proceeds. When the first row of struts is installed, the depth of excavation is small and no significant yielding of the soil mass will have taken place. As the depth of excavation increases, significant yielding of the soil occurs before strut installation but the first row of struts prevents yielding near the surface. Deformation of the wall, therefore, will be of the form shown in Figure 11.37(a), being negligible at the top and increasing with depth. Thus the deformation condition for active conditions in Figure 11.13 is not satisfied and active earth pressures cannot be assumed to act on such walls. Failure of the soil will take place along a slip surface of the form shown in Figure 11.37(a), only the lower part of the soil wedge within this surface reaching a state of plastic equilibrium, the upper part remaining in a state of elastic equilibrium. The limit states outlined earlier for propped walls must be met in design; however, because active conditions are not mobilised, free-earth support analysis should not be used to determine the strut/prop forces for ULS-8. An alternative procedure is described below. Additionally, because propping generally allows excavation to be made to greater depths, there will be a larger stress relief due to excavation which may result in heave of the soil at the bottom of the excavation in fine-grained soils (also termed **basal heave**). This limit state (denoted ULS-9) is essentially a reverse bearing capacity problem (involving unloading rather than loading of the ground), and is also described below.

Determination of strut forces

Failure of a braced wall is normally due to the initial failure of one of the struts (i.e. at ULS-8), resulting in the progressive failure of the whole system. The forces in the individual struts may differ widely because they depend on such random factors as the force with which the struts are lodged home and the time between excavation and installation of struts. The usual design procedure for braced walls is semi-empirical, being based on actual measurements of strut loads in excavations in sands and clays in a number of locations. For example, Figure 11.37(b) shows the apparent distributions of earth pressure derived from load measurements in the struts at three sections of a braced excavation in a dense sand. Since it is essential that no individual strut should fail, the pressure distribution assumed in design is taken as the envelope covering all the random distributions obtained from field measurements. Such an envelope should not be thought of as representing the actual distribution of earth pressure with depth but as a hypothetical pressure diagram from which the maximum likely characteristic strut loads can be obtained with some degree of confidence.

Based on 81 case studies in a range of soils in the UK, Twine and Roscoe (1999) presented the pressure envelopes shown in Figure 11.37(c) and (d). For soft and firm clays an envelope of the form shown in Figure 11.37(c) is proposed for flexible walls (i.e. sheet pile walls and timber sheeting) and, tentatively, for stiff walls (i.e. diaphragm and contiguous pile walls, see Section 11.10). The upper and lower pressure values are represented by $a\gamma h$ and $b\gamma h$, respectively, where γ is the total unit weight of the soil and h the depth of the excavation, including an allowance for over-excavation. For soft clay with wall elements extending to the base of the excavation, $a = 0.65$ and $b = 0.50$. For soft clay with wall elements embedded below the base of the excavation, $a = 1.15$ and $b = 0.5$. For firm clay, the values of a and b are 0.3 and 0.2, respectively.

The envelopes for stiff and very stiff clay and for coarse soils are rectangular (Figure 11.37(d)). For stiff and very stiff clays, the value of b for flexible walls is 0.3 and for stiff walls is 0.5. For coarse soils $b = 0.2$, but below the water table the pressure is $b\gamma'H$ (γ' being the buoyant unit weight) with hydrostatic pressure acting in addition.

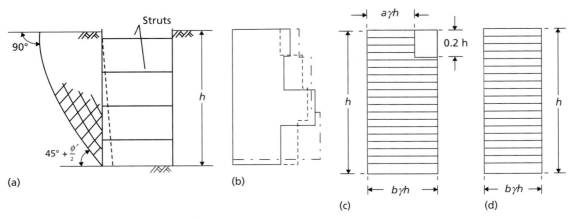

Figure 11.37 Earth pressure envelopes for braced excavations.

Applications in geotechnical engineering

In clays, the envelopes take into account the increase in strut load which accompanies dissipation of the negative excess pore water pressures induced during excavation. The envelopes are based on characteristic strut loads, the appropriate partial factor then being applied to give the design load for structural checks of the propping at ULS-8. The envelopes also allow for a nominal surface surcharge of 10 kPa on the retained soil on either side of the excavation.

Basal heave

Bearing capacity theory (Chapter 8) can be applied to the problem of base failure in braced excavations in fine-grained soils under undrained conditions (ULS-9). The application is limited to the analysis of cases in which the bracing is adequate to prevent significant lateral deformation of the soil adjacent to the excavation (i.e. when the other ULS conditions are satisfied). A simple failure mechanism, originally proposed by Terzaghi (1943), is illustrated in Figure 11.38, the angle at a being 45° and bc being a circular arc in an undrained material; therefore, the length of ab is $(B/2)/\cos 45°$ (approximately $0.7B$).

Failure occurs when the shear strength of the soil is insufficient to resist the average shear stress resulting from the vertical pressure (p) on ac due to the weight of the soil ($0.7\gamma Bh$) plus any surcharge (σ_q) reduced by the shear strength on cd ($c_u h$). Thus,

$$p = \gamma h + \sigma_q - \frac{c_u h}{0.7B} \tag{11.77}$$

The problem is essentially that of a bearing capacity analysis in reverse, there being zero pressure at the bottom of the excavation and p representing the overburden pressure. The shear strength available along the failure surface, acting in the opposite direction to that in the bearing capacity problem, can be expressed as $p_f = c_u N_c$ (Equation 8.17). Thus, for limiting equilibrium (i.e. to satisfy ULS-9),

$$p \leq p_f \tag{11.78}$$

Equation 11.78 may be solved for h to find the maximum depth of excavation for ULS-9 to be satisfied. In applying partial factors, p is the action, while p_f is the resistance, and these should be factored

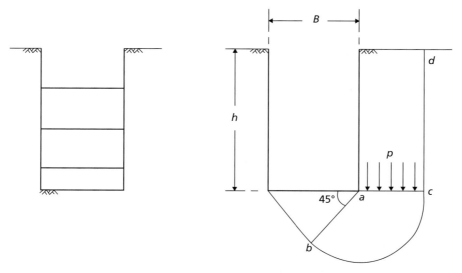

Figure 11.38 Base failure in a braced excavation.

accordingly. If a firm stratum were to exist at depth D_f below the base of the excavation, where $D_f<0.7B$, then D_f replaces $0.7B$ in Equation 11.77.

Based on observations of actual base failures in Oslo, Bjerrum and Eide (1956) concluded that Equation 11.77 gave reliable results only in the case of excavations with relatively low depth/breadth (h/B) ratios. In the case of excavations with relatively large depth/breadth ratios, local failure occurred before shear failure on cd was fully mobilised up to surface level, such that

$$p = \gamma h + \sigma_q \tag{11.79}$$

i.e. applying Equation 8.17 with $q_f = 0$.

Where there is a possibility that the base of the excavation will fail by heaving, this should be analysed before the strut loads are considered. Due to basal heave and the inward deformation of the clay, there will be horizontal and vertical movement of the soil outside the excavation. Such movements may result in damage to adjacent structures and services, and should be monitored during excavation; advance warning of excessive movement or possible instability can thus be obtained.

SLS design of braced excavations

Diaphragm or piled walls are commonly used in braced excavations. In general, the greater the flexibility of the wall system and the longer the time before struts or anchors are installed, the greater will be the movements outside the excavation. Settlement of the retained soil beside the excavation (SLS-1) is usually critical when braced excavations are made in urban areas, where differential settlement behind the wall may affect adjacent structures or services. Analysis of the serviceability limit state for retaining structures is difficult, typically requiring complex numerical analyses using a large body of soil parameters and requiring extensive validation. It is therefore preferable in design to use empirical methods based on observations of wall movements in successful excavations. Fortunately, databases of many tens of case histories now exist in a range of materials (for example Gaba *et al.* (2003); see Further reading), and limiting envelopes are summarised in Figure 11.39. In this figure, x is the distance behind the excavation (normalised by the excavation depth, h) and s_g is the settlement of the ground surface (again, normalised by h).

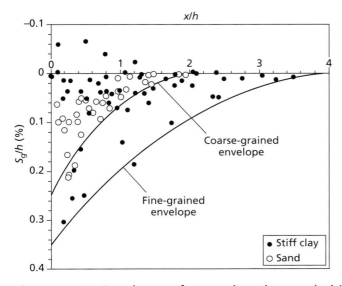

Figure 11.39 Envelopes of ground settlement behind excavations.

Applications in geotechnical engineering

The magnitude and distribution of the ground movements depend on the type of soil, the dimensions of the excavation, details of the construction procedure, and the standard of workmanship. Ground movements should be monitored during excavation and compared with the limits shown in Figure 11.39 so that advance warning of excessive movement or possible instability can be obtained. Assuming comparable construction techniques and workmanship, the magnitude of settlement adjacent to an excavation is likely to be relatively small in dense cohesionless soils but can be excessive in soft plastic clays.

11.10 Diaphragm walls

A **diaphragm wall** is a relatively thin reinforced concrete membrane cast in a trench, the sides of which are supported prior to casting by the hydrostatic pressure of a slurry of **bentonite** (a montmorillonite clay) in water. Stability of the trench during the excavation and casting phase is described in more detail in Section 12.2. When mixed with water, bentonite readily disperses to form a colloidal suspension which exhibits thixotropic properties – i.e. it gels when left undisturbed but becomes fluid when agitated. The trench, the width of which is equal to that of the wall, is excavated progressively in suitable lengths (known as **panels**) from the ground surface, as shown in Figure 11.40(a), generally using a power-closing clamshell grab; shallow concrete guide walls are normally constructed as an aid to excavation. The trench is filled with the bentonite slurry as excavation proceeds; excavation thus takes place through the slurry already in place. The excavation process turns the gel into a fluid, but the gel becomes re-established when disturbance ceases. The slurry tends to become contaminated with soil and cement in the course of construction, but can be cleaned and re-used.

The bentonite particles form a skin of very low permeability, known as the **filter cake**, on the excavated soil faces. This occurs due to the fact that water filters from the slurry into the soil, leaving a layer of bentonite particles, a few millimetres thick, on the surface of the soil. Consequently, the full hydrostatic pressure of the slurry acts against the sides of the trench, enabling stability to be maintained. The filter cake will form only if the fluid pressure in the trench is greater than the pore water pressure in the soil; a high water table level can thus be a considerable impediment to diaphragm wall construction. In soils of low permeability, such as clays, there will be virtually no filtration of water into the soil and therefore no significant filter-cake formation will take place; however, total stress conditions apply, and slurry pressure will act against the clay. In soils of high permeability, such as sandy gravels, there may be excessive loss of bentonite into the soil, resulting in a layer of bentonite-impregnated soil and poor filter-cake formation. However, if a small quantity of fine sand (around 1%) is added to the slurry the sealing mechanism in soils of high permeability can be improved, with a considerable reduction in bentonite loss. Trench stability depends on the presence of an efficient seal on the soil surface; the higher the permeability of the soil, the more vital the efficiency of the seal becomes.

A slurry having a relatively high density is desirable from the points of view of trench stability, reduction of loss into soils of high permeability, and the retention of contaminating particles in suspension. On the other hand, a slurry of relatively low density will be displaced more cleanly from the soil surfaces and the reinforcement, and will be more easily pumped and decontaminated. The specification for the slurry must reflect a compromise between these conflicting requirements. Slurry specifications are usually based on density, viscosity, gel strength and pH.

On completion of excavation, the reinforcement cage is positioned and the panel is filled with wet concrete using a **tremie pipe** which is dropped through the slurry to the base of the excavation. The wet concrete (which has a density of approximately twice that of the slurry) displaces the slurry upwards from the bottom of the trench, the tremie being raised in stages as the level of concrete rises. Once the wall (constructed as a series of individual panels keyed together) has been completed and the concrete has achieved adequate strength, the soil on one side of the wall can be excavated. It is usual for ground

Retaining structures

anchors or props to be installed at appropriate levels, as excavation proceeds, to tie the wall back into the retained soil (Sections 11.7–11.9). The method is very convenient for the construction of deep basements and underpasses, an important advantage being that the wall can be constructed close to adjoining structures, provided that the soil is moderately compact and ground deformations are tolerable. Diaphragm walls are often preferred to sheet pile walls because of their relative rigidity and their ability to be incorporated as part of the final structure.

An alternative to a diaphragm wall is the **contiguous pile wall**, in which a row of bored piles form the wall (Figure 10.40(b)), relying on arching between piles to support the retained soil. A thin concrete (or other) facing may be applied to the soil between piles to prevent erosion which could lead to progressive failure. A **secant pile wall** (Figure 10.40(c)) is similar to a contiguous wall though two sets of piles are installed to overlap with each other, forming a continuous structure. These are normally achieved by installing an initial row of cast-in-situ bored concrete piles (Chapter 9) in the ground at a centre-to-centre spacing of less than one diameter. These are known as 'female' piles. A lower strength concrete (or preferably, a higher strength concrete that has slow strength gain) is used in these piles. Once the row is complete, further 'male' piles are installed in the spaces between piles. As there is less than one pile diameter of spacing available, boring of the second set of piles will partially drill through the existing piles (hence the need for the concrete to have a low strength at this time). As the wall is formed from a series of overlapping circular piles, the resulting wall will have a ribbed surface; it may therefore be necessary to provide additional facing on the wall if it is to be used as part of the final structure.

The decision whether to use a triangular or a trapezoidal distribution of lateral pressure in the design of a diaphragm wall depends on the anticipated wall deformation. A triangular distribution (Section 11.7) would probably be indicated in the case of a single row of tie-backs or props near the top of the wall. In the case of multiple rows of tie-backs or props over the height of the wall, the trapezoidal distributions shown in Figure 11.37 might be considered more appropriate.

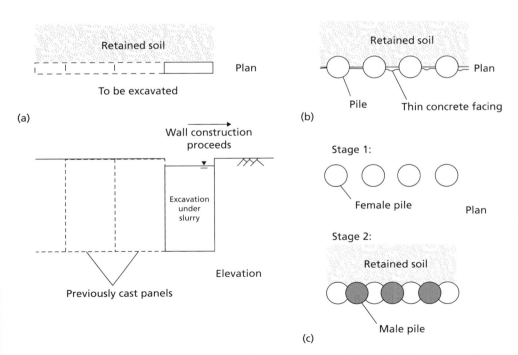

Figure 11.40 (a) Diaphragm wall, (b) contiguous pile wall, (c) secant pile wall.

Applications in geotechnical engineering

11.11 Reinforced soil

Reinforced soil consists of a compacted soil mass within which tensile reinforcing elements, typically in the form of horizontal steel strips, are embedded. (Patents for the technique were taken out by Henri Vidal and the Reinforced Earth Company, the term **reinforced earth** being their trademark.) Other forms of reinforcement include strips, rods, grids and meshes of metallic or polymeric materials, and sheets of **geotextiles**. The mass is stabilised as a result of interaction between the soil and the elements, the lateral stresses within the soil being transferred to the elements which are thereby placed in tension. The soil used as fill material should be predominantly coarse-grained, and must be adequately drained to prevent it from becoming saturated. In coarse fills, the interaction is due to frictional forces which depend on the characteristics of the soil together with the type and surface texture of the elements. In the cases of grid and mesh reinforcement, interaction is enhanced by interlocking between the soil and the apertures in the material.

In a reinforced-soil retaining structure (also referred to as a **composite wall**), a facing is attached to the outside ends of the elements to prevent the soil spilling out at the edge of the mass and to satisfy aesthetic requirements; the facing does not act as a retaining wall. The facing should be sufficiently flexible to withstand any deformation of the fill. The types of facing normally used are discrete precast concrete panels, full-height panels and pliant U-shaped sections aligned horizontally. The basic features of a reinforced-soil retaining wall are shown in Figure 11.41. Such structures possess considerable inherent flexibility, and consequently can withstand relatively large differential settlement. The reinforced soil principle can also be employed in embankments, normally by means of geotextiles, and in slope stabilisation by the insertion of steel rods – a technique known as soil nailing.

Both external and internal stability must be considered in design. The external stability of a reinforced soil structure is usually analysed using a limit equilibrium approach (Section 11.5). The back of the wall should be taken as the vertical plane (FG) through the inner end of the lowest element, the total active thrust (P_a) due to the backfill behind this plane then being calculated as for a gravity wall. The ultimate limit states for external stability are:

> ULS-1: bearing resistance failure of the underlying soil, resulting in tilting of the structure;
> ULS-2: sliding between the reinforced fill and the underlying soil; and
> ULS-3: development of a deep slip surface which envelops the structure as a whole.

ULS-1 is verified using the methods described in Section 11.4 for gravity walls; the methods required for verification of ULS-3 will be described in Chapter 12. The remainder of this section will focus on the verification of ULS-2. Serviceability limit states are excessive values of settlement and wall deformation as for the other classes of retaining structure considered previously.

In considering the internal stability of the structure, the principal limit states are tensile failure of the elements and slipping between the elements and the soil. Tensile failure of one of the elements could lead to the progressive collapse of the whole structure (an ultimate limit state). Local slipping due to inadequate frictional resistance would result in a redistribution of tensile stress and the gradual deformation of the structure, not necessarily leading to collapse (i.e. a serviceability limit state).

Consider a reinforcing element at depth z below the surface of a soil mass. The tensile force in the element due to the transfer of lateral stress from the soil is given by

$$T = K\sigma_z S_x S_z \tag{11.80}$$

Retaining structures

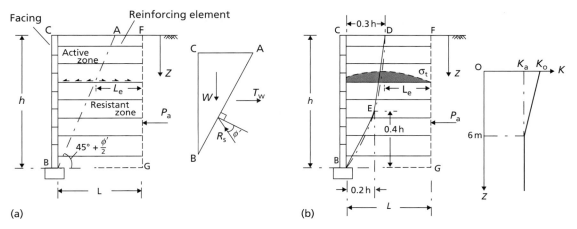

Figure 11.41 Reinforced soil-retaining structure: (a) tie-back wedge method, (b) coherent gravity method.

where K is the appropriate earth pressure coefficient at depth z, σ_z the vertical stress, S_x the horizontal spacing of the elements and S_z the vertical spacing. If the reinforcement consists of a continuous layer, such as a grid, then the value of S_x is unity, and T is the tensile force per unit length of wall. The vertical stress σ_z is due to the overburden pressure at depth z plus the stresses due to any surcharge loading and external bending moment (including that due to the total active thrust on the part of plane FG between the surface and depth z). The average vertical stress can be expressed as

$$\sigma_z = \frac{V}{L - 2e}$$

where V is the vertical component of the resultant force at depth z, e the eccentricity of the force and L the length of the reinforcing element at that depth. Given the design tensile strength of the reinforcing material, the required cross-sectional area or thickness of the element can be obtained from the value of T. The addition of surcharge loading at the surface of the retained soil will cause an increase in vertical stress which can be calculated from elastic theory.

There are two procedures for the design of retaining structures. One approach is the **tie-back wedge method** which is applicable to structures with reinforcement of relatively high extensibility, such as grids, meshes and geotextile sheets. This method is an extension of Coulomb's method, and considers the forces acting on a wedge of soil from which a force diagram can be drawn The active state is assumed to be reached throughout the soil mass because of the relatively large strains possible at the interface between the soil and the reinforcement; therefore the earth pressure coefficient in Equation 11.80 is taken as K_a at all depths and the failure surface at collapse will be a plane AB inclined at $45° + \phi'/2$ to the horizontal, as shown in Figure 11.41(a), dividing the reinforced mass into an active zone within which shear stresses on elements act outwards towards the face of the structure and a resistant zone within which shear stresses act inwards. The frictional resistance available on the top and bottom surfaces of an element is then given by

$$T_f = 2bL_e\sigma_z \tan\delta' \tag{11.81}$$

where b is the width of the element, L_e the length of the element in the resistant zone and δ' the angle of friction between soil and element. Slippage between elements and soil (referred to as **bond failure**) will not occur if T_f is greater than or equal to T. The value of δ' can be determined by means of direct shear tests or full-scale pullout tests.

Applications in geotechnical engineering

The stability of the wedge ABC is checked in addition to the external and internal stability of the structure as a whole. The forces acting on the wedge, as shown in Figure 11.41(a), are the weight of the wedge (W), the reaction on the failure plane (R) acting at angle ϕ' to the normal (being the resultant of the normal and shear forces), and the total tensile force resisted by all the reinforcing elements (T_w). The value of T_w can thus be determined. In effect, the force T_w replaces the reaction P of a retaining wall (as, for example, in Figure 11.21(a)). Any external forces must be included in the analysis, in which case the inclination of the failure plane will not be equal to $45° + \phi'/2$ and a series of trial wedges must be analysed to obtain the maximum value of T_w. The design requirement is that the factored sum of the tensile forces in all the elements, calculated from Equation 11.81, must be greater than or equal to T_w to satisfy ULS-2.

The second design procedure is the **coherent gravity method**, due to Juran and Schlosser (1978), and is applicable to structures with elements of relatively low extensibility, such as steel strips. Experimental work indicated that the distribution of tensile stress (σ_t) along such an element was of the general form shown in Figure 11.41(b), the maximum value occurring not at the face of the structure but at a point within the reinforced soil, the position of this point varying with depth as indicated by the curve DB in Figure 11.41(b). This curve again divides the reinforced mass into an active zone and a resistant zone, the method being based on the stability analysis of the active zone. The assumed mode of failure is that the reinforcing elements fracture progressively at the points of maximum tensile stress and, consequently, that conditions of plastic equilibrium develop in a thin layer of soil along the path of fracture. The curve of maximum tensile stress therefore defines the potential failure surface. If it is assumed that the soil becomes perfectly plastic, the failure surface will be a logarithmic spiral. The spiral is assumed to pass through the bottom of the facing and to intersect the surface of the fill at right angles, at a point approximately $0.3h$ behind the facing, as shown in Figure 11.41(b). A simplified analysis can be made by assuming that the curve of maximum tensile stress is represented by the bilinear approximation DEB shown in Figure 11.41(b), where $CD = 0.3h$. Equations 11.80 and 11.81 are then applied. The earth pressure coefficient in Equation 11.80 is assumed to be equal to K_0 (the at-rest coefficient) at the top of the wall, reducing linearly to K_a at a depth of 6 m. The addition of surface loading would result in the modification of the line of maximum tensile stress, and for this situation an amended bilinear approximation is proposed in BS 8006 (BSI, 1995), which provides valuable guidance on the deign of reinforced soil constructions to complement EC7.

Summary

1. Lower bound limit analysis may be used to determine the limiting lateral earth pressures acting on retaining structures in homogeneous soil conditions either directly (in undrained materials) or via lateral earth pressure coefficients K_a and K_p (in drained material). If the retaining structure moves away from the retained soil at failure, active earth pressures will be generated, while larger passive earth pressures act on structures which move into the retained soil at failure. These limit analysis techniques can account for a battered and/or rough soil–wall interface and sloping retained soil. Limit equilibrium techniques may alternatively be used which consider the equilibrium of a wedge of failing soil behind the retaining structure, and can additionally be used with flow-net/FDM techniques from Chapter 2 to analyse problems where seepage is occurring in the retained soil.

2. In-situ horizontal stresses within the ground are directly related to the vertical effective stresses (calculated using methods from Chapter 3) via the coefficient of lateral earth pressure at rest (K_0). This can be determined analytically for any soil based on its drained friction angle ϕ' and overconsolidation ratio (OCR). K_0 conditions apply when there is no lateral strain in the soil mass. Under extension, earth pressures will reduce to active values ($K_a < K_0$); under compression, earth pressures will increase to passive values ($K_p > K_0$).

3 Surcharge loading on the surface of retained soil may be accounted for by modifying the vertical total or effective stress used in lateral earth pressure calculations in undrained and drained materials, respectively. If the wall is constructed before the soil it retains is placed (a backfilled wall), compaction-induced lateral stresses may be induced along the wall.

4 Gravity retaining structures rely on their mass to resist lateral earth pressures. The key design criteria for these structures is maintaining overall stability (ULS). Earth pressure forces may be determined from the limiting earth pressures (point 1) and used to check the stability of the wall in bearing, sliding and overturning using the yield surface approach (Chapter 10). Embedded walls are flexible, and resist lateral earth pressures from a balance of lateral earth pressure forces from the soil behind the wall acting to overturn or translate the wall (active) and those resisting failure from the soil in front of the wall (passive). ULS design may be accomplished using the free-earth support method. Ground anchors or propping may be used to provide additional support, and this must also be designed to resist structural or pull-out failure. If a flexible wall is braced more heavily it is described as a braced excavation, the operative earth pressures on such a construction being much lower. These earth pressures are used to structurally design the bracing system avoiding progressive collapse. Basal heave may also occur in braced excavations, and must also be designed against. In flexible walls movements are also significant, imposing additional SLS criteria in design.

Problems

11.1 The backfill behind a smooth retaining wall, located above the water table, consists of a sand of unit weight $17\,\text{kN/m}^3$. The height of the wall is 6 m and the surface of the backfill is horizontal. Determine the total active thrust on the wall if $c'=0$ and $\phi'=37°$. If the wall is prevented from yielding, what is the approximate value of the thrust on the wall?

11.2 Plot the distribution of active pressure on the wall surface shown in Figure 11.42. Calculate the total thrust on the wall (active + hydrostatic) and determine its point of application. Assume $\delta'=0$ and $\tau_w=0$.

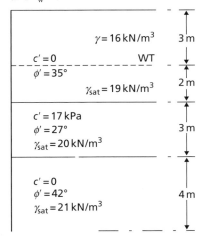

Figure 11.42 Problem 11.2.

Applications in geotechnical engineering

11.3 A line of sheet piling is driven 4 m into a firm clay and retains, on one side, a 3-m depth of fill on top of the clay. The water table is at the surface of the clay. The unit weight of the fill is 18 kN/m³, and the saturated unit weight of the clay is 20 kN/m³. Calculate the active and passive pressures at the lower end of the sheet piling (a) if $c_u = 50$ kPa, $\tau_w = 25$ kPa and $\phi_u = \delta' = 0$, and (b) if $c' = 0$, $\phi' = 26°$ and $\delta' = 13°$, for the clay.

11.4 Details of a reinforced concrete cantilever retaining wall are shown in Figure 11.43, the unit weight of concrete being 23.5 kN/m³. Due to inadequate drainage, the water table has risen to the level indicated. Above the water table the unit weight of the retained soil is 17 kN/m³, and below the water table the saturated unit weight is 20 kN/m³. Characteristic values of the shear strength parameters are $c' = 0$ and $\phi' = 38°$. The angle of friction between the base of the wall and the foundation soil is 25°. Check whether or not the overturning and sliding limit states have been satisfied to EC7 DA1b.

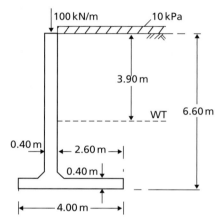

Figure 11.43 Problem 11.4.

11.5 The section through a gravity retaining wall is shown in Figure 11.44, the unit weight of the wall material being 23.5 kN/m³. The unit weight of the backfill is 19 kN/m³, and design values of the shear strength parameters are $c' = 0$ and $\phi' = 36°$. The value of δ' between wall and backfill and between base and foundation soil is 25°. The ultimate bearing capacity of the foundation soil is 250 kPa. Determine if the design of the wall is satisfactory with respect to the overturning, bearing resistance and sliding limit states, to EC7 DA1a.

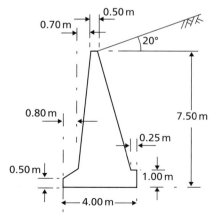

Figure 11.44 Problem 11.5.

11.6 The sides of an excavation 3.0 m deep in sand are to be supported by a cantilever sheet pile wall. The water table is 1.5 m below the bottom of the excavation. The sand has a saturated unit weight of 20 kN/m³ and a unit weight above the water table of 17 kN/m³, and the characteristic value of ϕ' is 36°. Determine the required depth of embedment of the piling below the bottom of the excavation if the excavation is to be designed to EC7 DA1b.

11.7 An anchored sheet pile wall is constructed by driving a line of piling into a soil for which the saturated unit weight is 21 kN/m³ and the characteristic shear strength parameters are $c' = 10$ kPa and $\phi' = 27°$. Backfill is placed to a depth of 8.00 m behind the piling, the backfill having a saturated unit weight of 20 kN/m³, a unit weight above the water table of 17 kN/m³, and characteristic shear strength parameters of $c' = 0$ and $\phi' = 35°$. Tie rods, spaced at 2.5-m centres, are located 1.5 m below the surface of the backfill. The water level in front of the wall and the water table behind the wall are both 5.00 m below the surface of the backfill. Determine the required depth of embedment to EC7 DA1b and the design force in each tie rod.

11.8 The soil on both sides of the anchored sheet pile wall detailed in Figure 11.45 has a saturated unit weight of 21 kN/m³, and a unit weight above the water table of 18 kN/m³. Characteristic parameters for the soil are $c' = 0$, $\phi' = 36°$ and $\delta' = 0°$. There is a lag of 1.5 m between the water table behind the wall and the tidal level in front. Determine the required depth of embedment to EC7 DA1a and the design force in the ties.

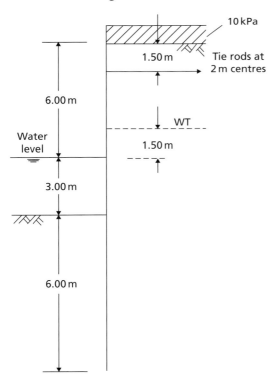

Figure 11.45 Problem 11.8.

11.9 A ground anchor in a stiff clay, formed by the gravel injection technique, has a fixed anchor length of 5 m and an effective fixed anchor diameter of 200 mm; the diameter of the borehole is 100 mm. The relevant shear strength parameters for the clay are $c_u = 110$ kPa and $\phi_u = 0$. What would be the expected characteristic ultimate load capacity of the anchor, assuming a skin friction coefficient of 0.6?

11.10 The struts in a braced excavation 9 m deep in a dense sand are placed at 1.5-m centres vertically and 3.0-m centres horizontally, the bottom of the excavation being above the water table. The unit weight of the sand is 19 kN/m³. Based on design shear strength parameters $c'=0$ and $\phi'=40°$, what load should each strut be designed to carry? (Use EC7 DA1a.)

11.11 A long braced excavation in soft clay is 4 m wide and 8 m deep. The saturated unit weight of the clay is 20 kN/m³, and the undrained shear strength adjacent to the bottom of the excavation is given by $c_u = 40$ kPa, ($\phi_u = 0$). Determine the factor of safety against base failure of the excavation.

11.12 A reinforced soil wall is 5.2 m high. The reinforcing elements, which are spaced at 0.65 m vertically and 1.20 m horizontally, measure 65 × 3 mm in section and are 5.0 m in length. The ultimate tensile strength of the reinforcing material is 340 MPa. Design values to be used are as follows: unit weight of the selected fill = 18 kN/m³; angle of shearing resistance of selected fill = 36°; angle of friction between fill and elements = 30°. Using (a) the tie-back wedge method and (b) the coherent gravity method, check that an element 3.6 m below the top of the wall will not suffer tensile failure, and that slipping between the element and the fill will not occur. The value of K_a for the material retained by the reinforced fill is 0.30 and the unit weight of this material is 18 kN/m³.

References

Berezantzev, V.G., Khristoforov, V.S. and Golubkov, V.N. (1961) Load bearing capacity and deformation of piled foundations, in *Proceedings of the 5th International Conference on Soil Mechanics and Foundation Engineering, Paris, France*, pp. 11–15.

Bjerrum, L. and Eide, O. (1956) Stability of strutted excavations in clay, *Géotechnique*, **6**(1), 32–47.

British Standard 8006 (1995) *Code of Practice for Strengthened Reinforced Soils and Other Fills*, British Standards Institution, London.

Coulomb, C.A. (1776). Essai sur une application des régeles des maximus et minimus a quelque problémes de statique rélatif à l'architecture, *Memoirs Divers Savants*, 7, Académie Sciences, Paris (in French).

EC7–1 (2004) *Eurocode 7: Geotechnical design – Part 1: General rules, BS EN 1997–1:2004*, British Standards Institution, London.

Ingold, T.S. (1979) The effects of compaction on retaining walls, *Géotechnique*, **29**, 265–283.

Jaky, J. (1944). The coefficient of earth pressure at rest, *Journal of the Society of Hungarian Architects and Engineers*, Appendix 1, 78(22) (transl.).

Juran, I. and Schlosser, F. (1978) Theoretical analysis of failure in reinforced earth structures, in *Proceedings of the Symposium on Earth Reinforcement, ASCE Convention, Pittsburgh*, pp. 528–555.

Mayne, P.W. (2007) *Cone Penetration Testing: A Synthesis of Highway Practice*, NCHRP Synthesis Report 368, Transportation Research Board, Washington DC.

Mayne, P.W. and Kulhawy, F.H. (1982) Ko-OCR (At rest pressure – Overconsolidation Ratio) relationships in soil, *Journal of the Geotechnical Engineering Division, ASCE*, **108**(GT6), 851–872.

Pipatpongsa, T., Takeyama, T., Ohta, H. and Iizuka, A. (2007) Coefficient of earth pressure at-rest derived from the Sekiguchi-Ohta Model, in *Proceedings of the 16th Southeast Asian Geotechnical Conference, Subang Jaya, Malaysia, 8–11 May*, pp. 325–331.

Potts, D.M. and Fourie, A.B. (1984) The behaviour of a propped retaining wall: results of a numerical experiment, *Géotechnique*, **34**(3), 383–404.

Potts, D.M. and Fourie, A.B. (1985) The effect of wall stiffness on the behaviour of a propped retaining wall, *Géotechnique*, **35**(3), 347–352.

Terzaghi, K. (1943) *Theoretical Soil Mechanics*, John Wiley & Sons, New York, NY.

Twine, D. and Roscoe, H. (1999) Temporary propping of deep excavations: guidance on design, *CIRIA Report C517*, CIRIA, London.

Further reading

Frank, R., Bauduin, C., Driscoll, R., Kavvadas, M., Krebs Ovesen, N., Orr, T. and Schuppener, B. (2004) *Designers' Guide to EN 1997–1 Eurocode 7: Geotechnical Design – General Rules*, Thomas Telford, London.

This book provides a guide to limit state design of a range of constructions (including retaining walls) using Eurocode 7 from a designer's perspective and provides a useful companion to the Eurocodes when conducting design. It is easy to read and has plenty of worked examples.

Gaba, A.R., Simpson, B., Powrie, W. and Beadman, D.R. (2003) Embedded retaining walls – guidance for economic design, *CIRIA Report C580*, CIRIA, London.

This report provides valuable practical guidance on the selection of design and construction methodologies for flexible retaining structures, including procedural flowcharts. It also incorporates a large collection of case history data to inform future design.

For further student and instructor resources for this chapter, please visit the Companion Website at www.routledge.com/cw/craig

Chapter 12

Stability of self-supporting soil masses

> **Learning outcomes**
>
> After working through the material in this chapter, you should be able to:
>
> 1 Determine the stability of unsupported trenches, including those supported by slurry, and design these works within a limit state design framework (Eurocode 7);
> 2 Determine the stability of slopes, vertical cuttings and embankments, and design these works within a limit state design framework;
> 3 Determine the stability of tunnels and the ground settlements caused by tunnelling works, and use this information to conduct a preliminary design of tunnelling works within a limit state design framework.

12.1 Introduction

This chapter is concerned with the design of potentially unstable soil masses which have been formed through either human activity (excavation or construction) or natural processes (erosion and deposition). This class of problem includes slopes, embankments and unsupported excavations. Unlike the material in Chapter 11, however, the soil masses here are not supported by an external structural element such as a retaining wall; rather, they derive their stability from the resistance of the soil within the mass in shear.

Gravitational and seepage forces tend to cause instability in natural slopes, in slopes formed by excavation and in the slopes of embankments. A vertical cutting (or trench, formed of two vertical cuttings) is a special case of sloping ground where the slope angle is 90° to the horizontal. Design of self-supporting soil systems is based on the requirement to maintain stability (ULS) rather than on the need to minimise deformation (SLS). If deformation were such that the strain in an element of soil exceeded the value corresponding to peak strength, then the strength would fall towards the ultimate value. Thus, it is appropriate to use the critical state strength in analysing stability. However, if a pre-existing slip surface were to be present within the soil, use of the residual strength would be appropriate.

Section 12.2 will apply both limit analysis and limit equilibrium techniques to the stability of vertical cuts/trenches. These methods will then be extended to consider how fluid support may be used to improve the stability of such constructions (e.g. drilling of bored piles or excavation of diaphragm wall piles under slurry). In Sections 12.3 and 12.4, the analytical methods will be further extended to the consideration of slope and embankment design, respectively. Finally, in Section 12.5, an introduction to the

Applications in geotechnical engineering

design of tunnelling works will be considered, where the stability of a vertical cut face deep below the ground surface governs the design. This final section will also consider how the stability of tunnel headings may be improved by pressurising the cut face (analogous to the use of drilling fluids in trench support).

12.2 Vertical cuttings and trenches

Vertical cuts in soil can only be supported when soil behaves in an undrained way (with an undrained strength c_u) or in a drained soil where there is some cohesion (c'). As, in the absence of chemical or other bounding between soil particles, $c'=0$, vertical cuts and trenches cannot normally be supported under drained conditions. This is because, from the Mohr–Coulomb strength definition (Equation 5.11), a drained cohesionless soil will always fail when the slope angle reaches ϕ'. Under undrained conditions, however, vertical cuts may be kept stable up to a certain limiting depth/height which depends on the undrained strength of the soil. This is very useful during temporary works in fine-grained soils (typically clays) which are fast enough that undrained conditions can be maintained. The excavation of trial pits/trenches and bored pile construction techniques are two examples of where this is used in engineering practice.

Limiting height/depth using limit analysis

Figure 12.1 shows a simple upper bound failure mechanism UB-1 for a vertical cut in undrained soil having a unit weight γ. As the mechanism develops and the soil fails into the excavation/cut, work is input to the system from the potential energy recovered as the weight of the sliding block moves downwards with gravity. The vertical force due to the weight of the block (W) is given by

$$W = \frac{\gamma h^2}{2 \tan \theta} \quad (12.1)$$

per metre length of the cut. If the block slides along the slip plane at velocity v, then the component in the vertical direction (in the direction of W) is $v \sin \theta$. From Equation 8.2, the work input is then

$$\sum W_i = \frac{\gamma h^2 v \sin \theta}{2 \tan \theta} = \frac{1}{2} \gamma h^2 v \cos \theta \quad (12.2)$$

As in Chapter 8, energy is dissipated in shear along the slip plane; the length of the slip plane $L_{OA} = h/\sin \theta$ per metre length of cut, the shear stress at plastic failure is c_u, and the slip velocity is v. Then, from Equation 8.1, the energy dissipated is

$$\sum E_i = \frac{c_u h v}{\sin \theta} \quad (12.3)$$

By the upper bound theorem, if the system is at plastic collapse, the work done by the external loads/pressures must be equal to the energy dissipated within the soil, so from Equation 8.3:

$$\sum W_i = \sum E_i$$
$$\frac{1}{2} \gamma h^2 v \cos \theta = \frac{c_u h v}{\sin \theta} \quad (12.4)$$
$$h = \frac{2 c_u}{\gamma \sin \theta \cos \theta}$$

Stability of self-supporting soil masses

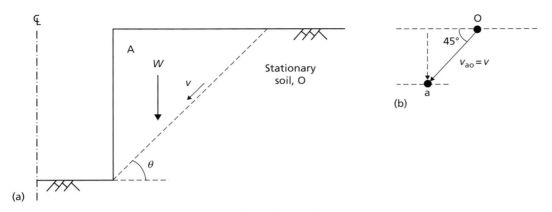

Figure 12.1 (a) Mechanism UB-1, (b) hodograph.

Equation 12.4 is a function of the angle θ. The cutting will fail when h is a minimum. The value of θ at which the minimum h occurs may be found from solving $dh/d\theta=0$, giving $\theta=\pi/4$ (45°); substituting this value into Equation 12.4 gives $h \leq 4c_u/\gamma$ for stability.

A simple lower bound stress field for the vertical cutting is shown in Figure 12.2(a). At plastic failure, the soil will move inwards towards the cutting, so the major principal stresses in the retained soil are vertical (active condition). Figure 12.2(b) shows the Mohr circle for the soil; for undrained conditions, $\sigma_3 = \sigma_1 - 2c_u$ and $\sigma_1 = \sigma_v = \gamma z$. The horizontal stresses on the vertical boundary must then sum to zero for equilibrium, i.e.

$$\int_0^h (\gamma z - 2c_u) dz = 0$$
$$\frac{1}{2}\gamma h^2 - 2c_u h = 0 \tag{12.5}$$

Solution of Equation 12.5 gives $h \leq 4c_u/\gamma$ for stability. This is the same as the upper bound, and therefore represents the true solution.

Limiting height/depth using limit equilibrium (LE)

The limit equilibrium (Coulomb) method considering a wedge of soil presented in Section 11.5 may also be used for assessing the stability of a vertical cut. Figure 12.3 shows a wedge at an angle θ in undrained soil. An additional force S is also included in this analysis to model the support provided by drilling fluid within a trench. The unit weight of the slurry is γ_s and that of the soil is γ, while the depth of the slurry is nh. The resultant resistance force along the slip plane (R_s from Section 11.5) is here split into normal and tangential components, denoted N and T, respectively. Considering force equilibrium

$$S + T\cos\theta - N\sin\theta = 0 \tag{12.6}$$

$$W - T\sin\theta - N\cos\theta = 0 \tag{12.7}$$

The resultant thrust from the slurry arises from the hydrostatic pressure distribution within the trench, i.e.

$$S = \int_0^{nh} (\gamma_s z) dz = \frac{1}{2}\gamma_s (nh)^2 \tag{12.8}$$

Applications in geotechnical engineering

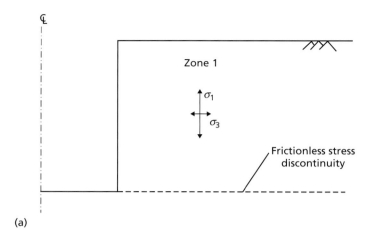

(a)

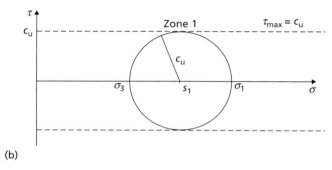

(b)

Figure 12.2 (a) Stress field LB-1; (b) Mohr circle.

The weight of the wedge is given by Equation 12.1 as before. The tangential force at failure is the shear strength of the soil multiplied by the slip plane area ($h/\sin\theta$ per metre length), i.e.

$$T = c_u \cdot \frac{h}{\sin\theta} \tag{12.9}$$

Substituting for S and T in Equation 12.6 and rearranging gives

$$N = \frac{S}{\sin\theta} + \frac{c_u h}{\tan\theta \sin\theta} \tag{12.10}$$

Then, substituting for W, T and N in Equation 12.7 and rearranging gives

$$h = \frac{2c_u}{\gamma \sin\theta \cos\theta} \cdot \left[\frac{1}{1 - \left(\dfrac{\gamma_s}{\gamma}\right) n^2 \tan\theta} \right] \tag{12.11}$$

If $\gamma_s = 0$ (i.e. there is no slurry in the trench), Equation 12.11 reduces to Equation 12.4.

Stability of self-supporting soil masses

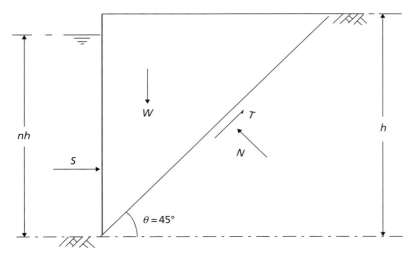

Figure 12.3 Stability of a slurry-supported trench in undrained soil.

As mentioned in Section 11.10, bentonite slurry forms a filter cake on the surface of the excavation allowing full hydrostatic pressures to be maintained, even against drained materials with no cohesion. Under drained conditions, the limit equilibrium analysis presented above may be modified to perform an effective stress analysis, accounting for the presence of the water table (at a height mh above the bottom of the trench) as shown in Figure 12.4.

Equations 12.6–12.8 may be used unchanged; however, the total shear resistance force T is now based on the effective stress along the slip plane, i.e.

$$T = c'\left(\frac{h}{\sin\theta}\right) + (N - U)\tan\phi' \qquad (12.12)$$

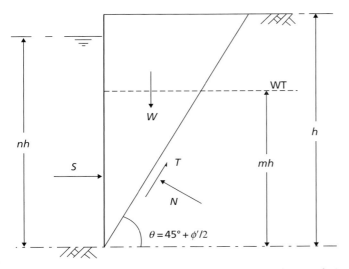

Figure 12.4 Stability of a slurry-supported trench in drained soil.

Applications in geotechnical engineering

where U, the boundary water force on the failure plane, is given by

$$U = \frac{1}{2}\gamma_w (mh)^2 \csc\theta \qquad (12.13)$$

When the wedge is on the point of sliding into the trench, i.e. the soil within the wedge is in a condition of active limiting equilibrium, the angle θ can be assumed to be $45° + \phi'/2$. The equations are solved using the same procedure as before; however the closed-form solution for the drained case is more complex than Equation 12.11. Instead, the equations may be straightforwardly programmed into a spreadsheet.

Figure 12.5(a) plots the normalised safe depth of excavation h as a function of slurry unit weight for excavation in an undrained soil. For the case of no slurry (unsupported excavation, $\gamma_s = 0$), $h = 4c_u/\gamma$ as before. In order to maintain workability for excavation, fresh slurry will typically have a density of 1150 kg/m³ ($\gamma_s = 11.3$ kN/m³). The data points in Figure 12.5(a) represent the maximum depths of excavation for some of the clays described in Chapters 5 and 7, namely the NC organic clay at Bothkennar, the glacial till at Cowden and the fissured clay at Madingley. The value of γ_s/γ for each of these clays is based on typical unit weights of 15.5, 21.5 and 19.5 kN/m³, respectively, and that of fresh slurry given above. It will be seen that excavation under slurry is particularly beneficial in NC soils, where $h = 14c_u/\gamma$ may be achieved (i.e. three and a half times the depth of an unsupported excavation). Even in the heavier clays, the excavation depth can be at least doubled by using slurry support.

Figure 12.5(b) plots the minimum slurry density required to avoid collapse as a function of the normalised water table height m in drained soil. It can be seen that excavation in such soils will only be problematic for situations where the water table is close to the ground surface. As a result, when installing bored piles in drained materials (e.g. sands) it is common to use steel casing towards the top of the excavation to prevent collapse.

12.3 Slopes

The most important types of slope failure are illustrated in Figure 12.6. In **rotational slips**, the shape of the failure surface in section may be a circular arc or a non-circular curve. In general, circular slips are associated with homogeneous, isotropic soil conditions, and non-circular slips with non-homogeneous

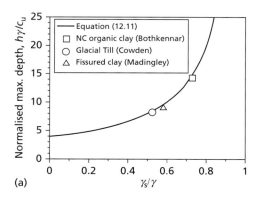
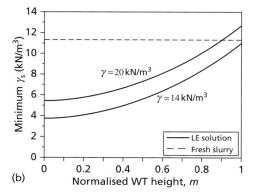

Figure 12.5 Slurry-supported excavations: (a) maximum excavation depth in undrained soil, (b) minimum slurry density to avoid collapse in drained soil ($\phi' = 35°$, $n = 1$).

Stability of self-supporting soil masses

conditions. **Translational** and **compound** slips occur where the form of the failure surface is influenced by the presence of an adjacent stratum of significantly different strength, most of the failure surface being likely to pass through the stratum of lower shear strength. The form of the surface would also be influenced by the presence of discontinuities such as fissures and pre-existing slips. Translational slips tend to occur where the adjacent stratum is at a relatively shallow depth below the surface of the slope, the failure surface tending to be plane and roughly parallel to the slope surface. Compound slips usually occur where the adjacent stratum is at greater depth, the failure surface consisting of curved and plane sections. In most cases, slope stability can be considered as a two-dimensional problem, conditions of plane strain being assumed.

An example of a rotational slip occurred 3–5 June 1993 at Holbeck, Yorkshire. Pore water pressure build-up as a result of heavy rain, coupled with drainage problems, was thought to be the cause of the failure, which involved approximately 1 million tonnes of glacial till as shown in Figure 12.7. The landslide caused catastrophic damage to the Holbeck Hall Hotel situated at the crest of the slope, as shown in Figure 12.7.

Limiting equilibrium techniques are normally used in the analysis of slope stability in which it is considered that failure is on the point of occurring along an assumed or a known failure surface. To check stability at the ultimate limiting state, gravitational forces driving slip (e.g. the component of weight acting along a slip plane) are considered as actions and factored accordingly; the forces developed due to shearing along slip planes are treated as resistances, along with any gravitational forces resisting slip, and factored down using a partial factor γ_{Rr}. In Eurocode 7, the normative value of $\gamma_{Rr} = 1.00$ for sets R1 and R3 and 1.10 for set R2.

Having analysed a given failure surface, the calculations should be repeated for a range of different positions of the slip surface. The failure surface that is closest to ULS is then the critical slip surface along which failure will occur. This process is normally automated using a computer.

Rotational slips in undrained soil

This analysis, in terms of total stress, covers the case of a fully saturated clay under undrained conditions, i.e. for the condition immediately after construction. Only moment equilibrium is considered in the analysis. If the soil is homogeneous, the failure surface can be assumed to be a circular arc in section. A trial failure surface (centre O, radius r and length L_a) is shown in Figure 12.8. Potential instability is due to the total weight of the soil mass (W per unit length) above the failure surface. The driving

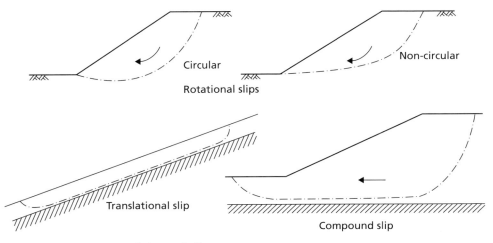

Figure 12.6 Types of slope failure.

Applications in geotechnical engineering

Figure 12.7 Rotational slope failure at Holbeck, Yorkshire.

(clockwise) moment about O is therefore $M_A = Wd$ (an action). The soil resistance is described by an anticlockwise moment $M_R = c_u L_a r$ about O. If the slip surface is a circular arc, then $L_a = r\theta$ from Figure 12.8. The criterion of stability at ULS is then described by

$$M_A \leq M_R \tag{12.14}$$

The moments of any additional forces (e.g. surcharge) must be taken into account in determining M_A. In the event of a tension crack developing, the arc length L_a is shortened and a hydrostatic force will act normal to the crack if it fills with water. It is necessary to analyse the slope for a number of trial failure surfaces in order that the most critical failure surface can be determined. In the analysis of an existing slope, M_A will be less than M_R (as the slope is standing) and the safety of the slope is usually expressed as a factor of safety, F where

$$F = \frac{M_R}{M_A} \tag{12.15}$$

From Equation 12.15 it can be seen that a stable slope will have $F > 1$, while an unstable slope will have $F < 1$. If $F = 1$, the slope is at the point of failure and the ULS stability criterion is regained.

Stability of self-supporting soil masses

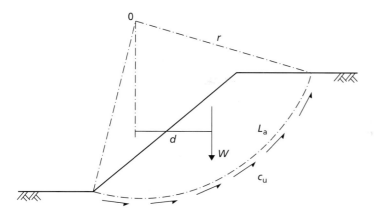

Figure 12.8 Limit equilibrium analysis in undrained soil.

In the design of a new slope, the aim is normally to find the maximum height h to which a slope of a given angle β can be constructed. To achieve this, the parameters W and d can be expressed in terms of the slope properties h and β and the properties describing the slip surface geometry (r and θ), though the derivation is not trivial. Fortunately, design charts have been published by Taylor (1937) for the case of c_u uniform with depth and an underlying rigid boundary, and by Gibson and Morgenstern (1962) for the case of c_u increasing linearly with depth ($c_u = Cz$). Both of these solutions express the critical conditions leading to slope failure as a non-dimensional **stability number** N_s, where

$$N_s = \frac{c_u}{F\gamma h} \tag{12.16}$$

In the case of increasing undrained shear strength with depth, $N_s = C/F\gamma h$. Values of N_s as a function of slope angle β are shown in Figure 12.9. Rearranging Equation 12.16 to give $F = c_u/N_s\gamma h$ and comparing this to Equation 12.15, it can be seen that the numerator of Equation 12.16 represents the resistance of the soil to failure while the denominator represents the sum of the driving actions for applying partial factors. These should be factored accordingly, along with the material properties (c_u and γ). In limit state design, the value of N_s is therefore determined from Figure 12.9 for a given slope angle; F is set to 1.0 such that the resulting equation governing ULS is

$$h \leq \frac{1}{N_s} \left[\frac{\left(\dfrac{c_u}{\gamma_{cu}}\right)}{\gamma_{Rr}\gamma_A \left(\dfrac{\gamma}{\gamma_\gamma}\right)} \right] \tag{12.17}$$

Equation 12.17 is then solved using the known soil properties and N_s to determine the maximum height of the slope.

Equation 12.16 may also be used to analyse existing slopes in place of the moment equilibrium method (Equation 12.15), in which case the slope height h is known and the equation is rearranged to find the unknown factor of safety. A three-dimensional analysis for slopes in clay under undrained conditions has been presented by Gens et al. (1988).

Applications in geotechnical engineering

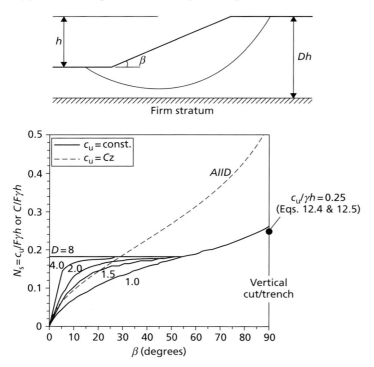

Figure 12.9 Stability numbers for slopes in undrained soil.

Example 12.1

A 45° cutting slope is excavated to a depth of 8 m in a deep layer of saturated clay of unit weight 19 kN/m³: the relevant shear strength parameters are $c_u = 65$ kPa.

a Determine the factor of safety for the trial failure surface specified in Figure 12.10.
b Check that no loss of overall stability will occur according to the limit state approach (using EC7 DA1b).
c Determine the maximum depth to which the slope could be excavated if the slope angle is maintained at 45°.

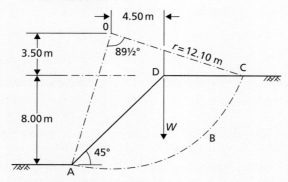

Figure 12.10 Example 12.1.

Solution

a In Figure 12.10, the cross-sectional area ABCD is 70 m². Therefore, the weight of the soil mass $W = 70 \times 19 = 1330$ kN/m. The centroid of ABCD is 4.5 m from O. The angle AOC is 89.5° and radius OC is 12.1 m. The arc length ABC is calculated (or may be measured) as 18.9 m. From Equation 12.15:

$$F = \frac{c_u L_a r}{Wd}$$
$$= \frac{65 \cdot 18.9 \cdot 12.1}{1330 \cdot 4.5}$$
$$= 2.48$$

This is the factor of safety for the trial failure surface selected, and is not necessarily the minimum factor of safety. From Figure 12.9, $\beta = 45°$ and, assuming that D is large, the value of N_s is 0.18. Then, from Equation 12.16, $F = 2.37$. This is lower than the value found previously, so the trial failure plane shown in Figure 12.10 is not the actual failure surface.

b For DA1b, $\gamma_\gamma = 1.00$, $\gamma_{cu} = 1.40$, $\gamma_A = 1.00$ (slope self-weight is a permanent unfavourable action) and $\gamma_{Rr} = 1.00$, hence

$$M_A = \gamma_A \left(\frac{W}{\gamma_\gamma}\right) d = 1.00 \cdot \left(\frac{1330}{1.00}\right) \cdot 4.5 = 5985 \text{ kNm/m}$$

$$M_R = \frac{\left(\frac{c_u}{\gamma_{cu}}\right) L_a r}{\gamma_{Rr}} = \frac{\left(\frac{65}{1.40}\right) \cdot 18.9 \cdot 12.1}{1.00} = 10520 \text{ kNm/m}$$

$M_A < M_R$, so the overall stability limit state (ULS) is satisfied to EC7 DA1b.

c The maximum depth of the cutting is given by Equation 12.17, the partial factors being as given above (same design approach as in part (b)) and $N_s = 0.18$ for $\beta = 45°$:

$$h \leq \frac{1}{0.18} \left[\frac{\left(\frac{65}{1.40}\right)}{1.00 \cdot 1.00 \cdot \left(\frac{19}{1.00}\right)} \right]$$

$$\leq 13.58 \text{ m}$$

Rotational slips in drained soil – the method of slices

In this method, the potential failure surface, in section, is again assumed to be a circular arc with centre O and radius r. The soil mass (ABCD) above a trial failure surface (AC) is divided by vertical planes into a series of slices of width b, as shown in Figure 12.11. The base of each slice is assumed to be a straight line. For the i-th slice the inclination of the base to the horizontal is α_i and the height, measured on the centre-line, is h_i. The analysis is based on the use of factor of safety (F), defined as the ratio of the available shear strength (τ_f) to the shear strength (τ_{mob}) which must be mobilised to maintain a condition of limiting equilibrium along the slip surface, i.e.

$$F = \frac{\tau_f}{\tau_{mob}}$$

Applications in geotechnical engineering

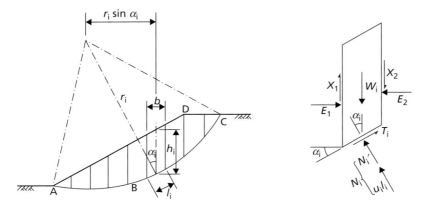

Figure 12.11 The method of slices.

The factor of safety is taken to be the same for each slice, implying that there must be mutual support between slices, i.e. inter-slice forces must act between the slices (E_1, X_1, E_2 and X_2 in Figure 12.11). The forces (per unit dimension normal to the section) acting on a slice are:

1. the total weight of the slice, $W_i = \gamma b h_i$ (γ_{sat} where appropriate);
2. the total normal force on the base, N_i (equal to $\sigma_i l_i$) – in general this force has two components, the effective normal force N'_i (equal to $\sigma'_i l_i$) and the boundary pore water force U_i (equal to $u_i l_i$), where u_i is the pore water pressure at the centre of the base and l_i the length of the base;
3. the shear force on the base, $T_i = T_{mob} l_i$;
4. the total normal forces on the sides, E_1 and E_2;
5. the shear forces on the sides, X_1 and X_2.

Any external forces (e.g. surcharge, pinning forces from inclusions) must also be included in the analysis.

The problem is statically indeterminate, and in order to obtain a solution assumptions must be made regarding the inter-slice forces E and X; in general, therefore, the resulting solution for factor of safety is not exact.

Considering moments about O, the sum of the moments of the shear forces T_i on the failure arc AC must equal the moment of the weight of the soil mass ABCD. For any slice the lever arm of W_i is $r_i \sin \alpha_i$, therefore at limiting equilibrium

$$\sum_i M_R = \sum_i M_A$$
$$\sum_i T_i r_i = \sum_i W_i r_i \sin \alpha_i$$

Now

$$T_i = \tau_{mob,i} l_i = \left(\frac{\tau_{f,i}}{F}\right) l_i$$

$$\therefore \sum_i \left(\frac{\tau_{f,i}}{F}\right) l_i = \sum_i W_i \sin \alpha_i$$

$$\therefore F = \frac{\sum_i \tau_{f,i} l_i}{\sum_i W_i \sin \alpha_i}$$

Stability of self-supporting soil masses

For an effective stress analysis (in terms of parameters c' and ϕ') $\tau_{f,i}$ is given by Equation 5.11, so

$$F = \frac{\sum_i (c'_i + \sigma'_i \tan \phi'_i) l_i}{\sum_i W_i \sin \alpha_i} \qquad (12.18\text{a})$$

Equation 12.18a can be used in the general case of c' and/or ϕ' varying with depth and position in the slope, the appropriate average values being used for each slice. For the case of homogeneous soil conditions, Equation 12.18a simplifies to

$$F = \frac{c' L_a + \tan \phi' \sum_i N'_i}{\sum_i W_i \sin \alpha_i} \qquad (12.18\text{b})$$

where L_a is the arc length AC (i.e. length of the whole slip plane). Equation 12.18b is exact, but approximations are introduced in determining the forces N'_i. For a given failure arc, the value of F will depend on the way in which the forces N'_i are estimated. In many cases, the critical state strength is normally appropriate in the analysis of slope stability, i.e. $\phi' = \phi'_{cv}$ and $c' = 0$, therefore the expression simplifies further to

$$F = \frac{\tan \phi'_{cv} \sum_i N'_i}{\sum_i W_i \sin \alpha_i} \qquad (12.18\text{c})$$

The Fellenius (or Swedish) solution

In this solution, it is assumed that for each slice the resultant of the inter-slice forces is zero. The solution involves resolving the forces on each slice normal to the base, i.e.

$$N'_i = W_i \cos \alpha_i - u_i l_i$$

Hence the factor of safety in terms of effective stress (Equation 12.18b) is given by

$$F = \frac{c' L_a + \tan \phi' \sum_i (W_i \cos \alpha_i - u_i l_i)}{\sum_i W_i \sin \alpha_i} \qquad (12.19)$$

The components $W_i \cos \alpha_i$ and $W_i \sin \alpha_i$ can be determined graphically for each slice. Alternatively, the values of W_i and α_i can be calculated. Again, a series of trial failure surfaces must be chosen in order to obtain the minimum factor of safety. It can be seen from the derivation of Equation 12.19 that the numerator represents the overall resistance, while the denominator represents the overall action driving failure. Therefore, Equation 12.19 may be used to verify the ULS by setting $F = 1$, factoring the numerator (resistance), denominator (action) and material properties appropriately as before, and ensuring that the numerator is larger than the denominator.

This solution is known to underestimate the true factor of safety due to the assumptions which are inherent in it; the error, compared with more accurate methods of analysis, is usually within the range 5–20%. Use of the Fellenius method is not now recommended in practice.

Applications in geotechnical engineering

The Bishop routine solution

In this solution it is assumed that the resultant forces on the sides of the slices are horizontal, i.e.

$$X_1 - X_2 = 0$$

For equilibrium, the shear force on the base of any slice is

$$T_i = \frac{1}{F}\left(c_i' l_i + N_i' \tan \phi_i'\right)$$

Resolving forces in the vertical direction:

$$W_i = N_i' \cos \alpha_i + u_i l_i \cos \alpha_i + \left(\frac{c_i' l_i}{F}\right)\sin \alpha_i + \left(\frac{N_i'}{F}\right)\tan \phi_i' \sin \alpha_i$$

$$\therefore N_i' = \frac{W_i - \left(\dfrac{c_i' l_i}{F}\right)\sin \alpha_i - u_i l_i \cos \alpha_i}{\cos \alpha_i + \left(\dfrac{\tan \phi_i' \sin \alpha_i}{F}\right)} \qquad (12.20)$$

It is convenient to substitute

$$l_i = b \sec \alpha_i \qquad (12.21)$$

Substituting Equation 12.20 into Equation 12.18a, it can be shown after some rearrangement that

$$F = \frac{1}{\sum_i W_i \sin \alpha_i} \cdot \sum_i \left\{ \left[c_i' b + (W_i - u_i b)\tan \phi_i'\right] \frac{\sec \alpha_i}{1 + \left(\dfrac{\tan \phi_i' \tan \alpha_i}{F}\right)} \right\} \qquad (12.22)$$

Bishop (1955) also showed how non-zero values of the resultant forces (X_1-X_2) could be introduced into the analysis, but this refinement has only a marginal effect on the factor of safety.

The pore water pressure can be expressed as a proportion of the total 'fill pressure' at any point by means of the dimensionless **pore water pressure ratio** (r_u), defined as

$$r_u = \frac{u}{\gamma h} \qquad (12.23)$$

Therefore, for the i-th slice

$$r_u = \frac{u_i b}{W_i}$$

Hence Equation 12.22 can be written as

$$F = \frac{1}{\sum_i W_i \sin \alpha_i} \cdot \sum_i \left\{ \left[c_i' b + W_i(1 - r_{u,i})\tan \phi_i'\right] \frac{\sec \alpha_i}{1 + \left(\dfrac{\tan \phi_i' \tan \alpha_i}{F}\right)} \right\} \qquad (12.24)$$

Stability of self-supporting soil masses

As the factor of safety occurs on both sides of Equations 12.22 and 12.24, a process of successive approximation must be used to obtain a solution, but convergence is rapid. Due to the repetitive nature of the calculations and the need to select an adequate number of trial failure surfaces, the method of slices is particularly suitable for solution by computer. More complex slope geometry and different soil strata can also then be straightforwardly introduced.

Again, the factor of safety determined by this method is an underestimate, but the error is unlikely to exceed 7% and in most cases is less than 2%. Spencer (1967) proposed a method of analysis in which the resultant inter-slice forces are parallel and in which both force and moment equilibrium are satisfied. Spencer showed that the accuracy of the Bishop routine method, in which only moment equilibrium is satisfied, is due to the insensitivity of the moment equation to the slope of the inter-slice forces.

Dimensionless stability coefficients for homogeneous slopes, based on Equation 12.24, have been published by Bishop and Morgenstern (1960) and Michalowski (2002). It can be shown that for a given slope angle and given soil properties the factor of safety varies linearly with r_u, and can thus be expressed as

$$F = m - nr_u \tag{12.25}$$

where m and n are the stability coefficients. The coefficients m and n are functions of β, ϕ', depth factor D and the dimensionless factor $c'/\gamma h$ (which is zero if the critical state strength is used).

A three-dimensional limit analysis for slopes in drained soil has been presented by Michalowski (2010).

> ### Example 12.2
>
> Using the Fellenius method of slices, determine the factor of safety, in terms of effective stress, of the slope shown in Figure 12.12 for the given failure surface: (a) using peak strength parameters $c' = 10\,\text{kPa}$ and $\phi' = 29°$; and (b) using critical state parameter $\phi'_{cv} = 31°$. The unit weight of the soil both above and below the water table is $20\,\text{kN/m}^3$.
>
>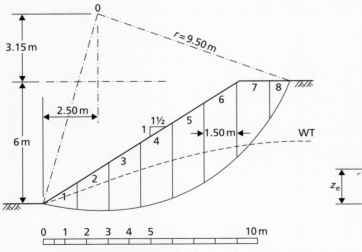
>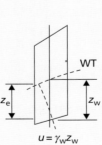
>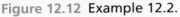
>
> **Figure 12.12** Example 12.2.

Solution

a The factor of safety is given by Equation 12.19. The soil mass is divided into slices 1.5 m wide. The weight (W_i) of each slice is given by

$$W_i = \gamma b h_i = 20 \cdot 1.5 \cdot h_i = 30 h_i \text{ kN/m}$$

The height h_i and angle α_i for each slice are measured from Figure 12.12 (which are drawn to scale), from which values of W_i are calculated using the expression given above, and values of l_i are calculated from Equation 12.21. The pore water pressure at the centre of the base of each slice is taken to be $\gamma_w z_w$, where z_w is the vertical distance of the centre point below the water table (as shown in the figure). This procedure slightly overestimates the pore water pressure, which strictly should be $\gamma_w z_e$, where z_e is the vertical distance below the point of intersection of the water table and the equipotential through the centre of the slice base. The error involved is on the safe side. The derived values are given in Table 12.1.

Table 12.1 Example 12.2

Slice	h_i (m)	α_i (°)	W_i (kN/m)	l_i (m)	u_i (kPa)	$W_i \cos\alpha_i - u_i l_i$ (kN/m)	$W_i \sin\alpha_i$ (kN/m)
1	0.76	−11.2	22.8	1.55	5.9	13.22	−4.43
2	1.80	−3.2	54.0	1.50	11.8	36.22	−3.01
3	2.73	8.4	81.9	1.55	16.2	55.91	11.96
4	3.40	17.1	102.0	1.60	18.1	68.53	29.99
5	3.87	26.9	116.1	1.70	17.1	74.47	52.53
6	3.89	37.2	116.7	1.95	11.3	70.92	70.56
7	2.94	49.8	88.2	2.35	0	56.93	67.37
8	1.10	59.9	33.0	2.15	0	16.55	28.55
						392.75	253.52

The arc length (L_a) is calculated/measured as 14.35 m. Then, from Equation 12.19,

$$F = \frac{c' L_a + \tan\phi' \sum_i (W_i \cos\alpha_i - u_i l_i)}{\sum_i W_i \sin\alpha_i}$$

$$= \frac{(10 \cdot 14.35) + (0.554 \cdot 392.75)}{253.52}$$

$$= 1.42$$

b Use of the critical state strength parameters only affects the values of c' and ϕ'; the calculations in Table 12.1 remain valid. Therefore,

$$F = \frac{(0) + (0.601 \cdot 392.75)}{253.52}$$

$$= 0.93$$

Despite $\phi'_{cv} > \phi'$, the factor of safety is lower in this case. This demonstrates that the (apparent) cohesion c' should not be relied upon in design.

Stability of self-supporting soil masses

Translational slips

It is assumed that the potential failure surface is parallel to the surface of the slope and is at a depth that is small compared with the length of the slope. The slope can then be considered as being of infinite length, with end effects being ignored. The slope is inclined at angle β to the horizontal and the depth of the failure plane is z, as shown in section in Figure 12.13. The water table is taken to be parallel to the slope at a height of mz ($0<m<1$) above the failure plane. Steady seepage is assumed to be taking place in a direction parallel to the slope. The forces on the sides of any vertical slice are equal and opposite, and the stress conditions are the same at every point on the failure plane.

In terms of effective stress, the shear strength of the soil along the failure plane (using the critical state strength) is

$$\tau_f = (\sigma - u)\tan\phi'_{cv}$$

and the factor of safety is

$$F = \frac{\tau_f}{\tau_{mob}} \tag{12.26a}$$

where τ_{mob} is the mobilised shear stress along the failure plane (see Chapter 11). The expressions for σ, τ_{mob} and u are

$$\sigma = \left[(1-m)\gamma + m\gamma_s\right]z\cos^2\beta$$
$$\tau_{mob} = \left[(1-m)\gamma + m\gamma_s\right]z\sin\beta\cos\beta$$
$$u = m\gamma_w z\cos^2\beta$$

giving

$$F = \frac{\left[(1-m)\gamma + m(\gamma_s - \gamma_w)\right]\tan\phi'_{cv}}{\left[(1-m)\gamma + m\gamma_s\right]\tan\beta} \tag{12.26b}$$

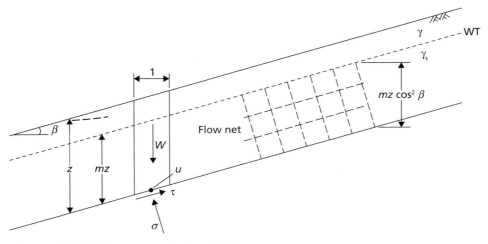

Figure 12.13 Plane translational slip.

Applications in geotechnical engineering

For a total stress analysis the $\tau_f = c_u$ is used, giving

$$F = \frac{c_u}{\left[(1-m)\gamma + m\gamma_s\right]z \sin\beta \cos\beta} \tag{12.26c}$$

As for rotational slips, the term in the numerator of Equation 12.26 represents the resistance of the soil to slip, while the denominator represents the driving action. For verification of the ULS, therefore, $F = 1.00$, and the numerator, denominator and material properties are factored appropriately.

Example 12.3

A long natural slope in an overconsolidated fissured clay of saturated unit weight $20\,\text{kN/m}^3$ is inclined at $12°$ to the horizontal. The water table is at the surface, and seepage is roughly parallel to the slope. A slip has developed on a plane parallel to the surface at a depth of 5 m. Determine whether the ULS is satisfied to EC7 DA1b using (a) the critical state parameter $\phi'_{cv} = 28°$, and (b) the residual strength parameter $\phi'_r = 20°$.

Solution

a The water table is at the ground surface, so $m = 1$. For DA1b, $\gamma_\gamma = 1.00$, $\gamma_{\tan\phi} = 1.25$, $\gamma_A = 1.00$ (slope self-weight is a permanent unfavourable action) and $\gamma_{Rr} = 1.00$. The resistance τ_f is

$$\tau_f = \frac{\left[(1-m)\left(\dfrac{\gamma}{\gamma_\gamma}\right) + m\left(\dfrac{\gamma_s - \gamma_w}{\gamma_\gamma}\right)\right] z \cos^2\beta \left(\dfrac{\tan\phi'_{cv}}{\gamma_{\tan\phi}}\right)}{\gamma_{Rr}}$$

$$= \frac{\left[0 + \left(\dfrac{20-9.81}{1.00}\right)\right] \cdot 5 \cdot \cos^2 12 \cdot \left(\dfrac{\tan 28}{1.25}\right)}{1.00}$$

$$= 20.7\,\text{kPa}$$

while the mobilised shear stress τ_{mob} (action) is

$$\tau_{mob} = \gamma_A \left[(1-m)\left(\dfrac{\gamma}{\gamma_\gamma}\right) + m\left(\dfrac{\gamma_s}{\gamma_\gamma}\right)\right] z \sin\beta \cos\beta$$

$$= \left[0 + \left(\dfrac{20}{1.00}\right)\right] \cdot 5 \cdot \sin 12 \cdot \cos 12$$

$$= 20.3\,\text{kPa}$$

As $\tau_f > \tau_{mob}$, the ULS is satisfied and the slope is stable.

b Using ϕ'_r in place of ϕ'_{cv} changes the resistance to $\tau_f = 14.2\,\text{kPa}$, while τ_{mob} remains unchanged. In this case $\tau_f < \tau_{mob}$, so the ULS is not satisfied (the slope will slip if residual strength conditions are achieved).

General methods of analysis

Morgenstern and Price (1965, 1967) developed a general analysis based on limit equilibrium in which all boundary and equilibrium conditions are satisfied and in which the failure surface may be any shape, circular, non-circular or compound. Computer software for undertaking such analyses is readily available. Bell (1968) proposed an alternative method of analysis in which all the conditions of equilibrium are satisfied and the assumed failure surface may be of any shape. The soil mass is divided into a number of vertical slices and statical determinacy is obtained by means of an assumed distribution of normal stress along the failure surface. The use of a computer is also essential for this method. In both general methods mentioned here, the solutions must be checked to ensure that they are physically acceptable. Modern computer-based tools are now available for analysing the ULS for slopes using limit analysis combined with optimisation routines (see the Companion Website for further details).

End of construction and long-term stability

When a slope is formed by excavation, the decreases in total stress result in changes in pore water pressure in the vicinity of the slope and, in particular, along a potential failure surface. For the case illustrated in Figure 12.14(a), the initial pore water pressure (u_0) depends on the depth of the point in question below the initial (static) water table (i.e. $u_0 = u_s$). The change in pore water pressure (Δu) due to excavation is given theoretically by Equation 8.55. For a typical point P on a potential failure surface (Figure 12.14(a)), the pore water pressure change Δu is negative. After excavation, pore water will flow towards the slope and drawdown of the water table will occur. As dissipation proceeds the pore water pressure increases to the steady seepage value, as shown in Figure 12.14(a), which may be determined from a flow net or by using the numerical methods described in Section 2.7. The final pore water pressure (u_f), after dissipation of excess pore water pressure is complete, will be the steady seepage value determined from the flow net.

If the permeability of the soil is low, a considerable time will elapse before any significant dissipation of excess pore water pressure will have taken place. At the end of construction the soil will be virtually in the undrained condition, and a total stress analysis will be relevant to verify stability (ULS). In principle, an effective stress analysis is also possible for the end-of-construction condition using the appropriate value of pore water pressure ($u_0 + \Delta u$) for this condition. However, because of its greater simplicity, a total stress analysis is generally used. It should be realised that the same factor of safety will not generally be obtained from a total stress and an effective stress analysis of the end-of-construction condition. In a total stress analysis it is implied that the pore water pressures are those for a failure condition (being the equivalent of the pore water pressure at failure in an undrained triaxial test); in an effective stress analysis the pore water pressures used are those predicted for a non-failure condition. In the long term, the fully drained condition will be reached and only an effective stress analysis will be appropriate.

On the other hand, if the permeability of the soil is high, dissipation of excess pore water pressure will be largely complete by the end of construction. An effective stress analysis is relevant for all conditions with values of pore water pressure being obtained from the static water table level or the steady seepage flow net.

Irrespective of the permeability of the soil, the increase in pore water pressures following excavation will result in a reduction in effective stress (and hence strength) with time such that the factor of safety will be lower in the long term, when dissipation is complete, than at the end of construction.

The creation of sloping ground through construction of an embankment results in increases in total stress, both within the embankment itself as successive layers of fill are placed, and in the foundation soil. The initial pore water pressure (u_0) depends primarily on the placement water content of the fill. The construction period of a typical embankment is relatively short, and, if the permeability of the

Applications in geotechnical engineering

compacted fill is low, no significant dissipation is likely during construction. Dissipation proceeds after the end of construction, with the pore water pressure decreasing to the final value in the long term, as shown in Figure 12.14(b). The factor of safety of an embankment at the end of construction is therefore lower than in the long term. Shear strength parameters for the fill material should be determined from tests on specimens compacted to the values of dry density and water content to be specified for the embankment (see Chapter 1).

The stability of an embankment may also depend on the shear strength of the foundation soil. The possibility of failure along a surface such as that illustrated in Figure 12.15 should be considered in appropriate cases.

Slopes in overconsolidated fissured clays require special consideration. A number of cases are on record in which failures in this type of clay have occurred long after dissipation of excess pore water pressure had been completed. Analysis of these failures showed that the average shear strength at failure was well below the peak value. It is probable that large strains occur locally due to the presence of fissures, resulting in the peak strength being reached, followed by a gradual decrease towards the critical state value. The development of large local strains can lead eventually to a progressive slope failure.

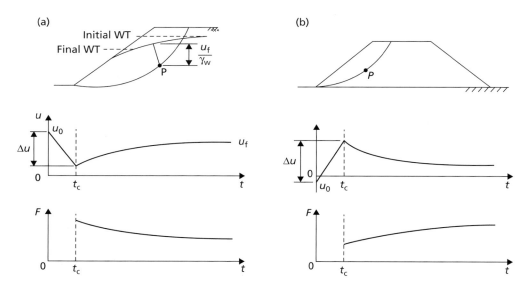

Figure 12.14 Pore pressure dissipation and factor of safety: (a) following excavation (i.e. a cutting), (b) following construction (i.e. an embankment).

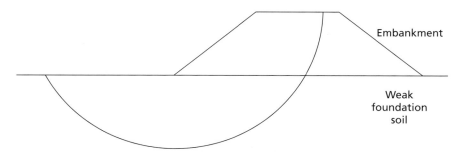

Figure 12.15 Failure beneath an embankment.

Stability of self-supporting soil masses

However, fissures may not be the only cause of progressive failure; there is considerable non-uniformity of shear stress along a potential failure surface, and local overstressing may initiate progressive failure. It is also possible that there could be a pre-existing slip surface in this type of clay and that it could be reactivated by excavation. In such cases a considerable slip movement could have taken place previously, sufficiently large for the shear strength to fall below the critical state value and towards the residual value.

Thus for an initial failure (a 'first time' slip) in overconsolidated fissured clay, the relevant strength for the analysis of long-term stability is the critical state value. However, for failure along a pre-existing slip surface the relevant strength is the residual value. Clearly it is vital that the presence of a pre-existing slip surface in the vicinity of a projected excavation should be detected during the ground investigation.

The strength of an overconsolidated clay at the critical state, for use in the analysis of a potential first time slip, is difficult to determine accurately. Skempton (1970) has suggested that the maximum strength of the remoulded clay in the normally consolidated condition can be taken as a practical approximation to the strength of the overconsolidated clay at the critical state, i.e. when it has fully softened adjacent to the slip plane as the result of expansion during shear.

12.4 Embankment dams

An embankment dam would normally be used where the foundation and abutment conditions are unsuitable for a concrete dam and where suitable materials for the embankment are present at or close to the site. An extensive ground investigation is essential, general at first but becoming more detailed as design studies proceed, to determine foundation and abutment conditions and to identify suitable borrow areas. It is important to determine both the quantity and quality of available material. The natural water content of fine soils should be determined for comparison with the optimum water content for compaction.

Most embankment dams are not homogeneous but are of zoned construction, the detailed section depending on the availability of soil types to provide fill for the embankment. Typically a dam will consist of a core of low-permeability soil with shoulders of other suitable material on each side. The upstream slope is usually covered by a thin layer of rockfill (known as **rip-rap**) to protect it from erosion by wave or other fluid actions. The downstream slope is usually grassed (again, to resist erosion). An internal drainage system, to alleviate the detrimental effects of any seeping water, would normally be incorporated. Depending on the materials used, horizontal drainage layers may also be incorporated to accelerate the dissipation of excess pore water pressure. Slope angles should be such that stability is ensured, but overconservative design must be avoided: a decrease in slope angle of as little as 2–3° (to the horizontal) would mean a significant increase in the volume of fill for a large dam.

Failure of an embankment dam could result from the following causes: (1) instability of either the upstream or downstream slope; (2) internal erosion; and (3) erosion of the crest and downstream slope by overtopping. (The third cause arises from errors in the hydrological predictions.)

The factor of safety for both slopes must be determined as accurately as possible for the most critical stages in the life of the dam, using the methods outlined in Section 12.3. The potential failure surface may lie entirely within the embankment, or may pass through the embankment and the foundation soil (as in Figure 12.15). In the case of the upstream slope, the most critical stages are at the end of construction and during rapid drawdown of the reservoir level. The critical stages for the downstream slope are at the end of construction and during steady seepage when the reservoir is full. The pore water pressure distribution at any stage has a dominant influence on the factor of safety of the slopes, and it is common practice to install a piezometer system (see Chapter 6) so that the actual pore water pressures can be measured and compared with the predicted values used in design (provided an effective stress analysis has been used). Remedial action could then be taken if, based on the measured values, the slope began to approach the ULS.

Applications in geotechnical engineering

If a potential failure surface were to pass through foundation material containing fissures, joints or pre-existing slip surfaces, then progressive failure (as described in the previous section) would be a possibility. The different stress–strain characteristics of various zone materials through which a potential failure surface passes, together with non-uniformity of shear stress, could also lead to progressive failure.

Another problem is the danger of cracking due to differential movements between soil zones, and between the dam and the abutments. The possibility of **hydraulic fracturing**, particularly within the clay core, should also be considered. Hydraulic fracturing occurs on a plane where the total normal stress is less than the local value of pore water pressure. Following the completion of construction the clay core tends to settle relative to the rest of the embankment due to long-term consolidation; consequently, the core will be partially supported by the rest of the embankment. Thus vertical stress in the core will be reduced and the chances of hydraulic fracture increased. The transfer of stress from the core to the shoulders of the embankment is another example of the arching phenomenon (Section 11.7). Following fracture or cracking, the resulting leakage could lead to serious internal erosion and impair stability.

End of construction and long-term stability

Most slope failures in embankment dams occur either during construction or at the end of construction. Pore water pressures depend on the placement water content of the fill and on the rate of construction. A commitment to achieve rapid completion will result in the maximisation of pore water pressure at the end of construction. However, the construction period of an embankment dam is likely to be long enough to allow partial dissipation of excess pore water pressure, especially for a dam with internal drainage. A total stress analysis, therefore, would result in an overconservative design. An effective stress analysis is preferable, using predicted values of r_u.

If high values of r_u are anticipated, dissipation of excess pore water pressure can be accelerated by means of horizontal drainage layers incorporated in the dam, drainage taking place vertically towards the layers: a typical dam section is shown in Figure 12.16. The efficiency of drainage layers has been examined theoretically by Gibson and Shefford (1968), and it was shown that in a typical case the layers, in order to be fully effective, should have a permeability at least 10^6 times that of the embankment soil: an acceptable efficiency would be obtained with a permeability ratio of about 10^5.

After the reservoir has been full for some time, conditions of steady seepage become established through the dam, with the soil below the top flow line in the fully saturated state. This condition must be analysed in terms of effective stress, with values of pore pressure being determined from a flow net (or using the numerical methods described in Section 2.7). Values of r_u up to 0.45 are possible in homogeneous dams, but much lower values can be achieved in dams having internal drainage. Internal erosion is a particular danger when the reservoir is full because it can arise and develop within a relatively short time, seriously impairing the safety of the dam.

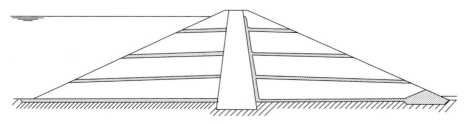

Figure 12.16 Horizontal drainage layers.

Stability of self-supporting soil masses

Rapid drawdown

After a condition of steady seepage has become established, a drawdown of the reservoir level will result in a change in the pore water pressure distribution. If the permeability of the soil is low, a drawdown period measured in weeks may be 'rapid' in relation to dissipation time and the change in pore water pressure can be assumed to take place under undrained conditions. Referring to Figure 12.17, the pore water pressure before drawdown at a typical point P on a potential failure surface is given by

$$u_0 = \gamma_w (h + h_w - h') \tag{12.27}$$

where h' is the loss in total head due to seepage between the upstream slope surface and the point P. It is again assumed that the total major principal stress at P is equal to the fill pressure. The change in total major principal stress is due to the total or partial removal of water above the slope on the vertical through P. For a drawdown depth exceeding h_w,

$$\Delta \sigma_1 = -\gamma_w h_w$$

From Equation 8.57, the change in pore water pressure Δu can then be expressed in terms of $\Delta \sigma_1$ by

$$\begin{aligned}\frac{\Delta u}{\Delta \sigma_1} &= \frac{B\left[\Delta \sigma_3 + A(\Delta \sigma_1 - \Delta \sigma_3)\right]}{\Delta \sigma_1} \\ &= B\left[1 - (1-A)\left(1 - \frac{\Delta \sigma_3}{\Delta \sigma_1}\right)\right] \\ &= \bar{B}\end{aligned} \tag{12.28}$$

Therefore the pore water pressure at P immediately after drawdown is

$$\begin{aligned}u &= u_0 + \Delta u \\ &= \gamma_w \left\{h + h_w (1 - \bar{B}) - h'\right\}\end{aligned}$$

Hence

$$\begin{aligned}r_u &= \frac{u}{\gamma h} \\ &= \frac{\gamma_w}{\gamma}\left[1 + \frac{h_w}{h}(1 - \bar{B}) - \frac{h'}{h}\right]\end{aligned} \tag{12.29}$$

The soil will be undrained immediately after rapid drawdown. An upper bound value of r_u for these conditions can be obtained by assuming $\bar{B} = 1$ and neglecting h'. Typical values of r_u immediately after drawdown are within the range 0.3–0.4.

Morgenstern (1963) published stability coefficients for the approximate analysis of homogeneous slopes after rapid drawdown, based on limit equilibrium techniques.

The pore water pressure distribution after drawdown in soils of high permeability decreases as pore water drains out of the soil above the drawdown level. The saturation line moves downwards at a rate depending on the permeability of the soil. A series of flow nets can be drawn for different positions of the saturation line and values of pore water pressure obtained. The factor of safety can thus be determined, using an effective stress analysis, for any position of the saturation line. Viratjandr and Michalowski (2006) published stability coefficients for the approximate analysis of homogeneous slopes in such conditions, based on limit analysis techniques.

Applications in geotechnical engineering

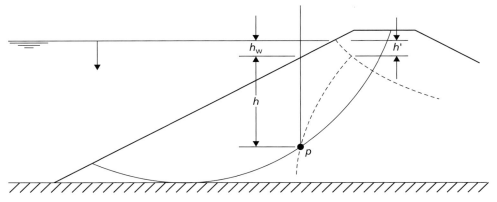

Figure 12.17 Rapid drawdown conditions.

12.5 An introduction to tunnels

Tunnels are the final class of problem that will be considered in this chapter, for which self-support of the soil mass controls the design. Shallow tunnels onshore may be constructed using the **cut-and-cover** technique; this is where a deep excavation is made, within which the tunnel is constructed, which is then backfilled to bury the tunnel structure. The design of such works may be completed using the techniques described in Chapter 11, and this class of tunnel will not be considered further here. In marine and offshore applications, sections of tunnel structure are floated out to site, flooded to lower them into a shallow trench excavated on the riverbed/seabed and connected underwater, followed by pumping out of the internal water. These are known as **immersed tube** tunnels. Some of the terminology related to tunnels is shown in Figure 12.18.

In this chapter, the design of deep tunnels which are formed by boring deep within the ground are considered. In certain conditions (namely undrained soil response and a shallow **running depth**) the tunnel may be self-supporting. For deeper excavations in undrained soil and for excavation in drained materials, the tunnel will need to be supported by an internal pressure to prevent collapse of the soil above the tunnel into the excavation (the ULS); this is known as **earth pressure balance** construction. Once the tunnel is complete this internal support pressure describes the structural loading which the tunnel lining must be able to resist and is used in the structural design of the tunnel lining. In addition to maintaining the stability of the tunnel (ULS condition), the design of tunnelling works also requires consideration of settlements at the ground surface which are induced by the tunnelling procedure to ensure that these movements do not damage buildings and other infrastructure (SLS).

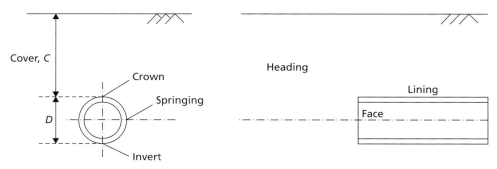

Figure 12.18 Terminology related to tunnels.

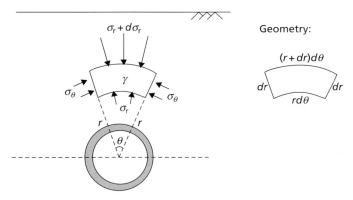

Figure 12.19 Stress conditions in the soil above the tunnel crown.

Stability of tunnels in undrained soil

Figure 12.19 shows an element of soil above the crown of the tunnel (i.e. within the cover depth, C). This element of soil is loaded similarly to that around a pressuremeter test (PMT, Figure 7.10(b)), though tunnel collapse involves the collapse of the cylindrical cavity (tunnel) rather than expansion in the PMT. As the stresses and strains are now in a vertical plane (rather than the horizontal plane for the PMT), the weight of the soil must also be considered. The volume of the soil element is given by

$$\left(\frac{(r+dr)d\theta + rd\theta}{2}\right)dr = rdrd\theta + \frac{d^2rd\theta}{2} \tag{12.30}$$

The second term in Equation 12.30 is very small compared to the first, and can be neglected. Resolving forces vertically then gives

$$(\sigma_r + d\sigma_r)(r+dr)d\theta + \gamma r dr d\theta = \sigma_r r d\theta + \sigma_\theta dr d\theta$$

$$\therefore r\frac{d\sigma_r}{dr} - (\sigma_r - \sigma_\theta) + \gamma r = 0 \tag{12.31}$$

Equation 12.31 is similar to Equation 7.15; the sign of the $(\sigma_r - \sigma_\theta)$ term has changed (cavity collapse instead of expansion) and there is an additional unit weight term. As in Chapter 7, the associated maximum shear stress is $\tau = (\sigma_r - \sigma_\theta)/2$, and in undrained soil at the point of failure, $\tau = c_u$. Substituting these expressions into Equation 12.31 and rearranging gives

$$d\sigma_r = \left[\left(\frac{2c_u}{r}\right) - \gamma\right]dr \tag{12.32}$$

At the tunnel (cavity) wall, $r=D/2$ and $\sigma_r=p$ (where p is any internal pressure within the tunnel); referring to Figure 12.18, at the ground surface $r=C+D/2$ and $\sigma_r=\sigma_q$ where σ_q is the surcharge pressure. Equation 12.32 may then be integrated using these limits to give

$$\int_p^{\sigma_q} d\sigma_r = \int_{D/2}^{C+D/2}\left[\left(\frac{2c_u}{r}\right) - \gamma\right]dr$$

$$\sigma_q - p = 2c_u \ln\left(\frac{2C}{D} + 1\right) - \gamma C \tag{12.33}$$

Applications in geotechnical engineering

Equation 12.33 may be used to determine the required support pressure based on the soil properties (c_u, γ), any external loading (σ_q) and geometric properties (C, D). Equation 12.33 is often expressed as a stability number N_t, where

$$N_t = \frac{\sigma_q - p + \gamma(C + D/2)}{c_u} \qquad (12.34)$$

For deep tunnels $C \gg D$, so that comparing Equations 12.33 and 12.34 gives an approximate expression for N_t, suitable for preliminary design purposes:

$$N_t = 2\ln\left(\frac{2C}{D} + 1\right) \qquad (12.35)$$

The foregoing analysis has considered collapse of the crown of a long tunnel (a plane-strain analysis was conducted). While this is appropriate for the finished tunnel, during construction there may additionally be collapse of soil ahead of the tunnel (the heading) into the face. This involves a more complex three-dimensional failure mechanism/stress field. In the case of undrained materials, Davis et al. (1980) presented stability numbers for use in Equation 12.34 for the case of a circular tunnel heading where

$$N_t = \min\left\{2 + 2\ln\left(\frac{2C}{D} + 1\right), 4\ln\left(\frac{2C}{D} + 1\right)\right\} \qquad (12.36)$$

Equations 12.35 and 12.36 are compared in Figure 12.20, which shows that the stability of the tunnel behind the excavation is usually critical (a lower N_t requires a higher support pressure to be supplied by the tunnel, from Equation 12.34).

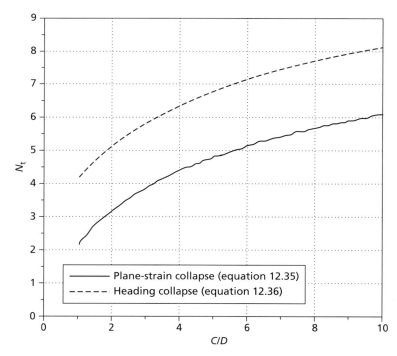

Figure 12.20 Stability numbers for circular tunnels in undrained soil.

Stability of tunnels in drained soil

Under drained conditions, the relationship between the radial and horizontal effective stresses is $\sigma'_q = K_p \sigma'_r$. Rewriting Equation 12.31 in terms of effective stresses and substituting then gives

$$\frac{d\sigma'_r}{dr} - \frac{\sigma'_r}{r}(K_p - 1) + \gamma' = 0 \qquad (12.37)$$

Equation 12.37 can be simplified for integration by substitution of $x = \sigma'_r/r$, so that $d\sigma'_r/dr = r(dx/dr) + x$, giving

$$r\frac{dx}{dr} + x = x(K_p - 1) - \gamma$$

$$\frac{dx}{(K_p - 2)x - \gamma} = \frac{dr}{r} \qquad (12.38)$$

Equation 12.38 is integrated between the same limits as before, but with $\sigma'_r = \sigma'_q$ where σ'_q is the effective surcharge pressure. This gives $x = 2p'/D$ at $r = D/2$ and $x = \sigma'_q/(C + D/2)$ at $r = C + D/2$, so that

$$\int_{2p'/D}^{\sigma'_q/(C+D/2)} \frac{dx}{(K_p - 2)x - \gamma} = \int_{D/2}^{C+D/2} \frac{dr}{r}$$

$$\frac{(K_p - 2)\left(\dfrac{2\sigma'_q}{\gamma(2C + D)}\right) - 1}{(K_p - 2)\left(\dfrac{2p'}{\gamma D}\right) - 1} = \left(\frac{2C + D}{D}\right)^{(K_p - 2)} \qquad (12.39)$$

Equation 12.39 may be rearranged to find the effective radial pressure applied by the overlying soil which the tunnel must support (p'). If the soil is dry, the total support pressure $p = p'$, and γ in Equation 12.39 is that for dry soil. If the soil is submerged, then $p = p' + u$, where u is the pore water pressure at the level of the springing and $\gamma = \gamma'$ (buoyant unit weight) in Equation 12.39 for a tunnel with an impermeable lining.

Equation 12.39 is plotted for various different values of ϕ' in Figure 12.21(a) for the case of $\sigma'_q = 0$. As expected, as the shear strength of the soil increases (represented by ϕ'), the required support pressure reduces. It can also be seen that for most common values of ϕ' (>30°) a maximum support pressure is reached even for shallow tunnels (low C/D). From Equation 12.39, this value is

$$\frac{p'_{max}}{\gamma D} = \frac{1}{2(K_p - 2)} \qquad (12.40)$$

Equation 12.40 is plotted in Figure 12.21(b), which demonstrates that long-term stresses (after consolidation is complete in the case of fine-grained soils) in tunnel linings are generally very small and independent of tunnel depth.

Further information regarding heading collapse in drained materials may be found in Atkinson and Potts (1977) and Leca and Dormieux (1990).

Serviceability criteria for tunnelling works

As material is excavated from the face of a tunnel, the soil ahead of the tunnel will slump towards the tunnel face under the action of its own self-weight. This will tend to lead to over-excavation of material,

Applications in geotechnical engineering

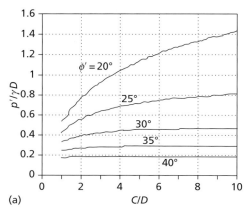
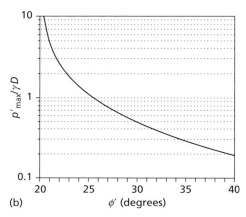

Figure 12.21 (a) Support pressure in drained soil for shallow and deep tunnels ($\sigma'_q = 0$), (b) maximum support pressure for use in ULS design ($\sigma'_q = 0$).

which will generate a settlement trough at the ground surface due to the loss of soil volume over and above that of the tunnel (Figure 12.22). This trough will have a maximum settlement immediately above the crown of the tunnel, reducing with radial distance from the tunnel. Any buildings or other infrastructure will therefore be subject to differential settlement as the ground beneath them subsides (see Section 10.2). The minimisation of damage to existing infrastructure is the main serviceability consideration in tunnelling in an urban environment.

Observation of tunnelling works has shown that the settlement trough can be described by

$$s_g = s_{max} e^{-x^2/2i^2} \tag{12.41}$$

where s_g is the settlement of the ground surface at a point defined by position x, s_{max} is the maximum settlement (above the crown of the tunnel), and i is the trough width parameter governing the shape of the

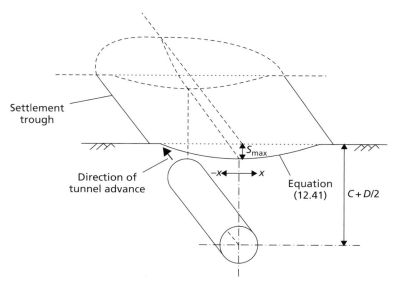

Figure 12.22 Settlement trough above an advancing tunnel.

curve, which is a function of soil type. In clays, $i=0.5(C+D/2)$; in sands and gravels, $i=0.35(C+D/2)$. The parameter i can also be expressed as a function of depth, so that the settlement profile can be found at any point between the ground surface and the tunnel; this can then be used to check differential movements of pipelines or other existing buried services.

The class of equation described by Equation 12.41 is also known as a **Gaussian curve**. The volume of over-excavated material per metre length of the tunnel (V_{soil}) can be found by integrating Equation 12.41 from $x=-\infty$ to $x=\infty$, giving

$$V_{soil} = \sqrt{2\pi} \cdot i s_{max} = 2.507 i s_{max} \tag{12.42}$$

(this is a standard result for a Gaussian curve). The volume of over-excavation in Equation 12.42 is usually normalised so that it can be expressed as a percentage of the tunnel volume, $V_t = \pi D^2/4$ per metre length. The resulting percentage is known as the **volume loss** (V_L):

$$V_L = \frac{2.507 i s_{max}}{0.25\pi D^2} = 3.192 \frac{i s_{max}}{D^2} \tag{12.43}$$

The volume loss depends on the tunnelling technique used and on the quality control that can be achieved during construction. A perfectly constructed tunnel would only excavate just enough soil for the tunnel so that $V_L=0$ and $s_{max}=0$ from Equation 12.43. In practice this is not possible, and volume loss is typically between 1% and 3% in soft ground. Volume loss may be minimised using modern **earth pressure balance (EPB) tunnel boring machines**. In these computer-controlled machines the cutting face is pressurised with the aim of matching the in-situ horizontal stresses within the ground; however, even these are not perfect. In design to SLS, a conservative (higher) value of V_L is selected based on previous experience in similar soils. Equation 12.43 is then used to determine s_{max}, from which the ground settlement profile is found using Equation 12.41. These settlements are then applied to infrastructure within the area affected by the settlement trough and checked for damage from differential settlement and tilt using the methods outlined in Section 10.2.

Summary

1. Trenches and open shafts may be excavated to a limited depth in fine-grained soils under undrained conditions (i.e. for temporary works only) and in bonded/cemented coarse-grained soils (having $c'>0$). These excavations may be extended in depth using fluid support in the trench (e.g. bentonite). Fluid support also allows for such excavations to be made in cohesionless soil. The overriding design criterion is preventing collapse of the excavation (ULS).
2. Limit analysis techniques may be applied to the stability of slopes and vertical cuttings in homogeneous soil. Limit equilibrium techniques may also be applied using the method of slices, which can also account for variable pore water pressure distribution and hence cases where seepage is occurring, and consider both rotational and translational slips. For both techniques, optimised failure surfaces must be found which give the most critical conditions. As for trenches, the overriding design criterion is preventing collapse of the slope and the occurrence of catastrophically large slip displacements (ULS).
3. In undrained materials, unsupported tunnels may be made to a limited depth in the short term. Under drained conditions and at deeper depths in cohesive material, internal support pressure must be applied along the axis of the tunnel

(from the tunnel lining) and at the face while excavation is proceeding. This information may be used to determine the earth pressures acting on a tunnel lining when the tunnel is completed, for structural design of the lining system at both ULS and SLS. In addition to preventing collapse of the tunnel (ULS), the design must also consider the ground settlement profile above the tunnel due to volume loss and the potential damage to surface or buried infrastructure due to gross or differential settlement in this region (SLS).

Problems

12.1 A diaphragm wall is to be constructed in a soil having a unit weight of $18\,kN/m^3$ and design shear strength parameters $c'=0$ and $\phi'=34°$. The depth of the trench is 3.50 m and the water table is 1.85 m above the bottom of the trench. Determine whether the trench is stable to EC7 DA1b if the unit weight of the slurry is $10.6\,kN/m^3$ and the depth of slurry in the trench is 3.35 m. Determine also the maximum depth to which the trench could be excavated if the slurry is maintained at the same level below the ground surface.

12.2 For the given failure surface, determine whether the slope detailed in Figure 12.23 is stable in terms of total stress to EC7 DA1a. The unit weight for both soils is $19\,kN/m^3$. The characteristic undrained strength (c_u) is 20 kPa for soil 1 and 45 kPa for soil 2. How would the answer change if allowance is made for the development of a tension crack?

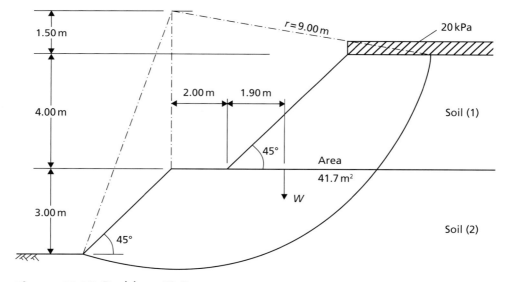

Figure 12.23 Problem 12.2.

12.3 A cutting 9 m deep is to be excavated in a saturated clay of unit weight $19\,kN/m^3$. The characteristic shear strength parameters are $c_u=30\,kPa$ and $\phi_u=0$. A hard stratum underlies the clay at a depth of 11 m below ground level. Determine the slope angle at which failure would occur. What is the allowable slope angle if the slope is to satisfy EC7 DA1b, and what is the overall factor of safety corresponding to such a design?

12.4 For the given failure surface, determine whether the slope detailed in Figure 12.24, is stable to EC7 DA1b using the Fellenius method of slices. The unit weight of the soil is 21 kN/m³, and the characteristic shear strength parameters are $c' = 8$ kPa and $\phi' = 32°$.

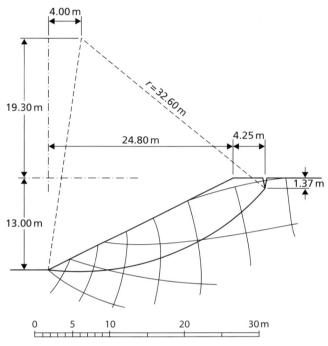

Figure 12.24 Problem 12.4.

12.5 Repeat the analysis of the slope detailed in Problem 12.4 using the Bishop routine method of slices.
12.6 Using the Bishop routine method of slices, determine whether the slope detailed in Figure 12.25 is stable to EC7 DA1-a in terms of effective stresses for the specified failure surface. The value of r_u is 0.20 and the unit weight of the soil is 20 kN/m³. Characteristic values of the shear strength parameters are $c' = 0$ and $\phi' = 33°$.

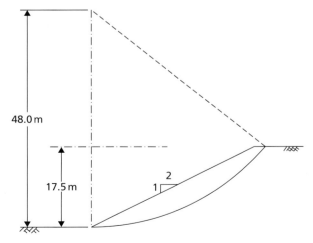

Figure 12.25 Problem 12.6.

12.7 A long slope is to be formed in a soil of unit weight $19\,kN/m^3$ for which the characteristic shear strength parameters are $c'=0$ and $\phi'=36°$. A firm stratum lies below the slope. It is to be assumed that the water table may occasionally rise to the surface, with seepage taking place parallel to the slope. Determine the maximum safe slope angle to satisfy EC7 DA1-b, assuming a potential failure surface parallel to the slope. Determine also the overall factor of safety for the slope angle determined above if the water table were well below the surface.

12.8 A circular tunnel of diameter $12\,m$ is to be bored in stiff clay with $c_u = 200\,kPa$ and $\gamma = 19\,kN/m^3$ (both constant with depth). A structure with a footprint of $16 \times 16\,m^2$ is situated directly above the tunnel, which is $16\,m$ high and has masonry load bearing walls. The tunnelling method can be assumed to induce a volume loss of 2.5%. Determine a suitable running depth of the tunnel (at the tunnel centerline).

References

Atkinson, J.H. and Potts, D.M. (1977) Stability of a shallow circular tunnel in cohesionless soil, *Géotechnique*, **27**(2), 203–215.

Bell, J.M. (1968) General slope stability analysis, *Journal ASCE*, **94**(SM6), 1253–1270.

Bishop, A.W. (1955) The use of the slip circle in the stability analysis of slopes, *Géotechnique*, **5**(1), 7–17.

Bishop, A.W. and Bjerrum, L. (1960) The relevance of the triaxial test to the solution of stability problems, in *Proceedings of the ASCE Research Conference on Shear Strength of Cohesive Soils, Boulder, Colorado*, pp. 437–501.

Bishop, A.W. and Morgenstern, N.R. (1960) Stability coefficients for earth slopes, *Géotechnique*, **10**(4), 129–147.

Davis, E.H., Gunn, M.J., Mair, R.J. and Seneviratne, H.N. (1980). The stability of shallow tunnels and underground openings in cohesive material, *Géotechnique*, **30**(4), 397–416.

EC7-1 (2004) *Eurocode 7: Geotechnical design – Part 1: General rules, BS EN 1997–1:2004*, British Standards Institution, London.

Gens, A., Hutchinson, J.N. and Cavounidis, S. (1988) Three-dimensional analysis of slides in cohesive soils, *Géotechnique*, **38**(1), 1–23.

Gibson, R.E. and Morgenstern, N.R. (1962) A note on the stability of cuttings in normally consolidated clays, *Géotechnique*, **12**(3), 212–216.

Gibson, R.E. and Shefford, G.C. (1968) The efficiency of horizontal drainage layers for accelerating consolidation of clay embankments, *Géotechnique*, **18**(3), 327–335.

Leca, E. and Dormieux, L. (1990) Upper and lower bound solutions for the face stability of shallow circular tunnels in frictional material, *Géotechnique*, **40**(4), 581–606.

Michalowski, R.L. (2002). Stability charts for uniform slopes, *Journal of Geotechnical and Geoenvironmental Engineering*, **128**(4) 351–355.

Michalowski, R.L. (2010) Limit analysis and stability charts for 3D slope failures, *Journal of Geotechnical and Geoenvironmental Engineering*, **136**(4), 583–593.

Morgenstern, N.R. (1963) Stability charts for earth slopes during rapid drawdown, *Géotechnique*, **13**(2), 121–131.

Morgenstern, N.R. and Price, V.E. (1965) The analysis of the stability of general slip surfaces, *Géotechnique*, **15**(1), 79–93.

Morgenstern, N.R. and Price, V.E. (1967) A numerical method for solving the equations of stability of general slip surfaces, *Computer Journal*, **9**, 388–393.

Skempton, A.W. (1970) First-time slides in overconsolidated clays (Technical Note), *Géotechnique*, **20**(3), 320–324.

Spencer, E. (1967) A method of analysis of the stability of embankments assuming parallel inter-slice forces, *Géotechnique*, **17**(1), 11–26.
Taylor, D.W. (1937) Stability of earth slopes, *Journal of the Boston Society of Civil Engineers*, **24**(3), 337–386.
Viratjandr, C. and Michalowski, R.L. (2006). Limit analysis of slope instability caused by partial submergence and rapid drawdown, *Canadian Geotechnical Journal*, **43**(8), 802–814.

Further reading

Frank, R., Bauduin, C., Driscoll, R., Kavvadas, M., Krebs Ovesen, N., Orr, T. and Schuppener, B. (2004) *Designers' Guide to EN 1997–1 Eurocode 7: Geotechnical Design – General Rules*, Thomas Telford, London.
This book provides a guide to limit state design of a range of constructions (including slopes) using Eurocode 7 from a designer's perspective and provides a useful companion to the Eurocodes when conducting design. It is easy to read and has plenty of worked examples.
Mair, R.J. (2008) Tunnelling and geotechnics: new horizons, *Géotechnique*, **58**(9), 695–736.
Includes some interesting case histories of tunnel construction, building on many of the basic concepts outlined in Section 12.5, and highlights current and future issues.
Michalowski, R.L. (2010). Limit analysis and stability charts for 3D slope failures. *Journal of Geotechnical & Geoenvironmental Engineering*, **136**(4), 583–593.
Provides stability charts similar to those in Section 12.3 for analysis of slopes under a range of soil conditions for the more complicated (but more realistic) cases of three-dimensional, rather than plane strain, failure (e.g. Figure 12.7).

For further student and instructor resources for this chapter, please visit the Companion Website at www.routledge.com/cw/craig

Chapter 13

Illustrative cases

Learning outcomes

After working through the material in this chapter, you should be able to:

1. Select characteristic values of engineering parameters from laboratory or in-situ data which are suitable for use in engineering design;
2. Understand the principal of operation of field instrumentation used to measure the response of geotechnical constructions, and be able to select appropriate instrumentation for verifying design assumptions;
3. Understand how the Observational Method may be used in geotechnical construction;
4. Apply the limit state techniques presented in Chapters 8–12 to the analysis and design of real geotechnical constructions in practice, to begin to develop engineering judgement.

13.1 Introduction

There can be many uncertainties in the application of soil mechanics in geotechnical engineering practice. Soil is a natural (not a manufactured) material, therefore some degree of **heterogeneity** can be expected within a deposit. A ground investigation may not detect all the variations and geological detail within soil strata, so the risk of encountering unexpected conditions during construction is always possible. Specimens of relatively small size, and subject to some degree of disturbance even with the most careful sampling technique, are tested to model the behaviour of large in-situ masses which may exhibit features which are not included in the test specimen (e.g. fissures in a heavily overconsolidated clay). Results obtained from in-situ tests can reflect uncertainties due to heterogeneity (e.g. values of N_k in Figure 7.22). Consequently, judgements must be made regarding the characteristic soil parameters which should be used in design. In clays, the scatter normally apparent in plots of undrained shear strength against depth is an illustration of the problem of selecting characteristic parameters (e.g. Figure 5.38). A geotechnical design is based on an appropriate theory which inevitably involves simplification of real soil behaviour and a simplified soil profile. In general, however, such simplifications are of lesser significance than uncertainties in the values of the soil parameters necessary for the calculation of quantitative results. Details of construction procedure and the standard of workmanship can result in further

Applications in geotechnical engineering

uncertainties in the prediction of the performance of geotechnical constructions. Section 13.2 discusses the interpretation of ground investigation data, and the selection of characteristic values for use with the design techniques outlined in Chapters 8–12.

In most cases of simple, routine construction, design is based on precedent/experience and serious difficulties seldom arise. In larger or unusual projects, however, it may be desirable, or even essential, to compare the actual performance to that predicted during design. Lambe (1973) classified the different types of prediction. Class A predictions are those made before the event. Predictions made during the event are classified as Class B, and those made after the event are Class C: in both these latter cases no results from observations are known before predictions are made, though further independent ground data may be available at these later stages to develop more reliable characteristic values of the soil parameters. If observational data are available at the time of prediction these types are classified as B1 and C1, respectively, with the observational data usually being used to infer what the values of the soil parameters must be to give the observed response (this procedure is also sometimes referred to as **back-calculation**).

Studies of particular projects (case studies), as well as showing whether or not a safe and economic design has been achieved, provide the raw material for advances in the theory and application of soil mechanics. Case studies normally involve the monitoring over a period of time of quantities such as ground movement, pore water pressure and stress. Comparisons are then made with theoretical or predicted values, e.g. the measured settlement of a foundation could be compared with the calculated value. If failure of a soil mass has occurred and the profile of the slip surface has been determined, e.g. in the slope of a cutting or embankment, the mobilised shear strength parameters could be back-calculated and compared with the results from laboratory and/or in-situ tests. Empirical design procedures are based on in-situ measurements from case studies, e.g. the design of braced excavations is based on measurements of strut loads in different soil types (see Figure 11.37).

The measurements required in case studies depend on the availability of suitable instrumentation (described in Section 13.3), the role of which is to monitor soil or structural response as construction proceeds so that decisions made at the design stage can be evaluated and if necessary revised. The use of measurements to continuously re-evaluate design assumptions (Class B1 analyses) and refine the design or modify/control construction techniques is known as the **Observational Method** (Section 13.4). Instrumentation can also be used at the ground investigation stage to obtain information for use in design (e.g. details of groundwater conditions, as outlined in Chapter 6). However, instrumentation is only justified if it can lead to the answer to a specific question; it cannot by itself ensure a safe and economic design and the elimination of unpredicted problems during construction. It should be appreciated that a sound understanding of the basic principles of soil mechanics is essential if the data obtained from field instrumentation are to be correctly interpreted.

Section 13.5 will introduce a set of case histories covering a range of different geotechnical constructions. A detailed evaluation is not given in the main text, but each may be found as a self-contained document which may be downloaded from the Companion Website. In these cases, the basic ground investigation data will be given, from which characteristic values will be interpreted using the methods outlined in Section 13.2. The appropriate limit states will then be verified to Eurocode 7 as fully worked examples, and the results compared with the observed performance to demonstrate that the limit state design procedures described within this book produce designs which are acceptable.

13.2 Selection of characteristic values

In practical situations, the material property data derived from laboratory and in-situ testing will exhibit scatter. In order to perform the calculations described in the earlier chapters, it will be necessary to idealise the data from the ground investigation by a adding a fitting line (normally linear to ease subsequent calculation) which will remove the scatter and describe the variation in the material properties with depth. In layered soils, separate fitting lines can be used to characterise the different layers, and by using a sub-layering approach even very complex property variations may be idealised. These idealisations will then represent the characteristic values which will be used in the design phase. In determining characteristic values, Eurocode 7 recommends that this should represent a 'cautious estimate of the values likely to affect the limiting state considered'.

If a layer of soil has uniform properties with depth (also known as homogeneous) then the soil test data may be analysed statistically, with each data point representing a measurement of the uniform strength. The simplest fit to such a set of data would be to determine the mean value of the test data (X_{mean}). By definition, there will be a significant number of measured datapoints below the mean value, so this is not often used for ULS calculations where it might be unsafe. Schneider (1999) proposed that, instead, characteristic values should be taken as the value at 0.5 standard deviations below the mean, i.e.

$$X_k = X_{mean} - 0.5 s_X \tag{13.1}$$

where s_X is the standard deviation of the soil parameter X. In practice, however, there will normally be only a limited amount of test data available due to the desire to keep ground investigation (GI) costs as low as possible. Under these circumstances the use of Equation 13.1 is questionable due to small sample size.

While the mean and standard deviation of a given parameter are likely to vary greatly from soil to soil, it has been shown that the coefficient of variation (COV) of a given property lies within narrow ranges over a wide range of soil types. The coefficient of variation is defined as

$$COV = \frac{s_X}{X_{mean}} \tag{13.2}$$

The implication of this is that the variation in the standard deviation is linked to the variation in mean value for different soils at different sites. Conservative values of COV for different mechanical properties based on studies of large databases of soil tests are given in Table 13.1 after Schneider (1999). Using Equation 13.2, Equation 13.1 may be rewritten in terms of X_{mean} and COV, replacing the arbitrary value of 0.5 with a coefficient k_n:

$$X_k = X_{mean}(1 - k_n COV) \tag{13.3}$$

The coefficient k_n is a function of the number of test data points (n) used to calculate X_{mean}, and is based on the assumption that the data are normally distributed. A robust estimate of the mean characteristic value corresponding to a 95% confidence level that this is below the true mean value of the soil (value assuming that an infinite number of tests could be done) is found using

$$k_n = 1.64 \sqrt{\frac{1}{n}} \tag{13.4}$$

Values of X_k calculated using Equations 13.3 and 13.4 will be almost identical to those from Equation 13.1 if n is relatively large. If n is small Equation 13.1 will give poorer results, as s_X can be heavily influenced by one or two individual data points. Equations 13.3 and 13.4 should be used instead, as the statistical variation about the mean (defined by COV) is based on a large database (values in Table 13.1), even though the mean value is based on the smaller amount of test data from the GI.

Characteristic values calculated using the aforementioned techniques are particularly suitable for use in SLS calculations, where the aim is to predict the actual response as closely as possible. For ULS calculations a lower value may be desirable, in which none (or only a tiny fraction) of the measured test data falls below the characteristic value selected. Such values may be determined using Equation 13.3, with an alternative expression for k_n given by

Applications in geotechnical engineering

Table 13.1 Coefficients of variation of various soil properties

Soil property	COV
ϕ'	0.1
c_u, c'	0.4
m_v	0.4
γ	0

$$k_n = 1.64\sqrt{\frac{1}{n}+1} \tag{13.5}$$

The values of X_k so derived represent a **5% fractile value**, implying that there is only a 5% probability that somewhere in the layer (perhaps at an un-measured location) there is an element of soil having a strength lower than X_k. It is recommended to use the 5% fractile value if there is significant variation within the data (i.e. if it only loosely approximates a uniform distribution); a value between the 95% mean and the 5% fractile values may be used when the variation is small, based on previous experience in similar soils.

If the soil has a non-uniform linear variation with depth, the mean variation may be found by fitting a linear trendline to the data over the depth range of interest, usually using a least-squares fitting procedure. This profile may be used for SLS calculations. The intercept of this line may then be reduced manually until most or all of the test data points lie above the line when determining a lower value for ULS calculations. Statistical techniques may be applied to fitting linear trendlines, as described in Chapter 2 of Frank *et al.* (2004), but are beyond the scope of this book.

To demonstrate the application of the foregoing techniques, Figure 13.1 shows test data for undrained shear strength from three different fine-grained soil sites, from which characteristic profiles have been derived. Figure 13.1(a) shows data for a layer of predominantly clay-based glacial till from Cowden, near Kingston-upon-Hull (data from Lunne *et al.*, 1997). Both CPTU and UU triaxial testing on retrieved samples was conducted. The CPTU data was processed using Equation 7.37 with $N_k=22$, representing the upper end of the range for glacial tills given in Figure 7.22(a), and therefore a lower bound (conservative) value for c_u. The use of Equations (7.38) and (7.39) is not appropriate in this case as the values of B_q within the layer are mostly negative, due to sand and gravel-sized material contained within the clay matrix. It can clearly be seen that c_u is approximately uniform with depth. Assuming that the triaxial data are more reliable (the CPTU data are derived from an empirical correlation), and basing the value of c_{uk} on this only for simplicity, the mean of the 40 triaxial tests is 110 kPa. For $n=40$, $k_n=0.26$ for the 95% mean value (Equation 13.4) and $k_n=1.66$ for the 5% fractile (low) value. Taking $COV=0.4$ for c_u from Table 13.1 gives $c_{uk,mean}=99$ kPa and $c_{uk,low}=37$ kPa. These values are shown in Figure 13.1(a). In this case it would be reasonable to use a characteristic value closer to the 95% mean value, as both the triaxial and CPTU data show that the distribution is very uniform, with any variation generally serving to increase the strength. The 5% fractile value appears to be unrealistically conservative in this case.

Figure 13.1(b) shows triaxial data from an overconsolidated clay (Oxford Clay) at Tilbrook Grange in Cambridgeshire (after Clarke, 1993). In this example, data are shown from three spatially separated boreholes and the aim is to derive a single characteristic strength profile for a ULS check of a shallow foundation. As such, the data from the separate boreholes are assumed to represent one large database. Initially the variation of properties appears more complex compared to that in Figure 13.1(a), though it is possible to distinguish three different regions of increasing mean strength, being 0–7 m, 7–20 m and 20–30 m depth within each of which a uniform profile may be derived. This is the sub-layering approach described earlier. The techniques detailed in the previous example are now applied separately to the test data in each of the sub-layers, the results being summarised in Table 13.2 and shown in Figure 13.1(b).

Illustrative cases

Table 13.2 Example calculations for sub-layering approach

Sub-layer	n	$c_{uk,mean}$ (kPa)	COV	k_n (95% mean)	$c_{uk,mean}$ (kPa)	k_n (5% fractile)	$c_{uk,low}$ (kPa)
0–7 m	15	342	0.4	0.42	284	1.69	110
7–20 m	36	440	0.4	0.27	393	1.66	148
20–30 m	31	562	0.4	0.29	496	1.67	188

Compared to the previous example, there is significantly more variation in properties within the sub-layers specified and greater scatter. In this case it would be more appropriate to use the 5% fractile value for a ULS calculation, as it can be seen that some of the test values fall close to this line (particularly around the sub-layer interfaces). This is only one possible interpretation; a more accurate variation may be determined by using more sub-layers. However, the advantage of using the distribution in Figure 13.1(b) is that for most small shallow foundations it would be possible to use the two-layer approach at the ULS (Figure 8.11). This example therefore demonstrates that it is important to bear in mind what the characteristic values will ultimately be used for, as this may influence their determination.

Figure 13.1(c) shows UU triaxial test data for a heavily overconsolidated fissured clay (Gault Clay) in central Cambridge, taken prior to piling work. At this particular site, both small 38-mm diameter samples and larger 98-mm ones were tested. As mentioned previously in Section 7.5, small triaxial samples are unlikely to account for the fissures, as they may happen to have been taken from intact clay between fissures. From Figure 13.1(c), the larger samples suggest lower strengths than the smaller samples, and demonstrate that it is important to consider site observations of the soil fabric and geological information about the unit in interpreting characteristic values. In this case, the 98-mm data would be used in preference to the 38-mm data. A mean line is fitted to the data by a least-squares fit, where $c_u = 3.73z + 58.4$

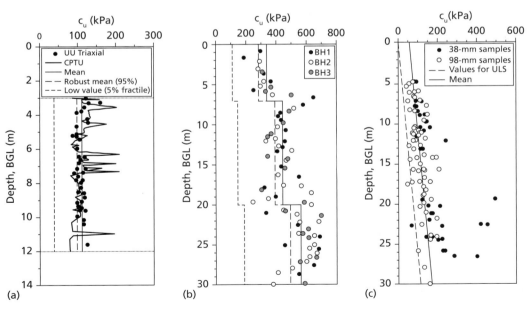

Figure 13.1 Examples of characteristic value determination (for undrained shear strength): (a) uniform glacial till, (b) layered overconsolidated clay, (c) overconsolidated fissured clay.

(kPa) where z is the depth below ground level in metres. Reducing the intercept to give $c_{uk} = 3.73z$ (kPa) would provide an appropriate characteristic profile for ULS design of a piled foundation.

A final note should be made regarding the selection of characteristic values of drained strength parameters. In most cases of design where the aim is to avoid failure, critical state strength properties should be used if available, with $\phi' = \phi'_{cv}$ and $c' = 0$. While ϕ'_{cv} is clearly representative of the ultimate strength of contractile ('loose') soil, if peak strength values are used for dilative ('dense') soils, failure will be characterised by a sudden failure (a brittle response). The use of lower critical state strength parameters for such soils therefore provides some additional margin of safety against collapse. Furthermore, laboratory and in-situ tests measure soil behaviour in a largely undisturbed state, where there may be apparent cohesion/interlocking due to the previous depositional history and/or true cohesion as a result of cementation/bonding. However, if the soil is subsequently worked during construction (e.g. excavation followed by recompaction), these destructive procedures will break down any pre-existing structural effects such that critical state parameters are most appropriate.

For some calculations, however, it may be desirable to select a high estimate of strength properties. An example would be determining the capacity of a pile so that a sufficiently sized piling rig can be specified. For such a calculation it is more conservative to over-predict the pile's capacity, to ensure that the piling rig will have sufficient capacity with a margin of safety (partial factors would also not be used for such an estimate). Under these conditions, it would be more appropriate to use peak strength values.

13.3 Field instrumentation

The most important requirements of a geotechnical instrument are reliability and sensitivity. In general, the greater the simplicity of an instrument, the more reliable it is likely to be. On the other hand, the simplest instrument may not be sensitive enough to ensure that measurements are obtained to the required degree of accuracy, and a compromise may have to be made between sensitivity and reliability. Instrumentation can be based on optical, mechanical, hydraulic, pneumatic and electrical principles; these principles are listed in order of decreasing simplicity and reliability. It should be appreciated, however, that the reliability of modern instruments of all types is of a high standard. The most widely used instruments for the various types of measurement are described below (N.B. measurement of pore water pressures is not described here, having been covered in Section 6.2).

Vertical movement

The most straightforward technique for measuring surface settlement or heave is **precise levelling**. A stable bench mark must be established as a datum, and in some cases it may be necessary to anchor a datum rod, separated by a sleeve from the surrounding soil, in rock or in a firm stratum at depth. For settlement observations of the foundations of structures, durable levelling stations should be established in foundation slabs or near the bottom of columns or walls. A convenient form of station, illustrated in Figure 13.2, consists of a stainless steel socket into which a round-headed plug is screwed prior to levelling. After levelling, the plug is removed and the socket is sealed with a perspex screw.

To measure the settlement due to placement of an overlying fill, a horizontal plate, to which a vertical rod or tube is attached, is located on the ground surface before the fill is placed, as shown in Figure 13.3(a). The level of the top of the rod or tube is then determined. The settlement of the fill itself could be determined from surface levels, generally using levelling stations embedded in concrete. Vertical movements in an underlying stratum can be determined by means of a **deep settlement probe**. One type of probe, illustrated in Figure 13.3(b), consists of a screw auger attached to a rod which is surrounded by a sleeve to isolate it from the surrounding soil. The auger is located at the bottom of a borehole, and anchored at the required level by rotating the inner rod. The borehole is backfilled after installation of the probe.

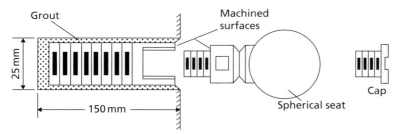

Figure 13.2 Levelling plug.

The **rod extensometer**, shown in Figure 13.3(c), is a simple and accurate device for measuring movement. The rods used are generally aluminium alloy tubes, typically 14 mm in diameter. Various lengths of rod can be coupled together as required. The lower end of the rod is grouted into the soil at the bottom of a borehole, a ragged anchor being threaded onto the rod, and the upper end passes into a reference tube grouted into the top of the borehole. The rod is isolated inside a plastic sleeve. The relative movement between the bottom anchor and the reference tube is measured with a dial gauge or displacement transducer operating against the top of the rod. An adjustment screw is fitted into a threaded collar at the upper end of the rod to extend the range of measurement. A multiple rod installation with rods anchored at different levels in the borehole enables settlement over different depths to be determined. The borehole is backfilled after installation of the extensometer. The use of rod extensometers is not limited to the measurement of vertical movement; they can be used in boreholes inclined in any direction.

Settlement at various depths within a soil mass can also be determined by means of a **multi-point extensometer**, one such device being the magnetic extensometer, shown in Figure 13.3(d), designed for use in boreholes in clay. The equipment consists of permanent ring magnets, axially magnetised, mounted in plastic holders which are supported at the required levels in the borehole by springs. The magnets, which are coated with epoxy resin as a guard against corrosion, are inserted around a plastic guide tube placed down the centre of the borehole. If necessary for stability, the borehole is filled with bentonite slurry. The levels of the magnets are determined by lowering a sensor incorporating a reed switch down the central plastic tube. When the reed switch moves into the field of a magnet, it snaps shut and activates an indicator light or buzzer. A steel tape attached to the sensor enables the level of the station to be obtained to an accuracy of 1–2 mm. Greater accuracy can be obtained by locating, inside the guide tube, a measuring rod to which separate reed switches are attached at the level of each magnet, each switch operating on a separate electrical circuit. The measuring rod, which consists of hollow stainless steel tubing with the electrical wiring running inside, is drawn upwards by means of a measuring head incorporating a screw micrometer, the level of each magnet being determined in turn. As a precaution against failure, two switches may be mounted at each level.

The settlement of embankments can be measured by means of steel plates with central holes which are threaded over a length of vertical plastic tubing and laid, at various levels, on the surface of the fill as it is placed. Subsequently, the levels of the plates are determined by means of an induction coil which is lowered down the inside of the tubing.

Hydraulic settlement gauges, which are used mainly in embankments, provide another means of determining vertical movement. In principle, the hydraulic overflow settlement cell (Figure 13.4) consists of a U-tube with limbs of unequal height: water overflows from the lower limb when settlement occurs, causing a fall in water level, equal to the settlement, in the higher limb. The cell, which is cast into a concrete block at the chosen location, consists of a sealed rigid plastic cylindrical container housing a vertical tube acting as an overflow weir. Polythene-coated nylon tubing leads from the weir to a vertical graduated standpipe remote from the cell. A drain tube leading from the base of the cell allows overflow water to be removed, and another tube acting as an air vent ensures that the interior of the cell is maintained at atmospheric

Applications in geotechnical engineering

pressure. The water tube is filled with de-aired water until it overflows at the weir. The standpipe is also filled and, when connected to the water tube from the cell, the water level in the standpipe will fall until it is equal to that of the weir. De-aired water is used to prevent the formation of air locks, which would affect the accuracy of the gauge. The use of back air pressure applied to the cell enables it to be located below the level of the manometer. The system gives the settlement at one point only, but can be used at locations which are inaccessible to other devices, and is free of rods and tubes which might interfere with construction.

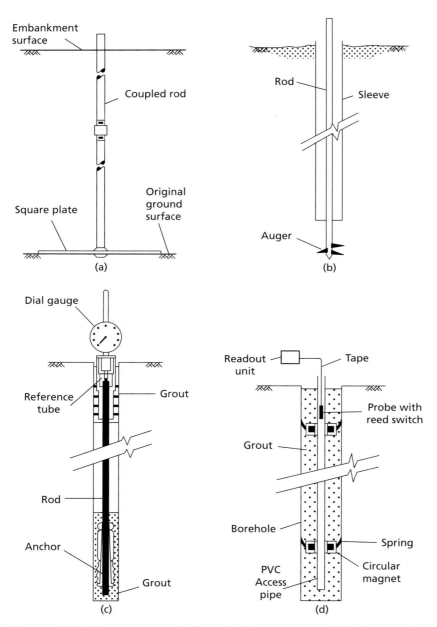

Figure 13.3 Measurement of vertical movement: (a) plate and rod, (b) deep settlement probe, (c) rod extensometer, and (d) magnetic extensometer.

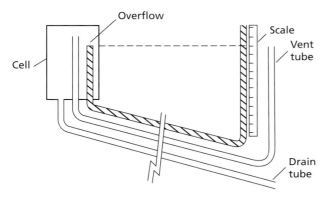

Figure 13.4 Hydraulic settlement cell.

Horizontal movement

In principle, the horizontal movement of stations relative to a fixed datum can be measured by means of a theodolite, using precision surveying techniques. However, in many situations this method is impractical due to site conditions. Movement in one particular direction can be measured by means of an **extensometer**, of which there are several types.

The tape extensometer (Figure 13.5(b)) is used to measure the relative movement between two reference studs with spherical heads, which may be permanent or demountable, as shown in Figure 13.5(a). The extensometer consists of a stainless steel measuring tape with punched holes at equal spacings. The free end of the tape is fixed to a spring-loaded connector which locates onto one of the spherical reference studs. The tape reel is housed in a cylindrical body incorporating a spring-tensioning device and a digital readout display. The end of the body locates onto the second spherical stud. The tape is locked by engaging a pin in one of the punched holes and tension applied to the spring by rotating the front section of the body, an indicator showing when the required tension has been applied. The distance is obtained from the reading on the tape at the pinned hole and the reading on the digital display. An accuracy of 0.1 mm is possible.

For relatively short measurements, a rod extensometer (Figure 13.5(c)) incorporating a micrometer can be used. The instrument consists of a micrometer head with extension rods of different lengths and two end pieces, one with a conical seating and another with a flat surface, which locate against spherical reference studs. The rods have precision connectors to minimise errors when assembling the device. An appropriate gauge length is selected and the micrometer is adjusted until the end pieces make contact with the reference studs. The distance between the studs is read from the graduated micrometer barrel. A typical rod extensometer can measure a maximum movement of 25 mm, to the accuracy of the micrometer.

The tube extensometer (Figure 13.5(d)) operates on a similar principle. One end piece has a conical seating as above; the other consists of a slotted tube with a coil spring inside. The slotted end piece is placed over one of the spherical reference studs, compressing the spring and ensuring firm contact between the second stud and the conical end piece at the other end of the extensometer. The distance between the outside end of the slotted tube and the reference stud is measured by means of a dial gauge, enabling the movement between the two studs to be deduced. Movements up to 100 mm can be measured using a tube extensometer.

The potentiometric extensometer consists of a rectilinear resistance potentiometer inside a cylindrical housing filled with oil. Contacts attached to the piston of the extensometer slide along the resistance wire. The piston position is determined by means of a null balance readout unit calibrated directly in displacement units. Extension rods coupled to the instrument are used to give the required gauge length. Movements of up to 300 mm can be measured, to an accuracy of 0.2 mm.

Applications in geotechnical engineering

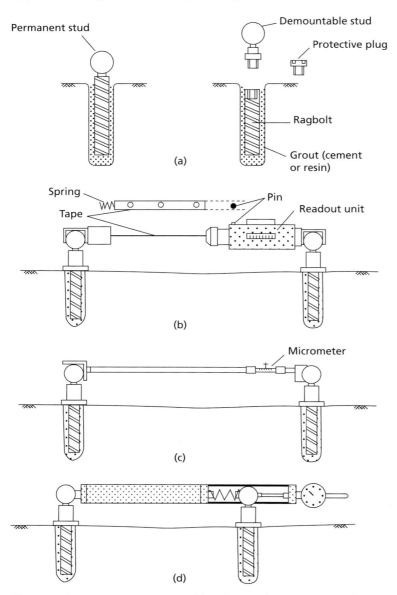

Figure 13.5 Measurement of horizontal movement: (a) reference studs, (b) tape extensometer, (c) rod extensometer, and (d) tube extensometer.

Various extensometers using transducers have also been developed. One such instrument uses a **vibrating wire strain gauge**, a diagram of which is shown in Figure 13.6. The wire is made to oscillate by means of an electromagnet situated approximately 1 mm from the wire. The oscillation induces alternating current, of frequency equal to that of the wire, in a second magnet. A change in the tension of the wire causes a change in its frequency of vibration, and a corresponding change in the frequency of the induced current. The meter recording the output frequency can be calibrated in units of length. One end of the device is fixed to the first reference point by a pinned connector, the other end consisting of a sliding head which moves with the other reference point. The vibrating wire is fixed at one end, and is

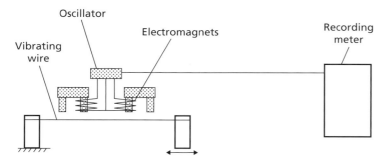

Figure 13.6 Vibrating wire strain gauge.

connected to the moving head through a tensioned spring. When a displacement occurs there is a change in spring tension, resulting in turn in a change in the frequency of vibration of the wire which is a function of the movement between the reference points. Movements of 200 mm can be measured to an accuracy of 0.1 mm. A number of extensometer units can be joined together by lengths of connecting rod to enable movements to be measured over a considerable length.

Horizontal movements at depth within a soil mass are often required in sloping ground to determine the moving mass and the likely location of the slip surface. Such measurements can be determined by means of the **inclinometer**, illustrated in Figure 13.7(a). The instrument operates inside a vertical (or nearly vertical) access tube which is grouted into a borehole or embedded within a fill, enabling the displacement profile along the length of the tube to be determined. The tube has four internal keyways at 90° centres. The inclinometer probe (known as the torpedo) consists of a stainless steel casing which houses a force balance accelerometer. The casing is fitted with two diametrically opposite spring-loaded wheels at each end, enabling the probe to travel along the access tube, the wheels being located in one pair of keyways, the accelerometer sensing the inclination of the tube in the plane of the keyways. The principle of the force balance accelerometer is shown in Figure 13.7(c). The device consists of a mass suspended between two electromagnets – a detector coil and a restoring coil. A lateral movement of the mass induces a current in the detector coil. The current is fed back through a servo-amplifier to the restoring coil, which imparts an electromotive force to the mass equal and opposite to the component of gravitational force which causes the initial movement. Thus, the forces are balanced and the mass does not, in fact, move. The voltage across a resistor in the restoring circuit is proportional to the restoring force and, in turn, to the angle of tilt of the probe. This voltage is measured, and the readout can be calibrated to give both angular and horizontal displacements. The vertical position of the probe is obtained from graduations on the cable attached to the device. Use of the other pair of keyways enables movements in the orthogonal direction to be determined. Readings are taken at intervals (δ) along the casing, and horizontal movement is calculated as shown in Figure 13.7(b). Depending on the readout equipment, movements can be determined to an accuracy of 0.1 mm.

In recent years, the **shape acceleration array** (SAA) has emerged as an alternative to the inclinometer. This device consists of a series of hinged rigid links which are connected together, each of which contains miniature accelerometers, to form a long jointed 'cable'. This can be fed from a spool into a borehole from the surface, requiring little specialist equipment other than the instrument itself. Changes in acceleration measured by each of these sensors can be integrated to determine the displacements along the length of the SAA using a computer. Such an instrument has become possible as a result of MEMS (micro electro-mechanical machines) acceleration technology, which provides highly miniaturised sensors at low cost. Compared to an inclinometer, an SAA is easier and quicker to install (as it has less separate components), cheaper, and more robust.

Applications in geotechnical engineering

The **slip indicator** is a simple device to enable the location of a slip surface to be detected (though not to accurately measure the magnitude of the deformation). A flexible plastic tube, to which a base-plate is attached, is placed in a borehole. The tube, typically 25 mm in diameter, is surrounded by a stiff sleeve. The tube is surrounded with sand, the sleeve being withdrawn as the sand is introduced. A probe at the end of a cable is lowered to the bottom of the tube. If a slip movement takes place, the tube will deform. The position of the slip zone is determined both by raising the probe and by lowering a second probe from the surface until they meet resistance.

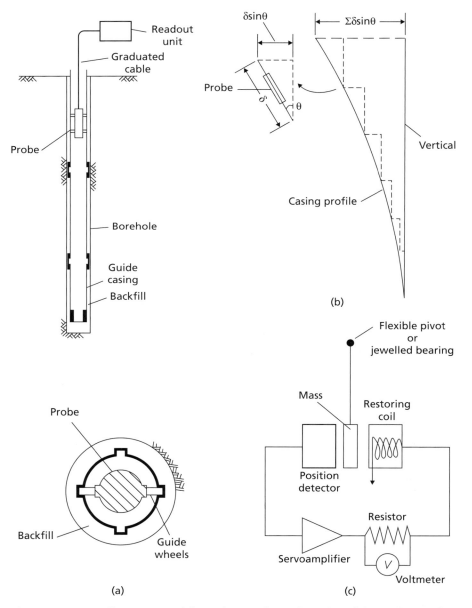

Figure 13.7 Inclinometer: (a) probe and guide tube, (b) method of calculation, and (c) force balance accelerometer.

Strain measurement

Vibrating wire strain gauges (described above) may also be attached to reinforcing bars in reinforced concrete elements including piles (where the axial strains may be used to infer the axial load carried by the pile) or retaining walls (where the difference in strain across the section is indicative of the induced bending moment in the wall). The measurement of bending strains in concrete foundation elements is important in determining the onset of cracking within the concrete, which may lead to water ingress, corrosion of reinforcement, and a consequent degradation in strength and stiffness.

In recent years, fibre-optic sensors have presented an alternative to traditional strain gauges. This involves installing a length of optical fibre within the concrete, with access provided to one end of the fibre. From this, the strain distribution along the length of the fibre can be determined using the **Brillouin optical time domain reflectometry technique (BOTDR)**. This involves attaching an analyser unit to one end of the fibre along which light is passed, the strains being determined from analysis of the reflection information received. A single optical fibre can be used in place of a number of strain gauges, saving both time and cost. A fibre-optic system is also more robust, requiring only the fibre itself to be embedded within the structure, with the sensitive electronics (the analyser) being required only when the instrument is to be read. An example of the use of a BOTDR sensor system within a tunnel lining may be found in Chueng et al. (2010).

Total normal stress

Total stress in any direction can be measured by means of pressure cells placed in the ground. Cells are normally disc-shaped, and house a relatively stiff membrane in contact with the soil. Such cells are also used for the measurement of soil pressure on structures. It should be appreciated that the presence of a cell, as well as the process of installation, modifies the in-situ stresses due to arching and stress redistribution. Cells must be designed to reduce stress modification to within acceptable limits. Theoretical studies have indicated that the extent of stress modification depends on a complex relationship between the aspect ratio (the thickness/diameter ratio of the cell), the flexibility factor (depending on the soil/cell stiffness ratio and the thickness/diameter ratio of the diaphragm) and the soil stress ratio (the ratio of stresses normal and parallel to the cell). The central deflection of the diaphragm should not exceed 1/5000 of its diameter, and the cell should have a stiff outer ring. Cell diameter should also be related to the maximum particle size of the soil.

There are two broad categories of instrument, known as the diaphragm cell and the hydraulic cell, represented in Figure 13.8. In the first category, the deflection of the diaphragm is measured by means of sensing devices such as electrical resistance strain gauges, linear transducers and vibrating wires, the readout in each case being calibrated in stress units. The hydraulic cell normally consists of two stainless steel plates welded together around their periphery, the interior of the cell being filled with oil. The oil pressure, which is equal to the total normal stress acting on the outside of the cell, is measured by a transducer (generally of the pneumatic or hydraulic type) connected to the cell by a short length of steel tubing. Thin spade-shaped cells, which can be jacked into the soil, are useful for the measurement of in-situ horizontal stress.

Load

The cells described above can be calibrated to measure load, the hydraulic type being the most widely used for this purpose. Such cells are used, for example, in the load testing of piles and the measurement of strut loads in braced excavations. Cells with a central hole can be used to measure tensile loads in tie-backs. The use of surface-mounted strain gauges (based on electrical resistance or vibrating wire principles) is an alternative method for the measurement of load in struts and tie-backs. Load cells based on

Applications in geotechnical engineering

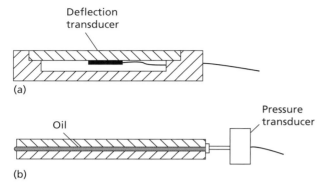

Figure 13.8 (a) Diaphragm pressure cell, and (b) hydraulic pressure cell.

the principle of photo-elasticity are also used. The load is applied to a cylinder of optical glass, the corresponding strains producing interference fringe patterns which are visible under polarised light. The number of fringes is observed by means of a fringe counter, the load being obtained by multiplying the fringe count by a calibration factor. Another method of determining the load in a structural member such as a tie-back is by using a **tell-tale**. A tell-tale is an unstressed rod attached to the member at one end, and running parallel to it. The change in length of the member under load is measured, using the free end of the tell-tale as datum. The load can then be determined from the change in length, provided the elastic modulus of the member is known.

13.4 The observational method

The main uncertainties in geotechnical prediction are the degree of variation and continuity of strata, pore water pressures (which may be dependent on macro-fabric features) and the values of soil parameters. Standard methods of coping with these uncertainties are the use of partial factors (which are useful only at ULS) and the making of assumptions related to general experience, which ignores the danger of the unexpected. The observational method, as described by Peck (1969), offers an alternative approach. The method is one of the design approaches listed in Eurocode 7 by which it can be verified that no relevant limit state is exceeded.

The philosophy of the observational method is to base the design initially on whatever information can be obtained, then to set out all conceivable differences between assumptions and reality. Calculations are then made, on the basis of the original assumptions, of relevant quantities which can be measured reliably in the field – e.g. settlement, lateral movement, pore water pressure. Predicted and measured values are compared as construction proceeds, and the design or the construction procedure is modified if necessary. The engineer must have a plan of action, prepared in advance, for every unfavourable situation that observations might disclose. It is essential, therefore, that the appropriate instrumentation to detect all such situations should be installed. If, however, the nature of the project is such that the design cannot be changed during construction if certain unfavourable conditions arise, the method is not applicable. In such circumstances, a design based on the least favourable conditions conceived must be adopted even if the probability of their occurrence is very low. If an unforeseen event were to arise for which no strategy existed, then the project could be in serious trouble. The potential advantages of the method are savings in cost, savings in time, and assurance of safety.

There are two distinct situations in which the method may be appropriate:

- Projects in which use of the method is envisaged from conception, known as 'ab initio' applications. It may be that an acceptable working hypothesis is not possible, and that use of the observational method offers the only hope of achieving the objective.
- Situations in which circumstances arise such that the method is necessary to ensure successful completion of the project, known as 'best-way-out' applications. It should be appreciated that projects do go wrong, and the observational method may then offer the only satisfactory way out of the difficulties encountered.

The comprehensive application of the method involves the following steps. The extent to which all steps can be followed depends on the type and complexity of the project.

1. Ground investigation to establish the general nature, sequence, extent and properties of the strata, but not necessarily in detail.
2. Assessment of the most probable conditions and the most unfavourable conceivable deviations from these conditions, geological conditions usually being of major importance.
3. Design process based on a working hypothesis of behaviour under the most probable conditions.
4. Selection of quantities to be observed as construction proceeds, and calculation of their anticipated values on the basis of the working hypothesis (e.g. using the principles in Chapters 8–12).
5. Calculation of values of the same quantities under the most unfavourable conditions compatible with the available data on ground conditions.
6. Decision in advance on a course of action or design modification for every conceivable deviation between observations and predictions based on the working hypothesis.
7. Measurement of quantities to be observed and evaluation of actual conditions.
8. Modification of design or construction procedure to suit actual conditions.

A simple example, given by Peck, of the use of the method planned from the design stage is the case of a braced excavation in clay, with struts at three levels. The excavation, to a depth of 14 m, was for the construction of a building in Chicago. The design of the struts was based on an assumed pressure distribution against the walls (see Section 11.9), based on an envelope of strut load measurements reported from a variety of sites. For a particular site, therefore, most of the struts could be expected to carry loads significantly less than those predicted from the diagram. For the site in question the struts were designed for loads equal to two-thirds of those given by the envelope, and the load factor used was relatively low. On this basis, a total of 39 struts was required. The probability then existed that a few struts would be overloaded and would fail. The procedure adopted was to measure the loads in every strut at critical stages of construction, and to have additional struts available for immediate insertion if necessary. Only three additional struts were required. The extra cost of monitoring the loads and in having additional struts available was low compared to the overall saving in using struts of smaller section loaded much closer to their capacity. The procedure also ensured that no strut became overloaded. If a much higher number of additional struts had been required, the overall cost might have been higher and construction might have been delayed; such a risk was judged to be minimal and, in any case, safety would still have been ensured.

13.5 Illustrative cases

On the Companion Website, seven detailed case histories may be found from projects in the UK and Northern Europe, as shown in Figure 13.9. These cover a range of different geotechnical constructions, including foundations (shallow and deep), excavations and slopes, in a variety of soil conditions as listed below:

Applications in geotechnical engineering

- Case study 1: Cellular raft on sand, Belfast, UK
- Case study 2: Raft on clay, Glasgow, UK
- Case study 3: Driven tubular pile in clay, Shropshire, UK
- Case study 4: Piled raft in clay, London, UK
- Case study 5: Deep excavation in clay, London, UK
- Case study 6: Failure of a cutting in clay, Oslo, Norway
- Case study 7: Motorway embankment construction, Belfast, UK.

In each case a short description of the construction is given, followed by the available ground investigation data (including both laboratory and in-situ test data where applicable). This is interpreted following the recommendations in Section 13.2 to determine simplified profiles of characteristic soil properties for use in design. Design calculations to Eurocode 7 at both ULS and SLS (as appropriate) are then detailed, to demonstrate how the methods described in the earlier chapters may be applied to real design situations. The resulting predictions are then compared to the observed performance to demonstrate that the design methods provided give acceptable designs, satisfying all limiting states.

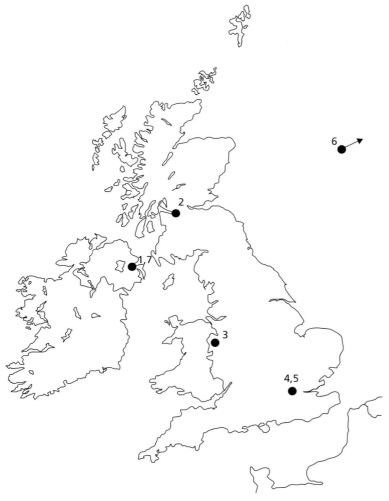

Figure 13.9 Location of illustrative cases.

> **Summary**
>
> 1 Characteristic values should always represent a cautious estimate of the values likely to affect the limiting state considered. Linearised distributions are normally most amenable to use in design, and their determination can be informed by statistical techniques. Complex ground conditions may be divided into a series of linearised sections using sub-layering. In the selection of characteristic values, the reliability of the test methods should be considered, as should the analytical techniques for which the characteristic values are desired.
> 2 A wide range of common field instrumentation exists for measuring vertical and horizontal movements, strains, total stress and load. Instrumentation of a project alone will not guarantee a successful outcome – it must always be installed with the aim of answering a particular question or verifying a particular design assumption.
> 3 The Observational Method involves the use of instrumentation to continuously measure performance during construction and modify/refine the design or alter construction procedures as construction proceeds, thereby ensuring that limiting states are always satisfied. This can be particularly useful for complex projects in difficult ground conditions where soil behaviour may be difficult to predict.
> 4 Seven illustrative cases (case studies) covering a range of geotechnical constructions are presented on the Companion Website, which demonstrate the selection of characteristic values (point 1) and the application of the techniques described earlier in the book to real-world situations. In this way, the reader can begin to develop engineering judgement (which is principally developed in engineering practice). It is also demonstrated that the limit state design procedures presented in this book give adequate designs and predictions of performance for use in routine geotechnical design.

References

Cheung, L.L.K., Soga, K., Bennett, P.J., Kobayashi,Y., Amatya, B. and Wright, P. (2010) Optical fibre strain measurement for tunnel lining monitoring, *Proceedings ICE – Geotechnical Engineering*, **163**(GE3), 119–130.

Clarke, M.J. (1993) *Large Scale Pile Tests in Clay*, Thomas Telford, London.

Lambe, T.W. (1973) Predictions in soil engineering, *Géotechnique*, **23**(2), 149–202.

Lunne, T., Robertson, P.K. and Powell, J.J.M. (1997) *Cone Penetration Testing in Geotechnical Practice*, E & FN Spon, London.

Peck, R.B. (1969) Advantages and limitations of the observational method in applied soil mechanics, *Géotechnique*, **19**(2), 171–187.

Schneider, H.R. (1999) Determination of characteristic soil properties, in *Proceedings of the 12th European Conference on SMFE*, Balkema, Rotterdam, Vol. 1, pp. 273–281.

Further reading

Frank, R., Bauduin, C., Driscoll, R., Kavvadas, M., Krebs Ovesen, N., Orr, T. and Schuppener, B. (2004) *Designers' Guide to EN 1997–1 Eurocode 7: Geotechnical Design – General Rules*, Thomas Telford, London.

This book contains further detail regarding the statistical derivation of characteristic values, with additional examples and further background on the statistical methods underlying the EC7 recommendations.

For further student and instructor resources for this chapter, please visit the Companion Website at www.routledge.com/cw/craig

Principal symbols

Roman

A	Air content
A	Pore pressure coefficient (related to major principal stress increment, axial symmetry)
A	Cross-sectional area
A_f	Plan area of foundation
A_p, A_{pi}	Area of pile cross-section
A_s	Pore pressure coefficient (related to major principal stress increment, plane strain)
B	Pore pressure coefficient (related to isotropic stress increment)
B	Footing width
B'	Effective footing width (accounting for uplift)
B_p	Width of square pile
B_q	Pore pressure parameter (CPT)
B_r	Width of raft or pile cap
C	Change in undrained shear strength per unit depth (undrained strength gradient)
C	Cover depth (e.g. to a tunnel)
C_1	Correction factor for footing depth (Schmertmann method)
C_b	Correlation factor between SPT and pile base resistance
C_c	Compression index
C_{cpt}	Correlation factor between CPT and pile base resistance
C_e	Expansion index
C_s	Correlation factor between SPT and pile shaft resistance
C_u	Coefficient of uniformity
C_z	Coefficient of curvature
C_A	Limiting value of an action effect (SLS)
C_N	Overburden correction factor (SPT)
COV	Coefficient of variation
D	Diameter
D_0	Outside diameter (e.g. of a pile)
D_{10}	Particle size (diameter) for which 10% of the particles are smaller
D_{50}	Mean particle size
D_b	Pile base diameter
D_s	Pile shaft diameter
E	Young's modulus
E'	Young's modulus of soil for drained conditions
E'_{oed}	Constrained modulus

Principal symbols

E_b	Operative Young's modulus below the base of a pile
E_p	Equivalent Young's modulus of a solid circular pile
E_{pi}	Young's modulus of pile material
E_u	Young's modulus of soil for undrained conditions
E_A	Action effect (SLS)
ER	Energy ratio (SPT)
F	Factor of safety
F_h	Normalised horizontal foundation stiffness factor
F_i	Intake factor
F_p	Pile slenderness correction factor
F_r	Friction ratio (normalised sleeve friction, CPT)
F_s	Shape factor (Burland and Burbidge method)
F_t	Time factor (Burland and Burbidge method)
F_z	Non-dimensional factor for bearing capacity in undrained soil with a strength gradient
F_I	Depth factor (Burland and Burbidge method)
F_α	Diffraction factor
F_ϕ	Normalised rotational foundation stiffness factor
G	Shear modulus
G_0	Small-strain shear modulus
G_s	Specific gravity (of soil particles)
$\overline{G}_{Lc}$	Median value of shear modulus over the critical length
H	Thickness of soil layer
H	Horizontal load (action)
H_{ult}	Sliding resistance of a foundation
H_w	Height of wall
I_c	Soil behaviour type index (CPT)
I_c	Compressibility index (Burland and Burbidge method)
I_{qr}	Influence factor (for stress, under rectangular loaded area)
I_s	Influence factor (for settlement)
I_z	Strain influence factor
I_{zp}	Influence factor (Schmertmann's method)
I_D	Relative density
I_L	Liquidity index
I_P	Plasticity index
I_Q	Influence factor (for stress, under point load)
J	Seepage force
K	Lateral earth pressure coefficient
K_0	Lateral earth pressure coefficient at rest (in-situ)
$K_{0,NC}$	Lateral earth pressure coefficient at rest for normally consolidated soil
K_a	Active earth pressure coefficient
K_{bi}	Pile base stiffness
K_f	Overall (foundation) vertical stiffness of a piled raft
K_h	Horizontal stiffness of shallow foundation
K_p	Passive earth pressure coefficient
K_r	Raft vertical stiffness
K_{rs}	Normalised raft–soil stiffness
K_{pg}	Pile group vertical stiffness
K_{pile}	Overall pile-head stiffness of a single pile
K_{si}	Pile shaft stiffness

Principal symbols

K_v	Vertical stiffness of shallow foundation
K_ϕ	Rotational stiffness of shallow foundation
L	Length
L_0	Length of pile not contributing to shaft load transfer
L_a	Length along slip plane
L_c	Critical length
L_p	Pile length
L_r	Length of raft or pile cap
L_{si}	Length of pile shaft
L_w	Length of wall
M	Gradient of the critical state line in q–p' space
M	Moment (action)
M_{max}	Maximum bending moment
M_p	Plastic moment capacity of a pile (yield moment)
M_s	Shaft–soil flexibility factor
M_A	Driving moment (an action)
M_R	Restoring moment (a resistance)
N	Specific volume on an unload–reload line at $p' = 1\,\text{kPa}$
N'	Corrected SPT blowcount accounting for driving-induced pore water pressure (Burland and Burbidge method)
$\overline{N}$	Average SPT blowcount
$N_{(60)}$	SPT N-value (corrected for energy transfer)
$(N_1)_{60}$	Normalised SPT blowcount (effects of overburden pressure removed)
N_c	Bearing capacity factor
N_d	Number of equipotential drops (in a flow net)
N_f	Number of flow channels (in a flow net)
N_k	Calibration factor for determining c_u from CPT data
N_{kt}	Calibration factor for determining c_u from CPTU data
N_q	Bearing capacity factor
N_s	Stability number (slope)
N_t	Stability number (tunnel)
N_γ	Bearing capacity factor
OCR	Overconsolidation ratio
P_a	Total active thrust
P'_a	Effective active thrust
P_p	Total passive thrust
P'_p	Effective passive thrust
Q	Applied point/concentrated load (or action)
Q_{bu}	Base resistance (of a deep foundation)
Q_f	Total load carried by piled raft foundation
Q_{pg}	Load carried by pile group
Q_r	Load carried by raft
Q_{su}	Shaft resistance (of a deep foundation)
Q_t	Normalised cone resistance (CPT)
R	Bearing resistance
R_a	Asperity height (roughness)
R_{avg}	Mean value of characteristic resistance from a population
R_k	Characteristic resistance
R_{min}	Lowest value of characteristic resistance from a population

Principal symbols

R_s	Resultant reaction force on a failure plane
R_Ω	Resistivity
S	Spacing (centre-to-centre)
S_e	Spacing (edge-to-edge)
S_r	Saturation ratio
S_t	Sensitivity
S_x	Horizontal spacing
S_z	Vertical spacing
T	Tension force
T_a	Traction force (active)
T_{bi}	Pile base load
T_f	Pull-out resistance (of anchorage)
T_p	Traction force (passive)
T_r	Time factor (consolidation, pore water drainage in radial direction)
T_s	Surface tension force
T_{si}	Pile shaft load
T_v	Time factor (consolidation, pore water drainage in vertical direction)
T_w	Total tensile force resisted by all reinforcing elements (reinforced soil)
U	Resultant pore water pressure thrust
U_r	Degree of consolidation (pore water drainage in radial direction)
U_v	Degree of consolidation (pore water drainage in vertical direction)
V	Vertical load (action)
V	Volume
V_s	Shear wave velocity
V_{soil}	Volume of over-excavated soil due to tunnelling works
V_t	Tunnel volume
V_L	Volume loss
W	Vertical force due to soil mass (i.e. weight)
X	Material property (generic)
X_{mean}	Mean value of a soil property
a	Area correction factor for a cone penetrometer
c'	Cohesion intercept
c_h	Coefficient of consolidation (pore water drainage in horizontal direction)
c_u	Undrained shear strength
c_{u0}	Undrained shear strength at the founding plane
c_{uFV}	Undrained shear strength measured in FVT (uncorrected)
c_v	Coefficient of consolidation (pore water drainage in vertical direction)
d	Embedment depth (foundations), Embedded length (retaining structures)
d_a	Anchor depth
e	Voids ratio
e_m	Eccentricity
e_{max}	Maximum possible voids ratio (loosest packing)
e_{min}	Minimum possible voids ratio (densest packing)
e_p	Exponent relating to pile group efficiency
f_s	Shaft friction measured along friction sleeve (CPT)
h	(pressure) Head
h	Horizontal displacement
h	Height of retained soil (alternatively, depth of excavation)
i	Hydraulic gradient

Principal symbols

i	Trough width parameter (settlement above tunnels)
i_c	Inclination factor (bearing capacity)
i_{cr}	Critical hydraulic gradient
i_e	Exit hydraulic gradient
i_q	Inclination factor (bearing capacity)
j	Seepage pressure
k	Coefficient of permeability (or hydraulic conductivity)
k_n	Parameter used in the statistical analysis of soil data
m_v	Coefficient of volume compressibility
n	Porosity
n	Number
p	Cavity pressure (PMT)/internal (support) pressure within a tunnel
p	Mean total stress (triaxial conditions)
p'	Mean effective stress (triaxial conditions)
p'_0	(Effective) Earth pressure at rest
p_a	Active earth pressure
p_c	Compaction-induced lateral pressure
p'_c	Preconsolidation pressure (isotropic compression in triaxial cell)
p_p	Passive earth pressure
p_y	Cavity pressure at soil yield (PMT)
p_L	Limit pressure (PMT)
q	Deviatoric stress (triaxial conditions)
q	Volumetric flow rate (seepage)
q	Bearing pressure (foundation)
q_c	Cone resistance
$\overline{q_c}$	Average cone resistance
q_f	Bearing capacity
q_n	Net pressure (foundation)
q_t	Cone resistance, corrected for pore water pressure effects (CPT)
r	Radius
r_c	Cavity radius
r_d	Drain radius
r_m	'Maximum' radius (pile settlement)
r_u	Pore water pressure ratio
s	Settlement
s	Parameter used in PMT interpretation in coarse-grained soil
s	Mean total stress (biaxial/plane strain conditions)
s'	Mean effective stress (biaxial/plane strain conditions)
s_b	Pile base settlement
s_c	Shape factor (bearing capacity)
s_c	Consolidation settlement (general term)
s_e	Elastic shortening (of a pile)
s_g	Ground surface settlement
s_i	Immediate settlement
s_{oed}	One-dimensional consolidation settlement (as measured in oedometer)
s_q	Shape factor (bearing capacity)
s_r	Pile-head settlement of a rigid pile
s_s	Pile shaft settlement
s_X	Standard deviation (of parameter X)

Principal symbols

s_γ	Shape factor (bearing capacity)
t	Time
t	Deviatoric stress (biaxial/plane strain conditions)
t_{90}	Time for 90% of primary consolidation to be complete
t_c	Construction period
t_r	Thickness of raft
u_0	Initial or in-situ pore water pressure (also initial back pressure in a triaxial test)
u_1	Pore water pressure measurement made on a CPT cone
u_2	Pore water pressure measurement made at the shoulder of a CPT cone
u_3	Pore water pressure measurement made at the upper end of the friction sleeve (CPT)
u_a	Pore air pressure
u_c	Suction pressure due to capillary rise
u_e	Excess pore water pressure
u_i	Initial excess pore water pressure
u_s	Static pore water pressure
u_{ss}	Steady-state seepage pore water pressure
$u_{(w)}$	Pore water pressure
v	Specific volume
v_0	Initial specific volume
v_d	Velocity (of pore water flow)
v_{fan}	Velocity within a shear fan
v_s	Seepage velocity
w	Batter angle (of retaining structure)
w	Water content
w_{opt}	Optimum water content
w_L	Liquid limit
w_P	Plastic limit
x	Distance behind excavation
y_c	Displacement at the cavity wall (PMT/tunnel)
z	Elevation head and depth
z_0	Depth of tension zone in active (undrained) soil
z_c	Critical depth
z_{f0}	Influence depth (Schmertmann's method)
z_{fp}	Depth of peak influence (Schmertmann's method)
z_I	Depth of influence (beneath a shallow foundation)

Greek

Δ	Differential settlement
Γ	Specific volume on the critical state line at $p' = 1$ kPa
α	Adhesion factor
α_j	Interaction factor (pile group settlement)
α_{rp}	Raft–pile interaction factor
α_{FV}	Empirical correlation factor for determining OCR from FVT data
β	Slope angle (to the horizontal)
β	Drained interface strength parameter
β_d	Angular distortion
χ	Parameter describing the effect of suction on effective stress in partially saturated soil

Principal symbols

δ'	Interface friction angle
δ'_{des}	Design value of interface friction angle (after applying partial material factor)
ε_a	Axial strain
ε_c	Cavity strain
ε_r	Radial strain
ε_s	Deviatoric shear strain (triaxial)
ε_t	Tensile strain
ε_v	Volumetric strain
$\varepsilon_{x,y,z}$	Normal strain (in x-, y- or z-direction)
ε_θ	Circumferential strain
η_g	Pile group efficiency
η_w	Viscosity of water
γ	Unit weight (weight density)
γ_w	Unit weight of water ($=9.81\,\mathrm{kN/m^3}$)
$\gamma_{(xz)}$	Shear strain
γ_y	Shear strain at soil yield
γ_c	Partial factor on cohesion intercept (c')
γ_{cu}	Partial factor on undrained shear strength
$\gamma_{A,dst}$	Partial factor on destabilising action
$\gamma_{A,stb}$	Partial factor on stabilising action
γ_A	Partial factor on action
γ_R	Partial factor on resistance
γ_{Ra}	Partial factor on pull-out resistance
γ_{Rb}	Partial factor on base resistance
γ_{Re}	Partial factor on earth pressures (when acting as a resistance)
γ_{Rr}	Partial factor on gravitational actions and shear resistance (slopes)
γ_{Rh}	Partial factor on sliding resistance
γ_{Rs}	Partial factor on shaft resistance
γ_{Rv}	Partial factor on vertical bearing resistance
γ_{RC}	Partial factor on total (pile) compressive resistance
γ_{RT}	Partial factor on total (pile) tensile resistance
γ_X	Partial factor on material property
$\gamma_{\tan\phi}$	Partial factor on $\tan\phi'$
γ_γ	Partial factor on unit weight (weight density)
κ	Slope of the isotropic unload–reload line
λ	Slope of the isotropic compression line
μ	Correction factor for FVT
μ_0	Elastic settlement factor for finite thickness layer
μ_1	Elastic settlement factor for finite thickness layer
μ_c	Settlement coefficient (Skempton-Bjerrum method)
ν	Poisson's ratio
ν'	Poisson's ratio for drained conditions
ν_u	Poisson's ratio for undrained conditions ($=0.5$)
θ	Rotation (angle)
θ_{fan}	Fan angle
ϕ'	Angle of shearing resistance
ϕ'_{cv}	Critical state angle of shearing resistance
ϕ'_{des}	Design value of angle of shearing resistance (after applying partial material factor)
ϕ'_{max}	Peak angle of shearing resistance

Principal symbols

ϕ'_{mob}	Mobilised angle of shearing resistance (within sloping ground)
ϕ'_u	Angle of shearing resistance under undrained conditions ($=0$ for saturated soil)
ϕ'_μ	True angle of friction (between soil particles)
ρ	Bulk (mass) density
ρ_c	Homogeneity factor (laterally loaded piles)
ρ_d	Dry density
ρ_{d0}	Zero-air voids dry density
ρ_s	Particle density
ρ_w	Bulk density of water ($=1000\,kg/m^3$)
σ	Total stress
σ'	Effective normal stress ($\sigma'=\sigma-u$)
$\sigma'_{1,2,3}$	Effective principal stresses
σ_a	Axial stress
σ_{h0}	In-situ horizontal total stress/lift-off pressure (PMT)
σ'_{max}	Preconsolidation pressure
σ_n	Total normal stress
σ'_n	Effective normal stress
σ_q	Total surcharge pressure (foundation)
σ'_q	Effective surcharge pressure (foundation)
σ_r	Radial stress
σ_{v0}	In-situ vertical total stress (total overburden pressure)
σ'_{v0}	In-situ vertical effective stress (effective overburden pressure)
σ_θ	Circumferential stress
$\Delta\sigma_z$	Change in vertical (total) stress
τ	Shear stress
$\bar{\tau}_0$	Average shear stress acting on pile shaft
τ_f	Shear stress at failure
τ_{int}	Interface shear strength
τ_{mob}	Mobilised shear stress (within a slope)
τ_w	Shear strength along wall
ξ_1	EC7 correlation factor (pile load test)
ξ_2	EC7 correlation factor (pile load test)
ξ_3	EC7 correlation factor (in-situ test)
ξ_4	EC7 correlation factor (in-situ test)
ψ	Dilation angle
ζ	SPT correction factor

Glossary

Chapter 1

Activity Ratio of the plasticity index to the clay fraction of a soil. Soils of high activity have a greater change in volume when the water content is changed.
Aggregation Group of particles acting as a single larger unit.
Air content (A) Ratio of the volume of air to the total volume of the soil.
A-line Boundary line dividing silts and clays on the plasticity chart for classifying soil (Figure 1.12).
Alluvium Soil deposit formed of material transported by fluvial processes (e.g. rivers).
Base exchange Replacement of cations in the water in the void space between clay mineral particles.
Bulk density/mass density Ratio of the total mass to the total volume of soil.
Clay fraction Percentage of clay-size particles in a soil.
Coarse-grained (soil) Soil deposit consisting mainly of sand and gravel size particles (single grain structure).
Coefficient of curvature (C_z) Coefficient describing the shape of a particle size distribution curve.
Coefficient of uniformity (C_u) Coefficient describing the slope of a particle size distribution curve.
Cohesion Strength of a soil sample when it is unconfined (true cohesion). Not to be confused with the shear strength parameter c' which is often termed apparent cohesion.
Colloidal Description of soil particles of size <0.002 mm.
Compactive effort Amount of energy supplied by compaction equipment.
Delta Soil deposit formed by deposition of material at an estuarine location.
Dispersed (structure) Structural arrangement of clay mineral particles consisting of a face-to-face orientation of particles (Figure 1.9a).
Drift deposits Recently deposited sediment.
Effective size Alternative term for D_{10}, the particle size of a soil for which 10% of the material is finer.
End-product compaction Compaction specification where the required minimum dry density is specified.
Engineered fill Soil which has been selected, placed and compacted to an appropriate specification with the object of achieving a particular engineering performance.
Fine-grained (soil) Soil deposit consisting mainly of clay and silt size particles (clay minerals may exert a considerable influence over soil behaviour).
Flocculated (structure) Structural arrangement of clay mineral particles consisting of an edge-to-face or edge-to-edge orientation of particles (Figure 1.9b).
Gap-graded A coarse grained soil consisting of particles of both large and small sizes but with a relatively low proportion of particles of intermediate size.
Glacial till Soil deposit formed of material transported by glaciation.
Interbedded/inter-laminated Deposit consisting of layers of different soil types.
Interstratified Deposit with alternating layers of varying soil types or with bands or lenses of other materials.

Glossary

Isomorphous substitution   Partial replacement of silicon and aluminium by other elements in clay minerals.

Liquid limit (w_L)   Water content above which the soil flows like a liquid (slurry).

Liquidity index (I_L)   Water content relative to plasticity index, Equation 1.2 ($I_L = 0$ – soil at plastic limit; $I_L = 1$ – soil at liquid limit).

Lithification   The conversion of loose sediment (soil) into sedimentary rock.

Loess   Soil deposit formed of wind-blown sediment.

Method compaction   Compaction specification where the type and mass of equipment, the layer depth and the number of passes are specified.

Modified AASHTO test   Standard test for compaction characteristics (4.5-kg mass falling freely through 450 mm, 27 blows). Used in the specification of engineered fills, usually for heavier compaction equipment of higher compactive effort.

Moisture condition value (MCV)   Parameter describing soil suitability for compaction, determined using the Moisture condition test (Section 1.7).

Optimum water content (w_{opt})   Water content at which the maximum value of dry density is obtained.

Organic soil   Soil containing a significant proportion of dispersed vegetable matter (e.g. peat). As the strength of the vegetable matter decreases, the soil may be described as fibrous, pseudo-fibrous or amorphous.

Particle density   Density of solid particles.

Particle size distribution (PSD)   The relative proportions of differently-sized particles within a soil deposit.

Partings   Bedding surfaces within a soil deposit which separate easily.

Pediment   A broad, gently sloping rock surface, often covered in sediment at the base of a steeper slope such as a mountain in arid regions where there is little vegetation to hold the overlying soil.

Plastic limit (w_P)   Water content below which the soil is brittle.

Plasticity index (I_P)   Range of water content over which a soil exhibits plasticity ($= w_L - w_P$).

Playa   An area of flat, dried-up land, esp. a desert basin from which water evaporates.

Poorly graded   A coarse-grained soil having a high proportion of the particles within narrow size limits (also termed uniform soil).

Porosity (n)   Ratio of the volume of voids to the total volume of the soil (alternative to void ratio).

Proctor test   Standard test for compaction characteristics (2.5-kg mass falling freely through 300 mm, 27 blows). Used in the specification of engineered fills, usually for lighter compaction equipment of lower compactive effort.

Relative density (I_D)   Degree of packing i.e. current void ratio relative to minimum and maximum values, Equation 1.23 ($I_D = 0$ – soil at loosest state, $e = e_{max}$; $I_D = 1$ – soil at densest state, $e = e_{min}$).

Residual soil   Sediment formed of weathered rock which has not been transported appreciably from the site of weathering.

Saturation line   Relationship between zero air voids dry density and water content.

Saturation ratio (S_r)   Percentage of the pore space taken up by water.

Sedimentation   Process involving settling of fine soil particles through water in a sedimentation tube to determine a soil's particle size distribution (for finer particles).

Shrinkage limit   Water content at which the volume of a soil reaches its lowest value as it dries out.

Sieving   Process involving passing soil through standardised test sieves of increasingly smaller aperture size, normally by vibration to determine a soil's particle size distribution (limited to coarser particles).

Single grain (structure)   Structural arrangement of soil particles where each particle is in direct contact with adjoining particles without there being any bond between them.

Solid (geology)   Class of geological material including rocks and older soil deposits which have undergone significant consolidation.

Specific gravity (G_s)   Ratio of the density of the solid particles to the density of water.
Specific volume (v)   Total volume of soil containing a unit volume of solids.
Transported soil   Sediment formed of weathered rock which has been transported from the site of weathering by ice, water or wind.
Unit weight/weight density (γ)   Ratio of the total weight to the total volume of soil.
Vibrating hammer test   Alternative test used in the specification of engineered fills.
Void ratio (e)   Ratio of the volume of voids to the volume of solids (i.e. an indicator of the particle packing).
Water content (w)   Ratio of the mass of water in the soil to the mass of solid particles.
Well graded   A coarse-grained soil which has no excess of particles in any size range and no intermediate sizes are lacking.
'Zero air voids' dry density (ρ_{d0})   Maximum possible value of dry density.

Chapter 2

Aquiclude   A zone of impermeable material surrounding water-bearing ground (creating perched conditions).
Aquitard   A zone of low permeability material surrounding water-bearing ground (creating perched conditions).
Artesian   Pore water pressure governed not by the local water table level, but by a higher water table level at a distant location.
Constant-head (permeability) test   Laboratory test for determining the coefficient of permeability (hydraulic conductivity) in soils of relatively high permeability (e.g. coarse-grained).
Equipotential   Line of constant pressure head (perpendicular to flow lines).
Falling-head (permeability) test   Laboratory test for determining the coefficient of permeability (hydraulic conductivity) in soils of relatively low permeability (e.g. fine-grained).
Finite Difference Method (FDM)   Computational technique for solving complex problems by dividing the system of interest into a series of smaller parts. The value of a particular parameter within each part is related to the parts it is connected to, so that by applying suitable boundary conditions a complete solution may be obtained.
Flow channel   Area of soil between two adjacent flow lines in a flow net.
Flow line   Line of groundwater flow (perpendicular to equipotential lines).
Flow net   Graphical construction of flow lines and equipotentials which can be used to solve steady-state seepage problems.
Flow rate   Volume of pore water flowing per unit time.
Hydraulic conductivity   Soil parameter describing the ease of pore water flow through soil. A high hydraulic conductivity means water flows easily. Also called coefficient of permeability.
Hydraulic gradient (i)   Difference in pressure head between two points divided by the distance between them. A hydraulic gradient is required for pore water flow (seepage).
Perched (water table)   Water bearing material existing above the water table.
Phreatic surface   Location where the pore water is under atmospheric pressure (also called the water table).
Piping   The process whereby voids in the form of channels or 'pipes' are created due to erosion where high hydraulic gradients exist.
Seepage   Process of pore water flow.
Stratified   Soil deposit which is layered.
Underseepage   Seepage beneath a structure (as opposed to through it in the case of an earthen dam).
Water table   Location where the pore water is under atmospheric pressure (also called the phreatic surface).

Glossary

Chapter 3

Consolidation Reduction in volume and increase in effective stress as positive excess pore water pressure dissipates.

Critical hydraulic gradient (i_{cr}) hydraulic gradient corresponding to zero resultant body force (onset of liquefaction).

Dissipation The reduction of excess pore water pressure as drainage takes place.

Drained conditions Soil state in which any excess pore water pressure has fully dissipated (i.e. consolidation is complete). Also referred to as 'long-term'.

Dynamic/seismic liquefaction Process by which the critical hydraulic gradient is reached (i.e. soil becomes liquefied) due to a reduction in volume of the soil skeleton during dynamic loading (including earthquakes).

Effective normal stress (σ') Represents the overall stress transmitted through the soil skeleton only (due to interparticle forces).

Excess pore water pressure (u_e) Pore water pressure increment induced by an increment of total stress, before seepage (consolidation) has started.

Heave Upward movement of soil.

Liquefied (soil) Soil in which the critical hydraulic gradient has been reached due to seepage, meaning that effective stress is zero, is said to be liquefied.

Net stress Effective stress, compensated for pore air pressure in partially saturated soil. In fully saturated soil (wet or dry), net stress = effective stress (σ')

Pore water pressure (u) Pressure of the water filling the void space between the solid particles.

Principle of Effective Stress Total stress is equal to the sum of effective stress (reaction of soil skeleton) and pore pressure, $\sigma = \sigma' + u$.

Resultant body force Resultant force acting on a soil mass due to the combined effects of gravity and seepage.

Seepage force (J) Frictional drag, acting in the direction of flow on the solid particles during seepage.

Static liquefaction Process by which the critical hydraulic gradient is reached (i.e. soil becomes liquefied) due to seepage.

Static pore water pressure (u_s) Initial pore water pressure before the application of an increment of total stress, when seepage is not occurring. In the absence of artesian conditions, this is the hydrostatic pressure due to the position of the water table.

Steady-state seepage pore water pressure (u_{ss}) Initial pore water pressure before the application of an increment of total stress, when seepage is occurring.

Swelling Increase in volume and reduction in effective stress as negative excess pore water pressure dissipates.

Total normal stress (σ) Force per unit area transmitted in a normal direction across a plane, imagining the soil to be a solid (single-phase) material.

Undrained conditions Soil state immediately after an increment of total stress is applied and before consolidation has started, in which the stress increment is entirely carried as an increment of excess pore water pressure. Also referred to as 'short-term'.

Chapter 4

Band drain Prefabricated vertical drain consisting of a flat plastic core indented with drainage channels, surrounded by a layer of filter fabric.

Coefficient of consolidation (c_v, c_h) Soil parameter governing the rate at which consolidation occurs. Subscript 'v' denotes pore water drainage in vertical direction, subscript 'h' denotes pore water drainage in horizontal direction.

Glossary

Coefficient of volume compressibility (m_v) Soil parameter describing the volume change per unit volume per unit increase in effective stress (alternatively, ratio of volumetric strain to applied stress).

Compression index (C_c) Slope of the virgin compression line (1DCL) on an e–log σ' plot.

Consolidation settlement Vertical displacement of the soil surface corresponding to the volume change at any stage of the consolidation process.

Constrained modulus (E'_{oed}) Elastic modulus calculated under one-dimensional loading conditions (i.e. zero lateral strain), as in the oedometer. Reciprocal of coefficient of volume compressibility.

Degree of consolidation Measure of the amount of consolidation which is complete at a given time – the ratio of current effective stress minus initial effective stress to the expected change in effective stress when consolidation is complete.

Expansion index (C_e) Slope of the unload–reload line on an e–log σ' plot.

Half-closed layer Layer of soil which is free to drain to only one of its boundaries (also called single drainage).

Heave Upward vertical displacement due to soil expansion.

Initial compression Compression of small quantities of air in the soil occurring before consolidation, resulting in a small amount of volume change.

Isochrone Curve showing the variation of a parameter with depth at a given instant in time.

Normally consolidated Description of stress history when the current effective stresses applied to an element of soil are also the maximum to which the soil has ever been subjected.

Open layer Layer of soil which is free to drain to both its upper and lower boundaries (also called double drainage).

Overburden pressure Alternative term for vertical stress.

Overconsolidated Description of stress history when the effective stresses at some time in the past have been greater than the present value.

Overconsolidation ratio (OCR) Numerical parameter quantifying the stress history of a soil, equal to the ratio of maximum effective vertical stress (the preconsolidation pressure) to current effective vertical stress. A normally consolidated soil has OCR=1; an overconsolidated soil has OCR>1. OCR can never be less than 1.

Preconsolidation pressure (σ'_{max}) The maximum effective vertical stress that has acted on a soil since it was deposited.

Pre-loading Application of a large overburden stress to soil to induce a state of overconsolidation, preparing soil for use in foundations. The applied stress is larger than that applied by the subsequent construction, so that volume changes during construction and operation will be along the unload–reload line and therefore small, reducing potential structural distress.

Primary consolidation Consolidation due to pore water drainage, following Terzaghi's theory of consolidation; the general term 'consolidation' is normally taken to mean the primary consolidation.

Sandwick drain Prefabricated vertical drain consisting of a filter stocking, usually of woven polypropylene, filled with sand.

Secondary compression Further slow compression of the soil occurring after primary consolidation is complete due to soil creep, continuing for an indefinite period of time.

Smear Reduction in the values of the soil properties for the soil immediately surrounding vertical drains (or any other inclusion) due to remoulding during installation.

Stress history The sequence of stresses (loading and unloading) that a soil deposit has been subjected to since it was initially deposited. This includes both natural (geological/hydrological) and man-made loadings.

Surcharge pressure Vertical stress applied on a given horizontal plane (usually the ground surface or founding plane) within a soil deposit.

Time factor Non-dimensional time during the consolidation process.

Virgin (one-dimensional) compression line (1DCL) Relationship between void ratio and (logarithm of) effective stress when the soil is in a normally consolidated state.

Glossary

Chapter 5

Air entry value Difference between the pore air and pore water pressures, below which air will not pass through a material.

Angle of dilation (ψ) Ratio of volumetric to shear strain, expressed as an angle (direction of movement) $= \tan^{-1}(d\varepsilon_v/d\gamma)$.

Angle of shearing resistance (ϕ') Line fitting constant used in the Mohr–Coulomb model.

Anisotropy Having different values of a given property or characteristic in different directions.

Apparent cohesion Description of the stress-independent shear strength (c') derived from a Mohr–Coulomb (straight line) failure envelope due to the line-fitting procedure (the soil in reality may be cohesionless).

Cohesion intercept (c') Line fitting constant used in the Mohr–Coulomb model.

Constitutive model Generalised relationship between stress and strain, linking the equations of equilibrium and compatibility.

Critical state angle of shearing resistance (ϕ'_{cv}) Equivalent angle of shearing resistance defining the slope of the critical state line on a τ–σ' plot.

Critical state line (CSL) Failure envelope defining all possible critical states for a soil in terms of void ratio and effective stresses.

Critical state Soil state (combination of stress and volume) in which the application of shear strain causes no further change of shearing resistance or soil volume.

Deviatoric stress invariant Stress parameter, which is a function of the principal stresses, causing only shear strain within an element of soil.

Dilatancy The increase in volume of a dense (coarse-grained) soil during shearing, usually due to particle interlocking and over-riding.

Elastic–perfectly plastic model Constitutive model in which soil strains as a function of applied stress (elasticity) until a limiting stress is reached, at which point the soil yields and unrestricted plastic flow occurs (perfect plasticity).

Elastic–strain hardening plastic model Constitutive model in which soil strains as a function of applied stress (elasticity) until a limiting stress is reached (yield point); following yield, further stress is required to cause further strain (i.e. the soil gets stronger – this is known as hardening).

Elastic–strain softening plastic model Constitutive model in which soil strains as a function of applied stress (elasticity) until a limiting stress is reached (yield point); if the applied shear stress is maintained following yield, then large strains will rapidly occur, resulting in catastrophic failure.

Equation of compatibility Equation describing the relationship between strain components within a soil element such that they are physically acceptable (compatible).

Equation of equilibrium Equation describing the equilibrium of a static element of soil under applied external stress (normal and shear) and self weight.

Extrasensitive Description of a clay with $8 < S_t < 16$.

Intermediate principal stress Under general triaxial conditions there are three principal stresses – the intermediate value lies between the major and minor values.

Isotropic compression line (ICL) Virgin compression line due to isotropic (rather than one-dimensional) consolidation.

Isotropic consolidation Consolidation under uniform confining pressure, equal strain occurring in all directions. This is not the same as one-dimensional consolidation, where lateral strain is prevented.

Isotropy Having the same value of a given property or characteristic in all directions.

Laboratory vane Simple and quick laboratory test for measuring the undrained shear strength of fine-grained soils.

Mean stress invariant Stress parameter, which is a function of the principal stresses, causing only volumetric strain within an element of soil.

Mohr circle   Graphical construction representing the stress states on all possible planes within an element of soil.
Mohr–Coulomb criterion   A specific example of a rigid-perfectly plastic constitutive model, which is commonly used for describing the strength of soil.
Poisson's ratio (v)   Elastic constant expressing the ratio of strains in the two perpendicular directions to strain in the loading direction under uniaxial loading.
Pore water pressure coefficients (A,B)   Coefficients relating to the changes in pore water pressure under increments of total stress.
Principal stress difference   Difference between major and minor principal stresses (=deviatoric stress invariant, q).
Principal stresses   Stresses acting on a plane within a soil element where there is no resultant shear stress. The smallest value is the minor principal stress; the largest value is the major principal stress. The principal stresses can be used to define the position and size of a Mohr circle.
Quick   Description of a clay with $S_t > 16$.
Rigid–perfectly plastic model   Constitutive model in which the initial elastic behaviour is ignored, i.e. when only failure of the soil is of interest. This is commonly used for analysis of geotechnical structures at the ultimate limiting state (ULS).
Sensitive   Description of a clay with $4 < S_t < 8$.
Sensitivity (S_t)   Ratio of the undrained strength in the undisturbed state to the undrained strength, at the same water content, in the remoulded state.
Shear modulus (G)   Elastic material parameter defined as the ratio of shear stress to shear strain.
Shearbox/direct shear apparatus (DSA)   Laboratory test apparatus for determining the shear strength of soil. Can also be used to find the (interface) shear strength of soil–structure interfaces.
Simple shear apparatus (SSA)   Laboratory test apparatus for determining the shear strength of soil under a state of simple shear. Cannot be used to measure interface shear strength.
Stress path cell   Computer-controlled triaxial apparatus in which the radial and axial stresses and sample pore water pressure can be independently controlled, allowing for any stress conditions to be applied to an element of test soil.
Stress path   Line plotting deviatoric stress against mean stress. If the effective mean stress is used, the effective stress path (ESP) is described; if the total mean stress is used, the total stress path (TSP) is described.
Stress ratio   Deviatoric (shear) stress normalised by mean effective (volumetric) stress, e.g. τ/σ'.
Triaxial apparatus   Laboratory test apparatus for determining the full constitutive behaviour of soil (both shear strength and shear stiffness).
Triaxial compression   Loading regime within a triaxial cell in which shear stress is induced in a sample of soil by increasing axial stress while holding radial stress (confining pressure) constant.
Triaxial extension   Loading regime within a triaxial cell in which shear stress is induced in a sample of soil by reducing axial stress while holding radial stress (confining pressure) constant.
Triaxial shear strain   =deviatoric shear strain, i.e. 2/3 of the difference between the major and minor principal strains.
True cohesion   Stress-independent shear strength (c') derived from actual cementation or bonding between soil particles.
True triaxial (conditions)   General stress state in which $\sigma'_1 > \sigma'_2 > \sigma'_3$ (i.e. all three principal stresses are distinct).
Unconfined compression test   Triaxial test in which the confining pressure $\sigma_3 = 0$ (i.e. the sample is not held within a water bath).
Unconfined compressive strength (UCS)   Major principal (axial) stress (=deviatoric stress) at failure in an unconfined compression test.
Undrained (shear) strength (c_u)   Shear strength of soil under undrained conditions (only).
Young's Modulus (E)   Elastic material parameter defined as the ratio of normal stress to normal strain.

Glossary

Chapter 6

(Continuous) Flight auger   Steel rod surrounded by a helix which traps soil between the flights as it is rotated into the ground. Used in rotary boring (ground investigation and bored pile construction). A continuous flight auger has a helix running along its entire length so that the borehole can be excavated to full depth without requiring intermediate emptying.

Borehole   Deep circular hole which is augered or drilled for the purposes of determining the location and thickness of soil and rock strata, taking soil samples for laboratory testing and undertaking in-situ testing.

Bucket auger   Rotary boring tool which traps clay within a large bucket as it is rotated into the ground.

Chisel   Tool used in percussion boring for breaking-up boulders, cobbles and hard strata.

Clay cutter   Tool used in percussion boring in fine-grained soils. Similar to a shell, but relies on the undrained strength of fine-grained soils to retain the spoil within the shell.

Cone penetrometer   Tubular device ending in a cone-shaped tip which is jacked into the ground. The probe is instrumented and data logged from these instruments can be used for soil identification, ground profiling and determination of mechanical properties.

Core-catcher   A short length of tube with spring-loaded flaps, fitted between a sampling tube and the cutting shoe to prevent loss of cohesionless (coarse-grained) soil.

CPT sounding   In-situ test using a cone penetrometer, the data from which can be used for soil identification and ground profiling (in addition to determining soil parameters).

Desk study   Collation and interpretation of previously existing material relating to a site, including details of topology (mapping), geology and geotechnics, services, prior construction, contamination etc.

Disturbed sample   Samples having the same particle size distribution as the in-situ soil but in which the in-situ structure has not been preserved.

Heading   A short tunnel, with temporary support, that may be excavated from the bottom of a deep shaft or into a hillside to investigate the ground conditions.

In-situ   In the soil at the site in question.

Inversion   Back-calculation procedure involving the determination of a ground model having specified characteristics (e.g. transfer function for the passage of shear waves).

Investigation point   Location on the ground surface at which one (or more) investigative techniques will be applied.

Multi-channel analysis of surface waves (MASW)   A geophysical technique for determining ground profile, derived from SASW but using multiple simultaneous measurements of wave arrival to provide greater reliability, particularly in 'noisy' urban areas.

Non-targeted sampling   Investigation points are distributed uniformly over a wide area in cases where sources of contamination or other hazards are unknown/unexpected.

Piezocones (CPTU)   Cone penetrometer incorporating one or more pore water pressure transducers.

Piezometer   Instrument used to measure pore water pressure.

Push rods   Hollow tubes inserted between a cone penetrometer (or any other device) and the thrust machine to connect the two.

Response time   Time taken for the water level in a borehole to stabilise following boring.

Shear wave velocity (V_s)   Speed at which shear waves travel through an elastic medium (such as soil). Can be used to determine small strain shear modulus, G_0.

Shell   Tool used in percussion boring in coarse-grained soil consisting of a hollow tube which is driven into the ground by percussive force. A one-way valve at the bottom of the tube allows the soil into the tube and closes when the shell is withdrawn.

Standard Penetration Test (SPT)   In-situ soil test involving driving a thin-walled sampling tube over a specified penetration and counting the number of percussive blows required to achieve this. The blowcount is an indicator of soil density/strength.

Targeted sampling Investigation points are targeted around a particular known (or suspected) source of contamination/hazard, with the aim of determining its extent in a systematic way.

Thrust machine Vehicle or mounting consisting of hydraulic jacking equipment and a means of providing a reaction, used to drive a cone penetrometer into the ground.

Trial pit Shallow excavation made for the purposes of investigating near-surface layers, and their lateral extent.

Undisturbed sample Samples in which the in-situ soil structure has been essentially preserved (making them suitable for laboratory strength and stiffness testing).

Chapter 7

Split-barrel sampler Standardised sampling tube that can be split in half along its length for sample removal. This is the tool used to penetrate the test soil in the Standard Penetration Test (SPT).

Drop hammer A known mass falling under gravity from a known height, used to drive the split-barrel sampler in the SPT.

SPT blowcount (N) Number of blows of the drop hammer required to achieve penetration of the sample tube by 300 mm in the SPT.

Normalised blowcount ($N_1)_{60}$ SPT blowcount which has been normalised to account for the overburden pressure (σ'_{v0}) at the test depth.

Vane Steel rod with a cruciform section which is rotated in fine-grained soils to measure the undrained shear strength of soft fine-grained soils (c_u).

Pressuremeter A device which is inserted into a borehole, capable of expanding the walls of the borehole radially. By continually measuring the internal pressure required and the resultant deformations, soil strength and stiffness data can be derived.

Cavity strain (ε_c) Radial strain at the borehole (cavity) wall during a pressuremeter test.

Self-boring pressuremeter (SBPM) Pressuremeter which includes a cutting head at its tip, designed to install itself into the test soil with minimal disturbance.

Plane strain Also described as biaxial or 2-D stress conditions in which the soil may only strain in a single plane (i.e. the problem in question is so long in the third direction that strains in this direction can be neglected).

Lift-off pressure (σ_{h0}) Cavity pressure within a pressuremeter at which non-zero cavity strain is first recorded, implying that the in-situ horizontal total stress has been reached.

Limit pressure (p_L) Theoretical cavity pressure within a pressuremeter test which would lead to infinite radial expansion in the soil (impossible to achieve in practice).

Coefficient of lateral earth pressure (at rest, K_0) The ratio of horizontal effective stress to vertical effective stress. The value depends on the amount of strain within the soil: if no strain has occurred, such as in the in-situ condition, the soil is said to be 'at rest'.

Chapter 8

Action effect The response of a construction to an applied action (e.g. foundation settlement under an applied load).

Action A load or displacement which is applied to a construction.

Analysis Characterisation of the response of an existing construction (of known size/properties).

Associative flow rule Rule for plastic flow in drained materials for which soil dilation is a maximum ($\psi = \phi'$) and the normality principle applies.

Bearing capacity (q_f) Pressure acting over the foundation area which would cause shear failure of the supporting soil immediately below and adjacent to a foundation.

Glossary

Bearing capacity factor Non-dimensional factor describing the bearing capacity of a foundation.

Bearing resistance (R) Maximum load which a footing can carry before collapse (i.e. defining the ULS).

Bulb of pressure Zone of soil beneath a foundation within which the vertical stresses are greater than 20% of the applied footing pressure, representing the zone of soil which is expected to contribute significantly to the settlement of the foundation.

Characteristic value Cautious estimate of the value of a property affecting a limiting state.

Creep effect Increase of strain or movement with time, under constant applied stress.

Deep foundation Any foundation whose depth is significantly greater than its width.

Design value Characteristic value which has been modified by the application of a partial factor.

Design Determination of the required size/properties of a construction to satisfy a limiting state.

Effective (buoyant) unit weight (γ') Saturated unit weight minus the unit weight of water, i.e. in a fully saturated soil with the water table at the surface, $\gamma' z$ will directly give the effective vertical stress.

Favourable An action is favourable if it reduces the total applied loading.

Flexible Having negligible bending stiffness.

Footing Generic term for an individual shallow foundation element which is usually part of a larger foundation system.

Founding plane The level of the underside of the foundation.

General shear failure Failure mode of a shallow foundation in which continuous failure surfaces develop between the edges of the footing and the ground surface, a state of plastic equilibrium being fully developed throughout the soil above the failure surfaces.

Hodograph Diagram showing relative velocities.

Immediate settlement (s_i) Settlement occurring prior to consolidation (i.e. under undrained conditions).

Influence factor Non-dimensional factor describing the stress distribution within an elastic medium

Kinematically admissible (mechanism) A valid mechanism in which the motion of the sliding soil mass must remain continuous and be compatible with any boundary restrictions.

Limit state Performance requirement at which a foundation (or other construction) will become unsuitable/unsafe.

Local shear failure Failure mode of a shallow foundation in which there is significant compression of the soil under the footing and only partial development of a state of plastic equilibrium.

Lower bound (theorem) If a state of stress can be found, which at no point exceeds the failure criterion for the soil and is in equilibrium with a system of external loads, then collapse cannot occur; the external load system thus constitutes a lower bound to the true collapse load.

Normality principle As applied to a slip line, when the direction of movement is perpendicular to the resultant force, so that there is no energy dissipated in shearing.

Pad Footing supporting a single column or pier within a structure.

Permanent action Action which always acts on a construction during its design life, e.g. dead load of a structure.

Pile A type of deep foundation acting like a column which is used to support large concentrated loads or when poor quality soil exists close to the ground surface.

Punching shear failure Failure mode of a shallow foundation in which there is relatively high compression of the soil under the footing, accompanied by shearing in the vertical direction around the edges of the footing.

Raft Shallow foundation which extends over the entire footprint of a structure as a single continuous element.

Rigid Infinitely stiff in bending.

Serviceability limit state (SLS) Limit state at or above which a construction will cease to be suitable for its intended function (but will be safe).

Glossary

Shallow foundations Any foundation whose width is significantly greater than its depth.
Shear fan/fan zone Rotational plastic zone within which all soil is undergoing relative slip and within which the principal stress directions are continuously rotating.
Soil–structure interaction Transmission of load between a structure and the supporting (or surrounding) soil.
Strip footing Footing which is very long compared to its width, usually supporting a row of columns.
Ultimate limit state (ULS) Limit state at or above which a construction will suffer a catastrophic failure.
Unfavourable An action is unfavourable if it will increase the total applied loading – e.g. a downwards vertical load acting on a foundation.
Upper bound (theorem) If, in an increment of displacement, the work done by a system of external loads is equal to the dissipation of energy by the internal stresses within a kinematically admissible mechanism of plastic collapse, then collapse must occur; the external load system thus constitutes an upper bound to the true collapse load.
Variable action Action which acts only intermittently, e.g. wind or other environmental loading.
Weight density Alternative term for unit weight used in Eurocode 7.

Chapter 9

Adhesion factor (α) Proportion of the undrained shear strength which can be mobilised along an interface.
Base resistance (Q_{bu}) Resistance of a deep foundation derived from end-bearing at the tip (base).
Block failure (of a pile group) Failure mode in which the whole block of soil beneath the pile cap and enclosed by the piles fails as a single large pier.
Cast-in-situ pile Pile formed by placing concrete as a driven steel tube is withdrawn.
Constant rate of penetration test (CRP) Displacement-controlled pile load test conducted at a constant penetration rate of 0.5–2 mm/min (in compression, slower in tension) until either a steady ultimate load is reached or the settlement exceeds 10% of the pile diameter/width.
Continuous Flight Auger (CFA) pile Specific type of bored pile in which a helical auger is drilled into the ground over the desired pile length in a single process. The plug of soil trapped between the flights is then withdrawn from the ground as concrete is pumped into the shaft through a tube running down the centre of the auger.
Correlation factor Empirical factor in Eurocode 7 used in defining a characteristic value of pile resistance from test data based on the amount of testing undertaken.
Design Verification Load (DVL) The expected working load carried by the pile in the final foundation.
Displacement pile Pile where installation involves displacement and disturbance of the soil around the pile.
Driving shoe Rigid cap placed over the tip of the pile to protect it from damage under high local driving stresses.
Dynamic (pile) test Non-destructive test of a pile in which a stress wave is passed along the pile.
(pile group) Efficiency (η_g) The ratio of the average load per pile in a group at failure to the resistance of a single pile.
Freestanding (pile group) When the pile cap is clear of the ground surface.
Hyperbolic method Empirical method used for predicting pile settlement based on load-deflection behaviour being well-approximated by a curve of hyperbolic shape.
Interface friction angle (δ') Angle of shearing resistance along a soil–structure interface.
Interface shear test Direct shear test in which half of the shearbox contains a structural material and the other half soil, the purpose being to measure the shear strength along the interface.

Glossary

Jacked (pile) Push-in method of installing a displacement pile, avoiding the noise associated with driving.

Kentledge Dead weight consisting of blocks (usually of precast concrete or iron).

Maintained Load Test (MLT) Load-controlled pile test in which load is applied to the pile over a series of stages, each of which is maintained for a period of time.

Model factor Empirical factor in Eurocode 7 functioning as an additional partial resistance factor accounting for unreliability/uncertainty in the correlations used when deriving a characteristic value of pile resistance from test data.

Negative skin friction Situation when a consolidating layer around a deep foundation settles further than the pile so that the direction of skin friction in this layer is reversed, exerting an additional action on the pile (rather than contributing to the resistance).

Neutral plane Depth at which soil and pile settlement are equal, demarcating zones of negative and positive skin friction.

Non-displacement (bored) pile Pile which is installed without soil displacement. Soil is removed by boring or drilling to form a shaft, concrete then being cast in the shaft to form the pile.

Pier/caisson Type of deep foundation which has a much smaller length to diameter ratio than a pile.

Pile cap A slab cast on top of the piles in a group which distributes load between the piles. It is often in contact with the underlying soil.

Plugged A hollow pile is plugged when the plug of soil within the pile has a resistance from enhanced interface friction along the interior walls which is higher than the base pressure acting upwards on the plug. The pile will subsequently have the resistance of a closed-ended pile of the same outside diameter.

Proof load 150% of the working load (i.e. $1.5 \times DVL$).

Randolph and Wroth method Simple linear elastic method for predicting pile settlement, considering purely elastic behaviour of the soil and idealised soil conditions.

Scour Erosion of soil from around a construction as a result of tidal, wave or flow actions.

Shaft resistance/skin friction (Q_{su}) Resistance of a deep foundation derived from interface friction between the foundation material and the soil along the shaft.

Shell pile Tubular pile filled subsequently with concrete after driving.

Slenderness ratio (L_p/D_0) Ratio of pile length to diameter (or width).

Statnamic (pile) test Rapid load test in which an explosive charge is used to reduce the reaction mass required in a load test.

Tapered pile A pile where the diameter or width decreases with depth.

Trial/test piles Piles which are constructed solely for the purposes of load testing, usually before the main piling works commence, and which can be taken to the ULS.

T–z method Numerical technique for predicting pile settlement utilising a finite difference scheme in the solution. Any type of soil–pile interaction may be modelled (linear or non-linear) allowing more complicated soil and pile geometry to be considered.

Under-ream Enlarged base in a bored pile, used to increase base resistance and tensile capacity.

Working/contract piles Piles that will be part of the final foundation and as such are loaded to only 150% of the working load that the pile will ultimately carry in a load test, allowing for the SLS to be verified.

Yield stress ratio (c_u/σ'_{v0}) Strength normalisation which incorporates the effects of overconsolidation in undrained soil. Normally consolidated soils which are fully saturated usually have $c_u/\sigma'_{v0} = 0.2-0.25$. Overconsolidated soils have higher values (as c_u is larger at the same depth).

α-method Determination of shaft resistance in undrained conditions (i.e. with interface friction characterised by αc_u).

β-method Determination of shaft resistance in drained conditions (i.e. with interface friction characterised by $\sigma' \tan \delta'$).

Chapter 10

Angular distortion, (β_d) Rotation occurring between two points due to differential settlement ($=\Delta/L$).
Cellular raft Thick raft within which voids are formed to reduce bearing pressure without dramatically reducing bending stiffness of the raft.
Critical length (L_c) Point along a pile below which the displacement of a pile under lateral load, applied at the head, is negligible.
Deep basement A hollow structure below ground level, providing usable space and also acting as a heavily embedded shallow foundation.
Differential settlement (Δ) Difference in settlement between two parts of a building (or other) structure.
Footprint The plan area covered by a building (or other) structure.
Inclination factor (i_c) Modification factor used in determining bearing capacity to account for the application of horizontal load alongside vertical load (the resultant of these forces is inclined to the vertical, hence the name).
Piled raft A very large pile group in which the cap (raft) is relatively flexible.
Plunge column A column from a building structure which is cast directly into the head of a concrete pile used to support it.
Raked (pile) A pile installed at an angle to the vertical so that part of the axial resistance may be used to carry a component of any horizontal loading.
Tilt Overall (gross) rotation of a structure.
Uplift (hydraulic) Failure mode in which a foundation or other underground structure becomes buoyant.
Uplift The process whereby moment loading causes part of the foundation to lift-off the soil, losing contact and reducing the area of the foundation contributing to bearing capacity.
Yield surface Surface representing combinations of actions (e.g V, H and M) which will result in a state of plastic equilibrium being reached.

Chapter 11

Active condition Condition where the value of horizontal stress decreases to a minimum value due to relative soil-structure movement such that a state of plastic equilibrium develops for which the major principal total and effective stresses are vertical.
Active earth pressure (p_a) Total stress acting normal to a retaining structure associated with active conditions within the soil mass. The minimum limit pressure.
Active earth pressure coefficient (K_a) Lateral earth pressure coefficient describing the minimum lateral (active) earth pressure that can occur within a soil deposit as a function of the vertical effective stress (associated with lateral extension of the soil).
Arching Redistributions of lateral pressure from yielding to unyielding sections of a soil mass.
Backfill Soil that is placed (usually compacted) behind a retaining structure after it has been constructed. Also, used to refer to any soil that is used to fill an excavation. The term is also used as a verb to describe the process of backfilling.
Basal heave Heave of the soil at the bottom of a deep excavation due to stress relief on excavation.
Bentonite A colloidal suspension (slurry) of montmorillonitic clay in water, used to provide temporary fluid support to the sides of an excavation.
Bond failure Slippage between reinforcement elements and soil.
Coherent gravity method Analysis method for reinforced soil retaining structures considering progressive failure of reinforcing elements.

Glossary

Composite wall   Reinforced-soil retaining structure, formed from alternating layers of soil and reinforcement.

Contiguous pile wall   Concrete retaining wall formed from a row of non-overlapping bored piles.

Diaphragm wall   A relatively thin reinforced concrete membrane used as a retaining structure, cast in a trench prior to excavation.

Earth pressure at-rest (p'_0)   Lateral earth pressure when the lateral strain in the soil is zero.

Filter cake   Skin of very low permeability (a few millimetres thick) formed by the deposition of bentonite particles on excavated soil faces.

Fixed anchor length (L)   The grouted length of tendon in a ground anchor, through which force is transmitted to the surrounding soil.

Flexible retaining wall   Flexible retaining structure which resists soil movement by bending.

Free anchor length   The length of tendon between the fixed anchor length and the bearing plate in a ground anchor, over which no force is transmitted to the soil.

Geotextiles   Family of polymeric materials used in sheet form. Geogrids having an open structure and high tensile strength are used as soil reinforcement; geomembranes are closed sheets used as impermeable drainage barriers.

Gravity retaining wall   Large, monolithic retaining structure which keeps the retained soil stable due to its mass.

Ground anchor   Anchorage consisting of a high-tensile steel cable or bar (tendon), one end of which is held securely in the soil by a mass of cement grout or grouted soil; the other end of the tendon is anchored against a bearing plate on the structural unit to be supported.

Limit equilibrium   Analysis technique which involves postulating a failure mechanism as in upper bound limit analysis. While the latter considers an energy balance during a virtual movement of the mechanism, limit equilibrium considers equilibrium of the limiting forces acting on the mechanism at the point of failure.

Limit pressures   General term for the active (minimum) and passive (maximum) lateral earth pressures.

Method of Characteristics   Lower bound limit analysis technique, involving the formation of a slip line field.

Overdig   Reduction in soil level in front of a retaining wall for design purposes to allow for the possibility of future planned or unplanned excavation in front of the wall.

Panel   Section of a diaphragm wall.

Passive condition   Condition where the value of horizontal stress increases to a maximum value due to relative soil–structure movement such that a state of plastic equilibrium develops for which the major principal total and effective stresses are horizontal.

Passive earth pressure (p_p)   Total stress acting normal to a retaining structure associated with passive conditions within the soil mass. The maximum limit pressure.

Passive earth pressure coefficient (K_p)   Lateral earth pressure coefficient describing the maximum lateral (passive) earth pressure that can occur within a soil deposit as a function of the vertical effective stress (associated with lateral compression of the soil).

Reinforced earth   General term describing soil containing geosynthetic or other reinforcement.

Resultant thrust   Force per unit length of wall representing the integrated effect of the lateral earth pressure distribution.

Retained soil   Soil mass supported by a retaining structure, that would not stand without support.

Secant pile wall   Concrete retaining wall formed from a row of overlapping bored piles.

Tie-back wedge method   Analysis method for reinforced soil retaining structures which is an extension of Coulomb's method (considering the forces acting on a wedge of soil from which a force diagram can be drawn).

Tie-backs   Ground anchors used to provide additional restraint to flexible retaining structures.

Tremie pipe Pipe through which concrete is pumped which can be lowered into an excavation.
Virtual back Vertical plane through the retained soil for a cantilever wall over which the lateral earth pressure is assumed to act.

Chapter 12

Compound slip Slope failure mode in which the failure surface consists of both curved and plane sections.
Cut-and-cover Method of constructing underground space involving making a deep excavation, within which the structure is constructed, backfilling being used to bury the construction.
Earth pressure balance (EPB) tunnel boring machine Computer-controlled tunnelling machine, cylindrical in shape, in which the excavation process is carefully controlled to maintain balanced earth pressures at the face and hence minimise volume loss.
Earth pressure balance construction Technique used for constructing tunnels in which a support pressure is applied to the face being excavated to prevent collapse into the tunnel and subsequent over-excavation (ground-loss).
Gaussian curve Mathematical expression which can be used to describe the shape of a settlement trough above a tunnel.
Hydraulic fracturing Process occurring when the total normal stress on a plane is less than the local value of pore water pressure, resulting in cracking and pore water leakage (due to the increased permeability along the cracks).
Immersed tube tunnel Underwater tunnel constructed in sections which are prefabricated onshore, floated out to site, flooded to lower them into a shallow trench excavated on the riverbed/seabed, connected underwater and, finally, pumped out of the internal water.
Pore water pressure ratio (r_u) Pore water pressure normalised by total vertical stress.
Rip-rap Thin layer of rockfill placed on the upstream slope of an embankment dam to protect it from erosion by wave or other fluid actions.
Rotational slip Slope failure mode in which the failure surface in section is a circular arc or a non-circular curve.
Running depth Depth below ground surface of the centreline of a tunnel.
Stability number Non-dimensional number describing slope or tunnel stability, analogous to a bearing capacity factor in foundation analysis.
Translational slip Slope failure mode in which the failure surface tends to be plane and roughly parallel to the slope surface.
Volume loss (V_L) Volume of over-excavated soil normalised by tunnel volume, expressed as a percentage.

Chapter 13

5% fractile value Characteristic value X_k for which there is only a 5% probability that somewhere in the layer (perhaps at an unmeasured location) there is an element of soil having a strength lower than X_k.
Back-calculation Determination of soil properties that would need to exist to give an observed response.
Brillouin Optical Time Domain Reflectometry (BOTDR) Analysis technique that can be used with optical fibres for measuring strains along long structures.
Deep settlement probe Instrument for measuring vertical settlement within a borehole.

Glossary

Extensometers   Family of devices for measuring displacement along a given axis (used for measuring horizontal displacement).

Heterogeneity   Non-uniformity (opposite of homogeneity).

Hydraulic settlement gauge   Device measuring at a single point only, but can be used at locations which are inaccessible to other devices and is free of rods and tubes which might interfere with construction.

Inclinometer   Device for measuring a profile of horizontal displacement over a vertical line (e.g. along a borehole).

Multi-point extensometer   Extensometer capable of making measurements at multiple depths within a borehole.

Observational Method   The use of measurements to continuously re-evaluate design assumptions and refine the design or modify/control construction techniques.

Precise levelling   Surveying technique for measuring vertical settlement of surface structures capable of higher accuracy than conventional levelling.

Rod extensometer   Simple and accurate device for measuring relative vertical movement within a borehole.

Shape acceleration array (SAA)   Modern alternative to the inclinometer which is more robust, cheaper and easier to install.

Slip indicator   Simple device to enable the location of a slip surface to be detected within a slope (though not the amount of displacement).

Tell-tale   Unstressed rod attached to a structural member at one end and running parallel to it. The change in length of the member under load is measured using the free end of the tell-tale as datum.

Vibrating wire strain gauge   Device used for measuring strain (and therefore displacement) based on the change in natural frequency of a wire with the instrument as it stretches.

Index

Italic page numbers indicate tables; **bold** indicate figures. Authors are not listed in the index; readers requiring complete lists of cited works and authors should consult the reference lists following each chapter.

5% fractile value 504, *541*

A-line 18, *527*
alpha (α)-method 333, *538*
action effects 322, *535*
actions 317, 318–19, *535*
active condition 404, *539*
active earth pressure 405, 408–9, 410, 421, *539*
active earth pressure coefficient 406, 412, 415, *539*
activity 10, **11**, *527*
adhesion factor 333, **334**, *537*
advanced foundation topics, problems 399
aerial photography 201
air content 23, 24, 26, *527*
air entry value 159, *532*
alluvium 4, *527*
anchored and propped walls 437–41
anchored sheet pile walls **438**, **440**
angle of dilation 168–9, *532*
angle of shearing resistance 154, *532*
angular distortion 366–7, **367**, *368*, *539*
anisotropic soil conditions 57–9
anisotropy 147, *532*
apparent cohesion 171, *532*
aquiclude 39, *529*
aquitard 39, *529*
arching 440–1, **440**, 488, *539*
artesian conditions 39–40, *529*
associative flow rule 285, *535*
augers 205–6, **205**

beta (β)-method 335, *538*
back-calculation 502, *541*
backfilling 434–6, 437, *539*
band drain 136, *530*

basal heave 453, 454–5, *539*
base exchange 8, *527*
base resistance 328, 331–3, *537*
basic parabola 67–9, **69**
batter angle 409–13
bearing capacity 270, *535*; drained materials 285–94; shallow foundations 271–3; undrained materials 273–85
bearing capacity factors 281–5, **282**, **283**, **284**, **292**, *536*
bearing resistance 270, 316, *536*
bedding features 18
bending 368–9, **373**
bending moments 440–1, 372, **373**
bentonite 456, *539*
Bishop routine solution 480–1
block failure 354, *537*
bond failure 459, *539*
bookhouse structure **9**, 10
bored piles 330
borehole logs 216–18, **217**
borehole (permeability) tests 45–6, **47**
boreholes 202, 204, *534*
Bothkennar clay, use of correlations to estimate strength 193
braced excavations 452–6, **453**, **454**
Brillouin optical time domain reflectometry technique (BOTDR) 513, *541*
bucket augers 205, 206, *534*
bulb of pressure 298, *536*
bulk density 23, 24, 25, *527*

caissons 327, *538*
calibration, cone penetration testing (CPT) 253–4, **255**
cantilever sheet pile wall **437**
cantilever walls 436–7

543

Index

Casagrande apparatus **12**
Casagrande method 12
case studies, overview 502
cast-in-situ piles 329, 330, 537
cation layer 8
cavity expansion, idealised soil response 243
cavity strain 240, 535
cellular raft 378, 539
characteristic values 317, 502–6, **505**, 517, 536
chemical weathering 7
Chin's method **352**
chisels 204, 534
classification and description 14–16, *20*
classification systems 18–22
clay cutter 203, 204, 534
clay, definition 16
clay fraction 10, 527
clay minerals 7–8, **9**, **11**
clay minerals: basic units **8**
clay structures **9**
coarse-grained soils 527; coefficient of permeability 46; composite types *17*; cone penetration testing (CPT) 252–3, **253**; definition 16; determination of peak angle of shearing resistance **236**; dilatancy **170**; effect of age on SPT data interpretation **235**; grading 14; interpretation of PMT data 248–9, **249**, **250**; interpretation of SPT data 233–5; shear strength 168–74, **169**
coefficient of consolidation 116, 121–6, **121**, **123**, 530
coefficient of curvature 14, 527
coefficient of lateral earth pressure 256, 535
coefficient of permeability 41, *41*, 42–6, 125–6
coefficient of uniformity 14, 527
coefficient of variation 503, **504**
coefficient of volume compressibility 105, 531
coefficients for vertical displacement **302**
coherent gravity method 460, 539
cohesion 13, 15, 171, 527
cohesion intercept 154, 532
cohesive soils, use of correlations to estimate strength **193**
colloidal 527
colloidal size 7
combined loading 387–9, 397–9; deep foundations 389–98; shallow foundations 380–9
common energy ratios *234*
compaction 26–34, **31–3**
compaction-induced earth pressures 434–6, **436**
compactive effort 26, 527
compactive state and stiffness *17*
complex variable theory 66
composite walls 458, 540
compound slips 473, 541

compressed air samplers 214, **215**
compressibility characteristics 104–5
compression index 105, 531
compression ratios 123
cone penetration testing (CPT) 218–22, **219**, 252–60; calibration 253–4, **255**; coarse-grained soils 252–3, **253**; determination of OCR 256; examples 257–60, 315–16; fine-grained soils 253–60, **256**; pile resistance 340–1, **340**; shallow foundations 313–16; soil behaviour type classification chart **219**, **221**; use of data **221**
cone penetrometer 218, 534
confining fluid pressure 162
conformal transformation 66–7, **67**
consistency limits, fine-grained soils **11**
consolidated–drained (CD) triaxial test 166, 175, 176, 178
consolidated–undrained and drained triaxial tests, typical results **176**
consolidated–undrained (CU) triaxial test 166, 175–6
consolidation 84, 530; assumed linear relationship **113**; coefficient of 116; compressibility characteristics 104–5; compression ratios 123; correction for construction period 131, **131**; degree of 112–14; element within a consolidating layer of soil **115**; examples 129–30, 132–5; finite difference method (FDM) 127–30; under increase in total stress **114**; oedometer test 102–9; one-dimensional depth-time Finite Difference mesh **128**; overview 101–2; piston analogy 85–6, **85**; pre-loading 140–1, **141**; preconsolidation pressure 105–6, **106**; problems 142–3; secondary compression 126–7; summary 142; Terzaghi's theory of one-dimensional consolidation 115–16; time factor for radial drainage **138**; vertical drains 136–8
consolidation equation 116–20
consolidation settlement 101, 109–12, **111**, 531
consolidation theory, shallow foundations 304–10
constant-head test 42, 45, 529
constant rate of penetration tests (CRP) 351–2, 537
constitutive model 147, 532
constrained modulus 105, 531
contact pressure, under rigid area **301**
contaminated ground 227–8
contiguous pile wall 457, **457**, 540
continuous flight auger (CFA) pile 330, **331**, 537
continuous flight auger piling rig **330**
continuous-flight augers 205–6, 534
continuous samplers 214, **215**
continuum mechanics 145–8
contract piles 351, 538
core-catcher 213, 534
correlation factors 341, *341*, 353, 537

Index

Coulomb's frictional model 152
Coulomb's theory of earth pressure 429–31, **429**, **431**, 469–72
CPT soundings 202, 534
creep effect 314, 536
critical hydraulic gradient 91, 530
critical length 395, 539
critical state 169, 532
critical state angle of shearing resistance 169, 190–1, **190**, 532
critical state framework 183–8
critical state line (CSL) 169, 184, 185–6, **185**, 532
critical state parameters 186–7
crushing 171
cut-and-cover technique 490, 541
cylindrical blocks (vertical drains) **137**

Darcy's law 42–3, 44, 46, 57
data, selection of characteristic values 502–6
decision-making 501
deep basements 378–9, 539
deep foundations 271, **328**, 536; capacity under combined loading 395–8; combined loading 389–98; determination of bearing capacity factor and shape factor **332**, **333**; examples 337–9, 356–8; load testing 350–3; negative skin friction 358–9, **358**; overview 327–9; pile groups 353–8; pile installation 329–31; pile resistance from in-situ test data 340–1; pile resistance under compressive loads 331–9; pile settlement 341–9; piles under tensile loads 349–50; problems 359–61; serviceability limit states (SLS) 341, 395–7; summary 359; ultimate limit states (ULS) 395
deep settlement probe 506, **508**, 541
degree of consolidation 112–14, **120**, 531
degree of saturation 26
Delft continuous sampler 214
delta 4, 527
deposition 3; desert **5**; fluvial, **5**; glacial **5**
depositional environments **5**
description and classification 14–16
design 316, 501–2, 536
design value 317, 536
design verification load 351, 537
desk studies 201, 534
deviatoric stress invariant 160, 532
diaphragm pressure cells 513, **514**
diaphragm walls 437, 456–7, **457**, 540
differential settlement 366–71, **372**, 373–4, **377**, 539
diffraction coefficient **356**
dilatancy 168, 532
dilatancy in coarse-grained soils **170**
dilatancy test 16

dilatometer test (DMT) 232, 260–1
direct shear apparatus **157**, 533
direct shear tests 156–7, 168, 173
direct wave 223
discontinuities 18
dispersed structure 9, **9**, 527
displacement piles 329, 330–1, 537
dissipation 84, 530
disturbed samples 210–11, 534
double drainage 117
double layer 8
drainage conditions, effect on shear strength 155
drained conditions 84, 530
drained soil: base resistance 332; bearing capacity 285–94; conditions along a slip plane **286**; lower bound stress field **289**; rotational slips 477–82; shaft resistance 334–5; stability of tunnels 493; upper bound mechanism **287**
drains 136–40
drift deposits 4, 527
driving shoes 329, 537
drop hammer 232, 535
dry density 27–9, **27**, **28**, **29**
dry strength test 16
Dupuit assumption 44
dynamic (pile) test 353, 537
dynamic/seismic liquefaction 96–7, 530

earth pressure at rest 415–18, 540
earth pressure balance 490; construction 490, 541; tunnel boring machines 495, 541
earthquake, liquefaction 97
effective (buoyant) unit weight 294, 536
effective size 14, 527
effective stress: example 81–2; finite difference method (FDM) 83; influence of seepage 87–91; interpretation **81**; overview 79; partially saturated soils 87; principle of 80–3; problems 98–9; response to change in total stress 83–6; summary 98
effective stress path (ESP) 170, **171**, **172**, **177**, **184**, **185**
effective vertical stress 81, 86, 94–5
efficiency, pile groups 353
elastic parameters, determination of: deep foundations 348; from laboratory tests 150–1; from pressuremeter test (PMT) 243–9, **247**, **249**; shallow foundations 302–3;
elastic–perfectly plastic model 147, **148**, 532
elastic–strain hardening plastic model 147, **148**, 532
elastic–strain softening plastic model 148, **148**, 532
elastic theory: deep foundations (combined loading) 39–7; deep foundations (vertical loading) 342–7; shallow foundations (combined loading) 385–7; shallow foundations (vertical loading) 300–4

545

Index

elasto-plastic soil behaviour, pressuremeter testing 244–6, **244**
electrical resistivity 225–7, **226**
element within a consolidating layer of soil **115**
elemental flow net field **59**
embankment dams 487–90; end of construction and long-term stability 488–90; horizontal drainage layers **488**; rapid drawdown 489, **490**; seepage control 71–3; seepage through 64–71; stability coefficients 489
end-product compaction 29, 527
engineered fill 26, 527
equation of compatibility 147, 532
equations of equilibrium 145–6, 532
equipotentials 48, **50**
erosion, in glacial regimes 3–4
Eurocode 7 103, 202, 269, 281, 292, 316, 336, 350, 352, 360, 361, 381, 383, 385, 416, 421, 447, 473, 503
excavated walls 437
excess pore water pressure 84, 113, 530
expansion index 105, 531
extensometers 509, 542
extrasensitive soils 180, 532

factor of safety: against heave 92–3; example 95; of slopes 474–5, 477–81, 483–4, **486**
failure envelopes and stress paths in triaxial tests **172**, **177**
fall cone **12**
falling-head test 42, **43**, 46, 529
fan zones 278–9, 537
favourable actions 319, 536
Fellenius solution 479
field compaction 29–34
field instrumentation 506–14, 517
field vane test (FVT) 192, 236–40, **238**, **239**
filter cake 456, 540
filter design, seepage control 73–4
fine-grained soils 527; cone penetration testing (CPT) 253–60, **256**; consistency limits **11**; definition 16; interpretation of SPT data 235–6, **237**; plasticity 10–13; pressuremeter testing 241–8, **247**; shear strength 174–80
finite difference method (FDM) 60–2, 529; consolidation 127–30; determination of head at an FDM node **61**; effective stress 83; seepage 60–2
fixed anchor length 448, **449**, 540
flexibility 300–1, 536
flexible retaining walls 403, 540
flight augers 205–6, 534
flocculated structure 9, **9**, 527
flow channels 51, 529
flow function 49

flow lines 49, 529
flow lines and equipotentials **50**
flow nets 51–6, **53**, **54**, **56**, **68**, 69–71, 529
flow rate 41, 529
footings 270, 536
footprint 366, 539
forces under seepage conditions **90**
foundation stiffness determination **387**
foundation systems **367**; deep basements 378–9; differential settlement 366–71; examples 373–4, 378–9; overview 365–6; pad/strip 371; piled 374–7; rafts 372–3; *see also* deep foundations; shallow foundations
founding plane 281, 536
free anchor length 448, **449**
freestanding piles 353, 537
frictional strength along a plane of slip **152**
fully saturated soil 24

gap-graded soil 14, 527
Gault clay, use of correlations to estimate strength **193**
Gaussian curve 495, 541
general shear failure 271, 536
geological history, understanding 5–6
geological maps and memoirs 201
geophones 223
geophysical methods 222–7
geotechnical materials: resistivities *227*; secondary compression characteristics *127*; shear wave velocities *223*; strength properties from index tests 189–94
geotextiles 458, 540
glacial till 4, 21, 527
gravel, definition 16
gravity retaining walls 403, 418–28, **419**, **420**, 540
grid rollers 31
ground anchorages 447–52, **448**, 540
ground anchors **448**, 449–50, **449**
ground investigation 190; borehole logs 216–18, **217**; cone penetration testing (CPT) 218–22, **219**; contaminated ground 227–8; cost 203; depth 202; electrical resistivity 225–7, **226**; geophysical methods 222–7; groundwater investigations 208–10; hand and portable augers 206; identification of soil layers by sounding **226**; intrusive methods 203–10; mechanical augers 205–6; objectives 201; overview 201–3; percussion boring 203–4; piezometers 208–10; precautions 227; procedure 202; rotary drilling 207–8; sample quality and end use 211; sampling 210–15; seismic refraction 222–4; selection of in-situ test methods 260–1; selection of laboratory test methods 215–16; shafts and headings 203; spectral analysis of surface waves (SASW) 225;

summary 228; trial pits 203; use of CPTU data **221**; wash boring 206–7
groundwater conditions, terminology **40**
groundwater investigations 208–10
group symbols 18

half-closed layer 117, 531
hand augers 206
headings 203, 534
heave 91, 101, 454–5, 530, 531
heterogeneity 501, 542
hodographs 274, **274**, **276**, **287**, 536
hogging 369–70
Holbeck landslide 473, **474**
homogeneous embankment dam section **65**
Hooke's Law 147
horizontal drainage layers, embankment dams **488**
horizontal movement 509–12; deep foundation 395–7; measuring instruments **510**; shallow foundation 385–7
hydration 8
hydraulic conductivity 41, *41*, 529
hydraulic fracturing 488, 541
hydraulic gradient 40, 529
hydraulic oedometer 124, **124**
hydraulic piezometer 210
hydraulic pressure cells 513, **514**
hydraulic settlement cell **509**
hydraulic settlement gauges 507, 542
hydraulic uplift 378, 539
hyperbolic method 342, 347–9, 537

illite 8, **9**, **11**
immediate settlement 302, 536
immersed tube tunnels 490, 541
in-situ 534
in-situ conditions, effective stresses **306**
in-situ tests 201, 231–66; cone penetration testing (CPT) 252–60; derivation of key soil properties *261*; dilatometer test (DMT) 232; field vane test (FVT) 236–40; overview 231–2; pile resistance 340–1, **341**; pressuremeter testing (PMT) 240–51; problems 262–4; selection 260–1; shallow foundations 311–16; standard penetration test (SPT) 232–6; summary 261–2
inclination factor 381, 383, 539
inclinometer 511, **512**, 542
influence factor 536; CPT 314; settlement 300, *301*; stress 295, *296*, 298, **298**
information, sources 202
initial compression 122, 531
initial variations of excess pore water pressure **120**
instrumentation, use of 502

intake factor 46
interbedded soils 18, 527
interface friction angles 334, **335**, 537
interface shear strength 333–4
interface shear test 334, 537
inter-laminated soils 18, 527
interlocking 152, 168–9
intermediate principal stress 160, 532
interstratified soils 18, 527
inversion 225, 534
investigation points 202, *202*, 534
isochrones 117, **118**, 531
isomorphous substitution 7, 8, 528
isotropic compression line (ICL) 175, **184**, **185**, 532
isotropic consolidation 159, 174–5, **175**, 532
isotropic elasticity 149
isotropy 532

jacked pile 331, 538

kaolinite 7, 8, **9**, **11**
kentledge 351, 538
kinematically admissible mechanism 272, 536
Kjellman sampler 214
Kozeny's basic parabola 67–9

laboratory compaction 26–9
laboratory shear tests 156–68
laboratory test methods, selection 215–16
laboratory vane 168, 532
lateral earth pressure coefficient at rest 255–6, 416–17, **417**
lateral strain and lateral pressure coefficient **417**
Leaning Tower of Pisa 368, **369**
levelling plug **507**
lift-off pressure 243, 247, 249, 535
limit equilibrium 429, 469–72, **475**, 540
limit pressures 415, 535, 540
limit state design 316–23, 335–7, 378, 418–21, 447, 455, 436–8, 473
limit states 269, 536, 316–23, 335–7, 378, 418–21, 447, 455, 436–8, 473
line load 297, **297**
linear elasticity 149–50, 243–4
liquefaction 91–7, 174, 530
liquefied soil 91, 530
liquid limit 10, 11–12, **22**, **193**, 528
liquidity index 10, **191**, **192**, 529
lithification 3, 528
load testing 350–3; correlation factors *353*; field instrumentation 513–14
local shear failure 271, 536
loess 4, 528

Index

log time method 121–2, **121**
lower bound (LB) theorem 272, 277–80, 288–90, 536
lower bound stress field 277–8, **278**, **279**, **381**, **384**, **405**, **410**, **470**

macro-fabric 14–15
magnetic extensometer **508**
maintained load tests (MLT) 351, 538
mass density 23, 527
maximum density 25
maximum negative pressure 40
mean stress invariant 160, 532
mechanical augers 205–6
Ménard pressuremeter **241**
method compaction 29, 528
Method of Characteristics 406, 540
method of slices 477–82, **478**
minimum density 25
model factors 341, 538
modified AASHTO test 27, 528
Mohr circles 153–5, 533
Mohr–Coulomb criterion 148, **153**, 155, 174, 533
Mohr–Coulomb model 153–5
moisture condition test 34, **34**
moisture condition value (MCV) 34, 528
moisture content 22
montmorillonite 8–9, **9**, **11**
moraine, transportation 4
multi-channel analysis of surface waves (MASW) 225, 534
multi-point extensometer 507, 542

National Geological Records Centre 202
natural clays, structure 9, **9**
nature of soils 6–10
negative skin friction 358–9, **358**, 538
net stress 87
neutral plane 359, 538
Nicoll Highway collapse **439**
Niigata earthquake **97**
non-displacement piles 330–1, **330**, 538
non-homogeneous soil conditions 18, 59–60, **59**
non-linear soil shear modulus **151**
non-targeted sampling 228, 534
normal consolidation 104
normalised blowcount 233, 535
normality principle 285, 536
normally consolidated soil 104, 531

observational method 502, 514–15, 517, 542
oedometer **102**; hydraulic 124, **124**
oedometer test 102–9, **102**, **103**
one-dimensional elastic modulus 105

open drive samplers 212–13
open layer 117, 531
open standpipe piezometer 209–10, **209**
optimum water content 27, 528
organic content, measuring 16
organic soils 16, 528
origins, of soils 3
overburden correction factor **234**
overburden pressure 105, 531
overconsolidated soil 104, 531
overconsolidation 104, 175
overconsolidation ratio (OCR) 104, 238, 255, **256**, 416, 531
overdig 420, 540
overturning 382

pad/strip 270, 536
pad/strip foundation systems 371
panels 456, 540
partially saturated soils 87, **88**
particle density 24, 528
particle size 3, **4**; analysis 13–14, 20–2; distribution curves 21; particle size distribution (PSD) 5–6, 13, **5**, 528
partings 18, 528
passive condition 404, 540
passive earth pressure 405, 408–9, 410, 421, 540
passive earth pressure coefficient 390, 406, 412, 540
peak strengths **171**, 172
peats 16
pediment 4, 528
perched water table 39, 529
percussion boring 203–4, **204**
permanent actions 318, 536
permeability and testing 41–6, *216*, *261*
permeability ellipse 58
phase diagrams **23**
phase relationships 22–6
phreatic surface 39, 529
piers 327, 538
piezocones 218, **220**, 252, 534
piezometers 124, 208–10, **210**, 534
pile caps 353, 355, 538
pile groups 353–8, **354**, 398
pile resistance 340, **340**
pile settlement 341–9, **343**, **345**, **348**
piled foundation systems 374–7
piled rafts 366, **376**, **377**, 539
piles 271, 327, 536; bearing capacity factors **332–3**; capacity under combined loading 395–8; capacity under horizontal and moment loading 389–94; constant rate of penetration tests (CRP) 351; critical failure mode **392**, **394**; critical length under lateral

loading **396**; displacement from elastic solutions 395–8; drained interface strength in drained fine-grained soils **336**; dynamic testing 353; examples 337–9; freestanding 353; installation 329–31; interpretation of capacity using Chin's method **352**; jacked 331, 538; lateral capacity **392**, **394**; lateral loading **391**, **393**; load testing 350–3; maintained load tests (MLT) 351; partial resistance factors for ULS design **336**; principal types **331**; resistance 328–9; resistance and limit state design 335–7; resistance from in-situ test data 340–1, **341**; resistance under compressive loads 331–9; resistance under tensile loads 349–50; serviceability limit states (SLS) 341; shaft friction in tension **350**; shape factors **332**; static load testing 351; Statnamic test 353; ultimate limit states (ULS) 336–7
piping 71, 529
piston samplers 213
plane strain 242, 404, 535
plastic limit 10, 11, 12–13, **193**, 528
plasticity 10–13, 15
plasticity chart 18, **19**
plasticity index 10, 528
plate anchors 447–8, **448**
plate loading tests (PLT) 232
playa 4, 528
plugged piles 329, 538
plunge column construction 366, 539
pneumatic-tyred rollers 30
point load 295, **296**, *296*
Poisson's ratio 149, 533
poorly graded soil 14, 528
pore pressure coefficient 165, 305, 533
pore water pressure 40, 80–1, 83–4, **120**, 208, 308–9, 441–2, 530
pore water pressure ratio 480, 541
porosity 23, 26, 42, 528
portable augers 206
potentiometric extensometer 509
power rammers 31
pre-loading 106, 140–1, **141**
precautions, ground investigation 227
precise levelling 506, 542
preconsolidation pressure 105–6, **106**, **175**, 531
preloading 140–1, **141**, 531
pressure cells 513
pressuremeter testing 240–51, **244**
pressuremeters 240, **241**, 535
primary consolidation 122, 531
principal stress difference 162, 533
principal stress rotation across a frictional stress discontinuity **279**, **289**
principal stresses 153, **153**, 160, **161**, 533

principle of effective stress 80–3, 530
problems: advanced foundation topics 399; basic characteristics 35–6; consolidation 142–3; deep foundations 359–61; effective stress 98–9; retaining structures 461–4; seepage 74–7; self-supporting soil masses 496–8; shallow foundations 324–5; shear 196–7; in-situ tests 262–4
Proctor test 26–7, 528
proof load 352, 538
punching shear failure 271, 536
push rods 218, 534

quick clays 180, 533

rafts 270, **372**, 373–4, **373**, 536
raked piles 389, 539
Randolph and Wroth method 342, 346–7, 538
Rankine's theory of earth pressure 406–8, **407**
rapid assessment procedures 15
rapid drawdown, embankment dams 489, **490**
rectangular area carrying uniform pressure 298, **298**
reference studs **510**
reinforced soil 458–60, **459**, 540
relative density 25, 528
residual friction angle 194, **194**
residual soil 3, 528
residual strength 188–9, **189**, 194
resistance 328–9, 331–9
resistivities, geotechnical materials *227*
response time 208, 534
resultant body force 88, 530
resultant thrust 421, **422**, **423**, 540
retained soil **404**, **405**, 540
retaining structures: anchored and propped walls 437–41; backfilling and compaction-induced earth pressures 434–6; braced excavations 452–6; cantilever walls 436–7; contiguous pile wall 457; Coulomb's theory of earth pressure 429–31; diaphragm walls 437, 456–7; earth pressure at rest 415–18; effect of wall properties 409–13; examples 408–9, 423–8, 432–3, 443–7, 451–2; free earth support analysis, anchored and propped walls 438–41; gravity retaining walls 418–28; ground anchorages 447–52; limiting lateral earth pressures 404–6; minimum deformation conditions **418**; Mohr circles for zone 1 soil, drained and undrained conditions **415**; Mohr circles for zone 2 soil, drained conditions **413**; Mohr circles for zone 2 soil, undrained conditions **411**; overview 403–4; pore water pressure 441–2; pressure distributions and resultant thrusts, drained soil **423**; pressure distributions and resultant thrusts, undrained soil **422**; problems 461–4; Rankine's theory of earth

Index

retaining structures *continued*
  pressure 406–8; reinforced soil 458–60, **459**; resultant thrust 421; rotation of principal stresses due to wall roughness and batter angle **410**; secant pile wall 457; sloping retained soil 413–15; stresses due to line load **435**; summary 460–1
rigid 536
rigid–perfectly plastic model 147, **148**, **272**, 533
rigidity 300–1
ring shear test 188–9, **189**
rip-rap 487, 541
Rissa landslide 180, **181**
rock cycle **4**
rod extensometer 507, **508**, 509, **510**, 542
root time method 122–3, **123**
rotary drilling 207–8, **207**
rotation 385, 396
rotational slips 472–82, 541
roughness 335, 409–13
running depth 490, 541

sagging 369–70
sampling 210–16, **215**, 228, 534, 535
sand, definition 16
sandwick drain 136, 531
satellite imagery 201
saturation line 28, 528
saturation ratio 23, 24, 528
scour 327, 538
scree 4
secant pile wall 457, **457**, 540
secondary compression 122, 126–7, *127*, 531
sedimentation 13, 528
seepage 39, 432, 529; adjacent to sheet piling **92**; anisotropic soil conditions 57–9; determination of head at an FDM node **61**; examples 54–6, 61–2, 69–71, 72–3, 94–5, 432–3; finite difference method (FDM) 60–2; flow lines and equipotentials **50**; flow nets 51–6; influence on effective stress 87–91; non-homogeneous soil conditions 59–60, **59**; permeability and testing 41–6; problems 74–7; soil water 39–40; summary 74; through embankment dams 64–71; transfer condition 63–4, **63**
seepage between two flow lines **50**
seepage control 71–4, **72**
seepage force 88, 530
seepage induced liquefaction 91–5
seepage pressure 90, 442
seepage theory 46–50
seepage through a soil element **48**
seismic cones (SCPTU) 253, 256
seismic refraction 222–4, **224**
seismic waves 222–3

seismographs 223
self-boring pressuremeter (SBPM) 240–1, **241**, 535
self-supporting soil masses: embankment dams 487–90; examples 476–7, 481–2, 484; limit equilibrium analysis **471**, **475**; overview 467; problems 496–8; rotational slips in drained soil 477–82; rotational slips in undrained soil 473–7; slope failure **473**; slopes 472–87; summary 495–6; translational slips 483–4; tunnels 490–5; vertical cuttings and trenches 468–72
sensitive soils 180, 533
sensitivity 180, 192, **192**, 533
serviceability limit states (SLS) 269, 419, 536; braced excavations 455–6; cantilever walls 437–8; deep foundations 341, 395–7; load testing 351–2; pile groups 354–8; shallow foundations 322–3, 385–7; tunnels 493–5
settlement 109–12, 131; above tunnels **494**; behind excavations **455**; consolidation theory 304–10; damage to walls **370**, **371**; differential 366–71, **372**; elastic theory 300–4; examples 303–4, 309–10, 356–8, 373–4; from in situ tests 311–6; limits 322; measuring 506–9; measuring instruments **508**; piles 341–9;
settlement coefficient, shallow foundations **308**, 309
settlement-reducing piles **377**
settlement trough 493, **494**
shaft resistance 328, **329**, 333–5, 538
shafts 203
shallow foundations 319–22, 537; analysis using CPT data 313–16; analysis using lower bound theorem 277–80, 288–90; analysis using SPT data 311–13; analysis using upper bound theorem 273–7, 285–8; bearing capacity and limit analysis 271–3; bearing capacity factors 281–5, 290–1; bearing capacity in drained materials 285–94; bearing capacity in undrained materials 273–85; bearing capacity under combined loading 380–9; combining upper and lower bounds 281; contact pressure distribution under rigid areas **301**; contours of equal vertical stress **299**; depth of influence and foundation width **312**; design 316–17, **270**; displacement from elastic solutions 385–7; distribution of strain influence factor **314**; elastic parameters 302–4; examples 294, 299–300, 303–4, 309–10, 315–16, 319–23, 387–9; limit state design 316–23; line load 297, **297**; lower bound (LB) theorem 272; overview 269–71; point load 295, **296**; problems 324–5; rectangular area carrying uniform pressure 298, **298**; serviceability limit states (SLS) 322–3, 385–7; settlement coefficient **308**, 309; settlement from consolidation theory 304–10; settlement from elastic theory 300–4; settlement from in-situ test data 311–16; shape

factors under drained conditions **293**; soil element under major principal stress increment **305**; stability from limit analysis 380–5; stress paths **306**; stresses beneath 295–300; strip area carrying uniform pressure 297, **297**; strip foundations 282–4; summary 323–4; ultimate limit states (ULS) 317–21, 380–5; upper bound (UB) theorem 272–7

shape acceleration array (SSA) 511, 542

shear: continuum mechanics 145–8; critical state framework 183–8; dense soil 168–9; direct shear test 156–7; effect of drainage conditions on shear strength 155; estimating soil strength 189–94; example 173; laboratory shear tests 156–68; linear elasticity 149–50; liquefaction 174; Mohr–Coulomb model 153–5; non-linear elasticity 150–1; problems 196–7; residual strength 188–9; soil as frictional material 151–2; stress ratio 170; summary 195; triaxial test 157–67; undrained shear strength and sensitivity 191–3; *see also* coarse-grained soils; fine-grained soils

shear fan 276, 287, 537
shear modulus 149, **151**, 533
shear stress 145
shear wave velocities *223*, 534
shearbox **157**, 533
shearing resistance 154, 190–1
sheepsfoot rollers 30
sheet piling 91–5, **92**
shell 203–4, 534
shell piles 329, 538
short-flight augers 205
shrinkage limit 11, 528
sieving 13, 528
silt, definition 16
simple shear apparatus (SSA) 168, 533
single drainage 117
single grain structure 6–7, **7**, 528
Single Particle Optical Sizing (SPOS) 13–14
Skempton–Bjerrum method 305, 309
skin friction 328, 538
slenderness ratio 334, 538
slip indicator 512, 542
slopes **473**, 476–7, 483–7, **483**, 486
sloping retained soil 413–15, **414**
slurry-supported excavations **469**, **470**, **471**, **472**
smear 139, 531
smooth-wheeled rollers 30
soil behaviour type classification charts (CPT) 219, 221, **222**
soil elasticity 149–51
soil element under isotropic stress increment **165**
soil element under major principal stress increment **305**
soil layers, identification by resistivity sounding **226**

soil masses, self-supporting *see* self-supporting soil masses
soil parameters, derivation of *see* coarse-grained soils; fine-grained soils, in situ tests, permeability and testing
soil plasticity models 151–5
soil strength, estimating from index tests 189–94
soil–structure interaction 269, 537
solid geology 4, 528
specific gravity 24, 529
specific volume 23, 529
spectral analysis of surface waves (SASW) 225
split-barrel samplers 213, 232, 535
SPT blowcount 232, 535
SPT correction factor *233*
stability coefficients, embankment dams 489
stability numbers 475, **476**, 492, **492**, 541
standard penetration test (SPT) 213, 232–6, **233**, **235**, **236**, 311–13, 340, **340**, 534
standards, laboratory testing of soil 11, 13, 16, 19, 27, 43, 102, 156, 166,
static liquefaction 91–5, 530
static load testing 351, **351**
static pore water pressure 83, 530
stationary piston samplers 213
Statnamic test 353, 538
steady-state seepage pore water pressure 84, 530
step-graded soil 14
strain hardening and strain softening elastic–plastic models **148**
strain measurement 513
stratified soils 41, 529
stress history 104, 531
stress path triaxial cell **167**, 416, 533
stress paths 170, **171**, **172**, **177**, **184**, **306**, 533
stress ratio 170, 533
stresses due to line load **435**
stresses, strip area carrying uniform pressure 297, **297**
strip footing 270, 282–4, **283**, 284, 537
strut forces 453–4
sub-layering 110, 302, 504–5, **505**
surcharge pressure 109, **273**, **405**, 531
surface wave 223
Swedish solution 479
swelling 84, 101, 530

T–z method 342, 344–7, **345**, 538
tamping rollers 30
tape extensometer 509, **510**
tapered piles 329, 538
targeted sampling 228, 535
tell-tale 514, 542
terminology, groundwater conditions **40**

Index

Terzaghi's theory of one-dimensional consolidation 115–20
test piles 350, 351, 538
Teton Dam, Idaho **65**
thin-walled samplers 213
thrust machine 218, 535
tie-back wedge method 459–60, 540
tie-backs 437, 540
tie rods *see* ground anchorages
till, glacial 4
tilt 367, **367**, *368*, 539
time factor 117, **120**, 138, 531
time–settlement curve 131
topographical maps 201
total stress path (TSP) 170
toughness test 15
transfer condition 63–4, **63**
transformation for embankment dam section 68
translational slips 473, 483–4, **483**, 541
transportation 3–4
transported soil 3, 529
tremie pipe 456, 541
trenches 468–72, **469**, **470**, **471**
trial piles 350, 351, 538
trial pits 202, 203, 535
triaxial apparatus **158**, 533
triaxial compression 157, 533
triaxial extension 163, 533
triaxial shear strain 164, 533
triaxial test 157–67; consolidated–drained (CD) 166, 175, 176; consolidated–undrained (CU) 166, 175–6; examples 178, 181, 187–8; failure envelopes and stress paths **172**, **177**; limitations and corrections 159–60; stiffness 163–4; strength 160–3; testing under back pressure 164–6; types 166–7; unconsolidated–undrained (UU) 166, 179–80
true cohesion 171, 533
true triaxial conditions 160, 533
tube extensometer 509, **510**
tunnels 490–5, **490**, **491**, **492**, **494**
turbostratic structure **9**, 10
two-dimensional induced state of strain in an element of soil **147**
two-dimensional state of stress in an element of soil **146**

ultimate limit states (ULS) 269, 317–21, 336–7, 537; anchored and propped walls 438–9; cantilever walls 437–8; deep basements 378; deep foundations 395; gravity retaining walls 418–21; load testing 350–1; partial factors *318*; pile groups 353–4; reinforced soil 458; shallow foundations 270–1, 380–5; tunnels 490

uncertainty 501–2
unconfined compression test 167–8, **167**, 533
unconfined compressive strength (UCS) 167
under-ream 330, 335, 538
underseepage 71, 529
undisturbed samples 156, 210–11, *216*, 535
undrained condition 84, 149, 530
undrained shear strength 155, 179–80, 191–3, 533
undrained strength of cohesive soils **193**
unfavourable actions 318, 537
Unified Soil Classification System (USCS) 19
unit weight 24, 25, 529
uplift 378, 382, 539
upper bound (UB) theorem 285–8, 537

V-H loading 380–2, **381**, **382**, **383**, **384**, **386**, **395**
V-H-M loading 382, **383**, 385, **386**
van der Waals forces 9
vane 168, 237, 535
variable actions 318, 537
velocity diagram *see* hodographs
vertical cuttings 468–72, **469**, **470**, **471**
vertical displacement **301**, *301*, **302**, 506–9, **508**
vertical drains 136–8, **136**
vibrating hammer test 27, 529
vibrating plates 31
vibrating wire strain gauge 510–11, **511**, 513, 542
vibratory rollers 31
virgin compression line (1DCL) 104, 531
virtual back 425, 541
void ratio 23, 26, 529
void ratio–effective stress relationship **104**
volume loss 495, 541
volumetric behaviour during shear 183–4, **184**

wall properties 409–13
wash boring 206–7, **207**
water content 10, 22, 25, 529
water table (WT) 39, 529
weathering, chemical 7
weight density 24, 529, 537
well graded soil 14, 529
well pumping tests 44–5, **45**
window sampler 214–15
working piles 351, 538

yield stress ratio 333, 538
yield surface 381, **382**, **383**, **386**, **395**, 539
Young's modulus 149, 533

'zero air voids' dry density 27, 28, 529

Civil Engineering Books from CRC Press

Soil Physics with HYDRUS: Modeling and Applications

David E. Radcliffe, Jiri Simunek

Co-authored by Dr. Jirka Simunek, creator of HYDRUS, this volume demonstrates two-and three- dimensional simulations and computer animations of numerical models using the popular software. Classroom-tested at the University of Georgia, this volume includes numerous examples and homework problems. It provides students with access to the base HYDRUS program as well as the Rosetta Database, which contains large volumes of information on the hydraulic properties of soils.

ISBN: 978-1-4200-7380-5
May 2010 | HB | £66.99 | 388 pp.

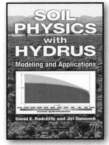

Reinforced Soil Walls and Slopes: Design and Construction

Mauricio Ehrlich, Leonardo Becker

When it comes to using reinforcements to grant better mechanical performance to soils, geosynthetics, one of the newest groups of building materials, have become mandatory in almost all works of infrastructure, draining applications, waterproofing, paving, erosion control and soil reinforcement.

ISBN: 978-85-797-5001-4
April 2011 | HB | £43.99 | 120 pp.

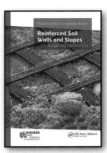

Soil Mechanics Fundamentals

Isao Ishibashi, Hemanta Hazarika

Comprehensive and introductory, this textbook covers all basic fundamental concepts of soil mechanics. It delineates how soil behaves, why it behaves that way, and the engineering significance of such behavior. The text explains the various behaviors of soils based on mathematics, physics, and chemistry in a simple but complete manner. The authors include essential engineering equations to emphasize the importance of fundamentals, and 180 exercise problems and solutions to aid with application.

ISBN: 978-1-4398-4644-5
December 2010 | HB | £49.99 | 344 pp.

New Techniques on Soft Soils

Edited by Marcio Almeida

New Techniques on Soft Soils covers a wide range of updated techniques on several topics, such as site investigation, vertical drains, surcharge, piled embankment, granular piles, deep mixing, monitoring and performance. An essential reference for designers and practitioners involved in soft soil construction, it provides a comprehensive view of current experiences and opinions of researchers and professionals from different parts of the world involving site investigation, design and construction on soft clays.

ISBN: 978-85-797-5002-1
April 2011 | HB | £93.00 | 340 pp.

Environmental Soil Properties and Behaviour

Raymond N. Yong, Masashi Nakano, Roland Pusch

This timely book examines changes in soil properties and behaviour caused by stresses from anthropogenic activities and environmental forces. It integrates soil physics, soil chemistry, and soil mechanics as vital factors in soil engineering. The book focuses on the use of soil as an environmental tool for management and containment of waste materials, as well as on the impact of ageing and weathering processes and soil contamination on the properties and behaviour of soils.

ISBN: 978-1-4398-4529-5
March 2012 | HB | £76.99 | 472 pp.

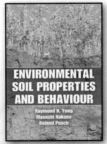

Geological Engineering

Luis Gonzalez de Vallejo, Mercedes Ferrer, Michael de Freitas

This extensively illustrated book examines the subject of geological engineering by focusing on four key areas: fundamentals, methods, applications and geohazards. As a clear guide to interpreting a geological setting for the purposes of engineering design and construction, it can be used both as a basic reference work for practising professionals as well as a standard introduction for students of engineering and related geological disciplines.

ISBN: 978-0-415-41352-7
January 2011 | HB | £76.99 | 700 pp.

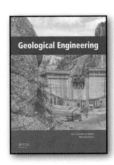

To view our full range of books and order online visit:

www.crcpress.com